ADVANCES IN ASYMMETRIC SYNTHESIS

Volume 3 • 1998

ADVANCES IN ASYMMETRIC SYNTHESIS

Editor: ALFRED HASSNER
Department of Chemistry
Bar-Ilan University
Ramat-Gan, Israel

VOLUME 3 • 1998

JAI PRESS INC.

Stamford, Connecticut *London, England*

100 Prospect Street
Stamford, Connecticut 06904-0811

JAI PRESS LTD.
38 Tavistock Street
Covent Garden
London WC2E 7PB
England

ISBN: 0-7623-0106-6

Transferred to digital printing 2005

CONTENTS

LIST OF CONTRIBUTORS

Naoki Asao	Department of Chemistry Graduate School of Science Tohoku University Sendai, Japan
Lucia Battistini	Università di Parma and Istituto per l'Applicazione delle Tecniche Chimiche Avanzate del CNR Sassari, Italy
Giovanni Casiraghi	Università di Parma and Istituto per l'Applicazione delle Tecniche Chimiche Avanzate del CNR Sassari, Italy
Robert E. Gawley	Department of Chemistry University of Miami Coral Gables, Florida
Oleg I. Kolodiazhnyi	Institute of Bioorganic Chemistry National Academy of Sciences of Ukraine Kiev, Ukraine
Marek Majewski	Department of Chemistry University of Saskatchewan Saskatoon, Canada
Gloria Rassu	Università di Parma and Istituto per l'Applicazione delle Tecniche Chimiche Avanzate del CNR Sassari, Italy

Hiroaki Sasai	Graduate School of Pharmaceutical Sciences The University of Tokyo Tokyo, Japan
Simon J. Sesay	Department of Chemistry Loughborough University Loughborough, England
Masakatsu Shibasaki	Graduate School of Pharmaceutical Sciences The University of Tokyo Tokyo, Japan
Naofumi Tsukada	Department of Chemistry Graduate School of Science Tohoku University Sendai, Japan
Jonathan M. J. Williams	School of Chemistry University of Bath Bath, England
Yoshinori Yamamoto	Department of Chemistry Graduate School of Science Tohoku University Sendai, Japan
Franca Zanardi	Università di Parma and Istituto per l'Applicazione delle Tecniche Chimiche Avanzate del CNR Sassari, Italy

PREFACE

Enantioselective synthetic methods are not only in the forefront of chemical and pharmaceutical research but activity in this area is constantly on the upswing. Some advances in this field have undoubtedly been stimulated by the urgency to obtain drugs or compounds of medicinal interest as single enantiomers, and the keenness to synthesize natural products in nonracemic form.

The prominent methodology used in asymmetric syntheses is still the utilization of chiral reagents, chiral catalysts and kinetic resolution. In the previous volume of this series (Vol. 2), emphasis was on the use of enzymes as biocatalysts in the synthesis of nonracemic functionalized molecules, as well as synthesis of nonracemic amines, β-dicarbonyl compounds, and ferrocenes.

The authors of the seven chapters in this volume are all authorities and pioneers in the development of enantioselective methodology.

Yamamoto and colleagues demonstrate how metal amides derived from chiral amines, in particular silylated derivatives, can be used in stereoselective Michael additions to unsaturated esters in the presence of mild Lewis acids. Ultimately the products of these reactions serve in the synthesis of nonracemic β-lactams.

Chiral lithium amides can also be employed to achieve high enantioselectivity in the α-deprotonation of cyclic ketones, as shown in the chapter by Majewski.

In the framework of reactions of aldehydes with chiral carbanions α to amino functions leading to the synthesis of alkaloids and a variety of functionalized molecules with high enantiomeric excess, Gawley presents an instructionally useful general mechanistic rationale of factors affecting the stereoselective addition of nucleophiles to faces of diastereotopic carbonyl compounds.

In an interesting chapter describing applications to the synthesis of diverse natural products, Casiraghi et al. demonstrate how siloxy derivatives of furans, pyrroles, or thiophenes can be powerful synthetic substrates in Lewis acid-catalyzed reactions with chiral electrophiles such as imminium ions or aldehydes.

The great versatility of metal derivatives of BINOL in particular heterobimetallic BINOL complexes, in the highly successful asymmetric catalysts of various reactions from nitro aldol condensations to Michael additions and epoxidations is beautifully summarized by Shibasaki and Sasai, pioneers in the field.

How enantiomerically pure phosphorus ligands and Schiff bases can be applied to the well-known palladium-catalyzed allylic substitution by nucleophiles is discussed in a chapter by Williams and Sesay.

Finally, Kolodiazhnyi summarizes how tri-, tetra-, penta-, and hexa-coordinate phosphorus containing compounds are particularly suited not only to provide synthetic routes to nonracemic P-compounds but also to chirality transfer from P to other centers.

I am indebted to the authors of these chapters for their excellent presentations and fine cooperation. This volume is dedicated to my son Lawrence whose interest in chemistry I might have kindled.

Alfred Hassner
Editor

ASYMMETRIC SYNTHESIS OF β-AMINO ACIDS AND β-LACTAM DERIVATIVES VIA CONJUGATE ADDITION OF METAL AMIDES

Yoshinori Yamamoto, Naoki Asao, and
Naofumi Tsukada

Advances in Asymmetric Synthesis
Volume 3, pages 1–37.

ISBN: 0-7623-0106-6

I. INTRODUCTION

Since the presence of a 1β-methyl substituent has been found to enhance the chemical and metabolic stability of synthetic carbapenem antibiotics (e.g. **1**),[1–3] a number of stereoselective syntheses of the key 1β-methyl intermediate **2** have been reported.[4,5] Many of these syntheses proceed from 4-acetoxy-2-azetidinones. Other methods for introduction of the β-methyl group include catalytic hydrogenation[5] and L-Selectride[5] or borane reduction[6] of olefinic precursors of **2**, reduction of a hexacarbonyldicobalt-stabilized propargyl cation,[5] β-lactam formation from components derived from either (*S*)- or (*R*)-methyl 3-hydroxy-2-methylpropionate, and use of lactone intermediates.[5,6]

We developed an entirely new approach to the asymmetric synthesis of the β-lactam framework via a three-component coupling (TCC) process using metal dialkylamides; the regioselective 1,4-addition of the metal dialkylamides reagent to certain α,β-unsaturated enoates, followed by aldol condensation with acetaldehyde and subsequent manipulation, gave the β-lactam with high diastereomeric and enantiomeric excess.[7] One of the key steps for the TCC process is asymmetric conjugate addition of metal amides to α,β-enoates. How can we accomplish high asymmetric induction in this type of reactions? There are three possible ways for the asymmetric induction in the metal amide conjugate addition, which enables asymmetric synthesis of β-amino acid derivatives: (1) auxiliary, (2) substrate, and (3) reagent control (Eqs. 1–3).

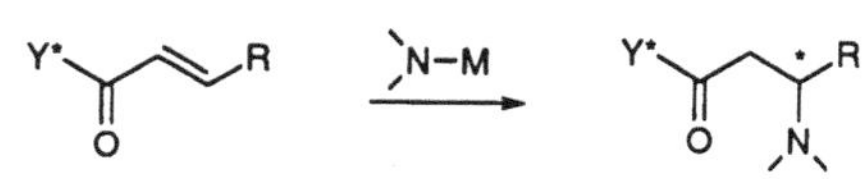

(1)

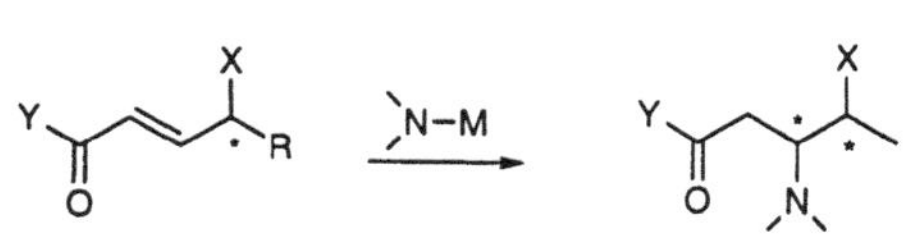

(2)

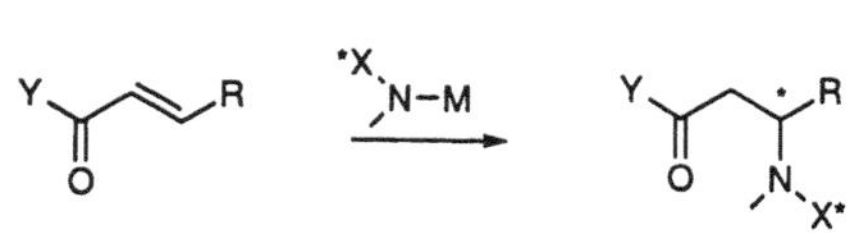

(3)

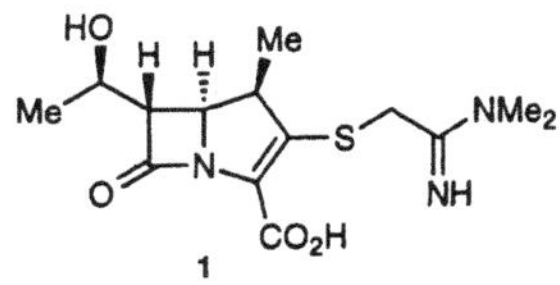
1

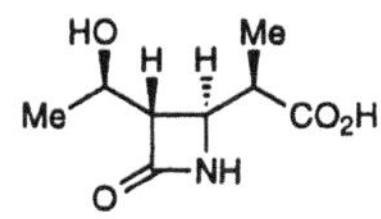
2

Before discussing these asymmetric syntheses, we had to investigate the conjugate addition of metal amides because systematic studies on the metal amide addition had not been carried out when we started this project. In this review, not only asymmetric synthesis of β-amino acids and β-lactams but also new findings related to the metal amide conjugate addition are described.

II. REACTIONS OF METAL AMIDES WITH α,β-UNSATURATED ESTERS

Metal dialkylamides R_2NM **3** are commonly used as strong bases for deprotonation of organic compounds. However, nucleophilic reactions of R_2NM, such as conjugate addition to α,β-unsaturated esters, have received little attention from a synthetic point of view.[8–10] The major reason is that nucleophilic reactions of R_2NM are always accompanied by deprotonation reactions; it has been difficult to control the reactivity of R_2NM. For example, the reaction of α,β-unsaturated esters with ordinary lithium dialkylamides gives a mixture of conjugate adducts (1,4-addition product **5**), deprotonation products at the γ-position, and substitution products at the ester group (1,2-adducts). We examined the conjugate addition of some lithium dialkylamides **3** to methyl crotonate (**4a**) for elucidation of regioselectivity (Eq. 4) and the results are summarized in Table 1.

Lithium *N*-benzyltrimethylsilylamide (LSA)[7] (**3a**) was an efficient reagent for the conjugate addition; by-products such as **6** and **7** were not detected (entry 1). It was reported that LDA gave **5** in good yields without contamination of **6** and **7**.[8–10] Our experiments revealed that LDA produced significant amounts of **7** as a by-product; precisely speaking, when the reaction was quenched with *n*-octyl iodide, methyl 2-vinyldecanoate was isolated since isolation of **7** itself was difficult owing to its volatile characteristics (entry 2). Further, the isopropyl group of LDA cannot be removed in the subsequent process and thus this reagent is not suitable for the preparation of β-amino acids. The amines substituted with silyl groups such as Bn(t-BuMe$_2$Si)NH, Bn(Ph$_2$MeSi)NH, Bn(Ph$_3$Si)NH, and Bn(t-BuMe$_2$Si)NH

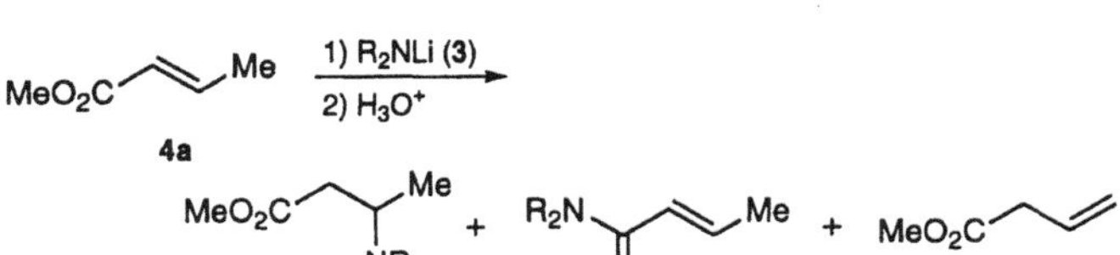

(4)

Table 1. Reactions of Methyl Crotonate (**4a**) with Lithium Amides

		Isolated Product (%)		
Entry	*R_2NLi (**3**)*	**5**	**6**	**7**
1	LSA [Bn(TMS)NLi] (**3a**)	88	—	—
2	LDA (iPr2NLi)	44	—	14[a]
3	Bn(*t*-$BuMe_2Si$)NLi	61	7	—
4	Bn(Ph_2MeSi)NLi	30	6	—
5	BnNHLi	20	60	—
6	Bn_2NLi	18	—	—
7	$(TMS)_2NLi$	—	—	9[a]

Note: [a]Instead of H_3O^+, the reaction was quenched with octyl iodide. Methyl 2-vinyldecanoate was isolated in the indicated yields.

were prepared from the reaction of benzylamine with the corresponding chlorosilanes (entries 3–4).[11] The conjugate adduct **5** was obtained along with **6** in the reaction with Bn(*t*-$BuMe_2Si$)NLi or Bn(Ph_2MeSi)NLi. When Bn(Ph_3Si)NLi or Bn(*t*-$BuPh_2Si$)NLi was used, several unidentified materials were produced. Other lithium amides such as BnNHLi and Bn_2NLi, in which the Bn group can be replaced by H in the subsequent process, also afforded both 1,4- and 1,2-adducts (entries 5–6). The deprotonation reaction leading to **7** took place with $(TMS)_2NLi$ (entry 7). Next we examined the reaction of LSA with other enoates. The results are summarized in Table 2.

Since the reactivity of ethyl acrylate (**4b**) and methyl methacrylate (**4c**) was too high, the addition of chlorotrimethylsilane was essential to prevent the formation of polymers (entries 1–2). The reaction of tiglic acid methyl ester **4d** was quite sluggish and the 1,4-adduct was obtained in low yield (entry 3). No desired 1,4-addition product was produced with ethyl 3,3-dimethylacrylate (**4e**), but 1,2-adduct (α,β-unsaturated amide) was obtained in low yield (entry 4). Interestingly, while the reaction of (*E*)-methyl 2-decenoate (**4g**) gave a conjugate adduct in good yield, desired 1,4-adduct was not obtained with (*Z*)-isomer **4h** (entries 6–7). Thus, an *E* geometry of α,β-unsaturated esters appears indispensable for this type of conjugate addition. Except for esters, a carboxylic amide and thioester can be used for the conjugate addition of LSA (entries 8–9). The reaction of ethyl phenypropiolate (**4k**) gave a 1,2-addition product exclusively instead of the desired 1,4-adduct (entry 10). Next, we examined the regioselectivities in the reaction of α,β,γ,δ-unsaturated esters **8** with metal dialkylamides (Eq. 5) and the results are summarized in Table 3.

The reactions of **8** with metal dialkylamides may provide a mixture of 1,4-(**9**), 1,2-(**10**), and 1,6-adducts **11** (Eq. 5). The reaction using LSA did not produce a good result; the 1,4-adduct **9a** was obtained in only 29% yield along with small

Table 2. The Conjugate Addition of LSA (**3a**) to Various Enoates

Entry	*Enone* **4**	*Yield*[a]	*Entry*	*Enone* **4**	*Yield*[a]
1[b]	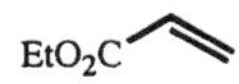 **4b**	38%	6	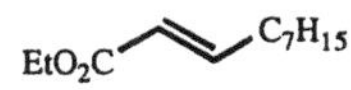 **4g**	86%
2[b]	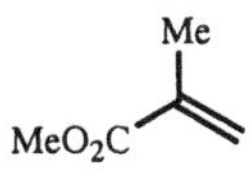 **4c**	60%	7[d]	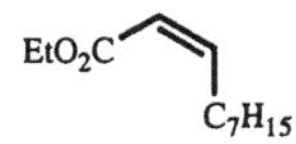 **4h**	0%
3	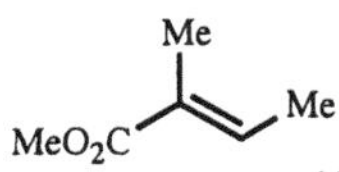 **4d**	15%	8	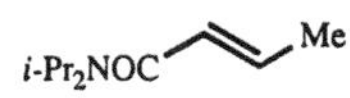 **4i**	44%
4[c]	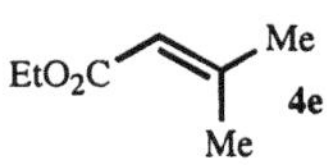 **4e**	0%	9	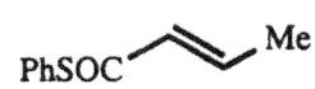 **4j**	68%
5	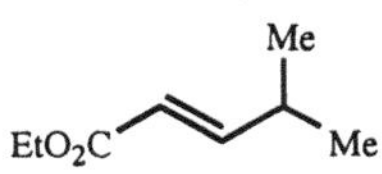 **4f**	67%	10[e]	 **4k**	0%

Notes: [a]Isolated yield of the corresponding β-amino acid derivative.
[b]A mixture of substrate and chlorotrimethylsilane was added to a LSA solution.
[c]1,2-Addition product was obtained in 15% yield.
[d](*E*)-Methyl 3-decenoate was obtained quantitatively.
[e]1,2-addition product was obtained in 77% yield.

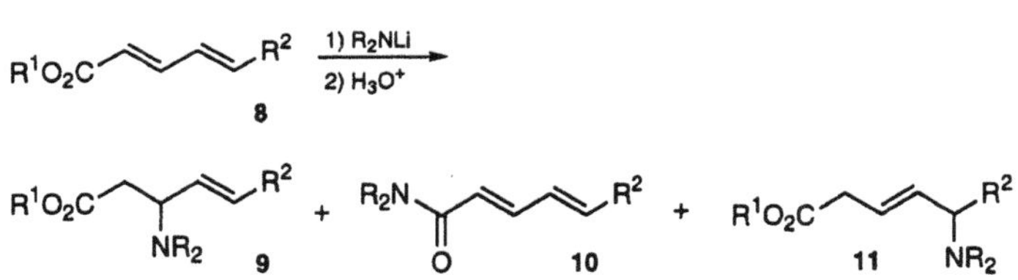

(5)

	R^1	R^2
8a:	Me	Me
8b:	*t*-Bu	Me
8c:	Me	Ph
8d:	*i*-Pr	Ph
8e:	*t*-Bu	Ph
8f:	*i*-Pr	Me

	R^1	R^2	R
9-11a:	Me	Me	Bn and H
9-11b:	Me	Me	Bn and Bn
9-11c:	*t*-Bu	Me	Bn and H
9-11d:	Me	Ph	Bn and H
9-11e:	*i*-Pr	Ph	Bn and H
9-11f:	*t*-Bu	Ph	Bn and H

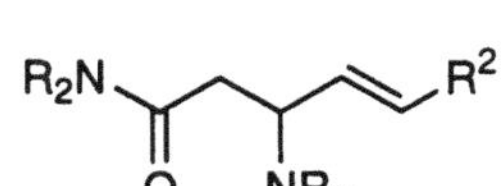

13

Table 3. Reactions of **8** with R_2NM[a]

Entry	**8**	R_2NLi	**9**	**10**	**11**
1	**8a**	Bn(TMS)NLi	29	8	0
2	**8a**	$\{[Bn(TMS)N]_2Cu\}Li$ (**12a**)	54	0	0
3	**8a**	Bn(TMS)NLi / cat. CuI	60	9	0
4	**8a**	$\{[Bn(TMS)N]_2Cu(CN)\}Li_2$ (**12b**)	48	trace	0
5[b]	**8a**	Bn_2NLi	0	0	0
6[c]	**8a**	$[(Bn_2N)_2Cu]Li$	32	0	0
7	**8a**	BnNHLi	0	37[d]	0
8	**8a**	$[(BnNH)_2Cu]Li$	0	87	0
9[e]	**8a**	$BnNH_2$	0	0	48
10	**8b**	$\{[Bn(TMS)N]_2Cu\}Li$	68	trace	0
11	**8c**	$\{[Bn(TMS)N]_2Cu\}Li$	60	10	0
12	**8d**	$\{[Bn(TMS)N]_2Cu\}Li$	85	0	0
13	**8e**	$\{[Bn(TMS)N]_2Cu\}Li$	78	trace	0

Notes: [a]Reactions were carried out at –78 °C except for entry 9.
[b]1,2-and 1,4-Adduct (**13**) was obtained in 54% yield.
[c]**13** was obtained in 68% yield.
[d]Several unidentified products were formed in addition to major products.
[e]The reaction was carried out at room temperature.

amounts of the 1,2-adduct **10a** (entry 1). The reaction of **8a** with copper reagents gave **9a** with relatively good 1,4-regioselectivities (entries 2–4); the cuprate-type **12a** (entry 2) and a higher order cyanocuprate-type **12b** (entry 4) reagent produced a similar regioselectivity. "Higher order" indicates that the stoichiometry of Bn(TMS)N, Cu, CN, and Li is 2 : 1 : 1 : 2; it does not mean that the copper species possesses the $\{[Bn(TMS)N]_2Cu(CN)\}Li_2$ structure.[12] CuI was used as source of Cu for the cuprate-type reagent, and CuCN was used for the higher order type reagent. In entries 1–4, the TMS group of R_2NM was removed during the work-up procedures; R_2N of **9** was BnNH. Lithium dibenzylamide gave the 1,2- and 1,4-adduct **13** (entry 5), and the corresponding cuprate reagent afforded a mixture of **9b** and **13** (entry 6). Lithium benzylamide (entry 7) and lithium bis(benzylamido)cuprate(I) (entry 8) produced the 1,2-adduct **10a**. The 1,6-adduct **11** was obtained in the reaction of **7a** with benzylamine (entry 9). These results suggested that the cuprate reagent **12a** might be promising for a regioselective 1,4-addition to α,β,γ,δ-dienoates. The effect of an ester group upon the regioselectivity of conjugate addition was examined using the cuprate reagent **12a**. The 1,4-adducts **9c**, **9e**, and **9f** were obtained exclusively in the case of a sterically bulky ester group (entries 10, 12, and 13), whereas the 1,2-adduct **10b** was afforded as a minor product in the case of methyl ester (entry 11, see also entry 2). Judging from the results of entries 12 and 13, the isopropyl ester seemed to give a better yield and regioselectivity than the *t*-butyl ester. Accordingly it is clear that: (1) a dienoate gives the 1,6-adduct **11** with $BnNH_2$, the 1,4-adduct **9** with the cuprate reagent of LSA **12a**, and a mixture of **9**, **10**, and **11** with the lithium reagents; and (2) the presence of an *i*-Pr group in R^1 diminishes the formation of the 1,2-adduct.

III. 1,2-DIASTEREOSELECTIVITY

The selectivities of the addition of amines,[13] alkoxides,[14] and carbon nucleophiles[15] to α,β-enoates bearing a chiral center at the γ-position had been reported. The *syn* selectivity was previously reported for the conjugate addition of benzylamine[13a,b] and various alkoxides[14] to acyclic α,β-enoates and their derivatives. A more complex situation holds for the addition of organometallic reagents (carbon nucleophiles) to acyclic γ-alkoxy-α,β-unsaturated enoates and enones;[15] *anti* selectivity has been observed frequently with organocopper reagents, but in certain cases organolithium and copper reagents have reacted with *syn* selectivity.

We first investigated the diastereoselectivity in the addition of lithium amides $LiNR^1R^2$ **3** to γ-silyloxy and alkoxy-α,β-unsaturated esters **14** (Eq. 6).[16] The results are summarized in Table 4. The reaction of *t*-butyldimethylsilyl (TBDMS) protected enoate **14a** with lithium benzyl(trimethylsilyl)amide **3a** gave an approximate 1:1 mixture of the *syn*-**15a** and *anti*-**16a** diastereomers (entry 1). The *syn* diastereoselectivity increased to 89~90% with sterically bulkier *t*-butyldiphenylsilyl **14b** and tri(isopropyl)silyl protected enoate **14c** (entries 2 and 3). Chemical yields were very high in the reactions with **14b** and **14c**. The *syn* diastereoselectivity reached to ~100% using the trityloxy derivative **14d** although the chemical yield decreased to 77% (entry 5). It is clear from these experiments that an increase of steric bulk at the OR group produces high *syn* diastereoselectivity.

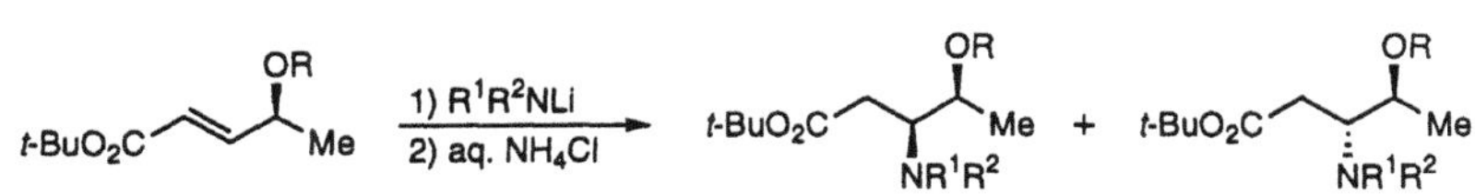

(6)

14

15a: R = *t*-BuMe$_2$Si, R^1 = H, R^2 = CH_2Ph — **16a:** R = *t*-BuMe$_2$Si, R^1 = H, R^2 = CH_2Ph

15b: R = *t*-BuPh$_2$Si, R^1 = H, R^2 = CH_2Ph — **16b:** R = *t*-BuPh$_2$Si, R^1 = H, R^2 = CH_2Ph

15c: R = *i*-Pr$_3$Si, R^1 = H, R^2 = CH_2Ph — **16c:** R = *i*-Pr$_3$Si, R^1 = H, R^2 = CH_2Ph

15d: R = *i*-Pr$_3$Si, R^1 = R^2 = CH_2Ph — **16d:** R = *i*-Pr$_3$Si, R^1 = R^2 = CH_2Ph

15e: R = Ph_3C, R^1 = H, R^2 = CH_2Ph — **16e:** R = Ph_3C, R^1 = H, R^2 = CH_2Ph

15f: R = Me, R^1 = R^2 = CH_2Ph — **16f:** R = Me, R^1 = R^2 = CH_2Ph

Table 4. 1,2-Asymmetric Induction in the Reaction of **14** with R^1R^2NLi

Entry	*Enoates* **14**	*R*	R^1R^2NLi	*syn*-**15** : *anti*-**16**	*Yield*
1	**14a**	Si(*t*-Bu)Me_2	Bn(TMS)NLi (**3a**)	54:46	91%
2	**14b**	Si(*t*-Bu)Ph_2	**3a**	89 : 11	99%
3	**14c**	Si(*i*-Pr)$_3$	**3a**	90 : 10	95%
4	**14c**	Si(*i*-Pr)$_3$	Bn_2NLi (**3b**)	30 : 70	79%
5	**14d**	CPh_3	**3a**	100 : 0	77%
6	**14e**	Me	**3b**	63 : 37	83%

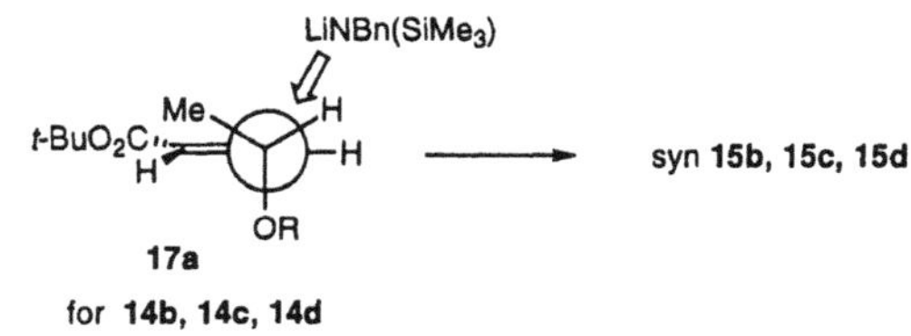

The *syn* selectivity for **14b–d** can be accounted for by a modified Felkin–Anh model **17a**, in which the OR group becomes *anti*, Me orients inside, and the nucleophile **3a** attacks from the less hindered outside position, as shown by the arrow. The coordination of lithium amide **3a** to an oxygen atom of the OR group followed by delivery of the nitrogen nucleophile from the side of OR would not take place; the chelation-controlled delivery of nucleophile **3a** is not conceivable due to the presence of sterically very bulky silyloxy groups[17] in addition to the soft nature of the nucleophile. The addition of lithium dibenzylamide **3b** to **14c** afforded a 30:70 mixture of *syn*-**15d** and *anti*-**16d** in 79% yield, indicating that the diastereoselectivity depended very much upon the structural difference of lithium amide reagents; either **3a** or **3b** (entry 3 vs. 4). Although the precise reason for the *anti* preference is not clear, the participation of a chelation model is conceivable in the

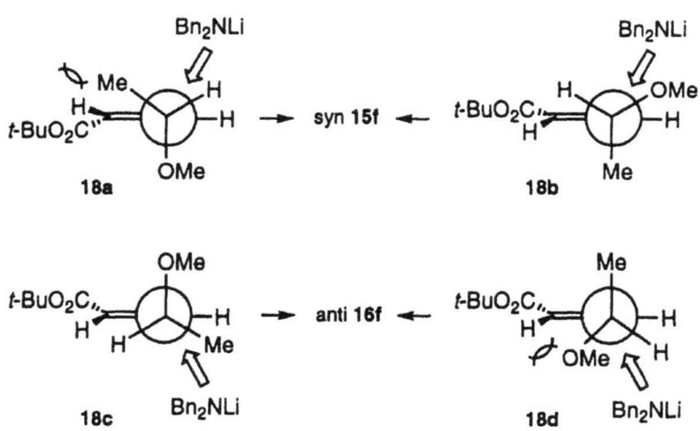

case of **3b** because of its hard characteristics in comparison with **3a**. Very interestingly, the reaction of γ-methoxy substituted enoate **14e** with lithium dibenzylamide **3b** gave a 63:37 mixture of the *syn*-**15f** and *anti*-**16f** diastereomers in 83% yield (entry 6). Among four possible transition-state geometries, **18d** would be most unfavorable due to the allylic strain between OMe and vinylic hydrogen group, and due to the presence of the Me group at the *anti* position.[18] Perhaps **18b** would be most favorable owing to both the chelation factor and minimum allylic strain, the presence of a small methyl group at the outside position, and the *anti*-OMe. In the case of **18a**, OMe is *anti* and this is stereoelectronically satisfactory, but there is allylic strain between the inside Me and an olefinic hydrogen. In the case of **18c** there is steric repulsion between the incoming reagent and Me group. Therefore, the reason for the *syn* preference in the case of **14e** is presumably due to both allylic strain and chelation factor.

To reduce ambiguous factors encountered when we considered probable transition-state geometries in **18**, the addition to the α-methyl-substituted enoate **19** was investigated. Interestingly, the reaction of **19** with **3b** gave the *syn*-diastereomer **20** exclusively in 85% yield (Eq. 7). No *anti*-isomer was detected. A 1:1 mixture of diastereoisomers concerning the α-position to ester group was obtained, although the stereochemistry at the α-position was not determined. It is now clear that the reaction proceeds through **21**, in which the allylic strain is minimized, leading to the *syn*-isomer **20** exclusively. Furthermore, the above result suggests that chelation-controlled delivery of **3b** is involved in the reaction of **19**.

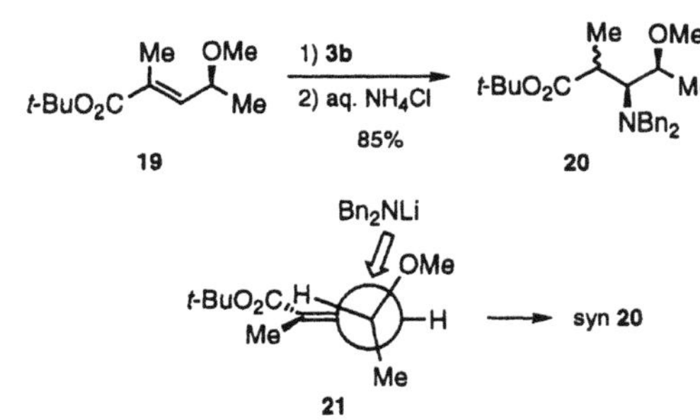

(7)

Next, we investigated the reaction of γ-phenyl-substituted enoate **22** in order to determine the effect of a substituent other than Me at the γ-position. The reaction of **22** with **3a** gave exclusively the *anti*-diastereoisomer **23** in 74% yield, with no detection of the *syn*-diastereoisomer (Eq. 8). Not only isopropyl ester **22**, but also ethyl and *tert*-butyl ester derivatives produced, upon treatment with **3a**, the corresponding *anti*-diastereoisomers exclusively although chemical yields were lower than reactions of **22**.

The *anti* diastereoselectivity can be explained by an allylic strain model **24a**. It should be noted that the reaction of **14a** (OTBDMS and Me substituent at the γ-position) with **3a** gave about 1:1 mixture of *syn*- and *anti*-diastereomers (entry 1,

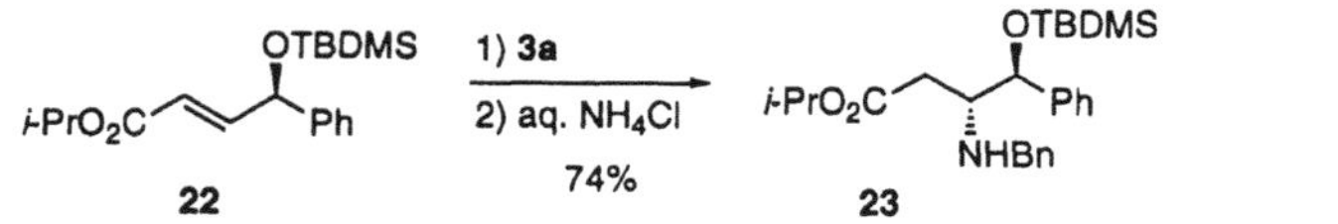

(8)

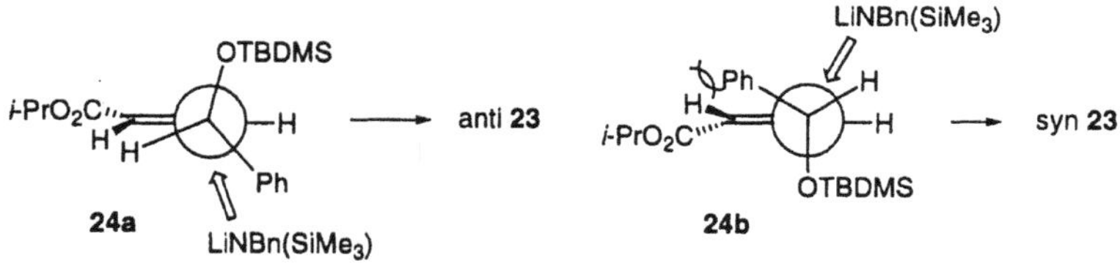

Table 4). The phenyl substituent enhanced the *anti* selectivity dramatically. A possible explanation for this difference is as follows. A serious steric repulsion (allylic strain) would arise between vinylic hydrogen and the phenyl group in the transition-state geometry **24b** which produces the *syn*-diastereomer. Accordingly, the reaction of **22** would proceed through **24a** to afford the *anti*-diastereomer **23**. On the other hand, the steric repulsion between the Me group and vinylic hydrogen would be not so serious, and thus the reaction of **14a** would proceed through both transition-state geometries **17b** and **17c**.

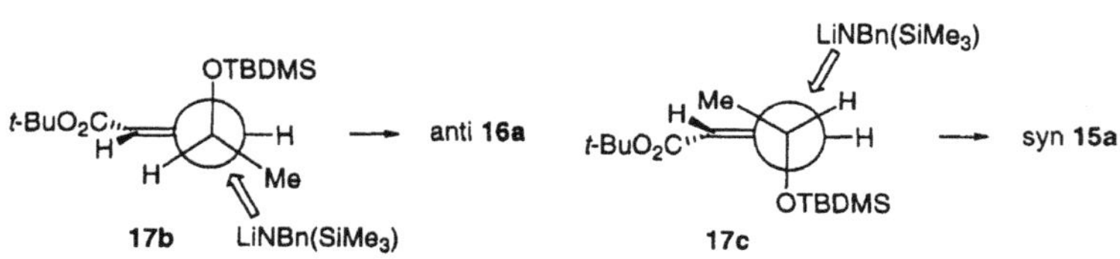

Finally, we investigated the diastereoselectivity in the reaction of enoates substituted with Ph and Me at the γ-position. The reaction of **25** with **3a** gave a 3:97 mixture of *syn*-**26** and *anti*-**26** diastereomers in 79% yield (Eq. 9). The very high *anti* selectivity is a reflection of a modified Felkin–Anh model **27** in which hydrogen orients outside, small Me group orients inside, and the bulky and electron

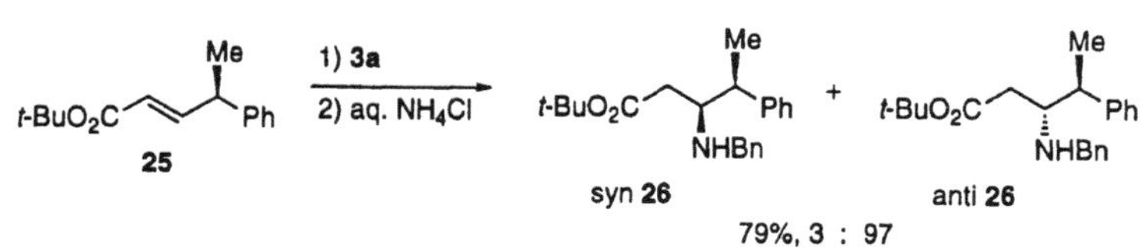

(9)

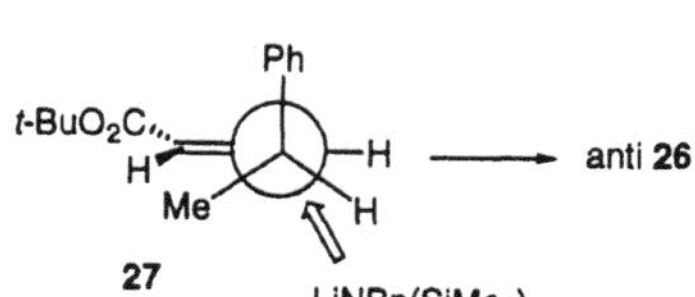

withdrawing phenyl group becomes *anti*. It is clear that not only an appropriate OR substituent at the γ-position but also a stereoelectronically suitable substituent (such as phenyl) produces high 1,2-asymmetric induction.

IV. STEREODIVERGENT SYNTHESIS OF THE (*Z*)- AND (*E*)-ENOLATES OF A β-AMINO ESTER[19]

Conjugate addition of metal amides to enoates produces metal enolates as intermediates. The stereochemistry of these enolate intermediates plays a key role for controlling the diastereoselectivities of further reactions, such as aldol condensations.[20] Fleming reported the stereodivergent synthesis of the β-silyl enolates via silyl cuprate addition to enoates.[21] Similarly, we examined the trapping of lithium enolates with trimethylchlorosilane. LSA was treated with **4a** in THF at –78 °C, and then chlorotrimethylsilane was added at this temperature. After additional stirring for several minutes, the solvents were removed in vacuo as soon as a cooling bath was removed. Evaporation was continued until the mixture was warmed to room temperature. Dry hexane was added to the residue, and then precipitated lithium chloride was separated by centrifuge under Ar atmosphere. Removal of hexane under reduced pressure, followed by Kugelrohr distillation gave a colorless oil in essentially pure form. The 400 MHz ^{1}H NMR analysis revealed that the silylketene acetal consists of only one stereoisomer. The nuclear Overhauser effect between the olefinic proton and methoxy group was observed (17%), indicating the *Z* geometry of the β-amino ketene acetal (**28**).

Next, we attempted to generate the (*E*)-isomer of the β-amino enolate. The (*Z*)-lithium β-amino enolate, derived from LSA and methyl crotonate (**4a**), was once protonated with methanol to prepare the corresponding β-amino ester **29**. Deprotonation of the resulting β-amino ester **29** with lithium diisopropylamide (LDA), subsequent addition of chlorotrimethylsilane, and a similar nonaqueous

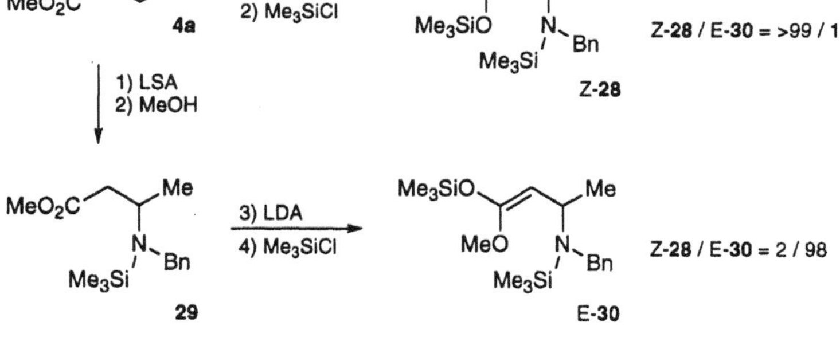

Scheme 1.

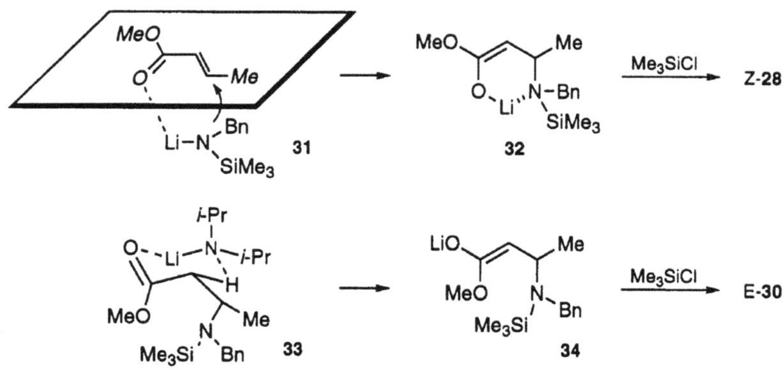

Scheme 2.

work-up gave a colorless oil. The ^{1}H NMR analysis indicated the *E*-form **30**, since the nuclear Overhauser effect between the olefinic proton and methoxy group was not observed. Consequently, we are now in a position to prepare both (*Z*)- and (*E*)-β-amino enolates in a stereodivergent way by merely changing the reagents and conditions (Scheme 1).

Very high *Z* selective formation of **28** is accounted for by chelation of the lithium to the ester group (Scheme 2). Presumably, the conjugate addition proceeds through the *s-cis*-form **31**, resulting in the stereoselective production of the (*Z*)-isomer. It is believed that the lithium enolate exists as a chelated species **32**. Exclusive formation of the (*E*)-enolate via the deprotonation procedure can be explained by the six-membered cyclic transition-state **33**, as is proposed in the related system.[22] The sterically bulky substituent takes an equatorial position in **33**, resulting in selective formation of the lithium (*E*)-enolate **34**.

V. CONJUGATE ADDITION OF LSA FOLLOWED BY ALKYLATION

We examined the reaction of the stereo-defined lithium enolates **32** and **34** with alkyl iodides in order to clarify the diastereoselectivity of the alkylation reaction (Eq. 10). The results are summarized in Table 5.[23] The alkylation proceeded in high to good yields. The silyl protective group of N–$SiMe_3$ was removed during the work-up process and we obtained finally the benzylamino derivatives. Although essentially no selectivity was observed via the (*Z*)-enolate **32**, moderate to good *anti* selectivity was obtained via the (*E*)-enolate **34**. In the chelated-form **32**, the nitrogen substituent adopts inside due to chelation, and thus RI can attack the α-carbon from both α- and β-faces with almost equal opportunity, since the stereoelectronic difference between Me and H is presumably not so distinct (**37** and **37a**). The *anti* selectivity via **34** is accounted for by the most stabilized transition-

Table 5. Alkylation of Lithium Enolates **32** and **34**

Entry	*Lithium Enolate*	*RI*	*syn-***35***/anti-***36**	*Yield (%)*
1	**32**	MeI	53/47	88
2	**32**	n-$C_8H_{17}I$	41/59	81
3	**34**	MeI	31/69	68
4	**34**	n-$C_8H_{17}I$	10/90	60

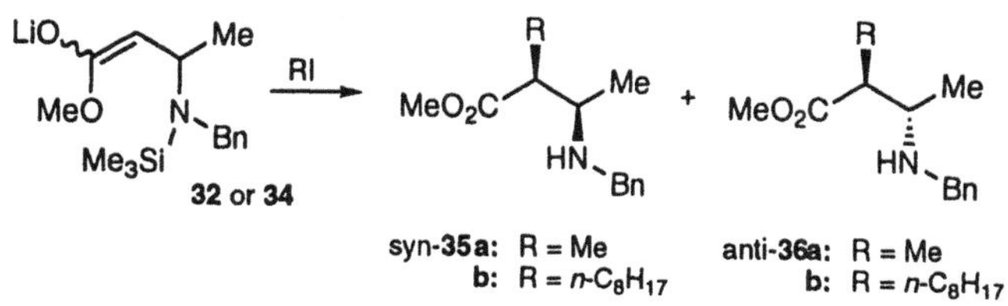

(10)

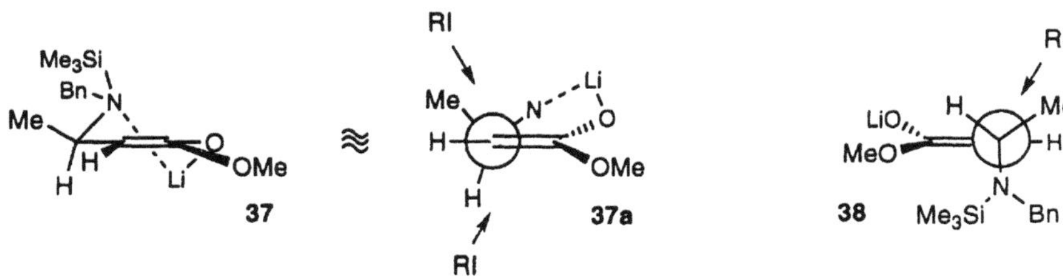

state geometry[24] **38**, in which the smallest H adopts inside, the nitrogen substituent adopts *anti* for stereoelectronic reasons, and Me adopts outside. The alkylation takes place from the less hindered Me side, giving the *anti* selectivity.

VI. PREPARATION OF α-ALKYLATED α,β-UNSATURATED ESTERS

The β-amino esters underwent β-elimination[25–27] via quarternization-base treatment to produce the corresponding α-alkylated α,β-unsaturated esters. *N*-Methylation of *syn*-**35b** with MeI/K_2CO_3 followed by treatment with silica gel produced a 91:9 mixture of the *trans*-**39a** and *cis*-**40a** isomers in 86% yield (Eq. 11).[28] Similar treatment of *anti*-**36b** gave a 86:14 mixture of *trans*-**39a** and *cis*-**40a** in 90% yield. The three-step sequence—conjugate addition of LSA to **4a**, enolate trapping with octyliodide, β-elimination—afforded a 88:12 mixture of *trans*-**39a** and *cis*-**40a** in 72% total yield. The conjugate addition of LDA to **4a**, followed by enolate trapping with octyl iodide and subsequent treatment with silica gel, gave a 93:7 mixture of *trans*-**39a** and *cis*-**40a** in 56% total yield. Therefore, a synthetic equivalent of the α-carbanion of an α,β-unsaturated ester can be generated by the conjugate addition of a nitrogen nucleophile (R_2NLi); LDA, as well as LSA, is also useful for this purpose.

Quite similarly, a 88:12 mixture of *trans*-**39b** and *cis*-**40b** was obtained in 52% yield via the three-step sequence with LDA. The α-benzylated derivative **39c** and **40c** (*trans*/*cis* = 87/13) was produced in 74% yield via LSA. Other α,β-unsaturated esters such as ethyl 4-methyl-2-pentenoate (**4l**) and ethyl 2-decenoate (**4m**) also underwent the conjugate addition of LDA-trapping with MeI-β-elimination, giving **39d**, **40d** (*trans*/cis = 95/5, 33%) and **39e**, **40e** (*trans*/*cis* = 88/12, 69%), respectively. Consequently, the three-step sequence via LSA or LDA provides a new synthetic procedure for trisubstituted enoates.

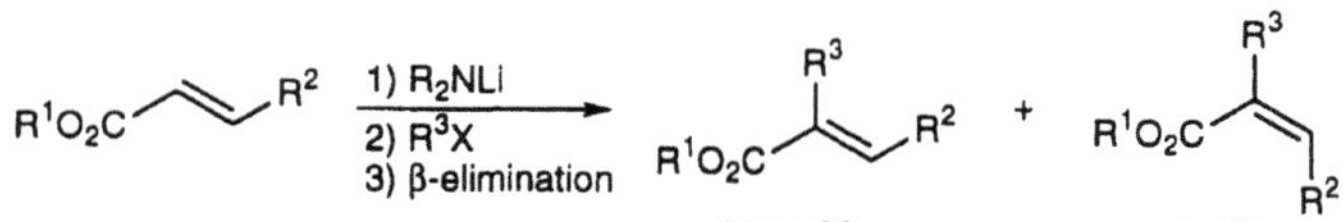

(11)

trans-**39** cis-**40**

4a: R^1 = Me, R^2 = Me
l: R^1 = Et, R^2 = *i*-Pr
m: R^1 = Et, R^2 = *n*-C_7H_{15}

a: R^1 = Me, R^2 = Me, R^3 = *n*-C_8H_{17}
b: R^1 = Me, R^2 = Me, R^3 = $CH_2CH{=}C(CH_3)_2$
c: R^1 = Me, R^2 = Me, R^3 = CH_2Ph
d: R^1 = Et, R^2 = *i*-Pr, R^3 = Me
e: R^1 = Et, R^2 = *n*-C_7H_{15}, R^3 = Me

VII. CYCLIZATION BASED ON THE CONJUGATE ADDITION–INTRAMOLECULAR ALKYLATION

Prior to our study, LDA had already been employed as a nitrogen nucleophile for the cyclization reaction.[9] The cyclization of ω-bromo ester **41**[29] with LSA in THF proceeded cleanly to give *trans*-2-aminocyclopentanecarboxylate **42** in 97% yield (Eq. 12).[30] When the conjugate addition–intramolecular alkylation of ω-bromo ester **43** with LSA was carried out in THF, *trans*-2-aminocyclohexanecarboxylate (**44**) was obtained in 67% yield.

(12)

41: n = 1
43: n = 2
45: n = 3

42: n = 1
44: n = 2
46: n = 3

A similar treatment of **43** with LSA in THF followed by addition of HMPA afforded **44** in 79% yield. *trans*-2-Aminocycloheptanecarboxylate **46** was derived in 73% yield as a single product from the reaction of ω-iodoester **45**[31] and LSA in the presence of HMPA. The stereostructures of these carbocycles were assigned on the basis of their ^{1}H NMR spectra. Thus, LSA is a useful nitrogen nucleophile for 5-, 6-, and 7-*tet*-cyclizations based on the conjugate addition–intramolecular alkylation that results in the thermodynamically more stable stereoisomer selectively.

VIII. CYCLIZATION BASED ON TANDEM CONJUGATE ADDITIONS

Because of the highly chemoselective nature of LSA, we expected that the enolate, generated by 1,4-addition of the amide to one of two enoate groups of an (*E*,*E*)-α,β,χ,ψ-unsaturated dioic acid ester, would add intramolecularly to the other enoate as shown in Eq. 13.[32] Reaction of dienedioate **47**[33] and LSA gave cyclopentanecarboxylates **48** in 92% yield as a 7:3 stereoisomeric mixture. After many unsuccessful attempts to separate the stereoisomers, amino esters **48** were convergently transformed into cyclopentenecarboxylate (**49**)[34] in 91% yield by treatment with iodomethane and K_2CO_3 in methanol. The ^{1}H NMR spectrum of **48** suggested that a major isomer was methyl *c*-3-(*N*-benzylamino)-*t*-2-(methoxycarbonyl)cyclopentane-1-acetate.

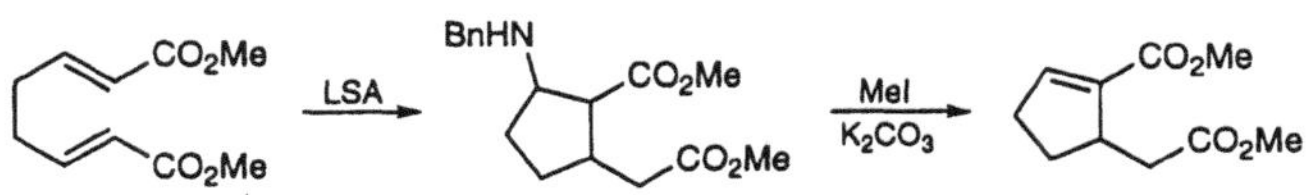

47 48 49 (13)

Cyclization of homologous dienedioate **50**[35] proceeded cleanly to give **51** as a sole product in 93% yield (Eq. 14). In contrast, no cyclization products were obtained from the higher homologue **52**[35] by treatment with LSA under similar conditions (Eq. 15). Now it is clear that we can use LSA as an initiator for the 5- and 6-*exo-trig* cyclizations. The stereoselective formation of **51** indicates that the reaction proceeds through a transition-state geometry **54**, with synclinal arrangement between the enolate double bond and the enoate double bond.

50 51 (14)

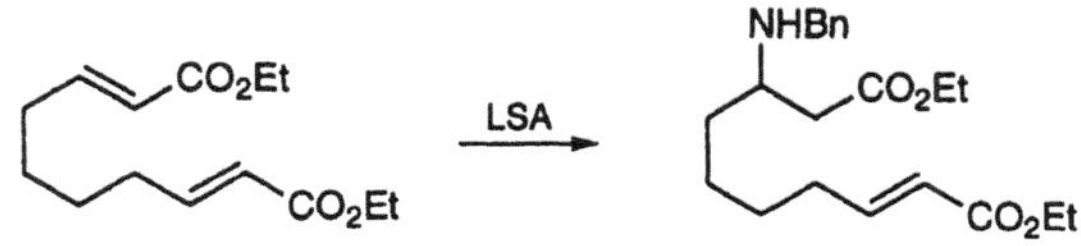

52 53 (15)

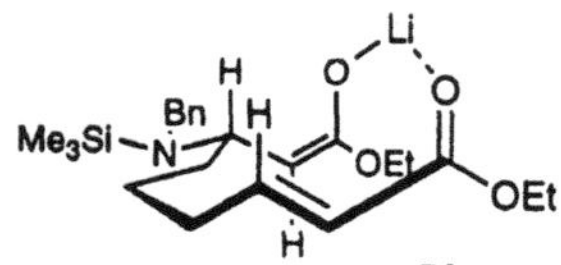

54

The cyclization of an unsymmetrical dienedioate by this type tandem conjugate addition was also investigated. Cyclopentenecarboxylate **49** and related compounds seemed to be candidates for synthetic precursors of cyclopenta[*c*]pyran monoterpenes. Therefore, we focused our attention on unsymmetrical dienedioates for 5-*exo-trig* cyclizations.

Dienedioate **55**[36] consists of crotonate and (*E*)-2-methyl-2-butenoate units. Reaction of **55** with LSA at –78 °C gave a complex stereoisomeric mixture of cyclopentanecarboxylates **56** (Scheme 3). Successive treatment of the mixture with iodomethane in the presence of K_2CO_3 and then with silica gel in boiling xylene gave a 72:28 mixture of unsaturated diesters **57** and **58** in 60% yield from **55**. Thus, LSA adds highly regioselectively to **55** in a 1,4-manner, and the reactivity of the (*E*)-2-methyl-2-butenoate part is less than that of the crotonate portion. The low reactivity of the tri-substituted enoate unit is presumably due to the increased electron density of the β-position due to the α-methyl group.

Unsymmetrical dienedioate **59** possesses crotonate and (*E*)-3-methyl-2-butenoate units. If conjugate addition of LSA to the crotonate unit proceeds more rapidly than deprotonation of Me group by LSA, 5-*exo-trig* cyclization was expected. In fact, cyclopentane **60** was derived from **59** in 28% yield (Eq. 16). The stereochemistry of **60** was determined by ^1H NMR spectroscopy. The NOE at the proton attached to C-2 upon irradiation of the methyl group attached to C-1, combined with the vicinal coupling constant, $J_{2,3} = 8.4$ Hz, indicated that the proton attached to C-2 and the benzylamino group was on the same face of the ring as the

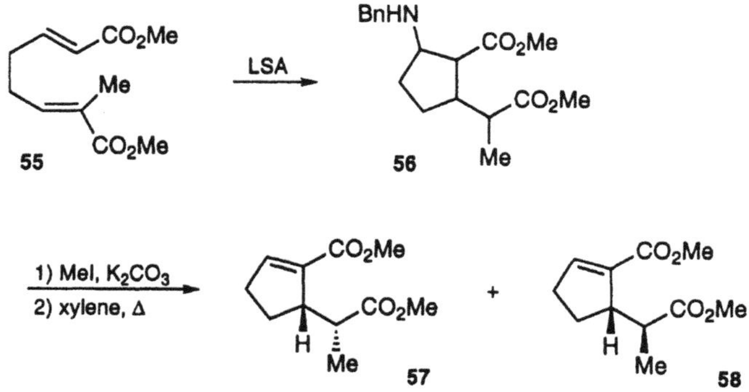

Scheme 3.

methyl group attached to C-1. The stereoselective cyclization to **60** may proceed through the transition state **61**, which would be more stable than **61a**.

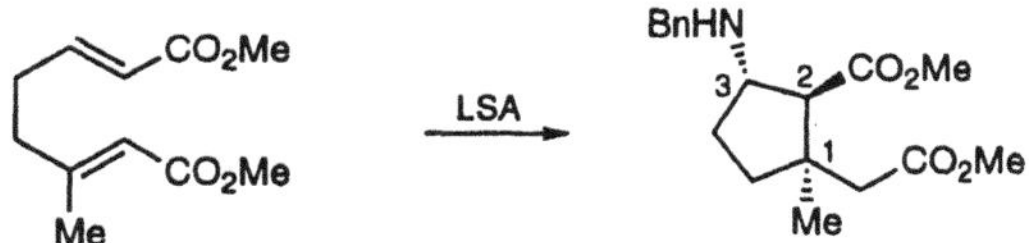

59 **60** (16)

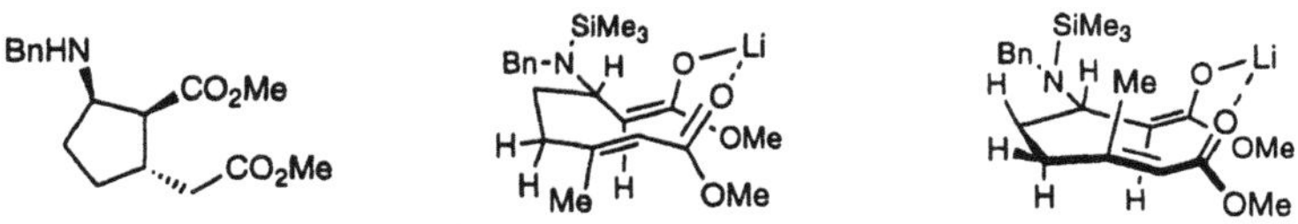

48a **61** **61a**

The stereostructure of **60** is not identical with that of a major product obtained from the cyclization of **47**, methyl *c*-3-(*N*-benzylamino)-*t*-2-(methoxycarbonyl)cyclopentane-1-acetate (**48a**). This means the methyl group at the β-position of **59** controls arrangement of the carbon framework in the transition state of the cyclization. In conclusion, 5-*exo-trig* cyclization of both symmetrical and unsymmetrical dienedioates can be achieved by treatment with LSA.

IX. TOTAL SYNTHESIS OF (±)-DIHYDRONEPETALACTONE AND (±)-ISODIHYDRONEPETALACTONE

These cyclopentene monoterpenes are physiologically active components for the *Felidae* aminals isolated from the leaves and galls of *Actinidia polygama* Miq. and from the essential oil of *Nepeta catraria*.[37,38] Our synthetic plan is shown in Scheme 4. Cyclopentenecarboxylates **57** and **58** are reasonable synthetic intermediates for Dihydronepetalactone (**64**) and Isodihydronepetalactone (**65**), respectively. We have already shown a synthetic route to obtain the compound **57** mainly (Scheme 3). In order to obtain **58** predominantly, we attempted *N*-methylation of the cyclized enoates derived from symmetrical dienedioate **47** and LSA. After *N*-methylation followed by deamination, **57** and **58** were obtained in a ratio of 32:68 in 64% yield. The third route to diesters **57** and **58** is via α-methylation of **49**. The reaction of the enolates, derived from the treatment of **49** with LDA, with iodomethane gave **57** and **58** in a 34:66 ratio in 76% yield.

Reduction of **57** with $LiAlH_4$ followed by oxidation with active manganese dioxide gave lactone **62** in 69% yield. A similar treatment of **58** gave lactone **63** in 58% yield. The stereostructures of **62** and **63** were defined on the basis of their ^1H NMR spectra, thus supporting the stereochemistry of **57** and **58** assigned previously. A highly stereoselective introduction of the C-7 methyl group of **64** was carried out

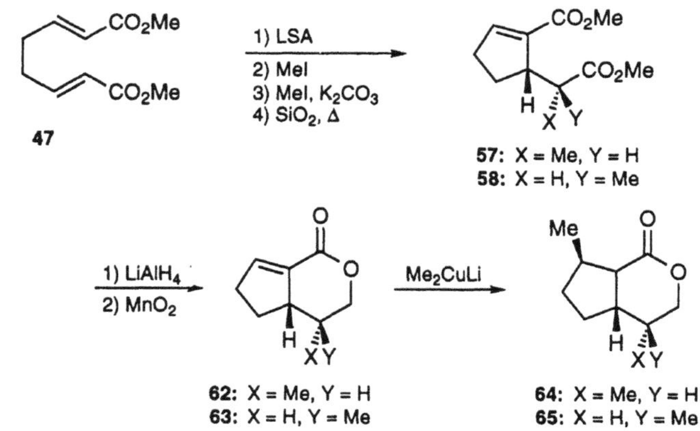

Scheme 4.

by treatment of **62** with dimethylcopper lithium. Similarly, **63** was converted into **65**. The spectral characteristics of **64** and **65** are identical with those of dihydronepetalactone and isodihydronepetalactone, respectively.

In conclusion, LSA is an efficient nucleophile for 5- and 6-*exo-trig* ring closures of α,β,χ,ψ-unsaturated dioic acid esters. The 5-*exo-trig* ring closures represent practical methodology for the synthesis of physiologically active cyclopentane monoterpenes.

X. ASYMMETRIC CYCLIZATION VIA TANDEM CONJUGATE ADDITION

Asymmetric carbocyclization is one of the most important process in organic synthesis. Among several useful methods, asymmetric Diels–Alder reactions have been extensively studied, and very elaborated chiral auxiliaries have been used to accomplish high diastereo- and enantioselectivity.[39] We examined the asymmetric carbocyclization by the tandem conjugate addition of certain metal amide reagents to nona-2,7-diene-1,9-dioic acid ester **66**, in which even a simple auxiliary such as menthyl group produces a high diastereo- and enantioselectivity (Eq. 17).[40]

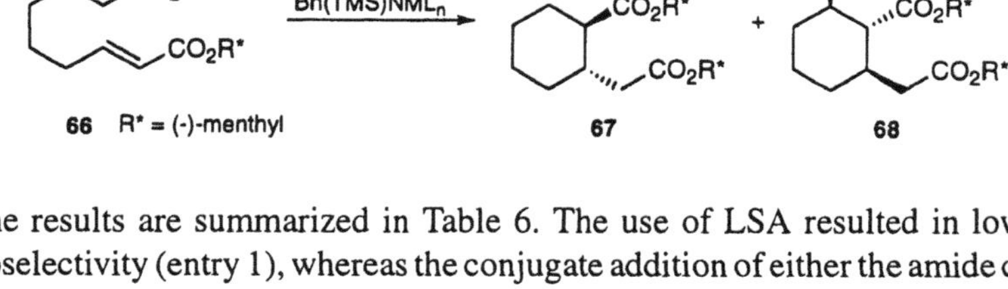 (17)

The results are summarized in Table 6. The use of LSA resulted in low diastereoselectivity (entry 1), whereas the conjugate addition of either the amide cuprate

Table 6. Asymmetric Cyclization of **66**[a]

Entry	*Metal Amide*	*Additive*	*Total Isolated Yield (%)*	*Diastereomer Ratio* **67 : 68**
1	Bn(TMS)NLi	—	70	61: 39
2	$[Bn(TMS)N]_2CuLi$	—	87	77: 23
3	$[Bn(TMS)N]_3ZnLi$	—	85	75: 25
4	$[Bn(TMS)N]_2CuLi$	$ZnCl_2$	60	87 : 13
5	$[Bn(TMS)N]_2CuLi$	$MgBr_2$ [b]	40	95 : 5
6	$[Bn(TMS)N]_2CuLi$	$MgBr_2$ [c]	56	91 : 9
7	$[Bn(TMS)N]_2CuLi$	$BF_3 \cdot OEt_2$	84	77 : 23
8	$[Bn(TMS)N]_3ZnLi$	$MgBr_2$ [b]	50	85 : 15
9	$[Bn(TMS)N]_3ZnLi$	$MgBr_2$ [c]	77	83 : 17

Notes: [a]Excess amounts of the amide reagent (3–9 equiv) were used. The diastereomer ratios were determined by ^1H NMR spectra.

[b]Commercially available $MgBr_2$ was crystallized from EtOH: *Purification of Lab. Chemicals*; Perrin, D. D., Armarego, W. L. F., Eds.; Pergamon: New York, 1988, p 331.

[c]The $MgBr_2$ was prepared in situ from the reaction of 1,2-dibromoethane with Mg in ether at 0 °C.

or zincate reagent enhanced both chemical yield and diastereoselectivity (entries 2 and 3). It should be noted that "cuprate" or "zincate" does not mean that the reagent possesses the structure $(R_2N)_2CuLi$ or $(R_2N)_3ZnLi$, respectively, but it indicates the stoichiometry of R_2N, Cu (or Zn), and Li. The amide cuprate was prepared from 2 equiv of LSA and 1 equiv of CuI in THF, and the zincate was prepared from 3 equiv of LSA and 1 equiv of $ZnCl_2 \cdot TMEDA$ complex in THF. The use of bidentate chelating reagents, such as $ZnCl_2$ and $MgBr_2$, as an additive further enhanced the diastereoselectivity (entries 4–6 and 8–9), whereas a monodentate Lewis acid $BF_3 \cdot OEt_2$ did not exert any significant influence upon the chemical yield and selectivity (entry 7 vs. 2).

It is widely accepted that a simple chiral auxiliary such as menthyl group does not produce high diastereoselectivity in the conjugate addition of nucleophiles to enoates. To accomplish high de, elaborated auxiliaries such as 8-phenylmenthyl,[41] camphor derivative,[42,43] pantolactone,[44] and oxazolidine[45] have been used. In fact, the zinc chloride mediated conjugate addition of the amide cuprate to menthyl monoenoates, **69a** and **b**, produced a 7:3 mixture of diastereomers **70** and **71** (Eq. 18). Comparison of this ratio with those of entries 2 and 4 in Table 6 suggested that the participation of another enoate moiety in the asymmetric conjugate addition of

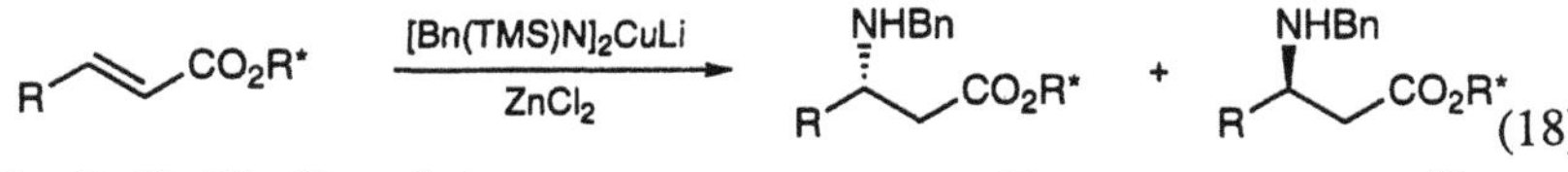

(18)

69a; R = Me, R* = (-)-menthyl
b; R = C_7H_{15}, R* = (-)-menthyl

70 **71**

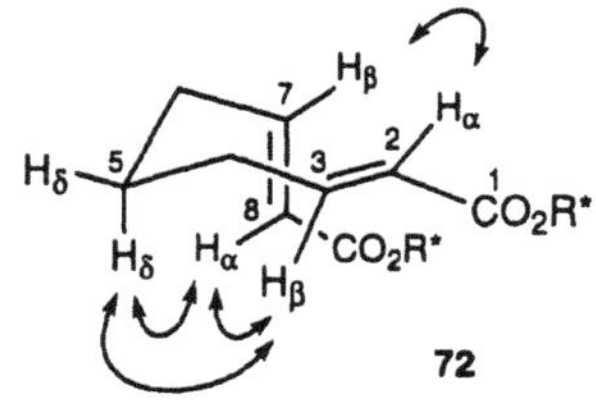

72

the amide cuprate to a double bond of **66** would cause the enhancement of the diastereoselectivity in the diendioate system. Even in the absence of Lewis acids such as $ZnCl_2$ and $MgBr_2$, NOEs were observed between H_α at C-2 and H_β at C-7, H_β at C-3 and H_α at C-8, H_β at C-3 and H_δ at C-5, and H_α at C-8 and H_δ at C-5. The NOEs were observed not only in $CDCl_3$ but also in THF-d_8, which was used as a solvent in the asymmetric cyclization. Accordingly, the diendioate **66** adopts a folded conformation (**72**) in solution, instead of a straightened structure. The bidentate Lewis acids such as $ZnCl_2$ and $MgBr_2$ chelate two oxygen atoms of the enoate groups, assisting **66** to take the folded structure. Boron trifluoride etherate, a monodentate Lewis acid, does not participate to fold the framework of the nonadiendioic acid ester.

Diastereoselective formation of **67** is accounted for by the nucleophilic addition to **73** (Scheme 5). The addition to either double bond of **73** produces **67**. The result of NOE experiments is in good agreement with this geometry. The addition to **74** from the direction shown by a solid arrow produces **67**, although the addition from the direction indicated by a dotted arrow affords **68**. Accordingly, the tandem conjugate addition to **74** would result in low diastereoselectivity. Furthermore, the geometry **74** is inconsistent with the observation obtained in the NOE measurement. It should be noted that the enoate geometry of **73** is a (*S*)-*cis* form whereas a

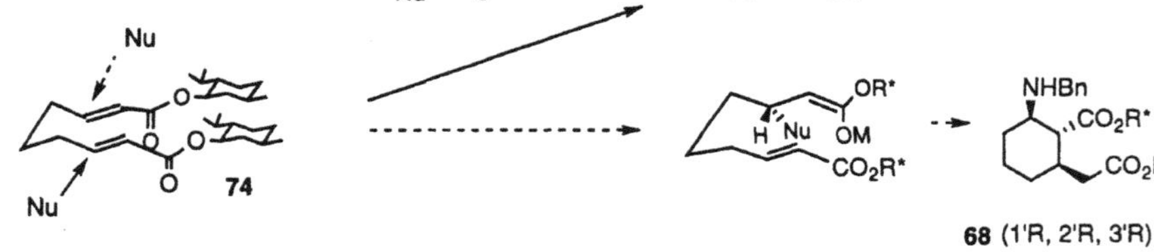

Scheme 5.

(*S*)-*trans* form is involved in the conjugate addition of organocopper-BF_3 reagents to 8-phenylmenthyl crotonate.[43]

XI. ALDOL CONDENSATION OF LITHIUM ENOLATES

The reactions of lithium enolates **32** and **34** with benzaldehyde and acetaldehyde are summarized in Table 7.[46–48] The reaction was normally quenched by water, but in some cases quenched with acetyl chloride to make isolation of products easy. The reaction proceeded in good yields except for the last case of Table 7. The (*Z*)-enolate **32** gave the *syn*(C-1 and C-2)–*anti*(C-2 and C-3) isomer **75** predominantly, while the (*E*)-enolate **34** produced the *anti*(C-1 and C-2)–*syn* (C-2 and C-3) isomer **76** preferentially (Eq. 19; C-numbers were shown in **75**). As expected, the diastereofacial selectivity was highly dependent upon the geometry of the enolates.

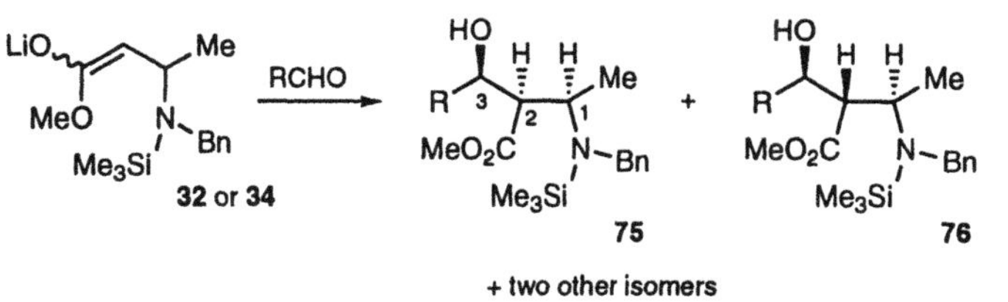

(19)

Concerning the diastereofacial selectivity between C-1 and C-2 of the aldol products, the (*Z*)-enolate **32** produces the *syn* selectivity predominantly, and the (*E*)-enolate **34** gives the *anti* selectivity preferentially. As mentioned above, the (*Z*)-enolate takes chelation structure (**37**). The aldehyde presumably approaches preferentially from the top side, since the approach from the bottom side is unfavorable owing to the presence of the pseudoaxial hydrogen. However, the following question may be raised. Why does the aldehyde attack selectively from the top side but not the alkyl halides (see Table 5)? The reason is not clear at present, but we assume that coordination of the lithium to the aldehyde oxygen produces the chelated transition state (**77**) and that such a π-π matching is more sensitive to the steric circumstance around the enolate face in comparison with the π-σ (RX)

Table 7. Aldol Reaction of **32** and **34** with Aldehydes

Entry	*Lithium Enolate*	*RCHO*	*Yield (%)*	**75**	:	*76*	:	*Two Other Isomers*
1	**32** (Z)	PhCHO	72	64	:	3	:	33
2	**32** (Z)	MeCHO	73	82	:	0	:	18
3	**34** (E)	PhCHO	64	6	:	80	:	14
4	**34** (E)	MeCHO	39	0	:	90	:	10

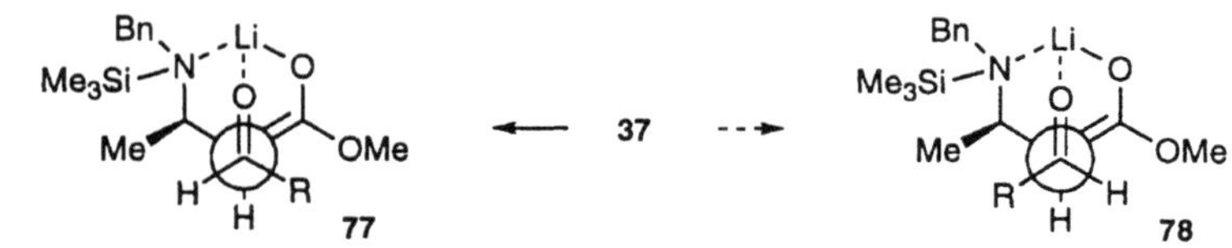

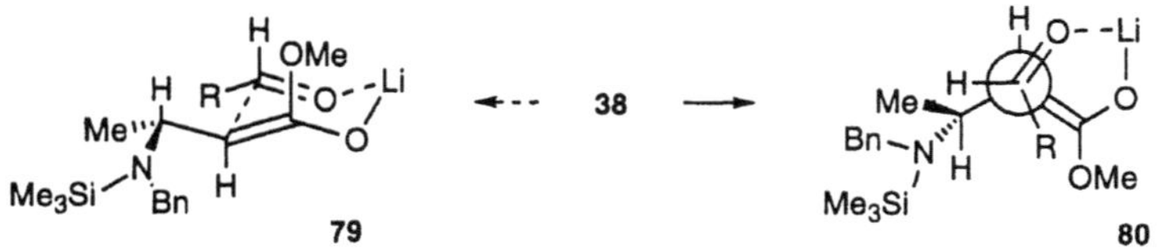

matching. More importantly, the model **77** can explain the *anti* selectivity between C-2 and C-3. An alternative model **78**, which produces the *syn* selectivity, is destabilized in comparison with **77** owing to the steric repulsion between the R and Me group.

The (*E*)-enolate gives predominantly C-1 and C-2 *anti*–C-2 and C-3 *syn* selectivity. The C-1 and C-2 *anti* selectivity is in good agreement with the *syn* selectivity of alkylation reactions (see Table 5), and can be explained similarly by the model **38**. The C-2 and C-3 *syn* selectivity cannot be explained by the ordinary proposed chair transition state model (**79**).[49] The *anti–anti*-isomer should be produced through **79**. Presumably, the distorted six-membered cyclic model (**80**), so-called skewed transition state, is involved in the aldol condensation owing to the presence of the very bulky silylamino group.

XII. ASYMMETRIC THREE-COMPONENT COUPLING PROCESS

A. Auxiliary Control

We examined the asymmetric synthesis of β-lactam framework with amidocuprate(I) $[Bn(TMS)N]_2CuLi$ (**12a**).[50] The conjugate addition of **12a** to **81a** gave **82a** in 80% yield with 72% de, and **81b** produced **82b** in 95% yield with 74% de (Eq. 20). The addition to **83** afforded **84** in 78% yield with 90% de (Eq. 21). The diastereomer ratios of **82a**, **82b**, and **84** were determined by their ^{1}H NMR spectra. It is noteworthy that the *R* configuration is produced in the amidocuprate(I) addition, whereas the opposite absolute stereochemistry is produced in high-pressure-induced reaction of R_2NH.[51] The "higher order" reagent $[Bn(TMS)N]Cu(CN)Li_2$ (**12b**) gave similar results. However, the aldehyde trapping reaction proceeded more smoothly and clearly with the higher order reagent **12b** than the trapping reaction using **12a**.

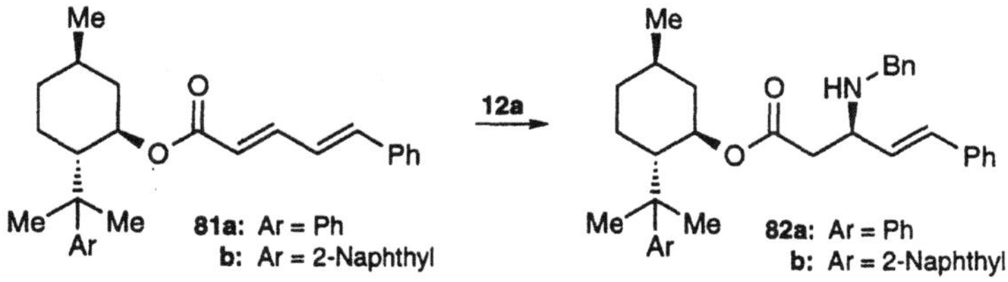

(20)

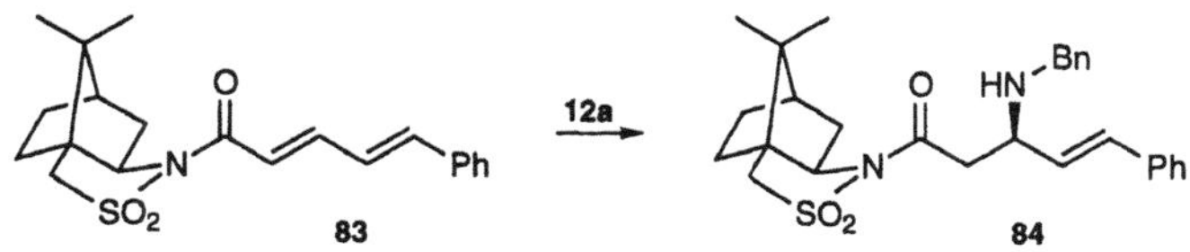

(21)

Finally, three-component coupling was carried out with **83**. The conjugate addition of the "higher order" reagent to **83**, followed by trapping with acetaldehyde, and subsequent protection of the hydroxyl group with *t*-butyldimethylsilyl chloride, gave **85b** as a single isomer in 71% overall yield from **83**. No other diastereomers were detected. Since the free hydroxyl form **85a** was unstable, protection was needed prior to isolation and purification. The reduction of **85b** with LAH led to deprotection of the bornanesultam X_N^* [52], and gave the corresponding alcohol in 60% yield. Protection of the alcohol with Et_3SiCl and the NH with $(Boc)_2O$ produced **86** in 60% yield. Selective deprotonation of the Et_3Si group followed by Swern oxidation and $NaClO_2$ oxidation afforded **87** in 68% yield. Removal of Boc with TFA followed by a standard cyclization procedure[53] gave **88** (**88a**: 47%, **88b**: 18%) in 65% yield (Scheme 6). Thus, three contiguous chiral centers can be precisely controlled in good yields by the three-component coupling process. Although the absolute stereochemistry of **88** does not correspond to natural β-lactams, a known technology[2i] can convert it into the correct configuration.

It is widely accepted that enoate–Lewis acid complexes prefer the *s-trans* conformation not only in the ground state, but also in the transition state of the reactions involving those complexes.[54] On the other hand, the relative populations of the *s-cis* and *s-trans* conformers of uncomplexed methyl cinnamate are almost equal in the isolated molecule at very low temperature and in solution at room and low temperatures. It has been clarified that the conjugate addition of metal amides to uncomplexed enoates proceeds predominantly through the *s-cis* conformation, and that most organocopper conjugate additions in the absence of Lewis acids or related metal salts take place preferentially in the *s-cis* conformation. Since **82a** (*R* at C3) and **84** (*R* at C3) are obtained from **81a** and **83**, respectively, the addition of the amide cuprate reagent takes place via the *s-cis* geometries of the (–)-8-phenyl-*p*-menth-3-yl ester **89** and the *N*-(2,4-pentadienoyl)-10,2-bornanesultam (**90**). The reagent attacks the β-carbon of **89** from the direction shown by an arrow, and that of **90** from the bottom side of a plane occupied by the dienoate (see an arrow at

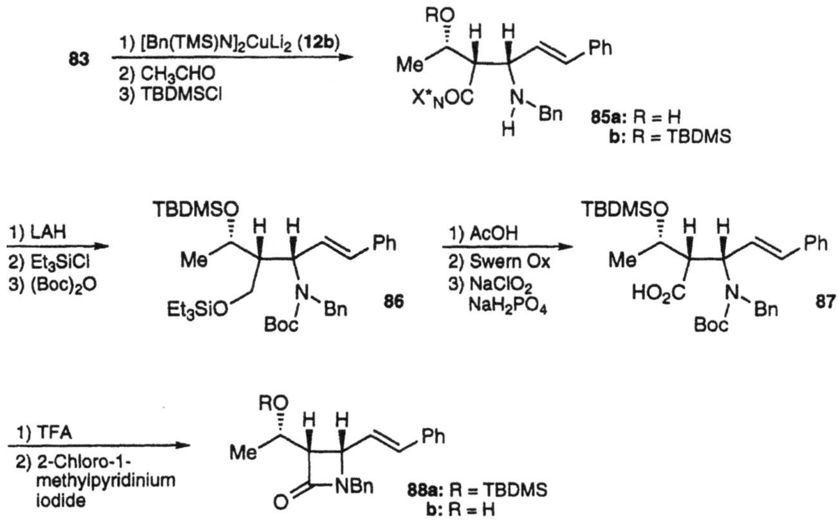

Scheme 6.

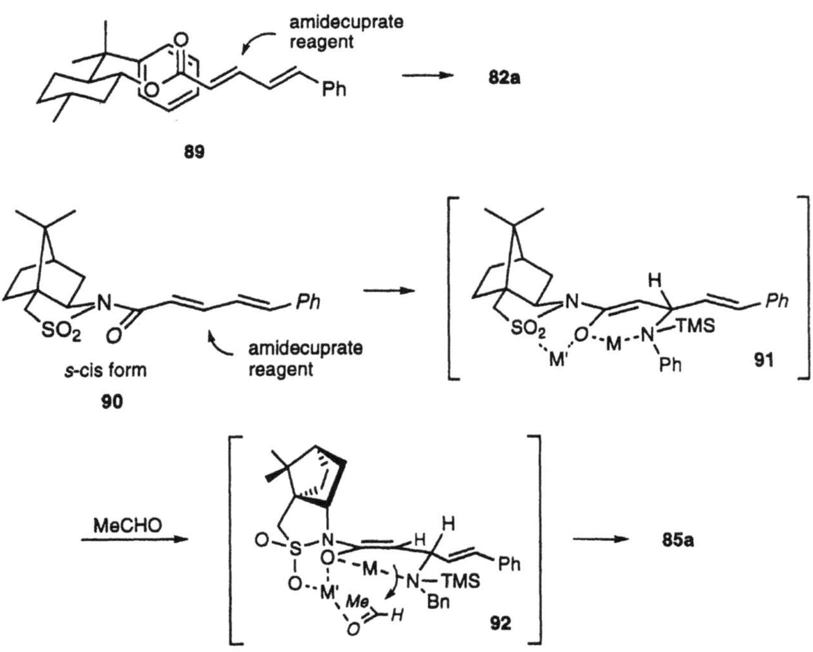

Scheme 7.

Scheme 7). The conjugate addition of $[Bn(TMS)N]_2Cu(CN)Li_2$ to the *s-cis*-form **90** would give the (*Z*)-type enolate **91**, in which the metal M (either Li or Cu) would chelate the nitrogen and oxygen atoms to form a six-membered ring. It is established that the conjugate addition of a lithium amide to an enoate produces the corresponding (*Z*)-enoate. The reaction of **91** with acetaldehyde would proceed as shown in **92**: A carbonyl oxygen would be chelated to M (and/or M'). This transition-state geometry would lead to the aldol condensation product **85a**.

B. Reagent Control

Hawkins and Lewis have reported the highly diastereoselective 1,4-addition of the chiral lithium amide of 3,5-dihydro-4*H*-dinaphth[2,1-*c*:1′,2′-*e*]azepine to α,β-unsaturated esters.[55] Davies and Ichihara have shown that the conjugate addition of homochiral lithium (*R*)-(α-methylbenzyl)-benzylamide (*R*)-**93** to certain enoates proceeds with very high diastereoisomeric excess.[56] Asymmetric conjugate addition of amines to α,β-unsaturated esters and nitriles has been reported.[13a,d,51,57] We examined the reaction of the chiral lithium amide (*R*)-**93** to α,β,γ,δ-unsaturated ester **8** (Eq. 22). The lithium amide **93** was found to be excellent reagent for asymmetric conjugate addition and regioselective 1,4-addition took place to give the corresponding β-amino esters in 81% isolated yield from **8c**, in 83% yield from **8d**, and in 98% yield from **8e**. In all cases only one diastereoisomer was produced. It should be noted that 1,6-addition does not take place; organocopper addition to dienoates often produces a mixture of 1,4- and 1,6-conjugate adducts.[58]

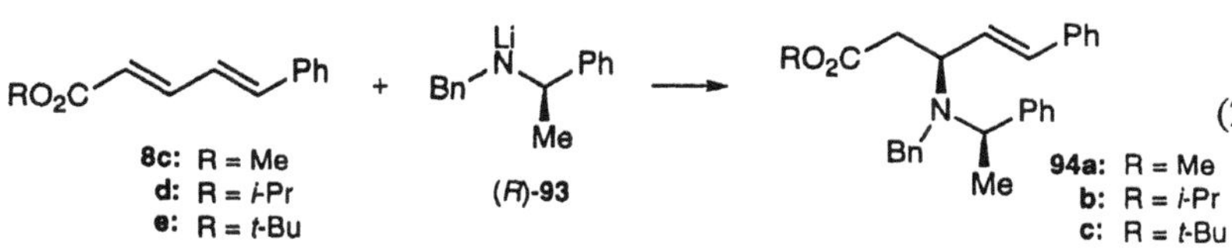

(22)

The β-amino ester **94c**, obtained from **8e** in 98% yield with >99% de, was treated with 3 equiv of LDA in THF at 0 °C and the resulting mixture was stirred for 2 h at this temperature. The mixture was cooled to −78 °C and then acetaldehyde (10 equiv) was added. Although the aldol products, **95** and its diastereomers, were obtained in quantitative yield, the diastereoisomer ratio was not high (entry 1, Table 8). To enhance the diastereoselectivity of the aldol process, we examined several additives (Eq. 23, Table 8). The use of trialkylboranes[59] and butyl borate as an additive did not give a satisfactory result (entries 2–4). $Bu_2BOSO_2CF_3$, Et_3Al, Bu_3SnCl,[50] $ZnCl_2$, and $(C_5H_5)_2ZrCl_2$[60] also gave unsatisfactory results. Finally we found that the use of trimethyl borate produced the highest de among the additives examined (entry 5). An attempt to generate in situ a boron enolate from **94c** upon treatment with dibutylboron trifluoromethanesulfonate and triethylamine[61] resulted in failure.

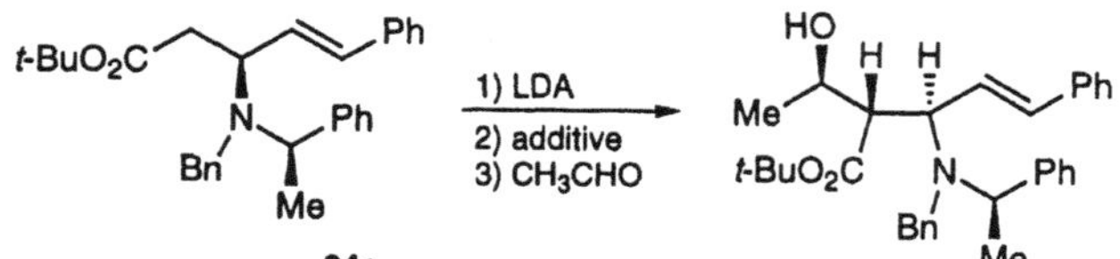

94c 95 (23)

Table 8. Aldol Reaction of the Enolate Derived from **94c**

Entry	*Additive*	*Product Ratio* **95** *: Other Diastereomers*	*Yield (%)*
1	none	78:22	100
2	BBu_3	81:19	72
3	BEt_3	86:14	82
4	$(BuO)_3B$	75:25	82
5	$(MeO)_3B$	91:9	89

Protection of the hydroxy group of **95** with TBDMSCl gave **96** in 90% yield. Hydrogenation in the presence of a catalytic amount of $Pd(OH)_2$ on carbon produced **97** in 60% yield. Treatment with trifluoroacetic acid in CH_2Cl_2 followed by cyclization with PPh_3–$(PyS)_2$–MeCN[62] gave **98** in 55% yield (Scheme 8).

It is clear that the modified three-component coupling process via the chiral lithium amide **93** provides the β-lactam framework **98** having correct absolute configurations at C-1, C-2, and C-3 positions (C-numbers were shown in **98**).

C. Substrate Control

Based on the previous results of 1,2-asymmetric induction we mentioned above, the reaction of LSA with *t*-butyl (4*S*)-4-trityloxy-2-pentenoate (**14d**) followed by

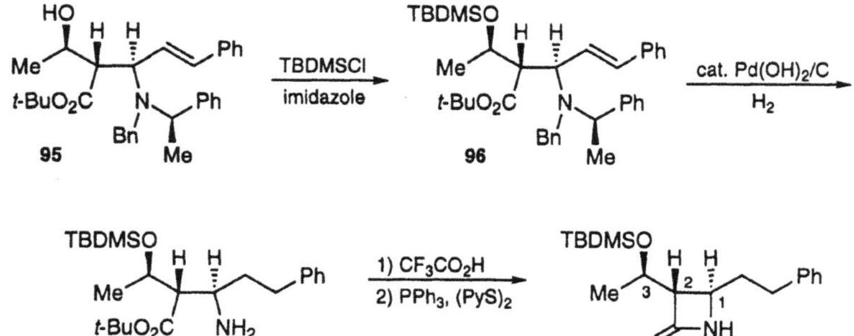

95 96 97 98

Scheme 8.

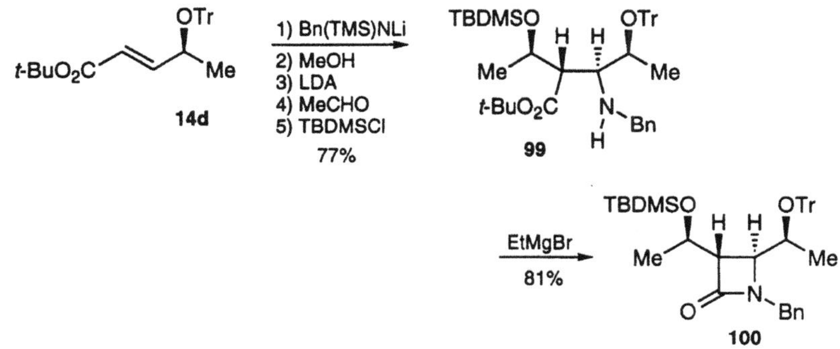

Scheme 9.

aldol condensation with acetaldehyde was examined. The desired diastereomer **99** was obtained as a single product in 77% yield (all in one pot). Conversion of **99** to the azetidinone **100** by EtMgBr[63] proceeded in 81% yield (Scheme 9).

Since (*E*)-enolates are formed stereoselectively from the reaction of β-amino esters with LDA, the (*E*)-isomer **101** is presumably a key intermediate for the aldol condensation of **15d**. The electrophilic attack of acetaldehyde to **101** would take place as shown in **102**; a hydrogen atom adopts inside due to severe 1,3-allylic strain by *t*-BuO group. The condensation would occur via a synclinal six-membered cyclic transition state **103**, giving **99** with essentially 100% de.

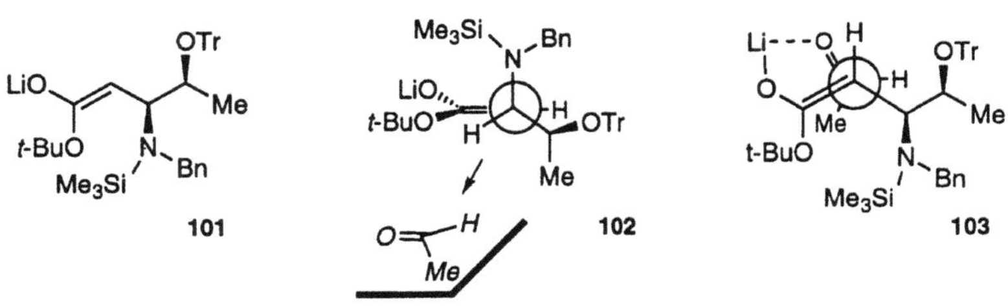

D. Reagent and Substrate Control

We examined the conjugate addition of several lithium amides to (4*R*)-γ-methyl-substituted α,β-unsaturated ester **104a**. The addition of LSA (Bn(TMS)NLi) gave a 73:27 mixture of **105a** and **106a** in 93% yield (Eq. 24). The conjugate addition of lithium dibenzylamide afforded a 73:27 mixture of **105b** and **106b** in 84% yield. The predominant formation of the *anti*-isomer **105** can be explained by a modified Felkin–Anh model **107** in which the largest silyloxymethyl group is in the *anti* position and the medium methyl group is in the inside and the lithium amide reagents attack the β-carbon of **104a** from the less hindered outside.

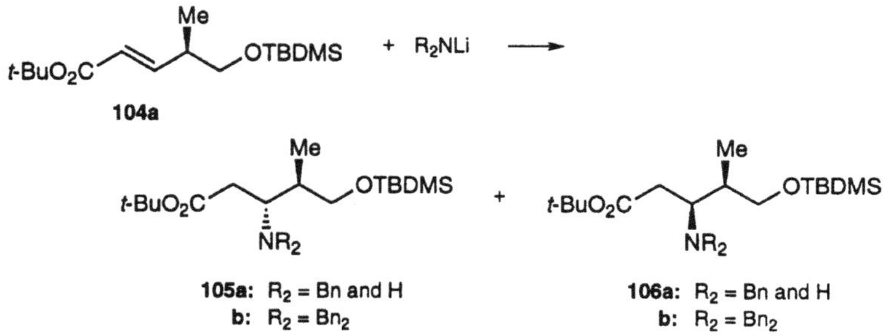

(24)

Bn(TMS)NLi	73	:	27
Bn_2NLi	73	:	27

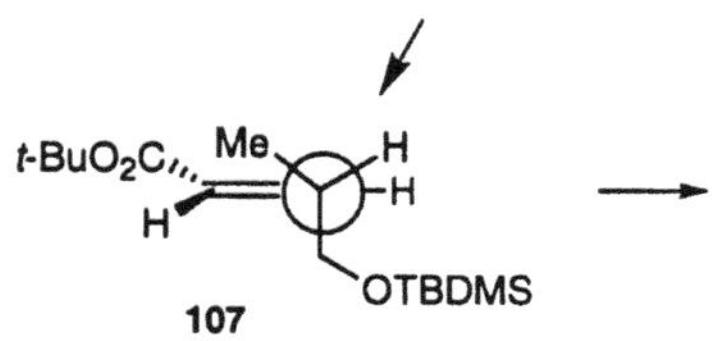

The conjugate addition of (*R*)-**93** to **104a** produced the *syn*-diastereoisomer **108** with essentially 100% de in 84% yield (Eq. 25). On the other hand, the addition of (*S*)-**93** to **104a** gave the *anti*-diastereoisomer **109** with essentially 100% de in 95% yield (Eq. 26). Accordingly, the asymmetric induction at the β-position of **104a** is controlled completely by the chirality of the lithium amide reagent, and the effect of the chirality of the γ-carbon upon the asymmetric induction is very small. The higher chemical yield in Eq. 23, in comparison with the yield in Eq. 22, suggests that the combination between (4*R*)-**104a** and (*S*)-**93** is a matched pair; this is supported by the predominant formation of the *anti*-isomer **105** in Eq. 24.

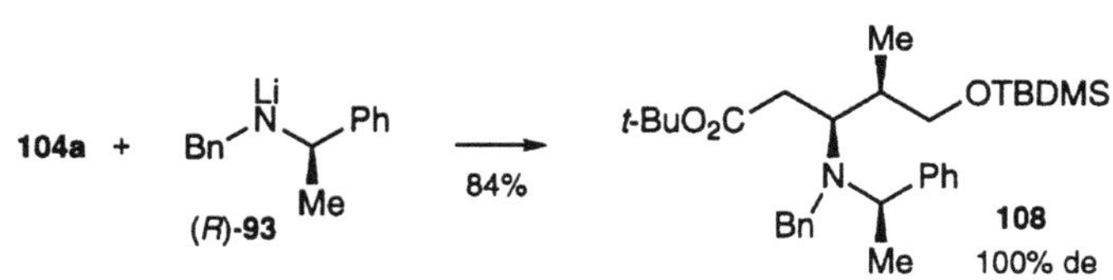

(25)

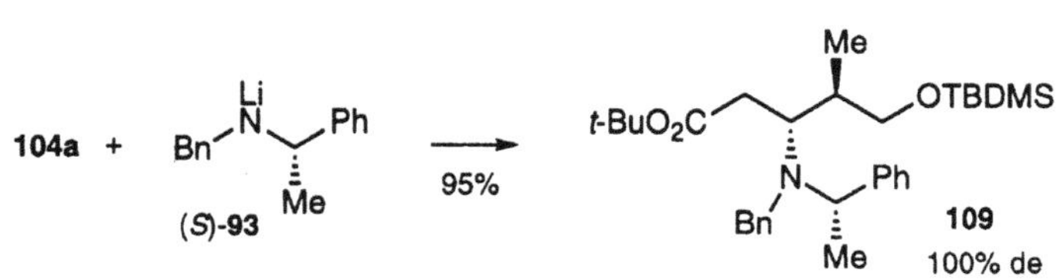

(26)

We next examined the conjugate addition to the (*Z*)-enoate **104b**, since it was known that the diastereoselectivity of the conjugate addition of organocopper reagents to γ-chiral α,β-unsaturated esters was dependent upon the geometry of the double bond.[64] The addition of (*R*)-**93** to **104b** gave the *syn*-isomer **108** with essentially 100% de in 77% yield, whereas the addition of (*S*)-**93** to **104b** afforded the *anti*-isomer **109** with essentially 100% de in 84% yield (Eqs. 27 and 28). Therefore, the double-bond geometry of **104** did not exert any influence upon the sense and extent of asymmetric induction. This observation is not in agreement with the previous result obtained from the conjugate addition of organocopper reagents to γ-methyl-substituted enoates. We carefully investigated the addition to **104a** and **104b** in order to clarify this difference.

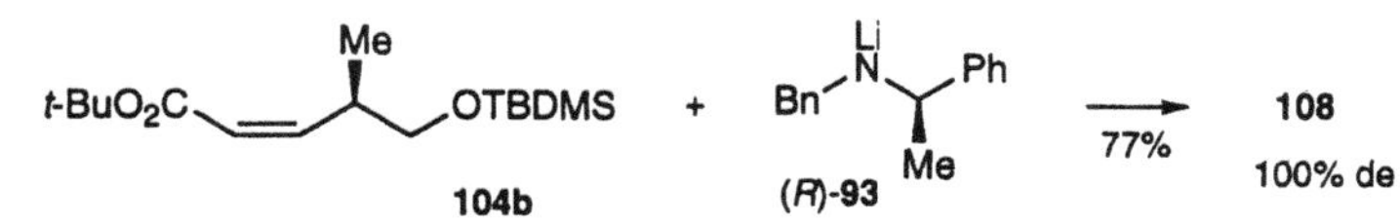

(27)

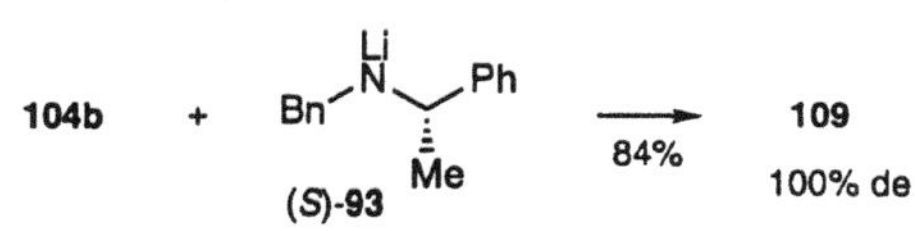

(28)

The reactions shown in Eqs. 25–28 were completed within 2 h at –78 °C. When the reaction of **104b** with (*R*)-**93** was stopped at an early stage, the formation of **104a** was observed along with the production of **108**. However, the formation of **104b** was not detected on the way of the reaction of **104a** with (*R*)-**93**. The time dependences of the yields of **104a**, **104b**, and **108** are shown in Figure 1. The progress of the reaction was followed by ^{1}H NMR spectra of the product mixture. It is now clear that the isomerization from **104b** to **104a** takes place in the reaction of **104b** (Eq. 27), whereas no isomerization occurs in the reaction of **104a** (Eq. 25).

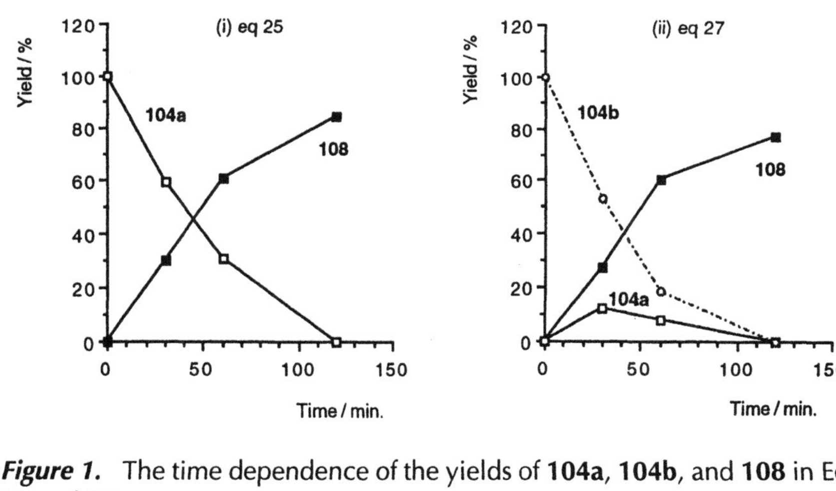

Figure 1. The time dependence of the yields of **104a**, **104b**, and **108** in Eqs. 25 and 27.

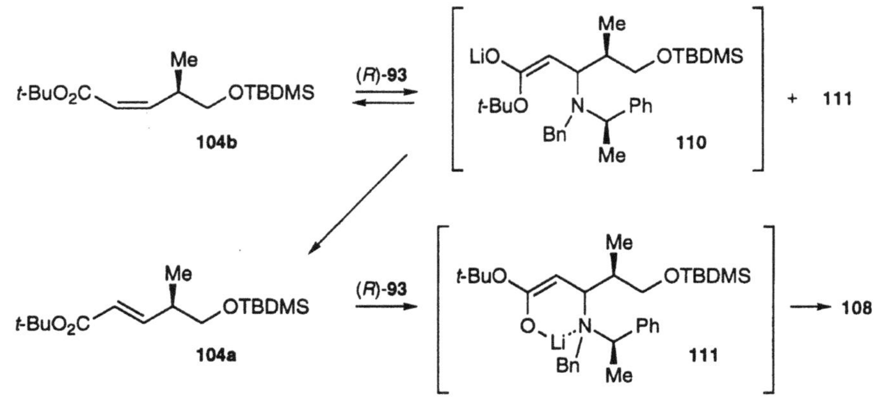

Scheme 10.

The *Z*/*E* isomerization might occur via a hydrogen abstraction at the γ-position by the lithium amide base. If this is the case, racemization at the γ-chiral carbon should be observed. Careful investigation indicated that only the β-amino ester isolable from the reaction mixture was **108** in which the absolute configuration at C-4 was *R*. Accordingly, no racemization takes place, indicating that a hydrogen abstraction at the γ-position does not occur.

A mechanistic rationale for the isomerization is shown in Scheme 10. The conjugate addition of (*R*)-**93** to **104b** presumably produces a mixture of the (*Z*)- and (*E*)-enolates **110** and **111**. The retroconjugate addition from **110** would take place under the reaction conditions, giving a mixture of **104b** and **104a**. The addition of (*R*)-**93** to **104a** would afford stereoselectively the (*E*)-enolate **111**, which does not undergo the retro-Michael addition due to the lithium chelation between a nitrogen and oxygen atom. The (*E*)-enolate **111** affords **108** upon hydrolysis. However, a possibility that the addition of (*R*)-**93** to **104b** produces **108** directly cannot be rigorously excluded.

Since the asymmetric induction at the β-position of **104** was controlled completely by the chirality of the Davies reagent, it occurred to us that the kinetic resolution of racemic **104** would take place and the desired diastereomer **108** might be obtained from the racemic substrate. The addition of 0.5 equiv of (*R*)-**93** to **104c** (racemic) gave 24% of the *anti*-diastereomer **112** and 56% of the recovered ester, whereas the addition of 0.5 equiv of (*S*)-**93** to **104c** afforded 27% of another *anti*-isomer **109** and 55% of the recovered ester (Eq. 29). The optical activity of the recovered ester was not determined. Here again, the asymmetric induction at the β-position of **104c** was dictated completely by the chirality of the chiral lithium amide. The *anti*-diastereoisomers **112** and **109** were stereoselectively obtained, and no *syn*-isomers were formed. This is reasonable because the attack of the reagents to **104c** proceeds primarily via the modified Felkin–Anh geometry **107** which

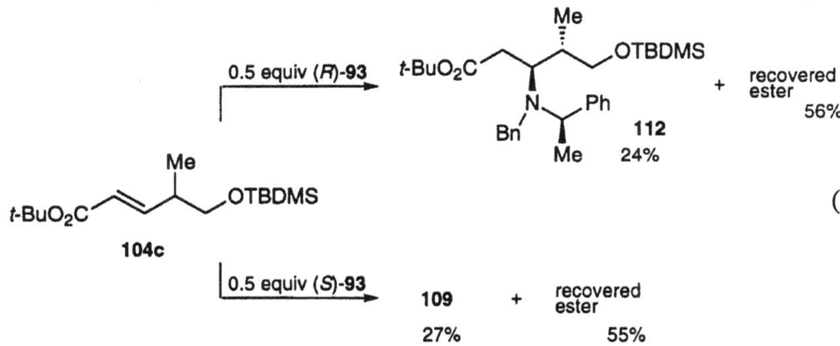

(29)

produces the *anti*-isomers **112** and **109**. Accordingly, the use of **104a** is essential to obtain the desired diastereomer **108** in which the absolute configurations at C-3 and C-4 (3*S*,4*R*) are in agreement with those of the 1β-methylcarbapenem key intermediate **2**.

XIII. ALDOL REACTION WITH ACETALDEHYDE: SYNTHESIS OF 1β-METHYLCARBAPENEM KEY INTERMEDIATE

Treatment of **108** with LDA in THF at 0 °C for 2 h, followed by addition of acetaldehyde at –78 °C either in the absence or presence of additives, gave the desired diastereomers (Eq. 30). The results are summarized in Table 9. The ratio of **113** to other diastereoisomers was 59:41 in the absence of additives (entry 1). The isomer ratio was not improved even using triethylborane and trimethyl borate as an

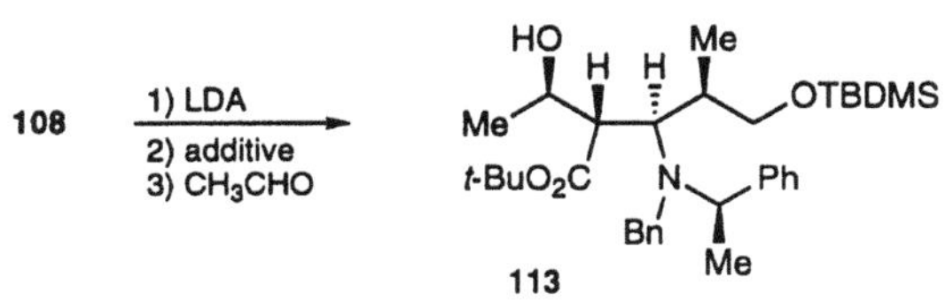

(30)

Table 9. Aldol Reaction of **108** in the Presence of Additives

Entry	*Additive*	*Product Ratio* **113** *: Other Diastereomers*	*Yield (%)*
1	none	59:41	87
2	BEt_3	53:47	74
3	$B(OMe)_3$	64:36	93
4	$EtAlCl_2$	77:23	78
5	$MeAlCl_2$	80:20	79

additive (entries 2 and 3); the use of trimethyl borate gave the highest diastereomeric ratio (91:9) in the condensation of acetaldehyde with the β-amino ester bearing no γ-methyl substituent as mentioned above. Instead of the boron additive, the use of aluminium dichloride gave the best result. The desired diastereomer **113** was separated easily from other diastereomers through silica gel column chromatography.

The hydrogen abstraction at the α-position of **108** by LDA presumably proceeds through a six-membered cyclic transition state **114**, as proposed by Ireland.[22] Actually, it is confirmed that treatment of a β-amino ester with LDA under the similar conditions produces the (*E*)-enolate with high stereoselectivity as mentioned above. The resulting (*E*)-enolate possesses a sterically demanding *tert*-butyloxy group at the *cis* position of the double bond, adopting conformation **115** as a stable conformer in order to avoid a severe 1,3-allylic strain. Acetaldehyde attacks from the less hindered side of the double bond, as shown in **115**, leading to the predominant formation of **113** (Eq. 31). It may be argued that **115** and **110** are the same (*E*)-enolates and **115** undergoes the aldol condensation without the retro-Michael addition which was observed in the case of **110**. Perhaps this difference is due to the difference of aggregation states of the lithium enolates and/or to the presence of $i\text{-}Pr_2NH$ in the reaction medium via **115**.

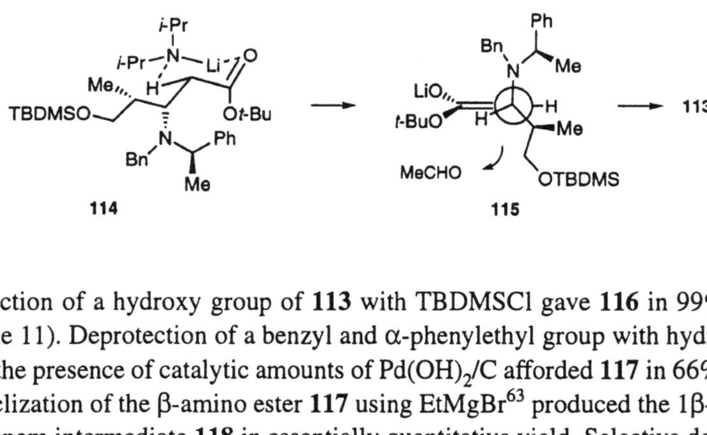

(31)

Protection of a hydroxy group of **113** with TBDMSCl gave **116** in 99% yield (Scheme 11). Deprotection of a benzyl and α-phenylethyl group with hydrogenation in the presence of catalytic amounts of $Pd(OH)_2/C$ afforded **117** in 66% yield. The cyclization of the β-amino ester **117** using EtMgBr[63] produced the 1β-methyl carbapenem intermediate **118** in essentially quantitative yield. Selective deprotection of a primary-OTBDMS group using NBS/aqueous DMSO[6c,65] afforded **119** in 73% yield. Treatment of **119** with $Me_2C(OMe)_2/BF_3\cdot OEt_2$ gave **120** in essentially quantitative yield. The stereostructure of **120** was confirmed unambiguously by comparing its ^1H NMR data with those of the authentic sample. Now it is clear that the modified TCC method provides efficiently a 1β-methyl carbapenem key intermediate **118**: (1) the conjugate addition of (*R*)-**93** to **104a** produces **108** with essentially 100% de (84% yield); (2) the conversion of **108** to **113** proceeds in 63% yield; and (3) **118** is obtained from **113** in 65% overall yield.

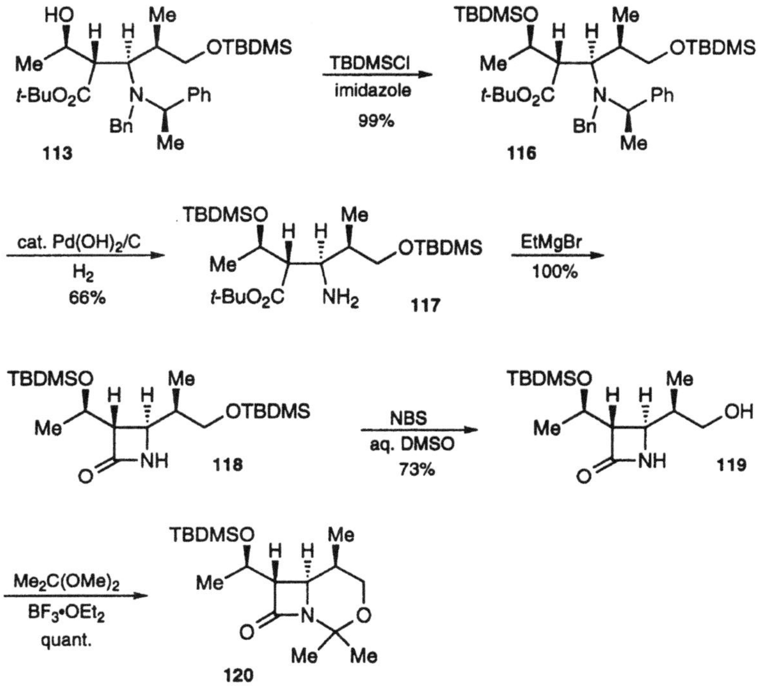

Scheme 11.

REFERENCES AND NOTES

1. (a) Berks, A. H. *Tetrahedron* **1996**, *52*, 331. (b) Ito, Y.; Terashima, S. In *Studies in Natural Products Chemistry*; Atta-ur-Rahman, Ed.; Elsevier: Amsterdam 1993, Vol. 12, pp. 145 (Chem. Abstr. 119:203186); (c) Ito, Y.; Terashima, S. *J. Synth. Org. Chem. Jpn.* **1989**, *47*, 606.
2. (a) Sunagawa, M.; Sasaki, A. *J. Synth. Org. Chem. Jpn.* **1996**, *54*, 761; (b) Hirai, K. *J. Synth. Org. Chem. Jpn.* **1992**, *50*, 112; (c) *Recent Progress in the Chemical Synthesis of Antibiotics*; Lukacs, G.; Ohno, M., Eds.; Springer-Verlag: Berlin, 1990, p. 533; (d) Nagahara, T.; Kametani, T. *Heterocycles* **1987**, *25*, 729; (e) Sugimura, M.; Hiraoka, T. *Yakugakuzasshi* **1987**, *107*, 175; (f) Kobayashi, S.; Gendaikagaku, Z. *Advanced Studies on Antibiotics*; Tokyo Kagaku Dojin: Tokyo 1987, p. 205; (g) Nagao, Y. *Kagaku* **1987**, *42*, 190; (h) Fukagawa, Y.; Shibamoto, N.; Yoshioka, T.; Ishikura, T. *β-Lactam Pharmaceutical Compounds*; Ueda, Y.; Shimizu, K., Eds.; Nankoudo: Tokyo, 1987, p. 664; (i) Nakai, T.; Chiba, T. *Pharmacia* **1986**, *22*, 612; (j) Drckheimer, W.; Blumbach, J.; Lattrell, R.; Scheunemann, K. H. *Angew. Chem., Int. Ed. Engl.* **1985**, *24*, 180; (k) Shibuya, M. *J. Synth. Org. Chem. Jpn.* **1983**, *41*, 62; (l) Kametani, T.; Fukumoto, K.; Ihara, M. *Heterocycles* **1982**, *17*, 463; (m) *Chemistry and Biology of β-Lactam Antibiotics*; Morin, R. B.; Gorman, M., Eds.; Academic Press: New York, 1982, p. 227; (n) Kametani, T.; Ihara, M. *J. Synth. Org. Chem. Jpn.* **1980**, *38*, 1025; (o) Hirai, K. *J. Synth. Org. Chem. Jpn.* **1980**, *38*, 97.
3. (a) Kahan, J. S.; Kahan, F. M.; Goegelman, R.; Currie, S. A.; Jackson, M.; Stapley, E. O.; Miller, T. W.; Hendlin, D.; Mochales, S.; Hernandez, S.; Woodruff, H. B. *The 16th Interscience Conference on Antimicrobial Agents and Chemotherapy;* Chicago, 1976, Abstract No. 227; (b) Kropp,

H.; Kahan, J. S.; Kahan, F. M.; Sandolf, J.; Darland, G.; Birnbaum, J. *The 16th Interscience Conference on Antimicrobial Agents and Chemotherapy;* Chicago; 1976, Abstract No. 228; (c) Schönberg, G. A.; Arison, B. H.; Hensens, O. D.; Hirshfield, J.; Hoogsteen, K.; Kaczka, E. A.; Rhodes, R. E.; Kahan, J. S.; Kahan, F. M.; Ratcliffe, R. W.; Walton, E.; Ruswinkle, L. J.; Morin, R. B.; Christensen, B. G. *J. Am. Chem. Soc.* **1978**, *100*, 6491; (d) Kahan, J. S.; Kahan, F. M.; Goegelman, R.; Currie, S. A.; Jackson, M.; Stapley, E. O.; Miller, T. W.; Miller, A. K.; Hendlin, D.; Mochales, S.; Hernandez, S.; Woodruff, H. B.; Birnbaum, J. *J. Antibiot.* **1979**, *32*, 1.

4. Shih, D. H.; Baker, F.; Cama, L.; Christensen, B. G. *Heterocycles* **1984**, *21*, 29.
5. (a) Mastalerz, H.; Menard, M. *J. Org. Chem.* **1994**, *59*, 3223; (b) Uyeo, S.; Itani, H. *Tetrahedron Lett.* **1994**, *35*, 4377; (c) Murayama, T.; Yoshida, A.; Kobayashi, T.; Miura, T. *Tetrahedron Lett.* **1994**, *35*, 2271; (d) Sakurai, O.; Ogiku, T.; Takahashi, M.; Horikawa, H.; Iwasaki, T. *Tetrahedron Lett.* **1994**, *35*, 2187; (e) Hirai, K.; Iwno, Y.; Mikoshiba, I.; Koyama, H.; Nishi, T. *Heterocycles* **1994**, *38*, 277; (f) Nagao, Y.; Nagase, Y.; Kumagai, T.; Matsunaga, H.; Abe, T.; Shimada, O.; Hayashi, T.; Inoue, Y. *J. Org. Chem.* **1992**, *57*, 4243; (g) Nagao, Y.; Kumagai, T.; Nagase, Y.; Tamai, S.; Inoue, Y.; Shiro, M. *J. Org. Chem.* **1992**, *57*, 4232; (h) Bender, D. R.; DeMarco, A. M.; Melillo, D. G.; Riseman, S. M.; Shinkai, I. *J. Org. Chem.* **1992**, *57*, 2411; (i) Tanner, D.; He, H. M. *Tetrahedron* **1992**, *48*, 6079; (j) Kobayashi, Y.; Ito, Y.; Terashima, S. *Tetrahedron* **1992**, *48*, 6079; (k) Kita, Y.; Shibata, N.; Tohjo, T.; Yoshida, N. *J. Chem. Soc., Perkin Trans.* **1992**, 1795; (l) Ito, Y.; Sasaki, A.; Tamoto, K.; Sunagawa, M. *Tetrahedron* **1991**, *47*, 2801; (m) Uyeo, S.; Itani, H. *Tetrahedron Lett.* **1991**, *32*, 2143; (n) Kitamura, M.; Nagai, K.; Hsiao, Y.; Noyori, R. *Tetrahedron Lett.* **1990**, *31*, 549; (o) Shirai, F.; Nakai, T. *Chem. Lett.* **1989**, 445; (p) Kawabata, T. Kimura, Y.; Ito, Y.; Terashima, S.; Sasaki, A.; Sunagawa, M. *Tetrahedron* **1988**, *44*, 2149; (q) Shirai, F.; Nakai, T. *J. Org. Chem.* **1987**, *52*, 5491; (r) Fuentes, L. M.; Shinkai, I.; King, A.; Purick, R.; Reamer, R. A.; Schmitt, S. M.; Cama, L.; Christensen, B. G. *J. Org. Chem.* **1987**, *52*, 2563; (s) Prasad, J. S.; Liebeskind, L. S. *Tetrahedron Lett.* **1987**, *28*, 1857; (t) Kim, C. U.; Luh, B.; Partyka, R. A. *Tetrahedron Lett.* **1987**, *28*, 507; (u) Hatanaka, M. *Tetrahedron Lett.* **1987**, *28*, 83. (v) Fuentes, L. M.; Shinkai, I.; Salzmann, T. N. *J. Am. Chem. Soc.* **1986**, *108*, 4675; (w) Nagao, Y.; Kumagai, T.; Tamai, S.; Abe, T.; Kuramoto, Y.; Taga, T.; Aoyagi, S.; Nagase, Y.; Ochiai, M.; Inoue, Y.; Fujita, E. *J. Am. Chem. Soc.* **1986**, *108*, 4673; (x) Iimori, T.; Shibasaki, M. *Tetrahedron Lett.* **1986**, *27*, 2149; (y) Deziel, R.; Favreau, D. *Tetrahedron Lett.* **1986**, *27*, 5687; (z) Shibata, T.; Iino, K.; Tanaka, T.; Hashimoto, T.; Kameyama, Y.; Sugimura, Y. *Tetrahedron Lett.* **1985**, *26*, 4739.
6. (a) Rao, A. V. R.; Gurjar, M. K.; Khare, V. B.; Ashok, B.; Deshmukh, M. N. *Tetrahedron Lett.* **1990**, *31*, 271; (b) Rao, A. V. R.; Gurjar, M. K.; Ashok, B. *Tetrahedron Asym.* **1991**, *2*, 255; (c) Ihara, M.; Takahashi, M.; Fukumoto, K.; Kametani, T. *J. Chem. Soc., Chem. Commun.* **1988**, 9; *Heterocycles* **1988**, *27*, 327; (d) Udodong, U. E.; Fraser-Reid, B. *J. Org. Chem.* **1988**, *53*, 2131; **1989**, *54*, 2103; (e) Bayles, R.; Flynn, A. P.; Galt, R. H. B.; Kirby, S.; Turner, R. W. *Tetrahedron Lett.* **1988**, *29*, 6345; (f) Kaga, H.; Kobayashi, S.; Ohno, M. *Tetrahedron Lett.* **1989**, *30*, 113.
7. (a) Asao, N.; Uyehara, T.; Yamamoto, Y. *Bull. Chem. Soc. Jpn.* **1995**, *68*, 2103; (b) Tsukada, N.; Shimada, T.; Gyoung, Y. S.; Asao, N.; Yamamoto, Y. *J. Org. Chem.* **1995**, *60*, 143; (c) Asao, N.; Shimada, T.; Tsukada, N.; Yamamoto, Y. *Tetrahedron Lett.* **1994**, *35*, 8425; (d) Asao, N.; Tsukada, N.; Yamamoto, Y. *J. Chem. Soc., Chem. Commun.* **1993**, 1660; (e) Yamamoto, Y.; Asao, N.; Uyehara, T. *J. Am. Chem. Soc.* **1992**, *114*, 5427; (f) Asao, N.; Uyehara, T.; Yamamoto, Y. *Tetrahedron* **1990**, *46*, 4563; (g) Uyehara, T.; Asao, N.; Yamamoto, Y. *J. Chem. Soc., Chem. Commun.* **1989**, 753; (h) Asao, N.; Uyehara, T.; Yamamoto, Y. *Tetrahedron* **1988**, *44*, 4173; (i) Uyehara, T.; Asao, N.; Yamamoto, Y. *J. Chem. Soc., Chem. Commun.* **1987**, 1410.
8. (a) Rathke, M. W.; Sullivan, D. *Tetrahedron Lett.* **1972**, 4249; (b) Herrmann, J. L.; Kieczykowski, G. R.; Schlessinger, R. H. *Tetrahedron Lett.* **1973**, 2433.
9. Little, R. D.; Dawson, J. R. *Tetrahedron Lett.* **1980**, *21*, 2609.
10. Hase, T. A.; Kukkola, P. *Synth. Commun.* **1980**, *10*, 451.
11. Diekman, J.; Thomson, J. B.; Djerassi, C. *J. Org. Chem.* **1967**, *32*, 3904; Narula, S.; Kapur, N. *Inorg. Chim. Acta* **1983**, *73*, 183.

12. For discussions on higher order cyano cuprates, see: Bertz, S. H. *J. Am. Chem. Soc.* **1990**, *112*, 4031; Lipshutz, B. H.; Sharma, S.; Ellsworth, E. L. *J. Am. Chem. Soc.* **1990**, *112*, 4032.
13. (a) Matsunaga, H.; Sakamaki, T.; Nagaoka, H.; Yamada, Y. *Tetrahedron Lett.* **1983**, *24*, 3009; (b) Dondori, A.; Boscarato, A.; Marra, A. *Synlett* **1993**, 256; (c) For the addition to 5-alkoxy-2(5*H*)-furanones, see: de Lange, B.; van Bolhuis, F.; Feringa, B. L. *Tetrahedron* **1989**, *45*, 6799; (d) For the addition to 2-hydroxyalkyl-propenoates, see: Perlmutter, P.; Tabone, M. *Tetrahedron Lett.* **1988**, *29*, 949.
14. Mulzer, J.; Kappert, M.; Huttner, G.; Jibril, I. *Angew. Chem. Int. Ed. Engl.* **1984**, *23*, 704.
15. (a)For the most recent article, see: Yamamoto, Y.; Chounan, Y.; Nishii, S.; Ibuka, T.; Kitahara, H. *J. Am. Chem. Soc.* **1992**, *114*, 7652, and references cited therein; (b) Yechezkel, T.; Ghera, E.; Ramesh, N.G.; Hassner, A. *Tetrahedron Asymm.* **1996**, *7*, 2423.
16. Asao, N.; Shimada, T.; Sudo, T.; Tsukada, N.; Yazawa, K.; Gyoung, Y. S.; Uyehara, T.; Yamamoto, Y. *J. Org. Chem.* **1997**, *62*, 6274.
17. Guanti, G.; Banfi, L.; Narisano, E. *Tetrahedron Lett.* **1991**, *32*, 6939, and references cited therein.
18. (a) Anh, N. T.; Eisenstein, O. *Nouv. J. Chem.* **1977**, *1*, 61; (b) Lodge, E. P.; Heathcock, C. H. *J. Am. Chem. Soc.* **1987**, *109*, 3353; (c) Wong, S. S.; Paddon-Row, M. N. *J. Chem. Soc., Chem. Commun.* **1990**, 456.
19. In this article, we defined that all metals including Li have higher priority than carbon for the describing the geometries of (*E*)- and (*Z*)-enolates.
20. Previous attempts to determine the stereochemistries of β-amino enolates, produced from the deprotonation of β-amino esters, resulted in failure. See: (a) Banfi, L.; Colombo, L.; Gennari, C.; Scolastico, C. *J. Chem. Soc., Chem. Commun.* **1983**, 1112; (b) Banfi, L.; Bernardi, A.; Colombo, L.; Gennari, C.; Scolastico, C. *J. Org. Chem.* **1984**, *49*, 3784.
21. (a) Fleming, I.; Hill, J. H. M.; Parker, D.; Waterson, D. *J. Chem. Soc., Chem. Commun.* **1985**, 318; (b) Fleming, I.; Kilburn, J. D. *J. Chem. Soc., Chem. Commun.* **1986**, 305.
22. Ireland, R. E.; Mueller, R. H.; Willard, A. K. *J. Am. Chem. Soc.* **1976**, *98*, 2868; Heathcock, C. H.; Pirrung, M. C.; Montgomery, S. H.; Lampe, J. *Tetrahedron* **1981**, *23*, 4087.
23. Seebach, D.; Estermann, H. *Tetrahedron Lett.* **1987**, *28*, 3103.
24. Paddon-Row, M. N.; Rondon, N. G.; Houk, K. N. *J. Am. Chem. Soc.* **1982**, *104*, 7162.
25. Yu, L. -C.; Helquist, P. *Tetrahedron Lett.* **1978**, 3423; *J. Org. Chem.* **1981**, *46*, 4536.
26. Still, W. C.; Schneider, M. J. *J. Am. Chem. Soc.* **1977**, *99*, 948.
27. Bernardi, A.; Beretta, M. G.; Colombo, L.; Gennari, C.; Poli, G.; Scolastico, C. *J. Org. Chem.* **1985**, *50*, 4442.
28. For the use of silica gel to eliminate an amino group, see: Snowden, R. L.; Wust, M. *Tetrahedron Lett.* **1986**, *27*, 699.
29. Vedejs, E.; Arnost, M. J.; Hogen, J. P. *J. Org. Chem.* **1979**, *44*, 3230; Oppolzer, W.; Gorrichon, L.; Bird, T. G. *Helv. Chim. Acta* **1981**, *64*, 486.
30. (a) Uyehara, T.; Shida, N.; Yamamoto, Y. *J. Chem. Soc., Chem. Commun.* **1989**, 113; (b) Uyehara, T.; Shida, N.; Yamamoto, Y. *J. Org. Chem.* **1992**, *57*, 3139.
31. Yamaguchi, M.; Tsukamoto, M.; Hirao, I. *Tetrahedron Lett.* **1985**, *26*, 1723.
32. The cyclization by tandem conjugate additions initiated by a carbon nucleophile was reported. See: Saito, S.; Hirohara, Y.; Narahar, O.; Moriwake, T. *J. Am. Chem. Soc.* **1989**, *111*, 4533.
33. Scheffer, R.; Wostradowski, T. *J. Org. Chem.* **1972**, *37*, 4317.
34. Kuritani, H.; Takaoka, Y.; Shingu, K. *J. Org. Chem.* **1979**, *44*, 452.
35. Anderson, J. P.; Baizer, M. M.; Patrovich, J. P. *J. Org. Chem.* **1966**, *31*, 3890.
36. Stork, G.; Winkler, J. D.; Saccomano, N. A. *Tetrahedron Lett.* **1983**, *24*, 465.
37. Sakan, T.; Isoe, S.; Hyeno, S.; Katsumura, R.; Maeda, T.; Wolinsky, J.; Dickerson, D.; Slabaugh, M.; Nelson, D. *Tetrahedron Lett.* **1965**, 3376 and references cited therein.
38. Syntheses of (±)-**53** and (±)-**54**: Wolinsky, J.; Euatace, E. J. *J. Org. Chem.* **1972**, *37*, 3376; Ficini, J.; d'Angelo, J. *Tetrahedron Lett.* **1976**, 6087.

39. Review articles: Paquette, L. A. In *Asymmetric Synthesis*; Morison, J. D., Ed.; Academic Press: New York, 1984, Vol. 3, Part B, Chap. 7; Oppolzer, W. *Angew. Chem. Int., Ed. Engl.* **1984**, *23*, 876.
40. Shida, N.; Uyehara, T.; Yamamoto, Y. *J. Org. Chem.* **1992**, *57*, 5049.
41. Whitesell, J. K.; Yaser, H. K. *J. Am. Chem. Soc.* **1991**, *113*, 3526. Corey, E. J.; Ensley, H. E. *J. Am. Chem. Soc.* **1975**, *97*, 6908.
42. For a review: Oppolzer, W. *Tetrahedron* **1987**, *43*, 1969. Helmchen, G.; Wegner, G. *Tetrahedron Lett.* **1985**, *26*, 6051.
43. Oppolzer, W.; Löher, H. J. *Helv. Chim. Acta* **1981**, *64*, 2808.
44. Poll, T.; Sobczak, A.; Hartmann, H.; Helmchen, G. *Tetrahedron Lett.* **1985**, *26*, 3095.
45. Evans, D. A. *Aldrichim. Acta* **1982**, *15*, 318.
46. Iwasawa, N.; Mukaiyama, T. *Chem. Lett.* **1986**, 637; Iwasawa, N.; Huang, H.; Mukaiyama, T. *Chem. Lett.* **1985**, 1045.
47. (±)-Thienamycin synthesis via a three component coupling method with silylcuprate. See: Palomo, C.; Aizpurua, J. M.; Urchegui, R. *J. Chem. Soc., Chem. Commun.* **1990**, 1390.
48. Conjugate addition-aldol condensation using titanium amides. See: Hosomi, A.; Yanagi, T.; Hojo, M. *Tetrahedron Lett.* **1991**, *32*, 2371.
49. For a review, see: Heathcock, C. H. In *Asymmetric Synthesis*; Morrison, J. D., Ed.; Academic Press: New York, 1984, Vol. 3, Chap. 2.
50. The copper (and/or lithium) enolate, produced in organocuprate conjugate additions, is not suitable for further aldol and alkylation reactions. Normally such an enolate is transmetallated to a zinc or tin enolate. See: Yamamoto, Y.; Yatagai, H.; Maruyama, K. *Si, Ge, Sn, and Pb Compds.* **1986**, *9*, 25; Suzuki, M.; Yanagisawa, A.; Noyori, R. *J. Am. Chem. Soc.* **1985**, *107*, 3348.
51. The high pressure induced addition of diphenylmethylamine to 8-(2-naphtyl)-*p*-menth-3-yl crotonate produced (*R*)-β-amino ester. See: d'Angelo, J.; Maddaluno, J. *J. Am. Chem. Soc.* **1986**, *108*, 8112.
52. Oppolzer, W.; Blagg, J.; Rodriguez, I.; Walther, E. *J. Am. Chem. Soc.* **1990**, *112*, 2767.
53. Huang, H.; Iwasawa, N.; Mukaiyama, T. *Chem. Lett.* **1984**, 1465.
54. Shida, N.; Kabuto, C.; Niwa, T.; Ebata, T.; Yamamoto, Y. *J. Org. Chem.* **1994**, *59*, 4068.
55. Hawkins, J. M.; Lewis, T. A. *J. Org. Chem.* **1992**, *57*, 2114; **1994**, *59*, 649.
56. (a) Davies, S. G.; Ichihara, O. *Tetrahedron: Asymmetry* **1991**, *2*, 183; (b) Davies, S. G.; Ichihara, O.; Walters, I. A. S. *Synlett* **1993**, 461; (c) Davies, S. G.; Garrido, N. M.; Ichihara, O.; Walters, I. A. S. *J. Chem. Soc., Chem. Commun.* **1993**, 1153; (d) Bunnage, M. E.; Davies, S. G.; Goodwin, C. J. *J. Chem. Soc., Perkin Trans. 1* **1993**, 1375; (e) Bunnage, M. E.; Davies, S. G.; Goodwin, C. J.; Walters, I. A. S. *Tetrahedron: Asymmetry* **1994**, *5*, 35. (f) Davies, S. G.; Walters, I. A. S. *J. Chem. Soc., Perkin Trans. 1* **1994**, 1129; (g) Davies, S. G.; Ichihara, O.; Walters, I. A. S. *J. Chem. Soc., Perkin Trans. 1* **1994**, 1141.
57. (a) Furukawa, M.; Okawara, T.; Terawaki, Y. *Chem. Pharm. Bull.* **1977**, *25*, 1319. (b) Estermann, H.; Seebach, D. *Helv. Chim. Acta* **1988**, *71*, 1824.
58. Yamamoto, Y.; Yamamoto, S.; Yatagai, H.; Ishihara, Y.; Maruyama, K. *J. Org. Chem.* **1982**, *47*, 119.
59. Yamamoto, Y.; Yatagai, H.; Maruyama, K. *Tetrahedron Lett.* **1982**, *23*, 2387.
60. Evans, D. A.; McGree, L. R. *Tetrahedron Lett.* **1980**, *21*, 3975; Yamamoto, Y.; Maruyama, K. *Tetrahedron Lett.* **1980**, *21*, 4607.
61. Mukaiyama, T.; Inoue, T. *Chem. Lett.* **1976**, 559; Inoue, T.; Mukaiyama, T. *Bull. Chem. Soc. Jpn.* **1980**, *53*, 174.
62. Kobayashi, S; Iimori, T.; Izawa, T.; Ohno, M. *J. Am. Chem. Soc.* **1981**, *103*, 2406.
63. Kano, S.; Ebata, T.; Shibuya, S. *J. Chem. Soc., Perkin Trans. 1* **1980**, 2105; Tufariello, J. J. *Tetrahedron Lett.* **1979**, *20*, 4359.
64. (a) Yamamoto, Y.; Nishii, S.; Ibuka, T. *J. Chem. Soc., Chem. Commun.* **1987**, 464; (b) Yamamoto, Y.; Nishii, S.; Ibuka, T. *J. Chem. Soc., Chem. Commun.* **1987**, 1572; (c) Yamamoto, Y.; Nishii, S.;

Ibuka, T. *J. Am. Chem. Soc.* **1988**, *110*, 617; (d) Yamamoto, Y.; Chounan, Y.; Nishii, S.; Ibuka, T.; Kitahara, H. *J. Am. Chem. Soc.* **1992**, *114*, 7652 and references cited therein.

65. Batten, R. J.; Dixon, A. J.; Taylor, R. J. K. *Synthesis* **1980**, 234.

ENANTIOSELECTIVE DEPROTONATION OF CYCLIC KETONES

Marek Majewski

Advances in Asymmetric Synthesis
Volume 3, pages 39–76.

ISBN: 0-7623-0106-6

I. INTRODUCTION

Perhaps no other area of organic chemistry has experienced as rapid growth during the last two decades as asymmetric synthesis. In 1971, Morrison and Mosher summarized the state of knowledge in the area in one volume of ca. 450 pages.[1] A similar compilation published in 1983–1984 required five volumes and almost 2000 pages of text and graphics.[2] Recently, a 10-volume series dealing with stereoselective synthesis comprising some 7000 pages has appeared.[3] Following the flow of ideas in this rapidly expanding field is a difficult task which could be facilitated by adopting a systematic approach to conceptual developments. A useful system of classification of stereoselective reactions, which involves analyzing the changes in the structure of the starting material and focuses on the reaction topicity was pioneered by Izumi and Tai and later has been clarified and popularized by Seebach.[4] According to this system, a stereoselective reaction can be termed either enantio- or diastereo-differentiating and the elements being differentiated by the reagent can be faces or groups (diastereotopic or enantiotopic) or even molecules (as in the case of enantiomers in a racemic mixture reacting with a chiral reagent at different rates). Seebach pointed out that there are essentially three approaches to the problem of synthesizing a chiral target compound in enantiomerically pure form (EPC synthesis): a researcher might start with a racemate and resolve it either by a classical resolution process (i.e. converting the mixture of enantiomers into the mixture of diastereoisomers and then taking advantage of different physical properties of the resulting species) or by using a chiral reagent in a kinetic resolution process (Figure 1, path a). Alternatively, one could start with a chiral substrate, available from natural sources, and convert it, hopefully without loosing the stereochemical information encoded in the substrate, into the target compound

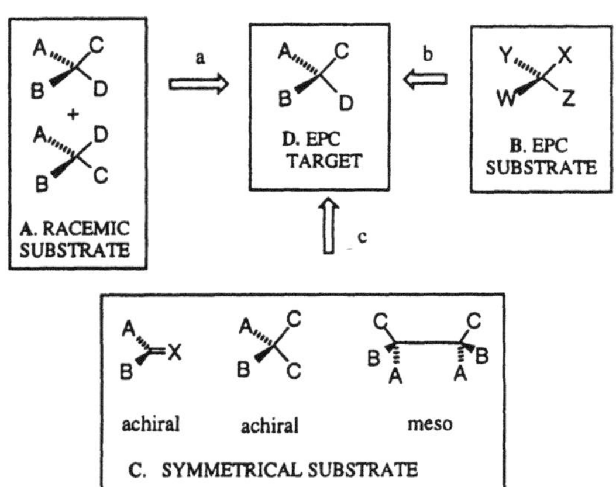

Figure 1. Synthesis of enantiomerically pure compounds (EPC).

(Figure 1, path b). Finally, the chemist could start with an achiral compound which might have stereogenic centers (a *meso* compound) and convert it selectively into the chiral target (Figure 1, path c). From the topicity point of view, the kinetic resolution approach is enantiomer-selective, the second pathway involves diastereotopic group- or face-selective reactions and the third method must utilize chiral reagents (or catalysts) to achieve enantiotopic face- or group-selective transformations. Combination moves on this scheme are, of course possible and thus one could start, e.g., with a racemic substrate and convert it into an achiral intermediate before attacking the target thus following the pathway **A** → **C** → **D**.

By the mid-1980s very few examples of enantiotopic group selective transformations were reported. In fact, not a single article in the five-volume *Asymmetric Synthesis* series[2] dealt with differentiation of enantiotopic groups—all reviews were concerned with enantiotopic face-selective transformations. This hole in the existing methodology apparently prompted a number of researchers around the world to initiate programs aimed at developing enantiotopic group-selective reactions.[5]

It is often perceived that a stereoselective reaction has a better chance of success (i.e. it is more likely to proceed with high selectivity) if a cyclic transition state, or a cyclic intermediate (often a complex), is involved.[6] Accordingly, reactions which could convert achiral compounds of C_s symmetry into chiral products having C_1 symmetry selectively (desymmetrization) probably should involve such cyclic transition states (or an intermediate) during the stereodifferentiating step. A schematic representation of this process is shown in Figure 2. It should be pointed out that the non-chiral starting material of C_s symmetry can be either achiral (no stereogenic centers present) or *meso* (stereogenic centers present) and that no new stereogenic elements need to be created at the actual reaction site—the overall change in the symmetry of the molecule is enough to reveal the latent chirality.

At present, essentially every major organic reaction has an enantioselective variant.[3] However, development of the stereoselective version of deprotonation of C–H acids, a fundamental reaction which occupies central place in organic synthesis, has been slow in coming. This is, perhaps, due to the fact that deprotonation typically converts an sp^3 carbon atom into an sp^2 carbon atom (as in enolate

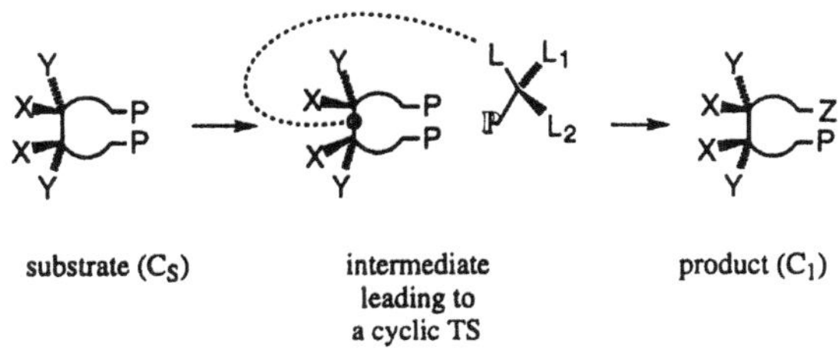

Figure 2. Desymmetrization with a chiral reagent.

formation) and thus thinking of this process in terms of stereoselectivity is counterintuitive, or at least contrary to the way organic chemistry has been taught at the undergraduate level. At this level, students were usually trained to regard a "chiral center" as an assembly comprising four different ligands connected to the central sp^3-hybridized carbon atom, and to think about synthesis of chiral molecules in terms of constructing the covalent bonds originating at the "chiral center." Implicitly, only reactions which produce a new sp^3-hybridized carbon atom were viewed as potentially useful in stereoselective synthesis.[7] Separating chirotopicity from stereogenicity, proposed by Mislow, was an important step in changing this way of thinking.[8]

The use of chiral lithium amides for enantioselective elimination, deprotonation and later for addition reactions started almost 20 years ago. Of special note here is the work of Whitesell on enantioselective opening of epoxides,[9] and the work of Duhamel's group on enantioselective protonation which also included the use of chiral lithium amides.[10] These reactions, albeit proceeding with low selectivity, opened the door for further development. In this account we will focus on the work done in the enantioselective deprotonation area by our group.

II. DEPROTONATION OF KETONES

A. Fundamentals

Deprotonation of ketones, to give the corresponding enolates, is the first step in many organic reactions, e.g. aldol, Michael, Robinson, Mannich, and other reactions involving alkylation, hydroxyalkylation, or acylation which are the most popular tools used for constructing carbon skeletons during synthesis of complex natural products.[11] Cyclic ketones were extensively investigated in this respect. Frequently, the deprotonation itself was treated rather sketchily and the overall process was described in an abbreviated form focusing on the new bond(s) being formed to the electrophilic reagent (Scheme 1—only one of two, or more, isomeric products possible for either **2** or **3** is shown in the scheme). However, many of the selectivity problems encountered during these reactions were a direct result of the deprotonation step.

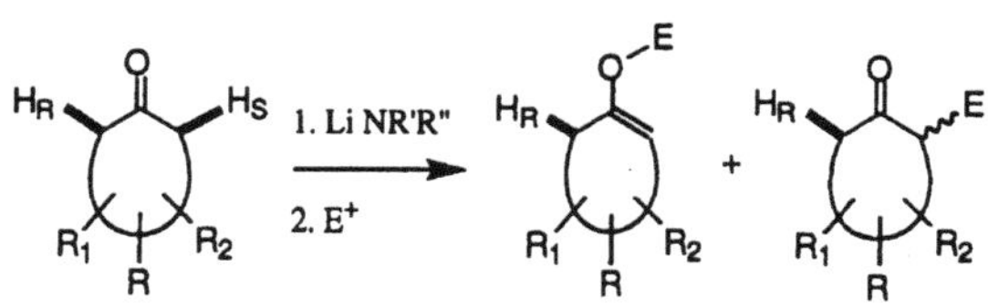

1 **2** **3**

Scheme 1.

Depending on the symmetry of the cyclic ketone **1**, which in turn is determined by the kind and disposition of the substituents R, R_1, and R_2, the two protons H_R and H_S, which can be abstracted by a base, could be enantiotopic (the ketone belongs to the C_s point group, e.g. cis-3,5-dimethylcyclohexanone), diastereotopic (the ketone belongs to the C_2 point group, e.g. *trans*-3,5-dimethylcyclohexanone), constitutionally heterotopic (the ketone belongs to the C_1 point group, e.g. 3-methylcyclohexanone), or homotopic (the ketone has an averaged symmetry C_{2v}, e.g. cyclohexanone).[12] The first case, i.e. deprotonation of cyclic ketones having C_s symmetry with chiral lithium amide bases, will be the focal point of this account. However, a chiral base could be used in other scenarios, e.g. to address the regioselectivity problem which arises when H_R and H_S are heterotopic.[13]

Ketone enolization with lithium amide bases is a complex process and it is now recognized that several steps are involved, complexation of the lithium amide to the carbonyl oxygen being, perhaps, the most important.[14,15] A more complete, although still simplistic, picture of the reaction is shown in Scheme 2.

The lithium amide behaves as a Lewis acid and complexes to the C=O group to form **4**. Proton transfer via a cyclic transition state follows and another complex **5** is generated (only one of the two different possible regio- or diastereoisomeric forms of **5** is shown; transfer of H_R would, of course, generate the other isomer). The presence of this complex is responsible for the internal proton return, a process where an electrophilic attack results in the reversal of deprotonation to give back the parent ketone **1**.[15] The schematic representation shown in Scheme 2 does not take into account aggregation of lithium amides and aggregation of lithium enolates.[16] However, the two complexes **4** and **5** afford some promise for control of stereoselectivity when a chiral lithium amide is used.

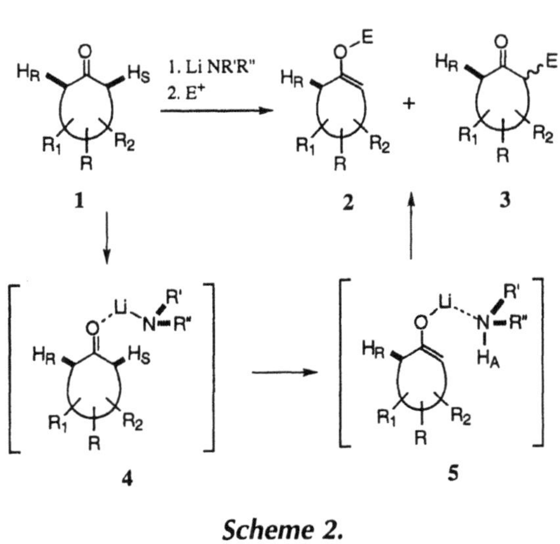

Scheme 2.

B. Enantioselective Deprotonation: A New Concept

Ability of chiral lithium amide bases to deprotonate ketones enantioselectively was first demonstrated in 1986 by the groups of Koga[17] (on 4-alkylcyclohexanone) and Simpkins[18] (on *cis*-2,6-dimethylcyclohexanone). At the same time we were investigating enantioselective deprotonation using other model compounds and especially *cis*-3,5-dimethylcyclohexanone.[19] The results of these early model studies are summarized in Scheme 3.

In all three systems non-racemic products were observed and the selectivity ranged from modest to good. As is common when developing a new enantioselective reaction the choice of the model compound(s) was critical. The three model ketones were clearly selected for their structural simplicity. Our model was the worst, because it required a nontrivial synthesis. The choice of the electrophile is also important and the electrophile should enter into a high- yielding reaction and the product should allow measurement of the enantiomeric excess by a reasonably simple protocol, preferably on a crude product, prior to purification which might affect the enantiomer ratio. In this respect the silylation reaction, chosen by Koga's group, proved problematic—the ee measurement relied on optical rotation and had to be reinvestigated later.[20] The enantiomer ratio of the nonracemic acetate **10** and the aldols **13** and **14** could be measured using NMR in the presence of optically active shift reagents.

Each of the three groups examined several lithium amides and the selectivity of deprotonation varied greatly with the structure of the amide. The most efficient base

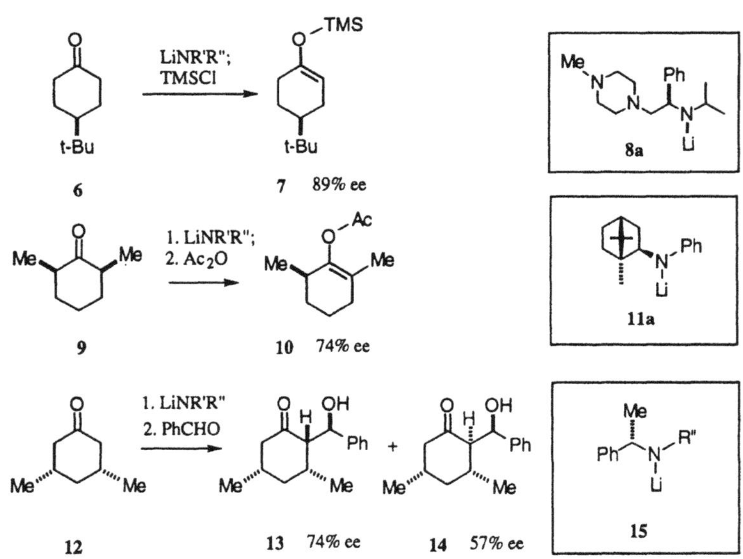

Scheme 3.

for deprotonation of 4-alkylcyclohexanones was the lithiated bidentate amine **8a**.[17] This, and similar lithium amides synthesized from bidentate derivatives of phenylglycine, proved later very efficient and versatile in selective deprotonation of different ketones (vide infra). The more sterically hindered 2,6-dimethylcyclohexanone was deprotonated most selectively by the chiral lithium amide **11a** derived from camphor.[18] Interestingly, this particular lithium amide was very inefficient in deprotonation of 3,5-dimethylcyclohexanone (11% ee).[19] We focused on chiral lithium amides derived from the parent amines of general structure **15**, which were easily obtained from the commercial α-methylbenzylamine (both enantiomers of this amine are available). The most selective was the benzyl derivative **15d**.[19] The two aldols **13** (the *cis-threo* isomer) and **14** (the *trans-erythro* isomer) were produced in the 1:1 ratio and the reaction was stopped at low conversion—under these conditions the ee values were different for both diastereoisomers. This was expected since the diastereoisomers are likely to be formed at different rates and hence there can be an element of kinetic resolution in low conversion reactions of this kind. We also briefly investigated the effects of temperature, solvents, and additives; these studies will be described in more detail later in this account. The structures of chiral lithium amides which were used in these and subsequent studies on ketone deprotonation are summarized in Figure 3.

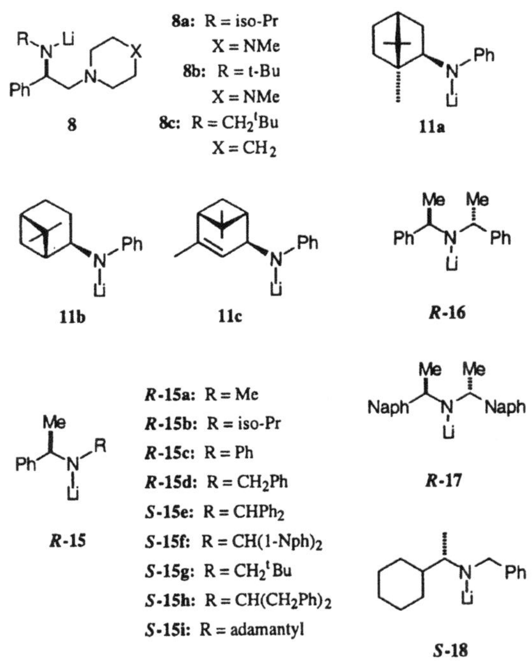

Figure 3. Chiral lithium amides used in enantioselective deprototation studies.

The preliminary model studies described above demonstrated that enantioselective deprotonation of ketones is possible. Three groups of chiral bases were used: lithium amides derived from amino acids, terpenes, and from α-methylbenzylamine. Not surprisingly, the level of selectivity varied greatly with the structure of the base and the ketone used. Also, based on Koga's report that 4-*t*-butylcyclohexanone was deprotonated much more selectively than 4-methyl- or 4-isopropylcyclohexanone,[17] it seemed that a fairly high level of conformational stabilization was necessary for a selective reaction.

C. Refinement of Enantioselective Deprotonation Methodology

At this stage there clearly was a lot room for improvements in the budding methodology. We decided to pursue enantioselective deprotonation studies on two parallel tracks: on one hand we would investigate the effect of reactants and conditions (solvents, temperature, additives, etc.) on the selectivity; on the other hand we would examine a number of different cyclic ketones of C_s symmetry, keeping in mind their potential synthetic utility. Structures of the ketones are shown in Figure 4. Protected 4-alkoxy- and/or 4-oxocyclohexanones (**19** and **20**) could be suitable starting materials for EPC synthesis of terpenoids, 2-substituted dioxanones (**21**) could be envisaged as synthetic equivalents of 1,3-dihydroxyacetone and could be utilized in synthesis of carbohydrates, and deprotonation of tropinone (**22**) might be a useful method in synthesis of alkaloids. The work on methodology development will be described first, followed by synthetic applications, although, in reality, these two aspects were being pursued simultaneously by our group.

Tropinone (**22**) provides a good model for studying deprotonation with chiral lithium amides. The compound has a bridged bicyclic skeleton, which is rather rigid and which normally exists with the pyrimidone ring in the chair conformation.[21] This renders the two hydrogens on the β-face axial, and the competition from the equatorial hydrogens for the reaction with the base should be minimized for steric reasons. Stereoelectronic arguments have been used in the past to postulate that in cyclic ketones axial protons should be abstracted preferentially to the equatorial

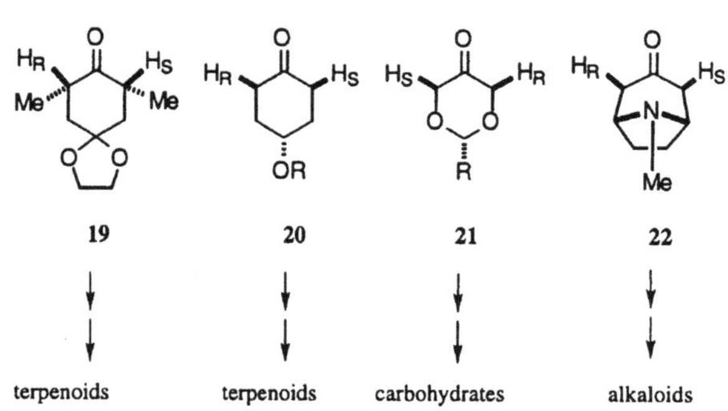

Figure 4. Cyclic ketones for enantioselective deprotonation studies.

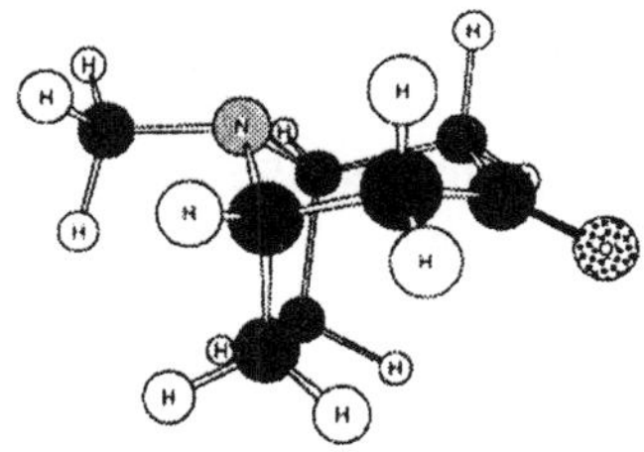

Figure 5. Ball and stick model of the most stable conformer of tropinone.

ones.[22] Although the stereoelectronic preference is difficult to measure and could be small, the β-face of tropinone should also be much more exposed to an attack by the base for steric reasons (Figure 5)—the notion that the β-face is less sterically hindered was confirmed later by results of electrophilic attack on tropinone enolate (vide infra).

Before addressing the enantioselective deprotonation the fundamentals of the tropinone lithium enolate chemistry had to be established, since little was known about generation of tropinone enolate under kinetic conditions. Tropinone was deprotonated with LDA and the resulting racemic enolate was subjected to reactions with several different electrophiles (Scheme 4).[23] The remarkable features of tropinone Li–enolate chemistry included a novel ring opening upon treatment of the enolate with chloroformates leading to compound **26**. It is noteworthy that the methyl group remains connected to the nitrogen atom, thus this is not a von Braun-type reaction. Use of methyl cyanoformate[24] gave the expected α-car-

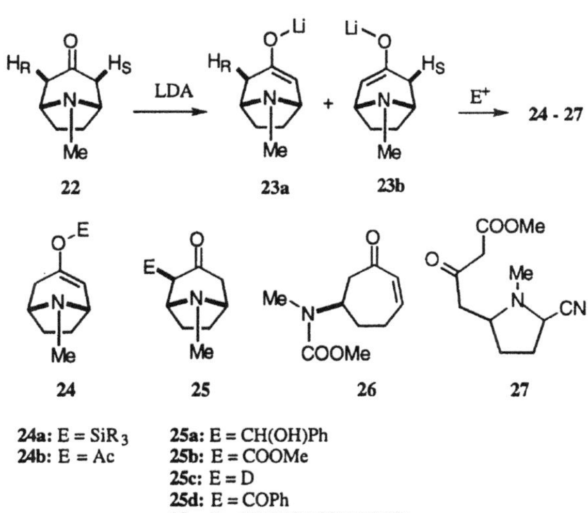

Scheme 4.

bomethoxy derivative **25b**; however, a substantial amount of a side product identified as **27** was also formed. Assuming that this side product originated via the cyanide ion attack on the initially formed **25b**, followed by a Grob fragmentation, we used silver salts to scavenge the cyanide ion—this eventually led to an efficient procedure for synthesis of compound **25b**.[25]

For an investigation of enantioselective deprotonation of tropinone we have chosen the aldol reaction with benzaldehyde as the model system. This reaction could, in principle, give four different diastereoisomeric aldols. Fortunately, however, the reaction turned out to be very diastereoselective and yielded only one product: the *exo-threo* isomer (Figure 6). This was rationalized by a much easier approach of the electrophile from the top face combined with the fact that the Zimmerman–Traxler cyclic transition state[26] in the like ("l") approach (note that hydrogen atoms were omitted in Figure 6 for clarity; for the "like"–"unlike" terminology cf. ref. 27) suffers from a severe steric hindrance arising from an interaction between the benzaldehyde phenyl and the tropinone N–Me group. The transition state resulting from the unlike approach ("ul") does not have this interaction. We also observed that compound **25a** was very stable and did not epimerize to the *endo* isomer upon treatment with bases or acids. Apparently the hydrogen bonding between the OH and the nitrogen stabilizes the structure.[28] The aldol reaction provided a good model for studies of enantioselectivity, because the enantiomer ratio could be measured by recording NMR spectra of **25a** in the presence of chiral shift reagents.[29]

Initially, the experiments with chiral lithium amides did not look very promising. A number of chiral lithium amides were tested in the aldol reaction and the

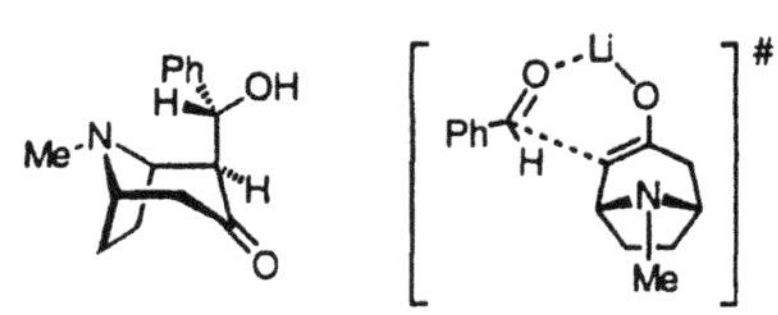

25a (exo-threo) TS

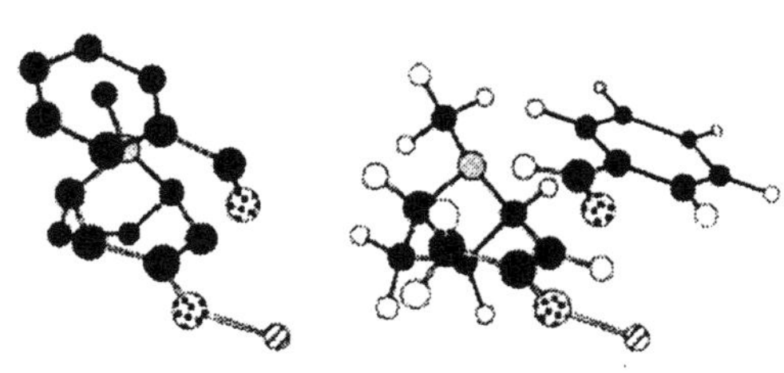

l *ul*

Figure 6. Diastereoselective formation of only one diasteroisomer of the tropinone aldol.

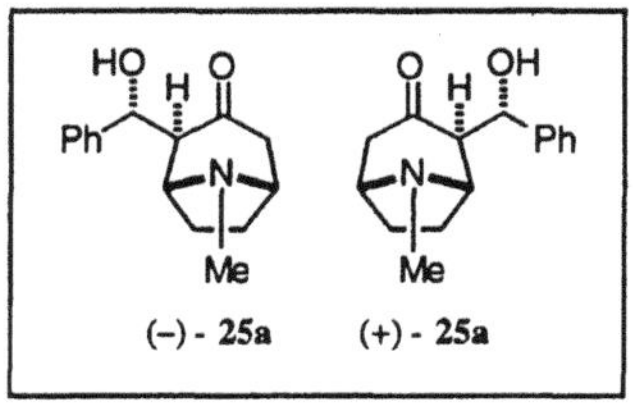

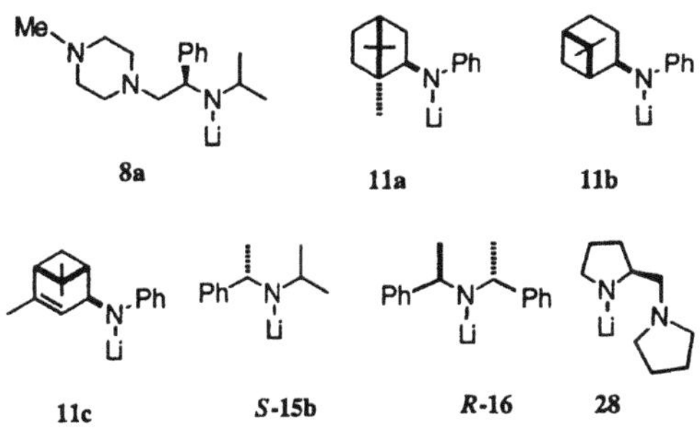

base	8a	11a	11b	11c	R-16	S-15b	28
ee	60 (+)	40 (–)	26 (–)	16 (–)	36 (–)	23 (+)	8 (+)
yield	64	75	70	97	91	93	85

Figure 7. Non-racemic aldols **25a** produced via enantioselective deprotonation and tropinone with chiral Li-amides.

enantiomeric excess of the non-racemic product **25a** was determined by NMR using a chiral shift reagent Eu(tfc)$_3$. The results are summarized in Figure 7.[25,30] Either the dextrorotatory or the levorotatory isomer was produced predominantly depending on the base used (this is denoted with the "+" or "–" sign in Figure 7). The absolute stereochemistry of deprotonation was determined by carbomethoxylation of the tropinone enolate and subsequent synthesis of anhydroecgonine methyl ester, the absolute configuration of which is known.[31] All the chiral lithium amides tested in this initial study showed low selectivity with the bidentate ligand developed by Koga being the best (compound **8a**). Interestingly, the proline-derived lithium amide **28**, which earlier proved quite enantioselective in epoxide opening reactions,[32] has shown very little selectivity. Similarly, the terpene-derived bases **11a**, which had been reported previously to be fairly selective in deprotonation of 2,6-dimethylcyclohexanone,[18] **11b** and **11c** were not selective.

Effects Caused by Additives

Certain lithium salts, and especially LiCl, are well known to affect the rates and selectivities of many diverse organic reactions.[33] We had previously reported a large detrimental effect of LiBr on diastereoselectivity of aldol addition reaction of cyclohexanone lithium enolate to benzaldehyde: addition of two or more equivalents of LiBr essentially destroyed diastereoselectivity (Scheme 5).[34] We had also noted an effect of LiBr on deprotonation of 3,5-dimethylcyclohexanone with chiral lithium amides (Scheme 5).[19] While lithium salt effects might be difficult to predict, they are often easy to determine experimentally. Deprotonation of tropinone with the chiral lithium amide ***R*-15b**, to which LiBr or LiCl were added, resulted in dramatic changes in the reaction enantioselectivity comparing to the reactions with the halide-free amide (Scheme 5).

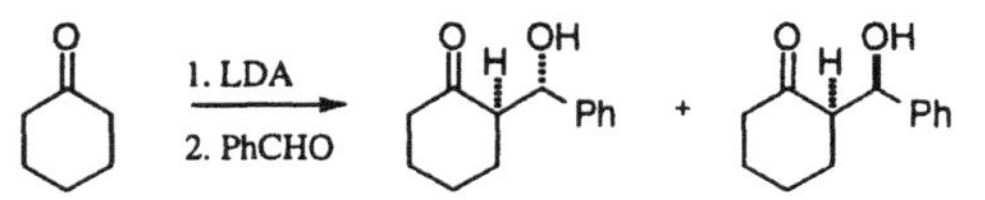

29 **30** (threo) **31** (erythro)

additive	threo : erythro
-	84 : 16
LiBr (1 eq.)	62 : 38
LiBr (2 eq.)	52 : 48

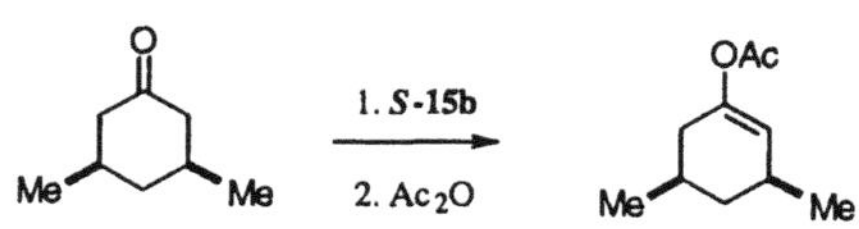

12 **32**

additive	ee (%)
-	30
LiBr (1 eq.)	64

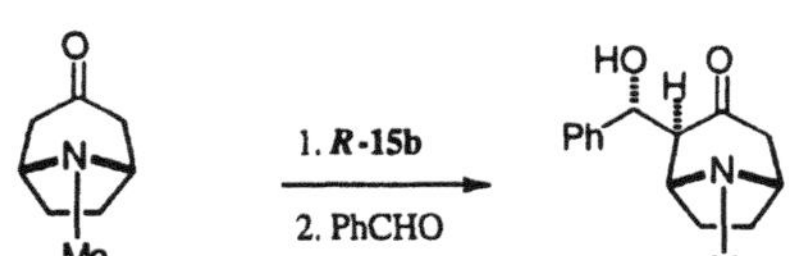

22 **25a**

additive	ee (%)
-	23
LiCl (1 eq.)	71

Scheme 5.

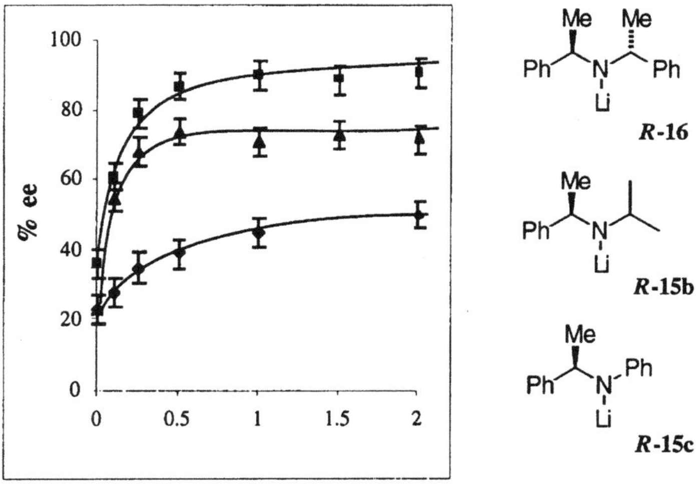

Figure 8. The effect of LiCl on enantioselectivity of deprotonation of tropinone with lithium amides ***R-16*** *(top)*, ***R-15b*** *(middle)* and ***R-15c*** *(bottom)*. The ee was measured on compound **25a**. Lines were fitted arbitrarily.

A more detailed investigation of the effects of lithium chloride on enantioselectivity of deprotonation of tropinone and other ketones revealed a pronounced and quite general effect. Figure 8 shows the influence of LiCl on deprotonation of tropinone with three different lithium amides (the enolate was trapped as the aldol product **25a**). Addition of small amounts of LiCl led to increases in stereoselectivity; the magnitude of this salt effect increased smoothly with the amount of the salt and leveled off at about one equivalent of LiCl. Deprotonation with the C_2 symmetrical amide ***R*-16** showed a remarkable enhancement of selectivity: in the absence of LiCl the reaction gave the aldol **25a** in low ee of ca. 30%; however addition of one, or more, equivalents of LiCl yielded the product in over 90% ee. Thus by the simple expedient of adding lithium chloride the stereoselectivity of the reaction was transformed from the synthetically useless to quite useful. Addition of even small amounts of LiCl caused a sharp rise in the level of stereoselectivity. A reasonable question which arose at this point was: how much lithium halide is there in the commercial BuLi solution? Concentration of halide ions (chloride, bromide, and iodide) in commercially available butyllithium was probed by cyclic voltammetry. The analysis indicated that the concentration of Cl^- was typically very low and the chloride content did not exceed 0.007 mole per 1 mole of BuLi.[30] Other halides were present in even smaller amounts.

The effect of lithium chloride proved to be quite general and, even though its magnitude depended on the structure of the ketone and the lithium amide, an increase in enantioselectivity was always observed. Scheme 6 summarizes the changes in enantioselectivity achieved by addition of LiCl to systems comprising different ketones and different lithium amide bases. These results were selected from several studies done by our group. Perhaps the most remarkable effect occurred in deprotonation of dioxanone **33** with the amide ***R*-15c**: in the absence of LiCl the reaction produced mainly the dextrorotatory isomer of the aldol **34** (one diastereoisomer only was formed; relative stereochemistry of this interesting product was determined by X-ray crystallography) and the stereoselectivity was poor (15% ee), whereas after addition of one equivalent of LiCl the levorotatory isomer was produced predominantly in much higher ee (54%).[46] This example underscores how difficult it is to draw a correlation between the structure of the chiral lithium amide and both the efficiency and the absolute stereochemistry of deprotonation.

Apart from LiCl, other additives which are often used in enolate chemistry to increase efficiency or selectivity include: other lithium halides, salts of other metals (e.g. $ZnCl_2$, $MgBr_2$), and polar co-solvents like HMPA, TMEDA, or DMPU.[36] We

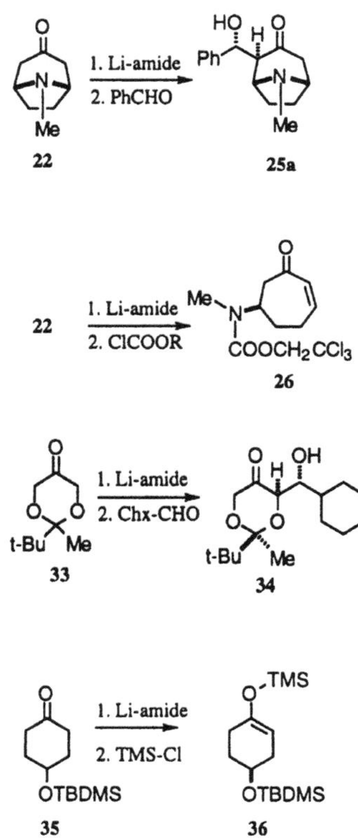

amide	LiCl	ee (%)
R-15b	-	23
	1 eq.	71
S-15c	-	23
	1 eq.	45
R-16	-	36
	1 eq.	90
S-15e	-	78
	1 eq.	87
S-16	-	44
	1 eq.	96
R-16	-	18
	0.5 eq	60
R-15c	-	15 (+)
	1 eq.	54 (-)
R-16	-	36
	1 eq.	70
R-17	-	74
	1 eq.	77

Scheme 6.

have investigated the effect of some of these additives using as the model reaction the ring opening with chloroformates mentioned above (e.g. **22** → **26**).[37] This reaction provided a convenient model because the enantiomeric composition of the products could be measured by HPLC using a column with a chiral phase. The results are summarized in Table 1 (numerically) and in Figures 9 and 10 (graphically).

Addition of even small amounts of LiCl or LiBr to the reacting system resulted in a marked increase of enantioselectivity in each case. The salt effects leveled off after one (LiCl) or two (LiBr) equivalents of the halide were added. The dependence of enantioselectivity on the number of equivalents of LiCl and LiBr is shown in Figure 10. Other lithium salts, i.e. iodide, fluoride, and perchlorate, showed no effect. Cerium trichloride and zinc chloride revealed salt effects similar to LiCl or LiBr but all our experiments with $ZnCl_2$ resulted in very low yields of compound

Table 1. Effect of additives on enantioselectivity of deprotonation of tropinone

Entry	*Additive (equiv)*		*ee (%)*	*Yield[a] of* ***26*** *(%)*
1	—		44	80
2	LiCl	(0.10)	49	84
3	LiCl	(0.25)	85	92
4	LiCl	(0.50)	95	90
5	LiCl	(1.0)	96	92
6	LiCl	(2.0)	95	94
7	LiBr	(0.10)	49	78
8	LiBr	(0.25)	54	82
9	LiBr	(0.50)	63	88
10	LiBr	(1.0)	73	93
11	LiBr	(2.0)	88	85
12	LiF	(1.0)	38	70
13	LiI	(2.0)	41	78
14	$LiClO_4$	(1.0)	45	85
15	$CeCl_3$	(1.0)	80	70
16	$ZnCl_2$	(1.0)	87	40
17	TMEDA	(1.0)	40	45
18	HMPA	(1.0)	39	71
19	DMPU	(1.0)	52	65
20	H_2O	(0.05)[b]	24	48
21	H_2O	(0.20)[b]	5	13

Notes: [a]Yield of the purified compound **4c** after column chromatography.
[b]Excess n-BuLi (5% or 20%) was used in these experiments.

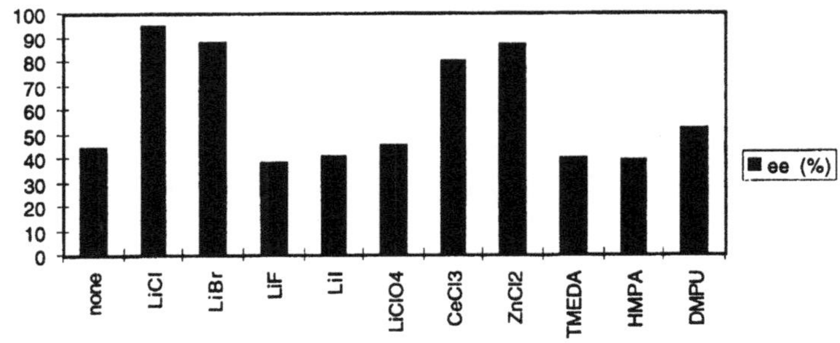

Figure 9. Maximum effects of additives on enantioselectivity of tropinone deprotonation (data from Table 1).

26. Polar organic co-solvents like TMEDA, HMPA, and DMPU did not affect noticeably the reaction selectivity. Interestingly, deliberate addition of water to the amide, followed by the addition of equimolar amount of *n*-BuLi (Table 1, entries 20 and 21; H_2O combined with *n*-BuLi should translate into addition of 0.05 or 0.20 molar equivalent of LiOH, respectively) essentially destroyed reaction selectivity and resulted in low yields. In all cases the yields were much lower than a simple assumption that a certain percentage of the base or the enolate was quenched would indicate. This demonstrates that thorough drying of the reagents is necessary. It is also not enough to titrate *n*-BuLi prior to use; if the reagent is contaminated with appreciable quantities of LiOH it will not work despite the reasonably high concentration of BuLi being present.

Overall, the amide **16** used in the presence of one molar equivalent of LiCl proved to be a very enantioselective reagent in tropinone deprotonation. A brief study of tropinone reaction with bidentate lithium amides of general structure **8**, which were

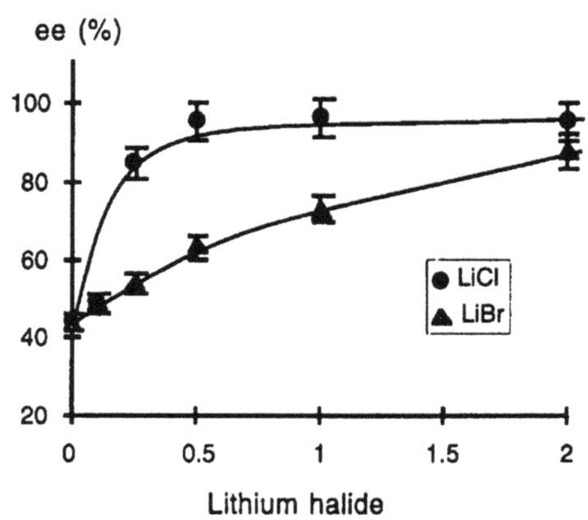

Figure 10. Effect of LiCl and LiBr on enantioselectivity of formation of **26**.

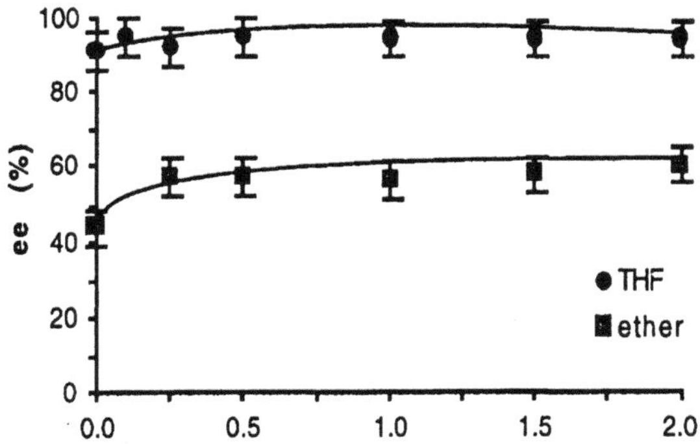

Figure 11. Effect of LiCl on enantioselectivity of deprotonation of tropinone with **8c** in Et_2O and THF.

developed by Koga's group and proved efficient, selective, and versatile reagents, revealed that some of these compounds also deprotonated tropinone very selectively. The selectivity was very much solvent-dependent but LiCl, and other additives, had little effect (Figure 11).

The origin of these remarkable effects of LiCl and other additives is not clear at this point. Other workers in the lithiation field, however, have convincingly demonstrated that lithium amides form mixed dimers with lithium halides (Figure 12).[38] Amides of general structure **8** are known to exists either as monomers or as dimers, depending on the solvent (Figure 12). These changes in aggregation clearly influence deprotonation selectivity.

Other Effects of Reaction Conditions

In most enantioselective reactions based on the $\Delta\Delta G^{\#}$ difference, lowering of the reaction temperature leads to higher selectivity.[39] We have taken advantage of this

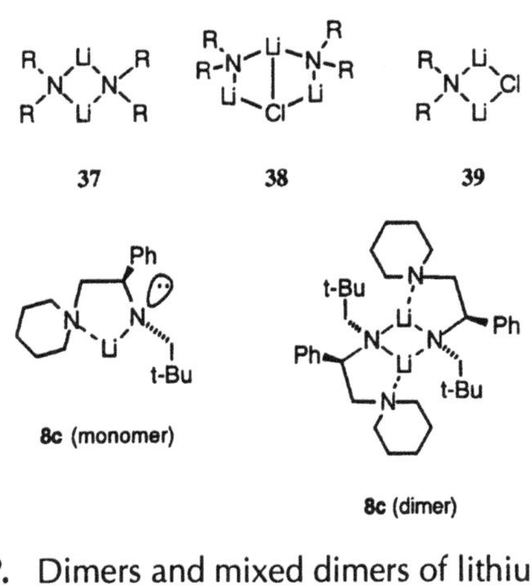

Figure 12. Dimers and mixed dimers of lithium amides.

effect in a synthetic study aimed at chiral butenolides (vide infra).[40] However, running the reaction at temperatures lower than the easily attainable –78 °C is often not practical. As far as solvents are concerned, THF seems to be the solvent of choice. Even though some stereoselective proton transfer reactions were demonstrated to work better in apolar solvents (hydrocarbons), our studies with deprotonation of cyclic ketones always afforded highest enantioselectivity in THF. The decrease in stereoselectivity caused by changing the solvent to diethyl ether when using lithium amide **8**, described above, is especially striking.

The picture which emerged at this point from our studies, and from the work of other groups, could be summarized as follows: cyclic ketones having C_s symmetry can be deprotonated enantioselectively by a variety of chiral lithium amides. There seem to be no straightforward correlation between the structure of the amide and the enantioselectivity. Bidentate amides having general structure **8** proved especially efficient and versatile. Much simpler monodentate lithium amides, and most notably the C_2 symmetrical amide **16** are capable of deprotonating ketones with high enantioselectivity when the reaction is conducted in the presence of one equivalent of LiCl. This led us to propose a procedure involving use of the hydrochloride of the parent amine and two equivalents of BuLi to generate the 1:1 Li–amide/LiCl mixture in situ.[30] This procedure offers some technical advantages, i.e. amine hydrochlorides are usually solids, easy to purify by crystallization, and very stable (long shelf life).

Structure–Activity Relationships. Ketone Structure

Relatively little can be said at this time about the relationships between the structures of the substrates (ketones and lithium amides) and the efficiency and also the absolute stereochemistry of deprotonation. A number of cyclic, symmetrical ketones have been used in studies with chiral lithium amides (Figure 13).[41] Overall, bicyclic bridged ketones afforded higher enantioselectivities in deprotonation with chiral lithium amides than monocyclic ketones. For some time we, and others, felt that reasonably high enantioselectivities can only be achieved when the ketone has a substantial level of conformational stabilization. This reasonable notion can be supported by pointing out that, assuming that in cyclic ketones the axial protons should be removed preferentially to the equatorial ones, due to stereoelectronic reasons,[22] the amide has only two choices and there is no competition involving the diastereotopic protons. After reinvestigation of Koga's early experiments on 4-alkylcyclohexanones,[20] and in view of the fact that 4-methyl-4-phenylcyclohexanone, which shows only a small preference (0.32 kcal/mole) for the phenyl group to stay axial,[42] was deprotonated enantioselectively (56% ee),[43] it seems that large conformational stability is not necessary.

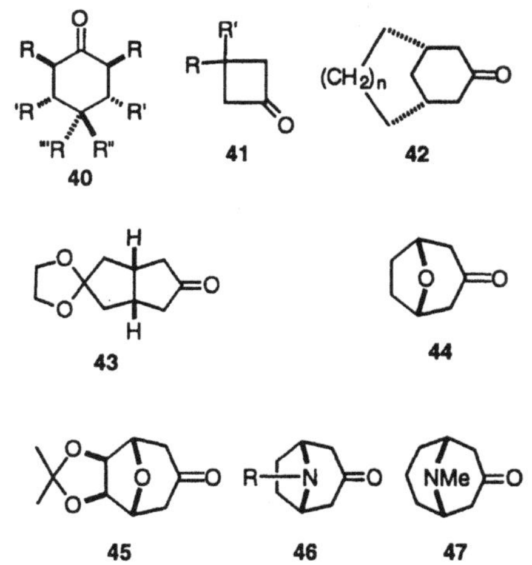

Figure 13. Symmetrical ketones used as substrates for enantioselective deprotonation.

Lithium Amide Structure

As mentioned earlier, three groups of chiral amines were used as sources for chiral lithium amide reagents: amines derived from terpenes (e.g. camphor), amines derived from either (*R*)- or (*S*)-α-methylbenzylamine, and bidentate amines derived from phenylglycine. The amides originating from terpenes seem to have been quickly abandoned by researchers, perhaps due to fairly long syntheses required for their production. The bidentate amines, developed by Koga's group, provided a number of lithium amides which have shown high enantioselectivity and versatility by working well with the number of diverse ketones. This family of chiral lithium amide bases includes perhaps the most elements of rational design.[44] It has been pointed out that the free electron pair on nitrogen is held is conformationally stable position and is strongly affected by the neighboring stereogenic center (cf. structure **8c** in Figure 12). Koga also carried out structure–selectivity relationship studies, the most important result of which was the demonstration that increasing the electron-withdrawing character of the group connected to the reactive nitrogen (i.e. by replacing the neopentyl in structure **8c** with the trifluoroethyl group) leads to significant increases in enantioselectivity.[44]

In order to gather more data and try to design better lithium amide reagents, we have investigated a number of diverse amides using deprotonation of dioxanone **33**, followed by addition to cyclohexanecarboxaldehyde to give the aldol **34**, as the model system.[45,46] The results of this study are presented in Scheme 7. Lithium amides of general structure **15** show some trends: increasing the steric bulk of the *R* group seems to result in greater selectivity (cf. compounds **15b** and **15e** or **15c**

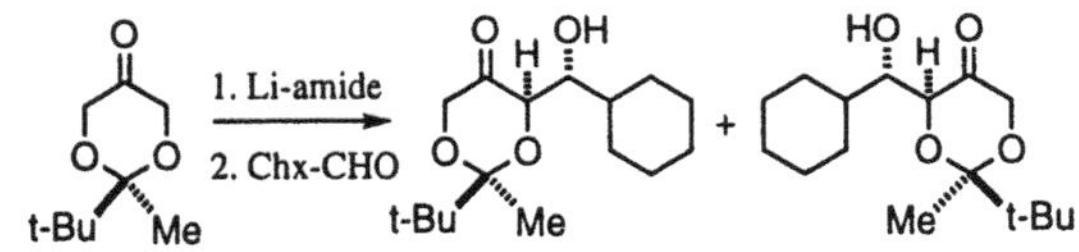

33 (+)-34 (–)-34

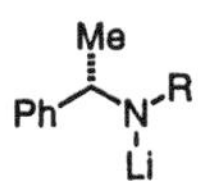

15

#	R	LiCl	ee (%)
R **- 15a**	Me	-	10 (–)
R **- 15a**	Me	0.5	4 (–)
R **- 15b**	i-Pr	0.5	16 (–)
R **- 15c**	Ph	-	15 (+)
R **- 15c**	Ph	1.0	54 (–)
S **- 15d**	CH_2Ph	-	13 (+)
S **- 15d**	CH_2Ph	0.5	39 (+)
S **- 15e**	$CHPh_2$	-	60 (+)
S **- 15e**	$CHPh_2$	1.0	72 (+)
S **- 15f**	$CHNph_2$	1.0	90 (+)
S **- 15g**	$CH_2{}^tBu$	1.0	19 (+)
S **- 15h**	$CH(CH_2Ph)_2$	1.0	60 (+)
S **- 15i**	adamantyl	1.0	80 (+)
S **- 15j**	CH_2CF_3	1.0	90 (+)
R **- 15k**	CH_2(p-OMePh)	-	19 (–)
R **- 15k**	CH_2(p-OMePh)	1.0	32 (–)
R **- 15l**	CH_2(p-FPh)	-	23 (–)
R **- 15l**	CH_2(p-FPh)	1.0	50 (–)

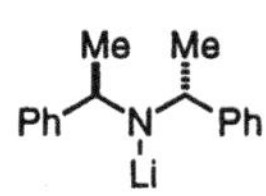

R **- 16**

no LiCl 18% ee (–)
1 eq. LiCl 60% ee (–)

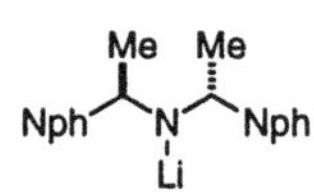

R **- 17**

no LiCl 50% ee (–)
0.5 eq. LiCl 60% ee (–)

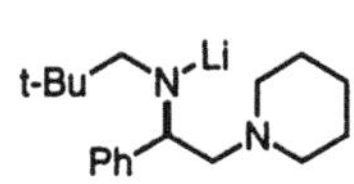

8c

no LiCl 23% ee (+)
0.5 eq. LiCl 29% ee (+)

Scheme 7.

and **15i**), and replacing a phenyl group or a *t*-butyl group with smaller but much more electronegative CF_3 resulted in a big increase in enantioselectivity (**15d, 15g,** and **15j**). Small changes in the electronic character of the *R* group had little effect (cf. **15d, 15k,** and **15l**). Bases **15k** and **15l**, having the methoxy group or the fluorine substituent, were tested in order to establish if the strong influence of the CF_3 group

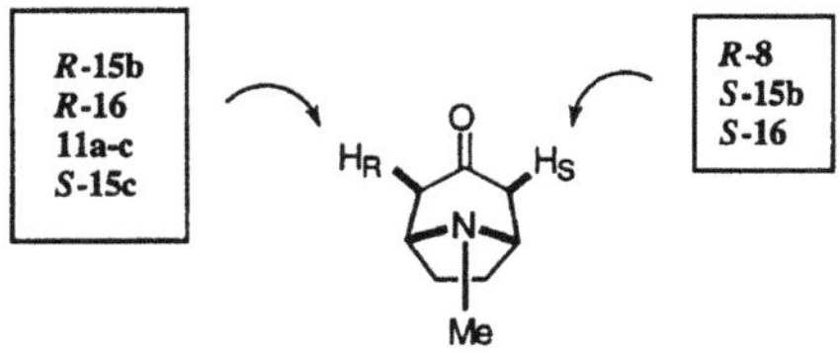

22

Figure 14. Absolute stereochemistry of tropinone deprotonation with different chiral Li-amides.

was electronic in nature. The changes in selectivity, although small, seem to support the notion that electron-withdrawing substituents in the amide are beneficial. Absolute configuration of the products of these studies deserves comment. The (+) and (–) enantiomers of compound **34** are believed to have absolute configurations as drawn on the basis of correlation of similar products of the aldol reaction of dioxanones with carbohydrate derivatives of known configuration. During the work on tropane alkaloids, described later in this account, we have correlated the structures of the products of some enantioselective deprotonation experiments, followed by carbomethoxylation, with anhydroecgonine, a tropane alkaloid of known absolute configuration.[25] Knowing the absolute stereochemistry of a single deprotonation involving tropinone and one chiral base allows, in principle, to assign absolute configuration to all products derived from tropinone by analyzing the NMR spectra taken in the presence of chiral shift reagents and determining which peak—the upfield one (representing one of the enantiomers) or the downfield one (representing the other enantiomer) is larger. The preferences of several lithium amides to attack either at the H_S or the H_R proton in tropinone are shown in Figure 14 and it is interesting to compare them with trends shown in Scheme 7. Broad generalizations could clearly be risky.

Mechanistic Considerations

Lithium diisopropylamide and other dialkyl lithium amides were studied in some detail during the last decade by experimental and theoretical methods and much progress has been made in unraveling their structure and properties. However, the mechanisms of fundamental reactions involving these amides, such as deprotonation of ketones, remain elusive. Surprisingly little is known, for example, about the kinetics of deprotonation of carbonyl compounds with lithium amides.[47] On the practical level, the conditions for deprotonation of ketones (amide, solvent, temperature, and time) are usually chosen arbitrarily, and the length of time allowed for deprotonation seems to depend greatly on the structure of both the ketone and the lithium amide.[48]

In order to gain some insight into the mechanism of ketone deprotonation we attempted a rate study which was based on enolization of the chiral ketone **48** with

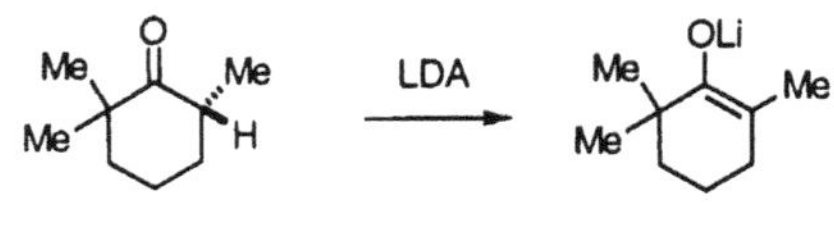

Scheme 8.

LDA (Scheme 8).[49] We hoped that a better understanding of the reaction mechanism would be useful in designing new chiral lithium amides. The enolate **49** is achiral and thus the decrease in optical rotation of the mixture of **48** with LDA provided a convenient way of monitoring the progress of deprotonation. The reaction was determined to be pseudo first order in ketone up to at least 75% conversion (Figure 15). Rate constants were measured at different concentration of LDA, which was the reagent used in excess (Table 2). The order in LDA was determined by the standard methods,[50] and proved to be 0.5 for the reaction in THF. Changing the solvent to diethyl ether resulted in a decrease of the rate constant,while addition of one molar equivalent of LiCl, which is known to affect the aggregation and reactivity of lithium amides (vide supra), had only slight effect on the rate constant, and excess of diisopropylamine resulted in a small increase in the rate constant.

The reacting system is approximated in Scheme 9 (well-known processes like aggregation of lithium enolates and diisopropylamine complexation to the enolate are ignored); lithium diisopropylamide exists primarily as the dimer **D** in equilibrium with the monomer. Either species could, in principle, form a complex with the

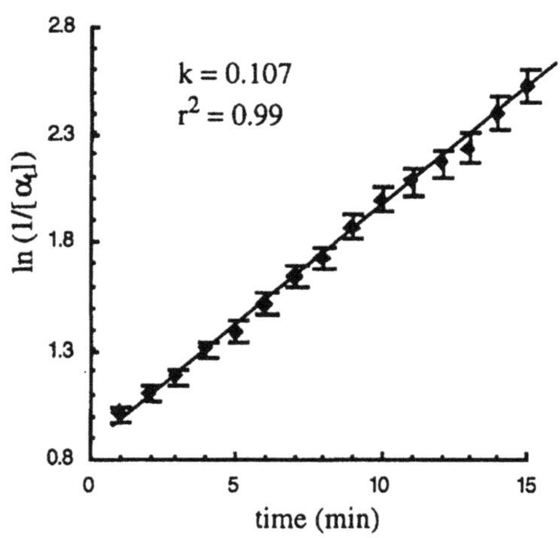

Figure 15. Plot of optical rotation data for the deprotonation of **48** (initial conc. 0.0186 M) with LDA (initial conc. 0.0186 M) in THF at –78 °C.

Table 2. Observed Rate Constants for Enolization of Ketone **48** with LDA

Entry	*[Ketone] (M)*	*[LDA] (M)*	*$k_\psi \times 10^3$ (s^{-1})*	*Solvent (additive)*
1	0.0186 ± 0.0006	0.557 ± 0.011	3.42 ± 0.30	THF
2	0.0186 ± 0.0006	0.372 ± 0.010	2.72 ± 0.24	THF
3	0.0186 ± 0.0006	0.186 ± 0.010	1.78 ± 0.09	THF
4	0.0186 ± 0.0006	0.093 ± 0.009	1.23 ± 0.06	THF
5	0.0190 ± 0.0006	0.381 ± 0.011	0.43 ± 0.02	Et_2O
6	0.0093 ± 0.0006	0.186 ± 0.010	1.15 ± 0.09	THF (LiCl)[a]
7	0.0376 ± 0.0007	0.372 ± 0.010	3.77 ± 0.33	THF ($^i Pr_2NH$)[b]

Notes: [a]Concentration of LiCl 0.1865 M.
[b]Concentration of diisopropylamine 0.093 M.

ketone and then undergo a proton transfer to form **49**. Our results were consistent with the pathway involving the monomer of the amide and the rate-determining proton transfer, analogous to the mechanism proposed by Collum for ester deprotonation.[47] The following rate equation was proposed:

$$\text{rate} = -d[\text{ketone}]/dt = k[\text{ketone}][\text{LDA}]^{1/2}$$

It should be pointed out that the results should be treated with caution; a combination of both the monomer and the dimer pathways is possible, especially for substrates less hindered sterically, and our data do not entirely rule out the dimer pathway being the major route.

The results of the kinetic study support the Ireland model[51] which provides a reasonable, albeit simple, description of deprotonation of carbonyl compounds by LDA. The question of which steric or stereoelectronic effects are responsible for enantioselectivity in deprotonation of ketones with chiral lithium amides remains unanswered.

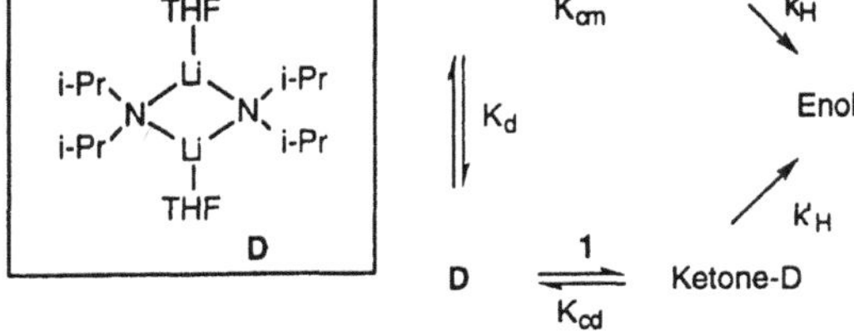

Scheme 9.

III. APPLICATIONS

A. Synthesis of Tropane Alkaloids

Tropane alkaloids are a group of natural products isolated from plants (mainly of *Solanaceae* family) and comprise over 200 compounds many of which are chiral.[52] Selected representative examples of chiral tropane alkaloids are shown in Figure 16 (it should be noted that the absolute configurations of both darlingine and knightinol in this figure are the opposite of the configurations published in other sources.[52] The changes in stereochemical assignments resulted from our studies as described below). Structurally, all tropane alkaloids contain the tropane (8-methyl-8-azabicyclo[3.2.1]octane) skeleton and a number of functional groups which, most often, are: a functionalized hydroxyl at C-3 (α or β), a side chain at C-2 or C-4 (α or β; the side chain is usually carbonyl-based or is either an alkyl or a hydroxyalkyl group) and a β-hydroxyl at C-6 or C-7. By the mid-1990s, even though a number of syntheses of these compounds were described,[52] the matter of enantioselective synthesis remained an unsolved general problem and the absolute configurations of many tropane alkaloids were not known.

Tropinone (**22**) could be envisaged as a convenient starting material for synthesis of diverse tropane alkaloids provided that several stereoselectivity problems could be solved. Enantioselective synthesis would require the selective delivery of a substituent (an alkyl, a hydroxyalkyl, a hydroxyl, or a carboalkoxy group) at C-2 or C-4 or, alternatively, at C-6 or C-7. Tropinone also makes a good starting material for methodology studies. As described in the preceding sections we established that tropinone lithium enolate **23** (Scheme 4), resulting from deprotonation of tropinone with LDA, can be hydroxyalkylated at the carbon terminus to give, diastereoselectively, the *exo-anti* aldol **25a**. The enolate can be carbomethoxylated using Mander's reagent to give **25b** and can also undergo a novel ring opening reaction to give

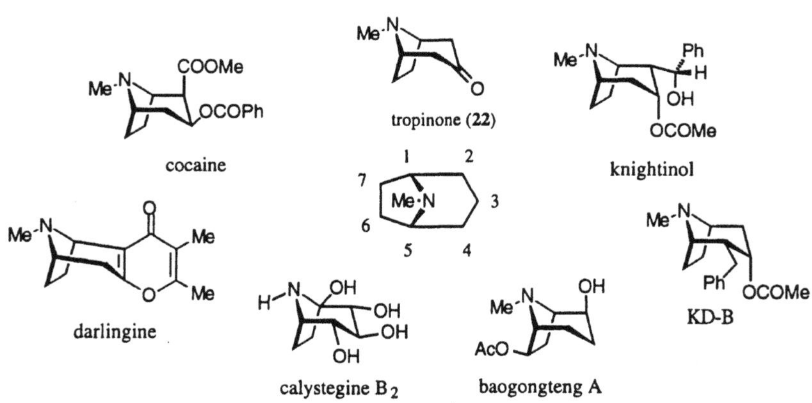

Figure 16. Tropane, tropinone, and selected tropane alkaloids.

26. This trio of reactions provides an attractive jump-off point for synthesis of tropane alkaloids. Since the enolate **23** can be generated enantioselectively using chiral lithium amide bases, compounds **25a, 25b**, and **26** can be easily obtained in enantiomerically "pure" form—either enantiomer of each compound can be produced in high ee. However, at the beginning of our studies we did not know the absolute stereochemistry of deprotonation. In order to establish which chiral bases abstract the H_S proton and which ones tend to attack at H_R (Figure 14) we synthesized anhydroecgonine, an alkaloid of the cocaine group and one of the few tropane alkaloids of known absolute configuration. The synthesis is shown in Scheme 10. Enantioselective deprotonation with the base **8c** followed by carbomethoxylation yielded the β-ketoester **25b**. A diastereoselective reduction on Adams' catalyst gave the corresponding alcohol **48** (in an earlier study we reported a mixture of α- and β-alcohols in this step,[25] eventually we were able to refine the conditions and obtain the α-alcohol as the only isolated product).[30] The alcohol **48** was easily dehydrated and the resulting unsaturated ester **49** was dextrorotatory. Since natural ecgonine, the structure of which had been correlated with cocaine (Scheme 10, inset), is levorotatory,[31] our product had to be *ent*-ecgonine and had to have the absolute configuration as drawn.

Acylation of tropinone with cyanoformates was used in enantioselective synthesis of chalcostrobamine, isobellendine, and darlingine. These alkaloids were described before as having the acyl side chain originating at C-4 of the tropane skeleton.[52] Our synthesis of darlingine is shown in Scheme 11. Enolization of

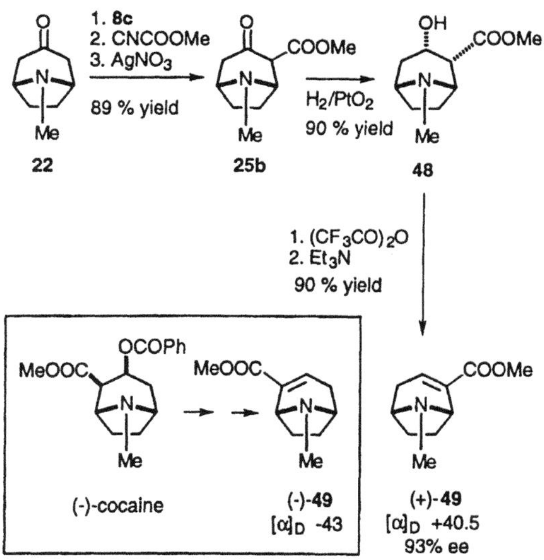

Scheme 10.

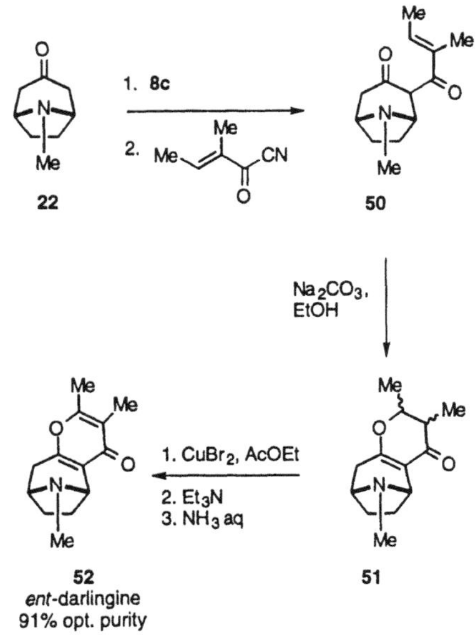

Scheme 11.

tropinone with the chiral base **8c**, followed by acylation using tigloyl cyanide, afforded compound **50** which readily cyclized under basic conditions to give the tricyclic species **51** (mixture of diastereoisomers), having the requisite pyranotropane skeleton. Introduction of the required double bond at the α,β position was done by bromination with $CuBr_2$, followed by elimination of HBr to give compound **52**.[53,54] This compound had all the spectral characteristics identical with darlingine, but was levorotatory. Since natural darlingine is dextrorotatory,[55] we have obviously synthesized the *ent* form, and the natural product must have the acyl substituent at C-2 (and not C-4, as in the structure **52**).

A number of tropane alkaloids contain the hydroxybenzyl or the benzyl group at either C-2 or C-4 of the tropane skeleton.[52] Examples include knightinol and alkaloid KD-B (Figure 16). Synthesis of knightinol seemed straightforward, since the method for efficient and enantioselective formation of the aldol **25a** had already been developed by our group. Surprisingly, however, changing the orientation of the side chain from the axial to the equatorial proved nontrivial. We could not find suitable conditions for isomerization of **25a** to the corresponding C-4 epimer. It seems that the configuration of **25a** is stabilized by intramolecular hydrogen bonding involving the OH and the nitrogen bridge; this had been established to be the case in the solid state.[28] In the end, protection of the aldol OH allowed the isomerization to the more stable isomer **54** having the side chain in the equatorial orientation (Scheme 12). Diastereoselective reduction using hydrogen on Adams'

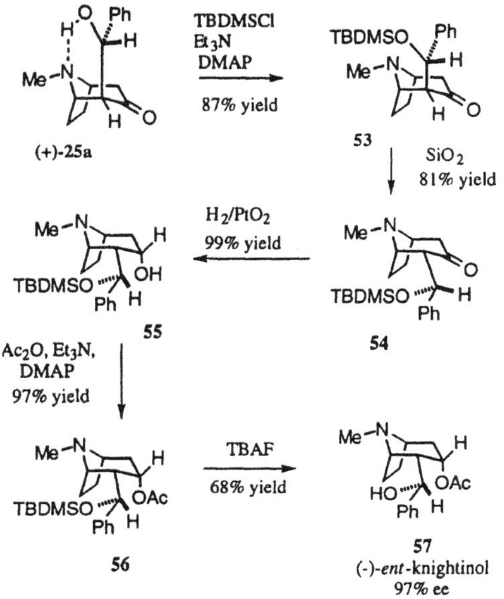

Scheme 12.

catalyst gave the *endo* alcohol **55**. Two standard functional group manipulation steps yielded compound **57** spectra of which were identical with these of natural knightinol but the optical rotation had the opposite sign.[53,54] Compound **57** was thus identified as *ent*-knightinol and the natural product must have the absolute configuration as shown in Figure 16.

Connecting a benzyl group to the carbon atom next to the tropinone carbonyl, in order to synthesize KD-B, looked like another simple task. However, all our attempts to alkylate the tropinone lithium enolate proved futile. Even with reactive alkyl halides like methyl iodide, allyl bromide, and benzyl bromide the alkylation either did not proceed at all or gave very low yields. Changing reaction conditions (higher temperatures, different solvents, using polar co-solvents), and adding the second equivalent of BuLi to prevent the internal proton return, did not help. Finally, we were forced to look for an indirect method. Aldol **25a** was used as the precursor to KD-B, the hydroxy group was removed by elimination and the resulting double bond proved easy to reduce chemoselectively. The reduction yielded the required benzyl substitutent (Scheme 13), and finally acetylation gave the levorotatory isomer of alkaloid KD-B in 62% overall yield (from tropinone).[54]

Tropinone-based synthesis of tropane alkaloids having a hydroxyl at either C-6 or C-7 requires a method to introduce the OH selectively, at one of these two enantiotopic atoms. Carbons 6 and 7 are not activated in the tropinone molecule and introduction of a functional group at one of these carbon atoms by a polar

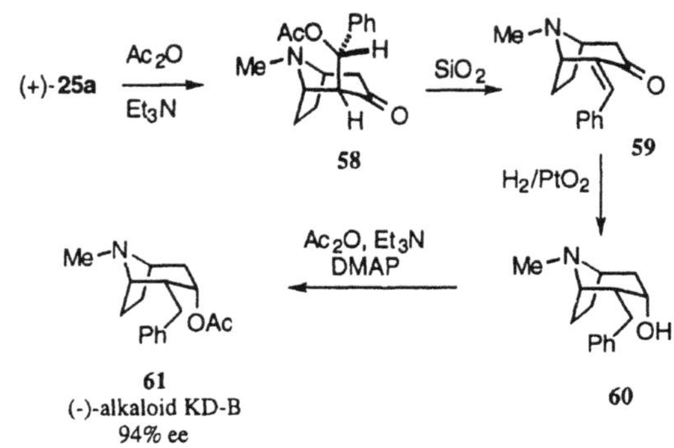

Scheme 13.

reaction would be difficult. However, in the product of the ring opening of tropinone by enolization followed by treatment with a chloroformate which was described earlier (cf. Scheme 4), either C-6 or C-7, depending on which enantiomer of the cycloheptenone was produced, is activated (allylic) and introduction of the OH at this position should be possible (Figure 17). If the nitrogen-bridged system could then be restored by removal of the COOR group from nitrogen and a Michael-type ring closure a way to the tropane skeleton hydroxylated at C-6 or C-7 would be open.

Functionalization at an allylic CH is, so to speak, easier said than done. Allylic bromination or hydroxylation of ketone **26** proved difficult and it seemed that the carbonyl functional group was the source of problems here. Simple protection of this group as an acetal did not improve the situation. Finally, a chemoselective reduction of the keto group using the Luche's method, followed by acetylation of the resulting alcohol **62**, yielded a derivative (**63**) which, after some experimentation, proved amenable to allylic hydroxylation (Scheme 12). The hydroxylation itself proved nontrivial and the detailed experimental procedure has been published elsewhere.[56] From the stereoselectivity point of view, the result was still disappoint-

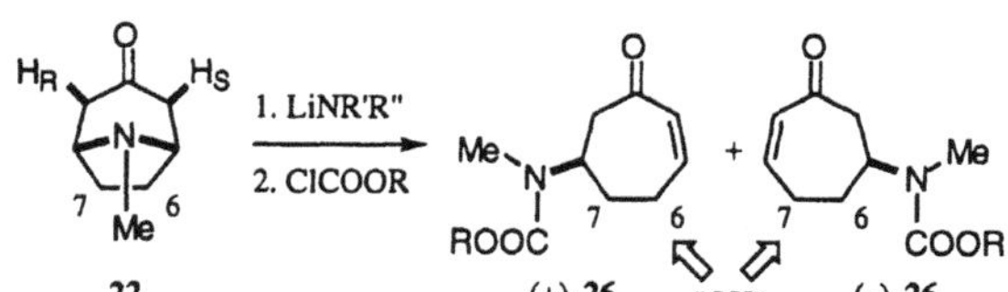

26a: R = Me
26b: R = CH_2Ph
26c: R = CH_2CCl_3

Figure 17. Activation of C6 or C7 of tropinone via the ring opening.

ing because two allylic alcohols, α- and β-isomers of **64**, were produced together with some ketone **65**. This did not look promising since only the β-alcohol seemed to be useful for synthesis of tropane alkaloids via Michael-type closure to C-5 (original tropinone numbering). However, compound **64** has some interesting latent symmetry, i.e. it should be noted that the β-isomer of **64** should yield a C_s symmetrical product upon acetylation. This observation led to the idea that if the nitrogen bridge of the tropane skeleton was to be restored by cyclization to C-4 (and not C-5, where it was connected originally), the 7β-acetoxytropinone would result, a useful entry into the 6(7)-β-hydroxytropanes. Thus, the strategy involving enantioselective deprotonation of tropinone, ring opening, 1,4-transposition of the carbonyl group, and ring reclosure was developed (Scheme 14).[37,56] Compound **66**

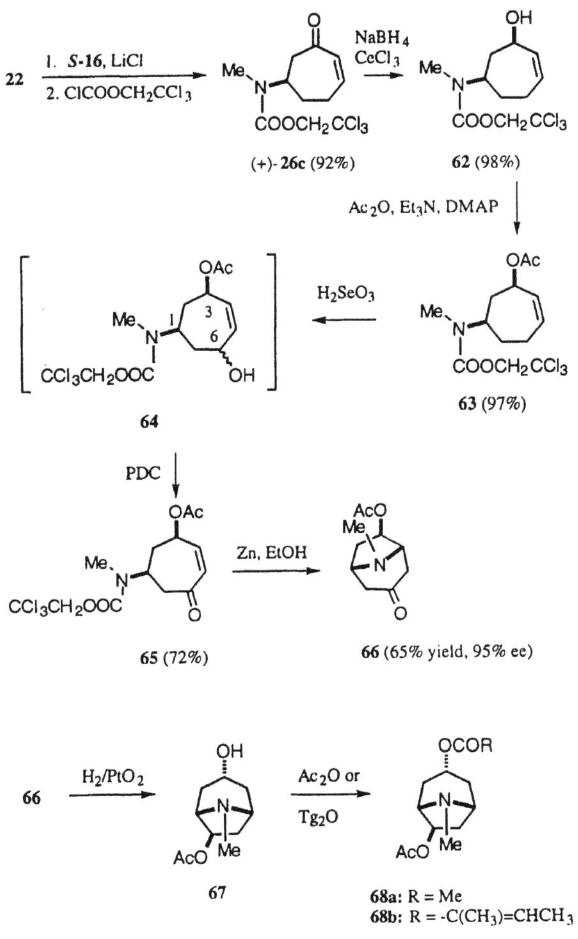

Scheme 14.

was used as starting material in synthesis of two natural products: (+)-3α,7β-diacetoxytropane **68a** and (–)-7β-acetoxy-3α-tigloyloxytropane **68b** (Scheme 14, second part).

A 1,3-transposition of the tropinone carbonyl group, via the Wharton rearrangement, in combination with the enantioselective chloroformate-promoted ring opening was used to synthesize physoperuvine (Scheme 15).[37,56] The bottleneck in the synthesis was the Wharton reaction, which proceeded only with a modest yield of 50%, but in the end, physoperuvine was synthesized in six steps, in over 95% ee, and 32% overall yield from tropinone.

In summary, enantioselective deprotonation of tropinone with chiral lithium amides, followed by hydroxyalkylation of the enolate with aldehydes, acylation using cyanoformates, and ring opening by chloroformates provided a key to development of a comprehensive synthetic strategy towards diverse chiral tropane alkaloids. An additional bonus was the assignment of absolute stereochemistry of these natural products. The alkaloids synthesized by this strategy are summarized in Figure 18 (yields were calculated from tropinone). Some of the compounds were synthesized in unnatural "*ent*" form which was mostly due to uncertainty as to their absolute configuration at the time the syntheses were started, but it should be noted that in each case the synthesis of the other enantiomer requires only using the mirror image of the chiral lithium amide. All syntheses were reasonably short, high yielding, and the target compounds were produced in high ee. In most cases, simple

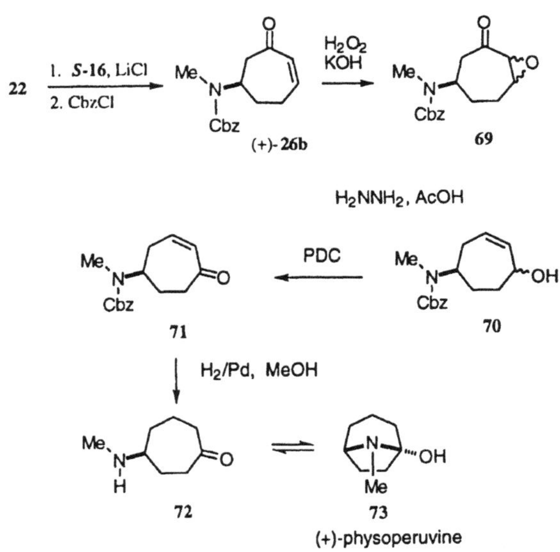

Scheme 15.

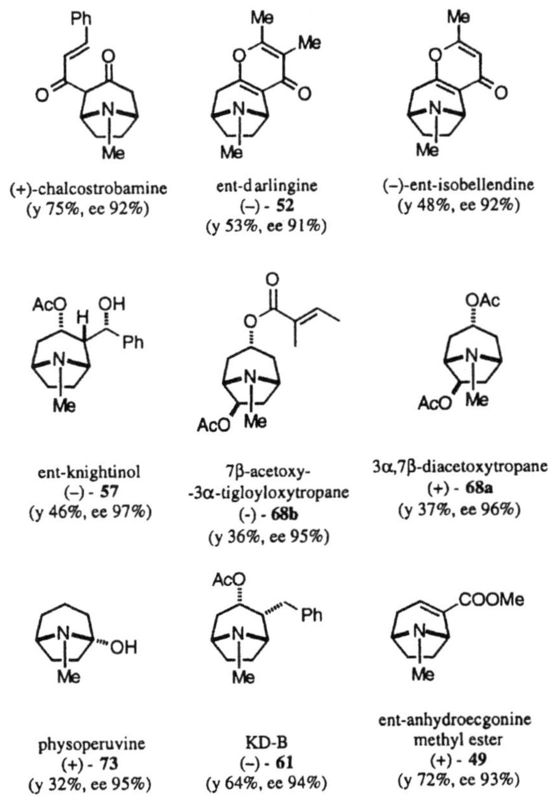

Figure 18. Tropane alkaloids synthesized via enantioselective deprotonation strategy.

purification, by e.g. one recrystallization, yielded optically pure compounds (the ee values in Figure 18 refer to nonpurified samples).

B. Synthesis of Terpenoids via Enantioselective Deprotonation

The initial studies on deprotonation with chiral lithium amides, described above, attempted to establish if an enantioselective reaction was possible and utilized alkyl-substituted cyclohexanones as simple models (cf. Scheme 3). Having another functional group on the cyclohexanone ring would increase the synthetic appeal of the methodology. With this idea in mind we studied at some length deprotonation of OH-protected 4-hydroxycyclohexanones with chiral lithium amides (cf. Figure 4).[57] The alkoxy group exerts much less conformational demand than the corresponding alkyl group; the A values for the methyl group and the methoxy group are 1.74 and 0.75 kcal/mol, respectively.[58] Thus the protected 4-hydroxycyclohex-

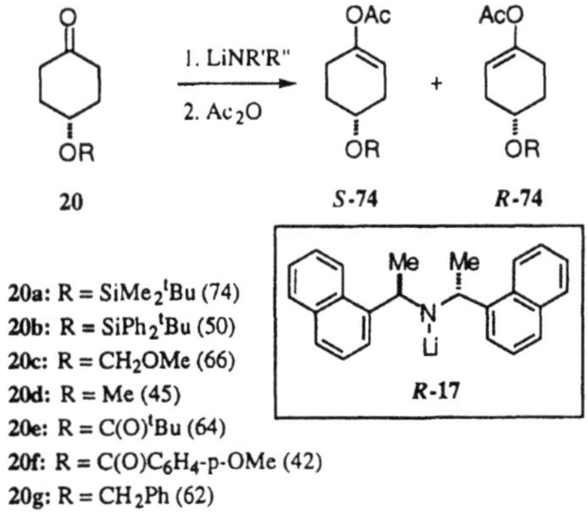

Scheme 16.

anone did not seem like a promising candidate for enantioselective deprotonation in view of Koga's original observations, suggesting that conformational stabilization effect is strongly tied to deprotonation enantioselectivity.[17] Initially, the examination of a number of 4-hydroxycyclohexanone derivatives of general structure **20** confirmed this pessimistic prediction (Scheme 16). Most of the lithium amides tested gave low or modest selectivities. The best amide was the C_2 symmetrical bis-naphthyl reagent ***R*-17**; the ee's obtained using this reagent are listed in Scheme 14 in parentheses. In general, the more sterically demanding was the protecting group the higher the enantioselectivity. Clearly, a certain measure of steric bias is necessary in cyclic ketones for deprotonation to proceed selectively. Manipulating the reaction conditions allowed, in the end, to deprotonate a 4-silyloxycyclohexanone with high enantioselectivity. Thus, when compound **20a** was very slowly added, at –100 °C, to a mixture of the lithium amide ***S*-17** and LiCl, which had been generated by mixing the hydrochloride of the amine and 2 equivalents of *n*-BuLi, and the resulting enolate was treated with TMS-Cl, the silyl enol ether **75** was produced in 90% ee (Scheme 17). The key elements here, as far as the reaction conditions were concerned, were the additive (LiCl), low temperature, and slow addition of the ketone to the base.

The results of this experiment were used in synthesis of two butenolide natural products.[40] The silyl enol ether **75** was oxidized to give a mixture of *cis*- and *trans*-hydroxyketones **76** and **77b** (Scheme 17). The reaction proceeded with low diastereoselectivity, but both product were useful. The *trans* isomer **76** was converted into (+)-dihydroaquilegiolide **79** by silylation, Peterson olefination, and removal of both silyl groups with concomitant lactonization using HF. The absolute stereochemistry was tentatively assigned by assuming that the base ***S*-17** abstracted the proton H_S, analogously to the deprotonation of tropinone by the same base.

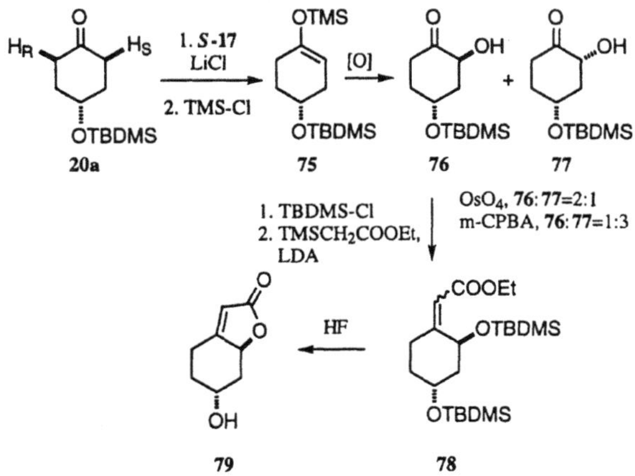

Scheme 17.

(–)-Dihydromenisdaurilide, the *cis* isomer of **79**, was synthesized by the same sequence of reactions from compound **77**.

C. Polyoxygenated Natural Products: Synthesis of Carbohydrate Derivatives

Heterocyclic ketones are promising starting materials for synthesis of diverse natural products via enantioselective deprotonation strategy. Carbohydrates and other polyoxygenated natural products could be envisaged as originating from 2-substituted-1,3-dioxa-5-ones. The aldol reaction of a dioxanone **80** with a chiral, 3-carbon aldehyde would produce a ketohexose derivative **81** (Scheme 18). Extend-

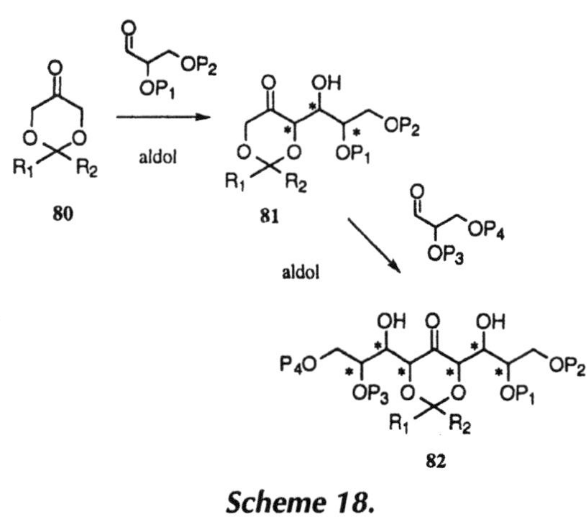

Scheme 18.

ing the carbon backbone the other way, by another aldol reaction with the same or a different aldehyde would lead to a carbohydrate derivative having nine carbon atoms in the skeleton. Some ketohexoses and many higher carbohydrates are not readily available and are interesting synthetic targets.

Dioxanones proved to be challenging compounds to work with. It took extensive experimental work to develop synthetic methodology required for efficient preparation of differently substituted dioxanones on a multigram scale.[45] Enolization of 1,3-dioxa-5-ones also was nontrivial and especially the heterocyclic ketones having only one alkyl group at the 2 position (acetals) suffered severe competition from the reduction with LDA[59] making them unsuitable starting materials for any enolate-based synthesis. Enolates of 2,2-dialkyl substituted dioxanones (ketals) could be generated efficiently, but their reactions with electrophiles were often unusual. Often the self-aldol condensation, leading to dioxanone dimers, interfered.[45,46] Clearly, the two oxygen substituents at the α-carbon atoms make the carbonyl group in a dioxanone very electrophilic with the reactivity towards nucleophilic attack approaching the reactivity of aldehydes. This behavior is precedented in other alkoxyketones.[61] We were also unable to alkylate any dioxanone lithium enolates using alkyl halides, usually the only products were the self aldols. The aldol reaction, on the other hand, worked well.[45] At this time the work on solving the many experimental problems of dioxanone chemistry is still ongoing. However, the preliminary results are interesting and demonstrate the potential for use of these compounds in synthesis of carbohydrates.[46]

The lithium enolate of 2,2-dimethyl-1,3-dioxa-5-one (**84**) reacted readily with the protected glyceraldehyde **85** and yielded a mixture of aldols as shown in Scheme 19. The two major products, which accounted for 88% of the crude product (12%

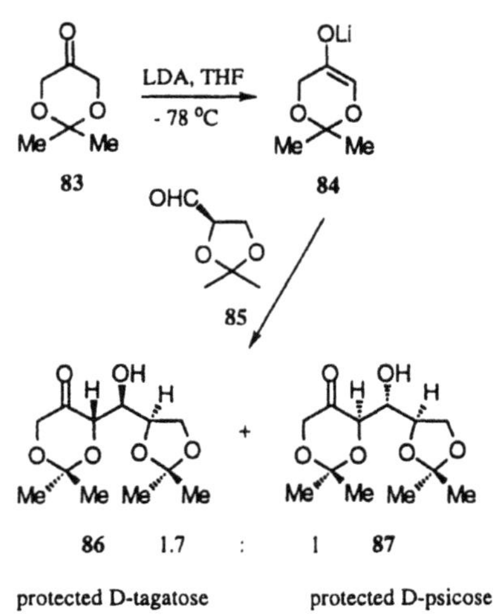

Scheme 19.

comprised a mixture of other aldol products which were discarded without separation during chromatographic purification of the two major products) were identified as derivatives of D-tagatose and D-psicose by reducing them to corresponding alcohols, deprotecting the OH groups, and comparing the products with D-tallitol and D-galacitol obtained by independent synthesis from D-talose and D-galactose. Thus the reaction shows fairly high aldol diastereoselectivity (*anti* aldols predominate) and low selectivity as far as discrimination between the two diastereotopic faces of the glyceraldehyde reagent is concerned. This behavior is precedented in glyceraldehyde chemistry.[62] Changing the reagent from the lithium enolate to the corresponding boron enolate resulted in a substantially improved diastereoselectivity. Thus, when the dioxanone **83** was treated with a mixture of dicyclohexylboron chloride and triethylamine, followed by addition of the aldehyde **85** and oxidative work up with H_2O_2 at pH 7, only the two aldols **86** and **87** were produced in a 6:1 ratio. And finally, double stereodifferentiation by using the chiral lithium enolate **89**, generated enantioselectively from the dioxanone **88** using the chiral lithium amide **90**, afforded a satisfactory route to derivatives of either D-psicose (**91**) or D-tagatose (**92**, Scheme 20). Both reactions with the chiral lithium amide were very diastereoselective; the aldol **91** accounted for 90% of the crude product with the other 10% comprising a mixture of other aldols and the final yield of the pure product **91** being 72%. The tagatose derivative **92** was produced somewhat less diastereoselectively with ca. 18% of the crude product mixture comprising other, unidentified aldols. Thus, although a lot of further experimental work is required, enantioselective deprotonation of dioxanones offers a promising synthetic strategy towards carbohydrates and other polyoxygenated natural products.

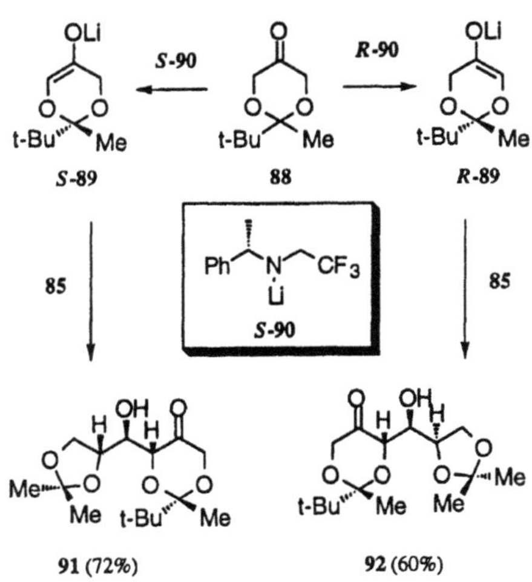

Scheme 20.

IV. SUMMARY AND CONCLUSIONS

One of the very attractive features of the new synthetic methodology utilizing chiral lithium amides is simplicity. Deprotonation of carbonyl compounds with lithium diisopropylamide is one of the most popular reactions in organic synthesis. Doing this reaction with a chiral lithium amide instead of LDA is technically simple and might improve the usefulness of a given reaction tremendously.

This account presents the work of our group on development and synthetic applications of enantioselective deprotonation of cyclic ketones with chiral lithium amides. Several other research groups were, and still are, active in this area with the groups of Koga and Simpkins deserving special recognition for advancing new concepts like catalytic enantioselective deprotonation, application of chiral lithium amides towards achieving regioselective reactions, and synthesis of new chiral lithium amide bases. The work progressed from the fundamental question ("is this possible?") through design and synthesis of new lithium amide reagents, elaboration of the reaction conditions, and application of the new methodology in synthesis of natural products. In the area of organic synthesis it was gratifying to see the chiral lithium amides becoming excellent tools for constructing a comprehensive synthetic strategy towards tropane alkaloids—we are close to the stage when an arbitrarily chosen enantiomer of any chiral tropane alkaloid could be synthesized by a short reaction sequence in reasonably high yield.

ACKNOWLEDGMENTS

I would like to thank many graduate and postgraduate collaborators which have done the experimental work described in this account. Professors Bick and Lounasmaa provided several samples of natural tropane alkaloids for which we are extremely grateful. The work was supported by a grant from National Science and Engineering Research Council of Canada.

REFERENCES

1. Morrison J. D.; Mosher, H. S. *Asymmetric Organic Reactions*; Prentice-Hall: Englewood Cliffs, NJ, 1971.
2. *Asymmetric Synthesis*; Morrison, J. D., Ed.; Academic Press: New York, 1983–1985, Vols. 1–5.
3. Helmchen, G.; Hoffman, R.W.; Mulzer, J.; Schaumann, E. (Eds.) In *Methods of Organic Synthesis (Houben-Weyl)*; Thieme: New York, 1996.
4. (a) Izumi, Y.; Tai, A. *Stereo-Differentiating Reactions;*, Academic Press: New York, 1977; (b) Seebach, D.; Hungerbuhler, E. In *Modern Synthetic Methods 1980*; Scheffold, R., Ed.; Salle: Frankfurt and Sauerlander: Aarau, 1980, p.107.
5. Reviews: (a) Ward, R. S. *Chem. Soc. Rev.* **1990**, *19*, 1; (b) Maier, M. *Nachr. Chem. Tech. Lab.* **1993**, *41*, 314.
6. Deslongchamps, P. *Aldrichimica Acta* **1984**, *17*, 59.
7. Wade, L. G. *Organic Chemistry*, 2nd ed.; Prentice-Hall: Toronto, 1991, pp. 266–267.
8. Mislow, K.; Siegel, J. *J. Am. Chem. Soc.* **1984**, *106*, 3319.

9. Whitesell, J. K.; Felman, S. W. *J. Org. Chem.* **1980**, *45*, 755.
10. (a) Duhamel, L. *C. R. Acad. Sci.* **1976**, *282C*, 125; (b) Duhamel, L.; Plaquevent J.-C. *Tetrahedron Lett.* **1977**, 2285; (c) Duhamel, L.; Duhamel, P.; Launay, J.-C.; Plaquevent J.-C. *Bull. Soc. Chim. Fr.* **1984**, II-421.
11. *Comprehensive Organic Synthesis;* Trost, B. M., Ed.; Pergamon: Oxford, 1991, Vols. 1–4.
12. Eliel, E. L.; Wilen, S. H.; Mander, L. N. *Stereochemistry of Organic Compounds*; Wiley-Interscience: Toronto, 1994.
13. House, H. O. *Modern Synthetic Reactions,* 2nd ed.; Benjamin: Menlo Park, CA, 1972.
14. (a) Williard, P. G.; Hintze, M. J. *J. Am. Chem. Soc.* **1987**, *109*, 5539; (b) *J. Am. Chem. Soc.* **1990**, *112*, 8602; (c) Williard, P. G.; Liu, Q.-Y.; Lochmann, L. *J. Am. Chem. Soc.* **1992**, *114*, 348.
15. Laube, T.; Dunitz, J. D.; Seebach, D. *Helv. Chim. Acta* **1985**, *68*, 1373.
16. (a) Collum, D. B. *Acc. Chem. Res.* **1993**, *26*, 227; (b) Gregory, K.; Schleyer, P. v. R.; Snaith, R. *Adv. Inorg. Chem.* **1991**, *37*, 47; (c) Willard, P. G.; Salvino, J. M. *J. Org. Chem.* **1993**, *58*, 1; (d) Williard, P. G.; Liu, Q.-Y. *J. Am. Chem. Soc.* **1993**, *115*, 3380; (e) Seebach, D. *Angew. Chem., Int. Ed. Engl.* **1988**, *27*, 1624; (f) Jackman, L. M.; Boriatynski, J. In *Advances in Carbanion Chemistry*; Snieckus V., Ed.; JAI Press, Greenwich, CT, 1992, Vol. 1, p. 45; (g) Seebach, D.; Amstutz, R.; Dunitz, J. D. *Helv. Chim. Acta* **1981**, *64,* 2622.
17. Shirai, R.; Tanaka, M.; Koga, K. *J. Am. Chem. Soc.* **1986**, *108*, 543.
18. Simpkins, N. S. *J. Chem. Soc. Chem. Commun.* **1986**, 88.
19. Majewski, M.; Gleave, D. M. *J. Org. Chem.* **1992**, *57*, 3599.
20. Aoki, K.; Nakajima, M.; Tomioka, K.; Koga, K. *Chem. Pharm. Bull.* **1993**, *41*, 994.
21. Zheng, G.-Z. M. Sc. Thesis, University of Saskatchewan, 1991.
22. Deslongchamps, P. *Stereoelectronic Effects in Organic Chemistry*; Pergamon: Toronto, 1983, pp. 274–284.
23. Majewski, M.; Zheng, G.-Z. *Synlett* **1991**, 173.
24. Crabtree, S. R.; Chu, W. L. A.; Mander, L. A. *Synlett* **1990**, 169.
25. Majewski, M.; Zheng, G.-Z. *Can. J. Chem.* **1992**, *70*, 2618.
26. C.f. ref. 11, Vol. 2, p. 155.
27. Seebach, D.; Prelog, V. *Angew. Chem., Int. Ed. Engl.* **1982**, *21*, 654.
28. Li, J.; Quail, J. W.; Zheng, G.-Z.; Majewski, M. *Acta Cryst.* **1993**, *C49*, 1410.
29. Eliel, E. L.; Wilen, S. H.; Mander, L. N. *Stereochemistry of Organic Compounds;* Wiley-Interscience: Toronto, 1994, Chap. 6.
30. Lazny, R. Ph. D.Thesis, University of Saskatchewan, 1996.
31. Campbell, H. F.; Edwards, O. E.; Kolt, R. *Can. J. Chem.* **1977**, *55*, 1372.
32. (a) Review: Mukayiama, T.; Asami, M. *Top. Curr. Chem.* **1985**, *127*, 133; (b) Asami, M.; Ishizaki, T.; Inoue, S. *Tetrahedron: Asymmetry* **1994**, *5*, 793.
33. Loupy, A.; Tchoubar, B. *Salt Effects in Organic and Organometallic Chemistry;* VCH: New York, 1991.
34. Majewski, M.; Gleave, D. M. *Tetrahedron Lett.* **1989**, *30*, 5681.
35. Majewski, R.; Lazny, P.; Nowak. *Tetrahedron Lett.* **1995**, *36*, 5465.
36. (a) Juaristi, E.; Beck, A. K.; Hansen, J.; Matt, T.; Mukhopadhyay, T.; Simson, M.; Seebach, D. *Synthesis* **1993**, 1271; (b) Hall, P. L.; Gilchrist, J. H.; Collum D. B. *J. Am. Chem. Soc.* **1991**, *113*, 9571; (c) Seebach, D.; Beck, A. K.; Studer A. In *Modern Synthetic Methods*; Ernst, B.; Leumann, C., Eds.; VCH: Weinheim, 1995,Vol. 7.
37. Majewski, M.; Lazny, R.; Ulaczyk, A. *Can. J. Chem.* **1997**, *75*, 754.
38. (a) Williard, P. G.; Hintze, M. J. *J. Am. Chem. Soc.* **1990**, *112*, 8602; (b) Mair, F. S.; Clegg, W.; O'Neil, P. A. *J. Am. Chem. Soc.* **1993**, *115*, 3388; (c) Jackman, L. M.; Rakiewicz, E. F. *J. Am. Chem. Soc.* **1991**, *113*, 1202; (d) Jackman, L. M.; Rakiewicz, E. F.; Benesi, A. J. *J. Am. Chem. Soc.* **1991**, *113*, 4101; (e) Glaser, R.; Hadad, C. M.; Wiberg, K. B.; Streitwieser, A. *J. Org. Chem.* **1991**, *56*, 6625.
39. Nogradi, M. *Stereoselective Syntheses*; VCH: New York, 1986.

40. Majewski, M.; Irvine, N. M.; MacKinnon, J. *Tetrahedron: Asymmetry* **1995**, *6*, 1837.
41. (a) Shirai, R.; Tanaka, M.; Koga, K. *J. Am. Chem. Soc.* **1986**, *108*, 543; (b) Aoki, K.; Noguchi, H.; Tomioka, K.; Koga, K. *Tetrahedron Lett.* **1993**, *34*, 5105; (c) Cousins, R. P. C.; Simpkins, N. S. *Tetrahedron Lett.* **1989**, *30*, 7241; (d) Cain, C. M.; Cousins, R. P. C.; Coumbarides, G.; Simpkins, N. S. *Tetrahedron* **1990**, *46*, 523; (e) Kim, H.-D.; Shirai, R.; Kawasaki, H.; Nakajima, M.; Koga, K. *Heterocycles* **1990**, *30*, 307; (f) Honda, T.; Kimura, N.; Tsubuki, M. *Tetrahedron: Asymmetry* **1993**, *4*, 21; (g) Muraoka, O.; Okumura, K.; Maeda, T.; Tanabe, G.; Momose, T. *Tetrahedron: Asymmetry* **1994**, *5*, 317; (h) Honda, T.; Kimura, N.; Sato, S.; Kato, D.; Tominaga, H. *J. Chem. Soc., Perkin Trans. 1* **1994**, 1043; (i) Underiner, T. L.; Paquette, L. A. *J. Org. Chem.* **1992**, *57*, 5438; (j) Izawa, H.; Shirai, R.; Kawasaki, H.; Kim, H.-D.; Koga, K. *Tetrahedron Lett.* **1989**, *30*, 7221; (k) Leonard, J.; Hewitt, J. D.; Ouali, D.; Raman, S. K.; Simpson, S. J. *Tetrahedron: Asymmetry* **1990**, *1*, 699; (l) Leonard, J.; Ouali, D.; Raman, S. K. *Tetrahedron Lett.* **1990**, *31*, 739; (m) Leonard, J.; Ouali, D.; Raman, S. K. *J. Chem. Soc., Perkin Trans. 1* **1992**, 1203; (n) Bunn, B. J.; Cox, P. J.; Simpkins, N. S. *Tetrahedron* **1993**, *49*, 207; (o) Bunn, B. J.; Simpkins, N. S. *J. Org. Chem.* **1993**, *58*, 533; (p) Momose, T.; Toyooka, N.; Seki, S.; Hirai, Y. *Chem. Pharm. Bull.* **1990**, *38*, 2072; (r) Momose, T.; Toyooka, N.; Hirai, Y. *Chem. Lett.* **1990**, 1319; (s) Muraoka, O.; Okumura, K.; Maeda, T.; Tanabe, G.; Momose, T. *Tetrahedron: Asymmetry* **1994**, *5*, 317.
42. Juaristi, E. *Introduction to Stereochemistry and Conformational Analysis*; Wiley: Toronto, 1991, p. 265.
43. Honda, T.; Kimura, N.; Tsubuki, M. *Tetrahedron: Asymmetry* **1993**, *4*, 1475.
44. Aoki, K.; Tomioka, K.; Noguchi, H.; Koga, K. *Tetrahedron* **1997**, *53*, 13641.
45. Majewski, M.; Gleave, D. M.; Nowak, P. *Can. J. Chem.* **1995**, *73*, 1616.
46. Nowak, P. Ph. D. Thesis, University of Saskatchewan, 1998.
47. Sun, X.; Kenkre, S.; Remenar, J. F.; Gilchrist, J. H.; Collum, D. B. *J. Am. Chem. Soc.* **1997**, *119*, 4765.
48. A typical deprotonation of a ketone with LDA is usually done in THF at –78°C for 15–30 min but a number of cases have been described involving either a very short or a very long deprotonation time, cf.: (a) Negishi, E.; King, A. O.; Tour, J. M. *Org. Synth. Coll. Vol. 7*, **1990**, 63; (b) Kende, A. S.; Fludzinski, P. *ibid.* p. 67; (c) Vedejs, E.; Larsen, S. *ibid.* p. 277; (d) Evans, D. A.; Gage, J. R.; Leighton, J. L. *J. Org. Chem.* **1992**, *57*, 1964; (f) Kuwajima, I.; Sato, T.; Arai, M. Minami, N. *Tetrahedron Lett.* **1976**, 1817.
49. Majewski, M.; Nowak, P. *Tetrahedron Lett.* **1998.** In press.
50. Espenson, J. H. *Chemical Kinetics and Reaction Mechanisms;* McGraw-Hill: Toronto, 1995.
51. Ireland, R. E.; Wipf, P.; Armstrong, J. D. *J. Org. Chem.* **1991**, *56*, 650.
52. Reviews: (a) Lounasmaa, M.; Tamminen, T. *The Alkaloids* **1993**, *44*, 1; (b) Lounasmaa, M. *ibid.* **1988**, *33*, 1.
53. Majewski, M.; Lazny, R. *Tetrahedron Lett.* **1994**, *35*, 3653.
54. Majewski, M.; Lazny, R. *J. Org. Chem.* **1995**, *60*, 5825.
55. Bick, I. R. C.; Gillard, J. W.; Leow, H. *Aust. J. Chem.* **1979**, *32*, 2523.
56. Majewski, M.; Lazny, R. *Synlett* **1996**, 785.
57. Majewski, M.; MacKinnon, J. *Can. J. Chem.* **1994**, 72.
58. Juaristi, E. *Introduction to Stereochemistry and Conformational Analysis*; Wiley: Toronto, 1991, p. 245.
59. Majewski, M.; Gleave, D. M. *J. Organomet. Chem.* **1994**, *470*, 1.
60. Gleave, D. M. Ph. D. Thesis, University of Saskatchewan, 1993.
61. (a) Heathcock, C. H. In *Comprehensive Organic Synthesis*; Trost, B. M., Ed., Pergamon Press: Oxford, **1991**,Vol. 2 p.140; (b) Thornton, E. R.; Das, G. *J. Am. Chem. Soc.* **1990**, *112*, 5360.
62. Jurczak, J.; Pikul, S.; Bauer, T. *Tetrahedron* **1986**, *42*, 447.

STEREOSELECTIVE ADDITION OF CHIRAL α-AMINOORGANOMETALLICS TO ALDEHYDES*

Robert E. Gawley

*Dedicated to Professor Waldemar Adam, on the occasion of his 60th birthday, July 26, 1997.

Advances in Asymmetric Synthesis
Volume 3, pages 77–111.

ISBN: 0-7623-0106-6

I. INTRODUCTION

In general, the steric course of addition of a nucleophile to a carbonyl can fall into one of several categories.[a] The most simple is the addition of an achiral nucleophile to a carbonyl whose faces are homotopic, in which case the product is achiral. The addition of phenyllithium to acetone (Scheme 1a) falls into this category. If the two carbonyl faces are enantiotopic, addition to the *Re* or *Si* faces affords transition states that are chiral and enantiomeric. The addition of phenyllithium to acetaldehyde (Scheme 1b) is an example of this type of reaction. In an achiral solvent, the two transition states will be isoenergetic (i.e. $\Delta\Delta G^{\ddagger} = 0$), and the two enantiomers of the product will be formed in equal amounts.[b] In order for the product ratio to be ≠ 1, the energies of activation must be different. To render the transition states diastereomeric ($\Delta\Delta G^{\ddagger} \neq 0$), there must be an additional element of chirality. This additional element could be in either of the reactants (organometallic or carbonyl), the solvent, or a catalyst, as illustrated by the following examples.

In the early 1950s, several groups began to analyze the factors affecting the stereoselective addition of a nucleophile to faces of carbonyl compounds that are diastereotopic by virtue of a proximate stereocenter. Principle among these were the efforts of Curtin,[3] Cram,[4] and Prelog.[5] These types of additions may be generalized as shown in Scheme 2. If the addition is irreversible (i.e. under kinetic control), the difference in free energy of activation, $\Delta\Delta G^{\ddagger}$, determines the product

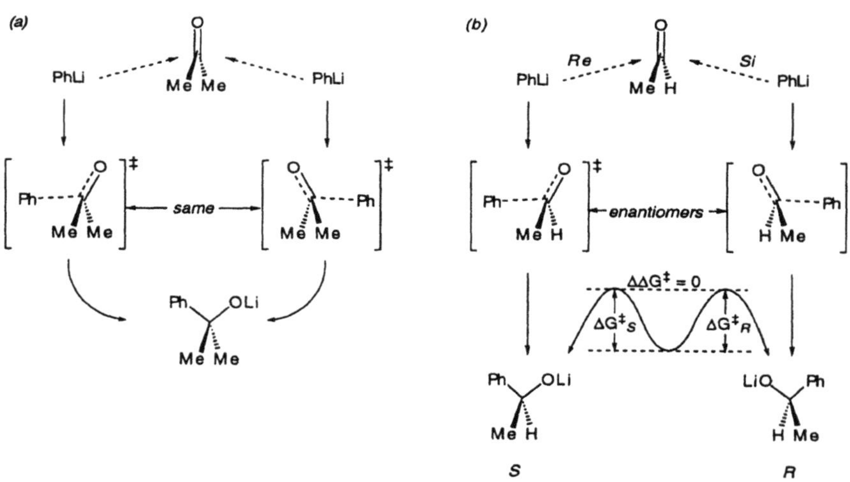

Scheme 1. (a) Addition of phenyllithium to either of the homotopic faces of acetone affords identical transition states. (b) Addition of phenyllithium to the enantiotopic faces of acetaldehyde affords enantiomeric transition states.

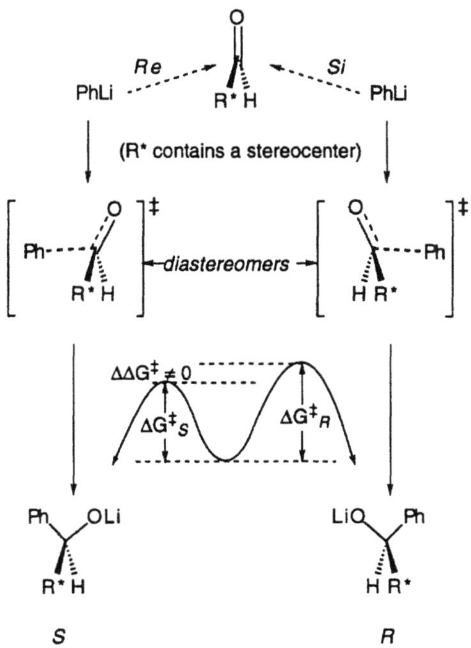

Scheme 2. Addition of phenyllithium to a chiral aldehyde affords diastereomeric transition states.

ratio. Often, the factors that influence the *Re/Si* face-selectivity are complex. For example, in the case of Cram's rule,[4] it has taken 40 years to derive a mechanistically sound rationale for the steric course of the addition.[c]

Another way to render the carbonyl faces diastereotopic is by complexation of a chiral Lewis acid to the carbonyl oxygen. This is the approach taken in, for example, the asymmetric diethylzinc reaction, as shown in Scheme 3.[6] For reviews of such processes, see refs. 7–11, chapter 5 in ref. 12, and pp. 137–141 of ref. 1.

Yet another way to render the transition states diastereomeric is to employ a chiral nucleophile (Scheme 4). One such class is organolithium and organomagnesium (Grignard) reagents in which the carbon bearing the metal is sp^3 hybridized and stereogenic. Although many secondary organolithium and Grignard reagents are chiral (e.g. *sec*-BuLi), much of the progress in stereoselective reactions of chiral organometallics has occurred in species containing a heteroatom in the position α to the metal. The heteroatom may be a first row element such as nitrogen or oxygen, or main group elements such as phosphorus, sulfur, selenium, or tellurium.[d] In this review, I will focus on nitrogen as the heteroatom.

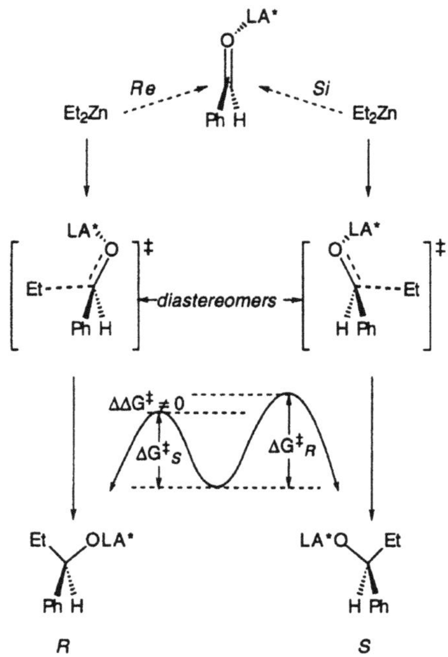

Scheme 3. Addition of diethylzinc to benzaldehyde, catalyzed by chiral Lewis acids.[6]

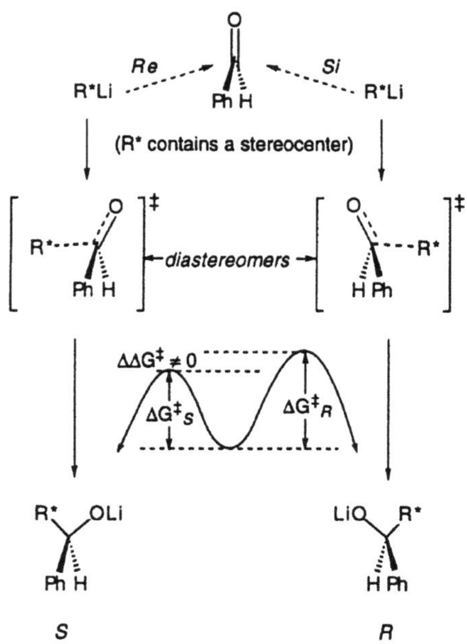

Scheme 4. Addition of a chiral organolithium to benzaldehyde affords diastereomeric transition states.

II. TRANSITION STATES AND MECHANISTIC RATIONALES

It is of interest to analyze the possible transition state assemblies of additions with the aim of rationalizing the steric course of the addition. In considering the mechanistic details of such additions, one must ask the following questions:

1. Is the organometallic configurationally stable? If so, what is the absolute configuration?
2. Does the addition occur by a polar pathway or a single electron transfer (SET) redox reaction, followed by a radical coupling?
3. If polar, does the addition to a carbonyl occur with retention or inversion at the carbanionic carbon?
4. Is there a preferred topicity for the addition? (i.e. Does the (*R*)-organometallic add to the *Re* face or *Si* face of the aldehyde?)

These considerations are illustrated in Scheme 5 for α-aminoorganometallics. Depending on the method of metallation, the configuration of the α-aminoorganometallic (boxed) may not be known, and its configurational stability may not be immediately obvious. Nevertheless, in reacting with an electrophile, there are two major pathways that can be followed: a polar addition and a radical mechanism involving oxidation of the carbanion by the electrophile (SET). Scheme 5 illustrates an aldehyde as an electrophile because analysis of the ratio of the four possible diastereomeric products may provide an opportunity to determine the mechanistic pathway(s) being followed. For example, if the organometallic is configurationally stable, a polar pathway proceeding with retention of configuration at the carbanionic carbon would give a mixture of the *R,R* and *R,S* addition products. If the configuration of the organometallic is known, then one can deduce the steric course at the carbanionic carbon. Conversely, a configurationally labile carbanion that adds with a given topicity would give a predominance of the *R,R* / *S,S* pair over the *R,S* / *S,R* pair, but the steric course (inversion vs. retention) may not be discernible.

For a radical pathway, two assumptions can be made: (1) pyramidal inversion (and translational motion in the solvent cage) of a free radical is faster than coupling to the aldehyde ketyl, and (2) radical coupling will show poor selectivity in adding to heterotopic carbonyl faces.[e] Under these conditions, an SET path would give an equimolar mixture of all four products.

Two final points should be made clear. First, the analysis outlined above is only complete if the α-aminoorganolithium is not racemic; otherwise, using the steric course of the reaction to elucidate mechanism is limited. The examples discussed below will illustrate the use of these principles. Second, the analysis outlined in Scheme 5 may be an oversimplification if the organometallic is not monomeric; in most of the examples discussed in this review, the aggregation state of the α-aminoorganometal is not known.

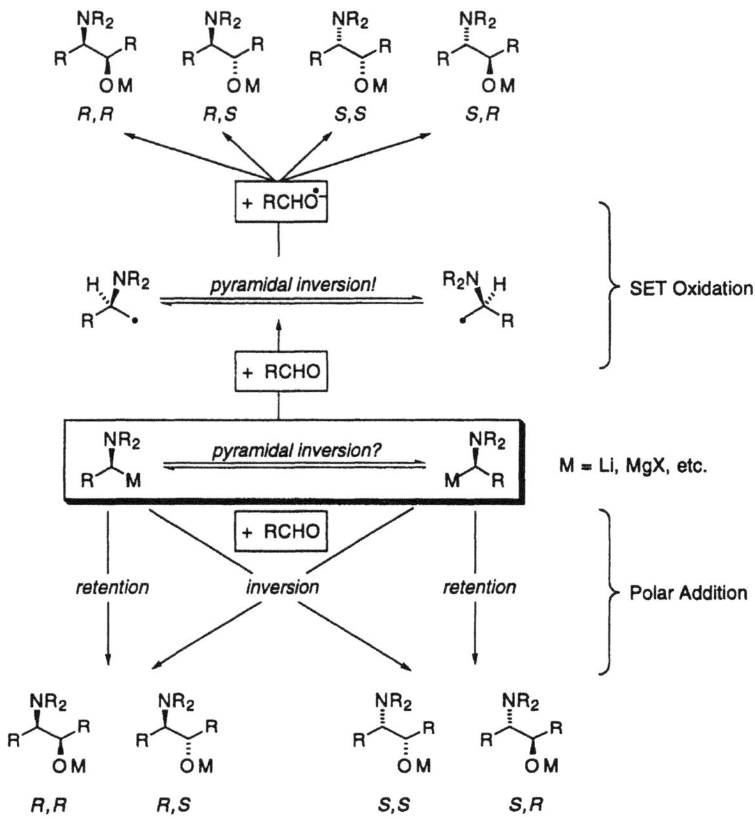

Scheme 5. Mechanistic possibilities for the polar addition of a chiral α-aminoorganometallic to an aldehyde.

α-Aminoorganolithium species have been known since the 1960s,[17–21] but examples of stereoselective reactions are a more recent development.[22–28] Analysis of the factors affecting the stereoselectivity of such additions is not always possible; nevertheless, synthetically useful processes have been developed.

III. TOPICITY AND TERMINOLOGY

Before reviewing the examples in detail, it is necessary to define the terminology that will be used to describe relative configuration and topicity in these reactions. In 1982, Prelog and Helmchen proposed the descriptors *l* and *u* (for *like* and *unlike*) to describe relative configuration,[29] and this terminology will be used here. Thus,

R,R and *S,S* pairs of stereocenters have the *l* relative configuration and *R,S* and *S,R* pairs are *u*. Following this precedent, Seebach and Prelog proposed that the steric course of reactions between two trigonal atoms could be classified topologically as *lk (like)* for reactions in which the *Re,Re or Si,Si* heterotopic faces are joined, and as *ul (unlike)* for reactions in which *Re,Si* or *Si,Re* faces were joined, as illustrated in Figure 1a.[30] These protocols are based on the CIP sequence rules[29,31] and are

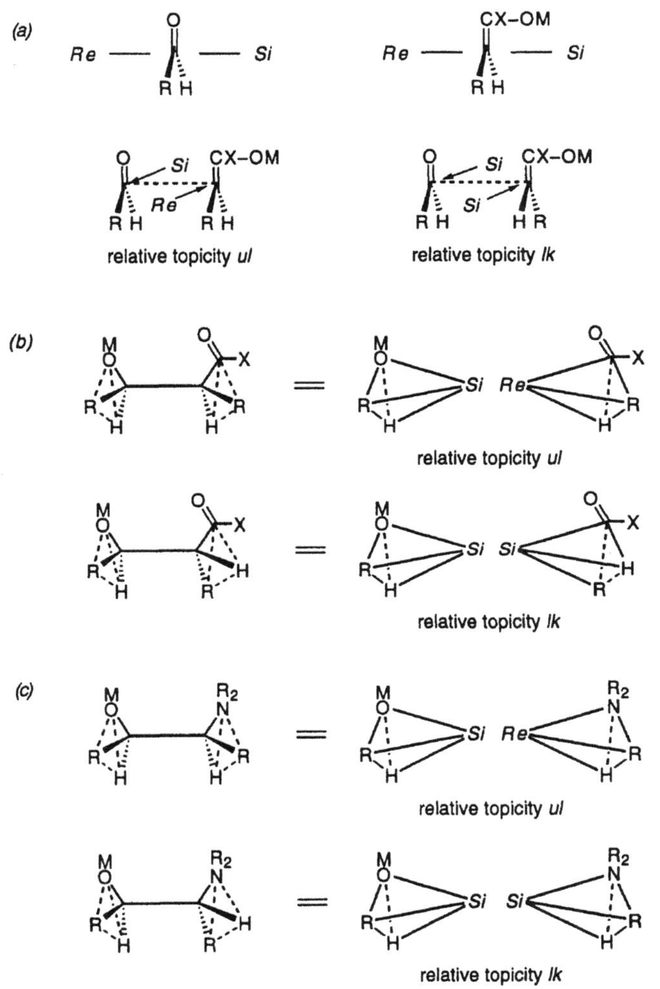

Figure 1. (a) Seebach–Prelog definition of relative topicity. (b) A second way to view the Seebach–Prelog concept. (c) Definition of relative topicity for β-amino alcohols based on relative configuration.

unambiguous in all respects when trigonal atoms are involved, because even in the transition state the reacting atoms are still only tetravalent. However, the same rules are not directly applicable to the reaction of a chiral, stereogenic nucleophile such as an organolithium, since the carbanionic carbon is tetrahedral in the ground state and pentavalent in the transition state and (with reference to Scheme 5) the reaction may occur with either retention or inversion of configuration (see also Figure 2 and the accompanying discussion, below). Nevertheless, examination of the products of the reaction of two trigonal atoms (Figure 1b) illustrates how the topicity may be defined based on the relative configuration of the products.[f] Thus, for an aldol addition, the three ligands of each of the former trigonal atoms form the bases of two tetrahedra, with the fourth vertex being the nucleophilic or electrophilic carbon of the other reactant. It can be easily seen that this fourth vertex is sitting on either a *Re* or *Si* face of a triangle and that these descriptors match the relative topicity according to the Seebach–Prelog definitions. Extending this concept to the β-amino alcohol product of addition of an α-aminoorganometallic to an aldehyde is straightforward, as shown in Figure 1c.

The illustrations in Figure 2 demonstrate why it would be impossible to try to define relative topicity for these reactions based on the reactants. The reaction at the metal-bearing carbon is an S_E2 process, which may occur with either retention or inversion of configuration. In reactions of chiral organolithiums with carbonyl compounds, both pathways are known.[33,34] Thus, the steric course of a reaction such as this must be specified with respect to both topicity *and* retention/inversion at the metal-bearing carbon. Note that the two structures in Figure 2 have the same topicity as defined in Figure 1c.

Assuming that the addition of a chiral organolithium to an aldehyde is a polar process, you might expect—based on the analysis outlined in Scheme 5—to see at least modest diastereoselectivities in additions of chiral organometals to aldehydes. The following tables list examples taken from the literature. Because the metal-bearing carbon in these examples is neither allylic nor benzylic, it is likely that these species are configurationally stable under the conditions listed.[35] Table 1 in Section IV lists examples of racemic carbanions, while Table 2 in Section V lists examples of chiral, nonracemic organometallics.

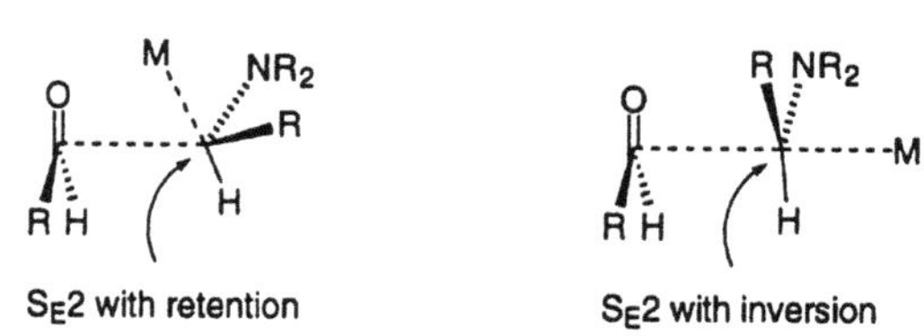

Figure 2. Transition structures of S_E2 reaction with retention and inversion at the metal-bearing carbon.

IV. EXAMPLES: CONFIGURATIONALLY STABLE, RACEMIC

The examples in Table 1 can only be evaluated in terms of topicity, since both enantiomers of the organometallic are present. Most of the examples listed are organolithiums, and the tendency is for *lk* topicity (*l* diastereomer). With lithiated pyrrolidine formamidines, 95–97% ds is possible (entries 3 and 4), although the results are highly variable as indicated by entries 1 and 2. Lithiated piperidine formamidines and urethanes exhibit very low (if any) *lk* selectivity (entries 5 and 6), although transmetallation of the formamidine to the magnesium derivative reverses the topicity to *ul* (entry 7). When the carbanionic carbon is acyclic (entry 9), there is also no selectivity. With the exception of entries 3 and 4, none of these diastereoselectivities are synthetically useful. Also, one wonders why the selectivity of addition of lithiated pyrrolidine formamidine is so variable (entries 1–3).

V. EXAMPLES: CONFIGURATIONALLY STABLE, NONRACEMIC

A few examples of organometallics that are configurationally stable and nonracemic have been reported, and are listed in Table 2. In every case, the organometal showed no facial selectivity (topicity preference) in additions to benzaldehyde. Nevertheless, the fact that these compounds are single epimers at the metal-bearing carbon allows us to probe the mechanism of addition to an extent not possible with the examples of racemic organometals from Table 1. For example, the lithiated pyrrolidine and piperidine in entries 1 and 2 exhibit extraordinary configurational stability. In 1993, Qianhui Zhang of my laboratory found that they are among the most stable organolithiums known, resisting racemization in THF/TMEDA mixtures at temperatures as high as –40 °C.[35,43,44] Although there is no *ul/lk* preference in their addition to benzaldehyde, analysis of enantiomer ratios in the products revealed that the addition occurred with 97% retention of configuration at the metal-bearing carbon.[45] The enantiomer ratios of the products were determined by rotation and comparison with literature values, and 97% retention may be within experimental error of 100%. On the other hand, if real, 4–5% of the product arises from inversion of configuration in these two examples. With reference to Scheme 5, there would be a number of mechanistic explanations:

1. The electrophile could catalyze the racemization of the organolithium.
2. There could be competition between polar pathways whereby the steric course at the metal-bearing carbon is a mixture of retention and inversion.
3. There could be a competitive stereorandom SET pathway that affords a mixture of all four possible stereoisomers.

The first option could be ruled out by experiment with catalytic amounts of electrophile, while the second and third possibilities might be distinguished with

Table 1. Simple Diastereoselectivity in the Addition of Configurationally Stable, Racemic, Chiral α-Aminoorganometallics to Benzaldehyde

Entry	Organometallic	Conditions	Products: *ul* Topicity		Products: *lk* Topicity	Reference
1	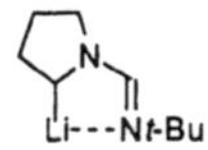	THF −78 °C	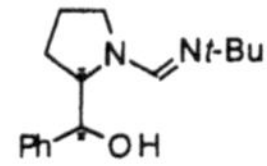 *rac-u*		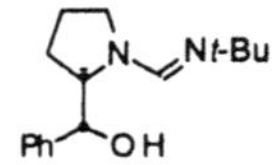 *rac-l*	36
			25	:	75	
2	"	"	50	:	50	37
3	"	"	5	:	95	38
4	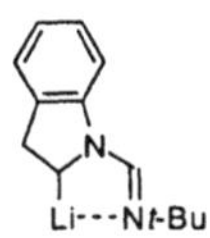	"	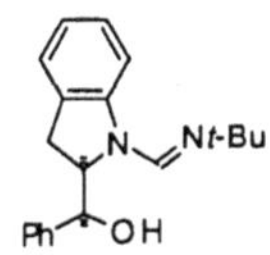 *rac-u*		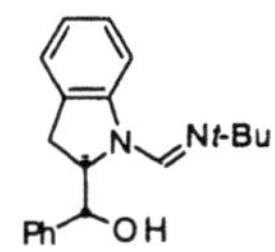 *rac-l*	39
			3	:	97	
5	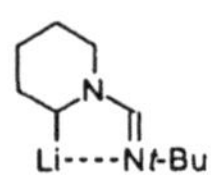	"	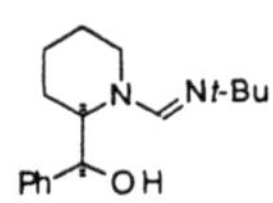 *rac-u*		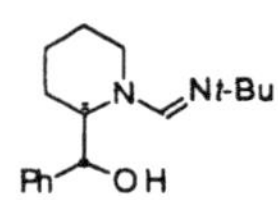 *rac-l*	37
			50-38	:	50-62	
6	"	80:20 ether/THF	45	:	55	38
7	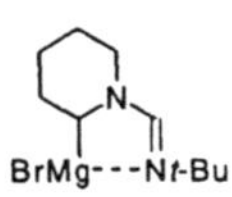	THF −78 °C	66	:	34	40
8	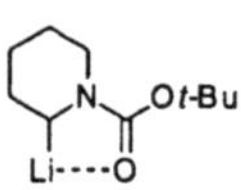	ether/TMEDA −78 °C →RT	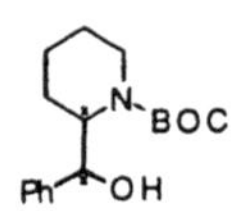 *rac-u*		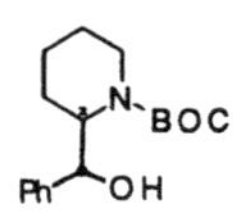 *rac-l*	41
			51	:	49	
9	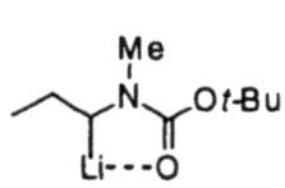	THF −78 °C	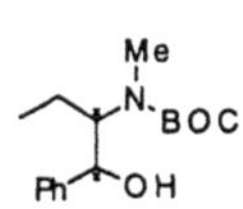 *rac-u*		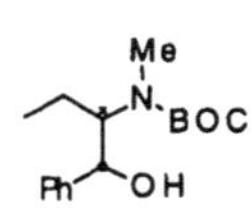 *rac-l*	42
			50	:	50	

Table 2. Simple Diastereoselectivity in the Addition of Configurationally Stable, Chiral Nonracemic α-Aminoorganometallics to Aldehydes

			Products		
Entry	*Organometallic*	*Conditions*	*ul Topicity*	*lk Topicity*	*Reference*
1	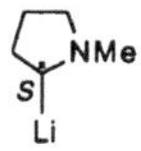 94% ee 97 : 3 er	THF –78 °C	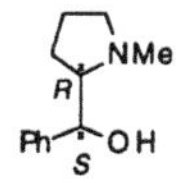 *u*, 89% ee 94.5 : 5.4 er	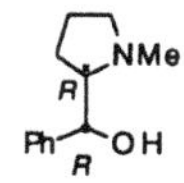 *l*, 89% ee 94.5 : 5.4 er	45
			50	: 50	
			97% retention of configuration at R*M		
2	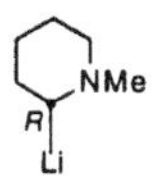 99% ee >99 : 1 er	"	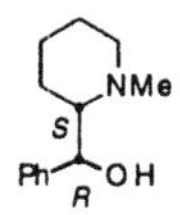 *u*, 92% ee 96 : 4 er	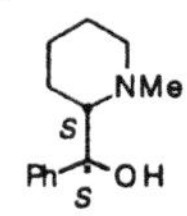 *l*, 92% ee 96 : 4 er	45
			50	: 50	
			97% retention of configuration at R*M		
3	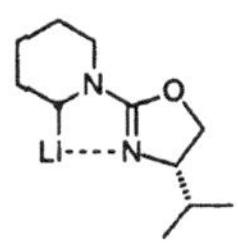	ether –100 °C →RT	after auxiliary removal: 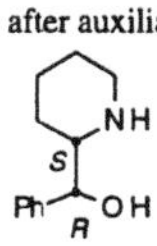 *u*, 0% ee 50 : 50 er	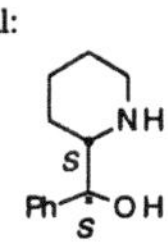 *l*, 0% ee 50 : 50 er	46
			50	: 50	
4	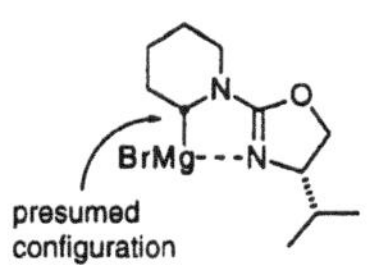	"	after auxiliary removal: 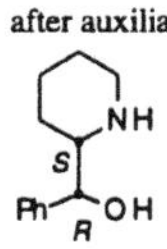 *u*, 100% ee >99 : 1 er	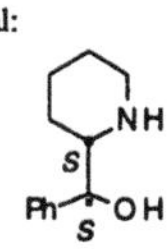 *l*, 50% ee 75 : 25 er	46
			50	: 50	
5	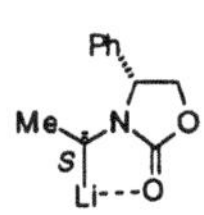 (similar with *R* diastereomer)	THF –78 °C	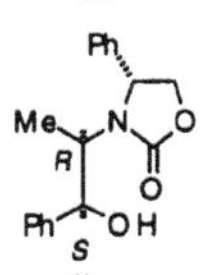	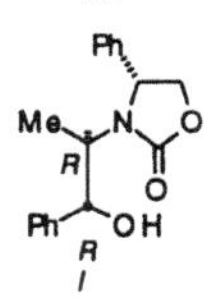	47
			50	: 50	
6	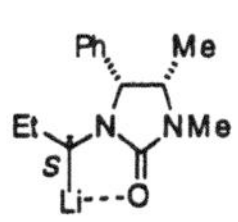 (similar with *R* diastereomer)	"	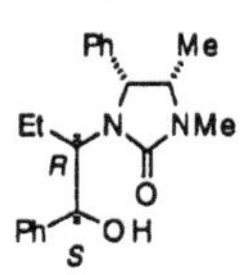	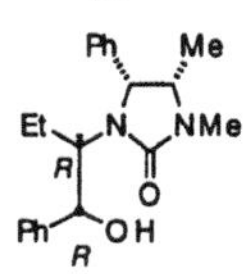	47
			50	: 50	

radical probe electrophiles. Such experiments have not been done, but note the following. In entry 3, the lithium is chelated by an amidine nitrogen (cf. entries 5 and 6 of Table 1). Not only is there no facial selectivity, but the enantiomer ratio of the amino alcohols formed after removal of the oxazoline auxiliary is 50:50.[46] Transmetallation from lithium to magnesium (entry 4; cf. entry 7 of Table 1) and addition to benzaldehyde afforded a mixture of two diastereomers (i.e. no *ul/lk* selectivity), but this time, the *u*-isomer was found to be enantiomerically pure after auxiliary removal. In contrast, the *l*-isomer was a 75:25 ratio of enantiomers.[46]

Although not shown in Table 2, addition of the organolithiums of entries 1–2 to benzophenone is stereorandom.[45] Thus, the random nature of reaction's course at C-2 of the piperidine in entry 3 suggests a mainly SET mechanism (also, recall entries 5 and 6 of Table 1). In contrast, the fact that the carbanionic carbon of the Grignard reagent (entry 4; cf. entry 7 of Table 1) retains its configuration in forming the *u* addition product suggests a purely polar mechanism. The presence of 25% of the minor *l*-enantiomer is revealing. If SET were the pathway by which it was formed, and if the assumption stated above (that SET additions are stereorandom) is true, then there should be an equal amount of the minor enantiomer of the *u*-diastereomer. The fact that there is not suggests that the products of *lk* topicity are formed by competing polar pathways, one with retention and one with inversion. Thus, the steric course of these additions suggests that:

1. For the lithiated *N*-methylpyrrolidine and *N*-methylpiperidine of entries 1 and 2, the addition follows a polar pathway.
2. For the metalated piperidinooxazolines of entries 3 and 4, the lithium compound adds by an SET mechanism and the Grignard follows a polar path.

Entries 5 and 6 of Table 2 were reported by the Pearson group, and differ from the previous examples in that the lithium-bearing carbon is not in a ring. Both of these organolithiums were obtained by transmetallation of the organostannane, and are the more stable of the two diastereomers at the lithium-bearing carbon.[47] Neither show any face-selectivity, consistent with the acyclic example reported by Chong, and shown in entry 9 of Table 1, but because of the thermodynamic bias favoring the *R* configuration at the nitrogen-bearing carbon in the products, no conclusions can be reached regarding the mechanism of addition.

VI. STEREOCHEMICAL RATIONALE (CHIRAL ORGANOLITHIUMS)

It is curious that, with reference to the analysis presented in Scheme 4, there is little or no stereoselectivity in the addition of these chiral nucleophiles to benzaldehyde! If the additions are polar, this means that $\Delta\Delta G^{\ddagger} = 0$ for the two competing transition states. Is this reasonable? The ab initio transition structures for addition of methyllithium monomer and dimer to formaldehyde are shown in Figure 3a, along with a Newman projection sighted along the forming C–C bond.[48] In both cases, a four-membered ring is postulated, and the configuration at the metal-bearing carbon is retained (or would be if the carbon was stereogenic). Note how one of the hydrogens of the methyl group projects into the H–C–H angle of the formaldehyde, clearly obvious in the Newman projection shown. Figures 3b and 3c show adaptations of this Newman projection to examples from the tables above. In Figure 3b,

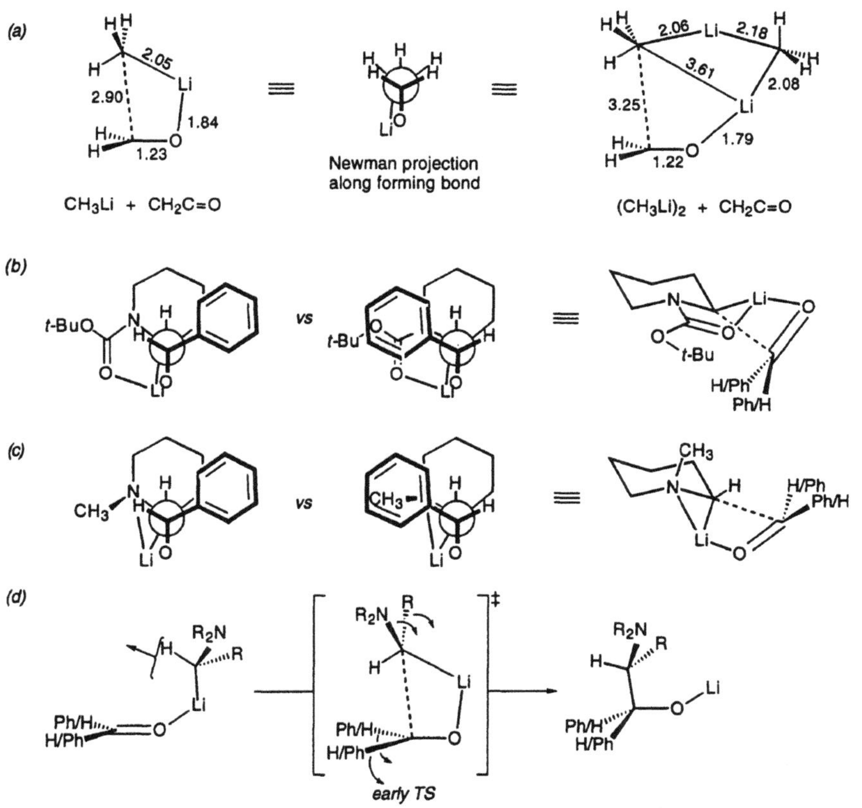

Figure 3. (a) Transition structure for the addition of methyllithium monomer and dimer to formaldehyde[48] (interatomic distances in angstroms). (b) Transition structures for the addition of 2-lithio-*N*-BOC-piperidine to benzaldehyde.[41] (c) Transition structures for the addition of 2-lithio-*N*-methylpiperidine to benzaldehyde.[45] (d) Analysis of the addition of α-aminoorganolithium monomer to benzaldehyde showing little interaction between the ligands on the carbonyl carbon and the carbanionic carbon in the transition structure.

the addition of 2-lithio-BOC-piperidine to benzaldehyde (Table 1, entry 8) is shown in Newman projection and in chair form.[41] In this organolithium, the metal is equatorial and is chelated by the carbonyl oxygen; the electrophile approaches from the equatorial direction.[49,50] Figure 3c shows the addition of 2-lithio-*N*-methylpiperidine to benzaldehyde (Table 2, entry 2). In this case, the lithium is shown bridging C-2 and the nitrogen;[51] in order for addition to occur with retention at the lithiated carbon,[45] the electrophile must approach from the pseudo axial direction of the chair conformation shown.

The lack of selectivity can be rationalized by the following analysis, based on the transition structures of Figure 3a. From either a monomeric or dimeric organolithium, the transition state is early, as evidenced by the 2.90–3.25 Å carbon–carbon bond distance. Figure 3d illustrates a monomeric α-aminoorganolithium adding to benzaldehyde. In the transition structure, the two non-hydrogen "carbanion" ligands (R and NR_2) are projected away from the direction of the forming bond. The arrows indicate the direction of motion of the various ligands as the initial coordination complex moves through the transition state and then towards product. Note that the carbonyl ligands (Ph and H) move away from the incoming nucleophile as the trigonal atom rehybridizes to sp^3, and that the NR_2 and R ligands move toward the lithium alkoxide. Although the carbonyl ligands (Ph and H) and the NR_2 and R ligands are synclinal in the final product (i.e. close enough to interact), the early transition state renders this fact irrelevant to the relative transition state energies. Even in the transition structure (Figure 3a) for addition of the methyllithium dimer, these structural features are preserved, so a similar analysis should hold for a dimeric organolithium nucleophile. On the other hand, this analysis neglects the interactions that exist between the lithium atom and either the nitrogen or the amide carbonyl (Figure 3b, c), so it is an oversimplification; the situation may be more complex than this analysis implies.

VII. EXAMPLES: CONFIGURATIONALLY LABILE, RACEMIC

When the metal-bearing carbon is benzylic or allylic, the barrier to inversion is lowered relative to the saturated systems. For example, an upper limit for the barrier to inversion of a lithiated isoquinolyloxazoline (vide infra Table 4, entry 6) has been placed at 8.2 kcal/mole and for the corresponding Grignard (vide infra Table 4, entry 7) at 9.8–10.1 kcal/mole.[52] With such barriers, inversion will be fast, even at low temperature, so (with reference to Scheme 5) the steric course of the reaction at the metal-bearing carbon cannot be determined.

Table 3 lists several examples of the addition of lithiated and magnesiated α-aminoorganometallics to aromatic aldehydes. Lithiated isoquinolines (entries 1,4,5), isoindoline (entry 6), and β-carboline (entry 7) show little or no selectivity in additions to aldehydes. But, in 1984, Seebach discovered that transmetallation of the lithium species to a Grignard reagent increases the simple diastereoselectivity

Table 3. Simple Diastereoselectivity in the Addition of Configurationally Labile, Racemic, Chiral α-Aminoorganometallics to Benzaldehydes (Unless Noted)

			Products		
Entry	*Organometallic*	*Conditions*	*ul Topicity*	*lk Topicity*	*Reference*
1	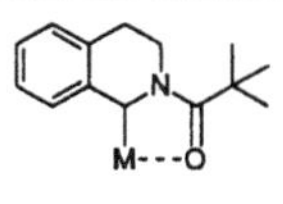 M = Li	THF –78 °C	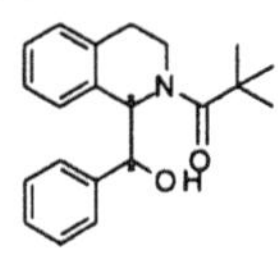 *rac-u* 66-78 :	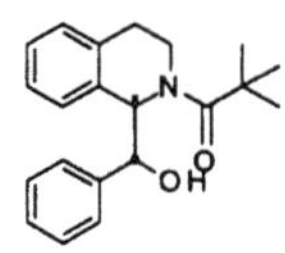 *rac-l* 34-22	40,53,56
2	" M = MgBr	THF –78 °C → RT	*rac-u* 100 :	*rac-l* 0	40,53
3	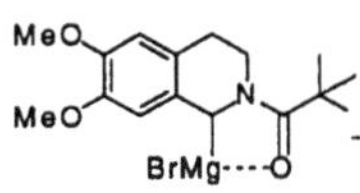	THF –78 °C → RT	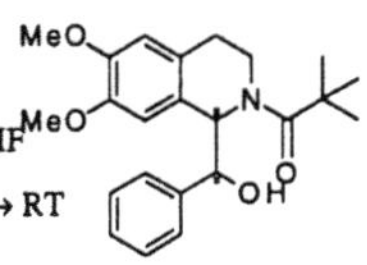 *rac-u* 100 :	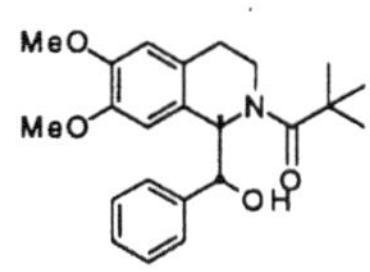 *rac-l* 0	57
4	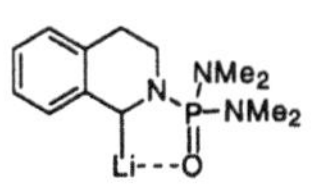		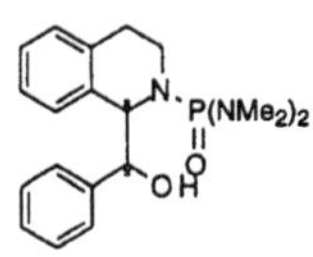 *rac-u* 50 :	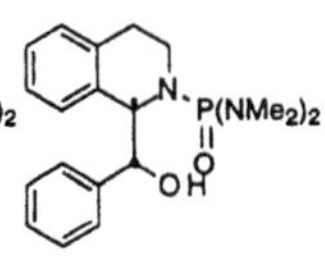 *rac-l* 50	58
5	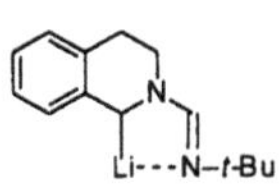 (3,4-dimethoxy-benzaldehyde used as electrophile)		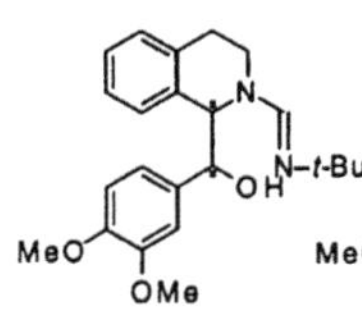 *rac-u* 50 :	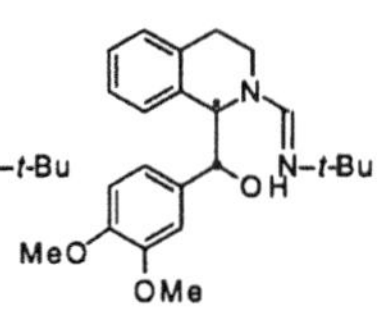 *rac-l* 50	59
6	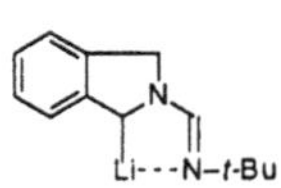		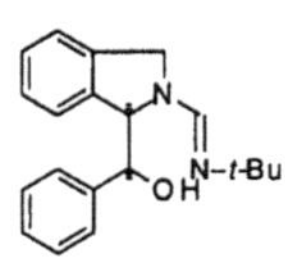 *rac-u* 50 :	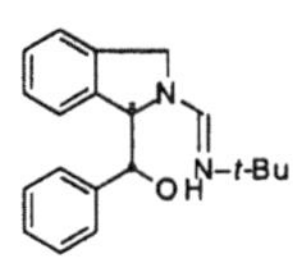 *rac-l* 50	60

(continued)

Table 3. (Continued)

Entry	Organometallic	Conditions	Products: *ul Topicity*	Products: *lk Topicity*	Reference
7	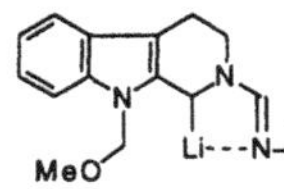	THF –78 °C → –30 °C	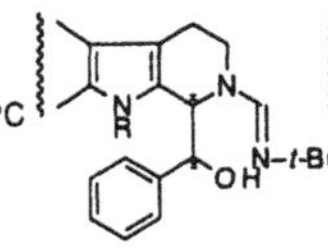 *rac-u* 50 :	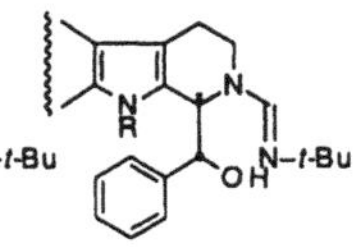 *rac-l* 50	61
8	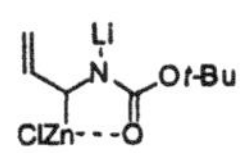 (structure postulated by present author)	THF –78 °C	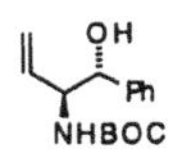 *rac-u* 94 :	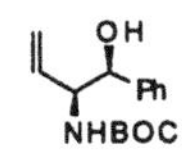 *rac-l* 6	54
9	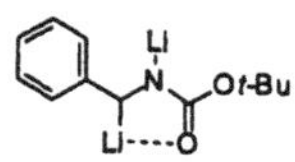 (structure postulated by present author; acrolein used as electrophile)	THF/TMEDA –78 °C	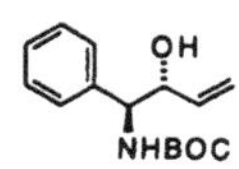 *rac-u* 14 :	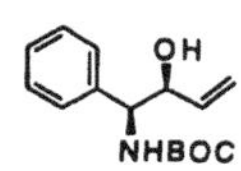 *rac-l* 86	55

(*ul* relative topicity) to 100%, as shown in entries 2 and 3.[40,53] As discussed in a later section, the Zürich group used this high selectivity in efficient syntheses of several (racemic) hydroxybenzyl isoquinoline alkaloids, a strategy that paved the way for a subsequent chiral auxiliary-based approach developed in Miami.

The examples shown in entries 8 and 9 show that dimetallated urethanes appear to have significantly higher diastereoselectivity than the isoquinoline additions, albeit with no predictable topicity sense. Beak found (entry 8) that having zinc as a counterion was necessary for regioselective addition at the α-position of the allyl group and afforded the *anti* addition product with 84% ds.[54] (The dilithio species gave a 1:1 mixture of α- and γ-addition with unspecified diastereoselectivity.) Greene used the dilithiated benzyl amine shown in entry 9 in a short synthesis of the taxotere side chain.[55] Unfortunately, several other electrophiles (benzaldehyde, crotonaldehyde, cyclohexyl glyoxalate, and furfural) were tested under similar conditions, but the low selectivity observed indicated that the moderately high selectivity of this example may be unique.

Table 4. Diastereoselectivity in the Addition of Configurationally Labile, Nonracemic, Chiral α-Aminoorganometallics to Aldehydes[a]

Entry	*Organometallic and Aldehyde*	*Conditions*	*Products*	*Reference*
1	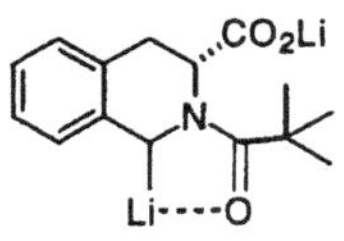 + PhCHO	THF −80 °C	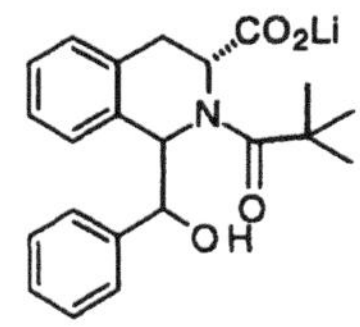 50:50 mixture of two unspecified diastereomers	62
2	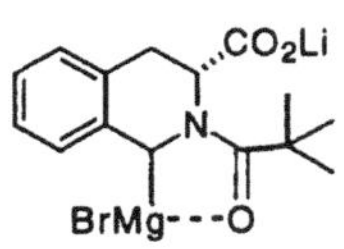 + PhCHO	THF −80 °C	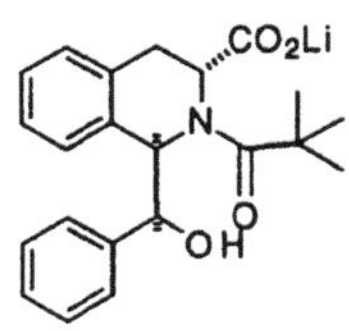 *ul* topicity 100% ds (presumed configuration)	62
3	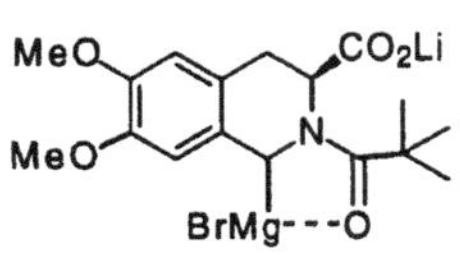 + 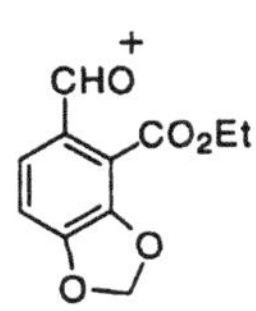	THF −80 °C	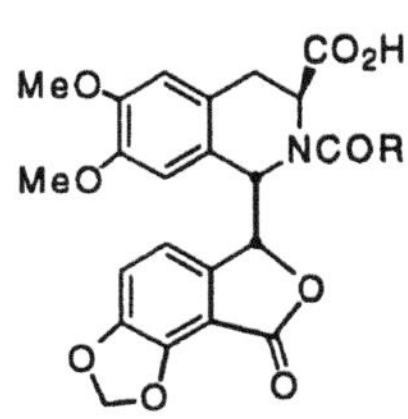 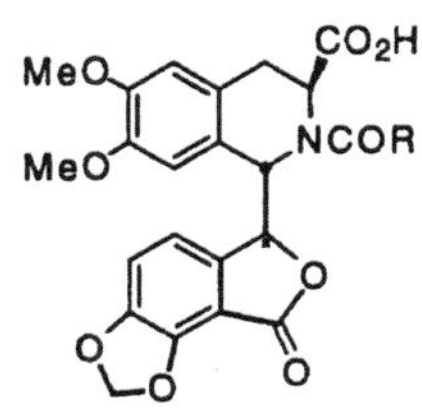 *ul* topicity : *lk* topicity 60 : 40	62
4	 + PhCHO R = Me	THF −78 °C	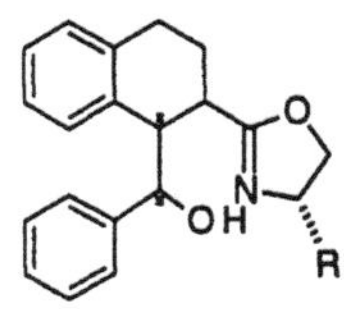 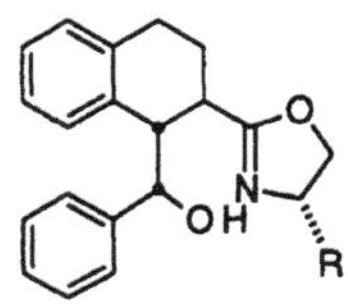 *ul* topicity : *ul* topicity 50 : 50	52
5	" R = Et	THF −78 °C	*ul* topicity : *ul* topicity 62 : 38	52

(continued)

Table 4. (Continued)

Entry	Organometallic and Aldehyde	Conditions	Products			Reference
6	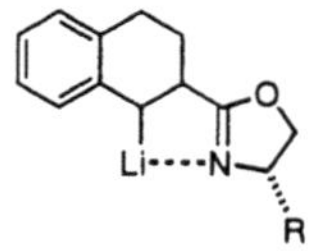 + PhCHO R = *i*-Pr	THF −78 °C	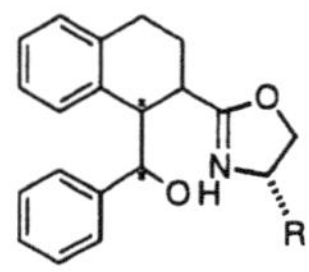 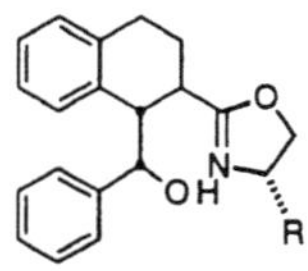 *ul* topicity 25:25	:	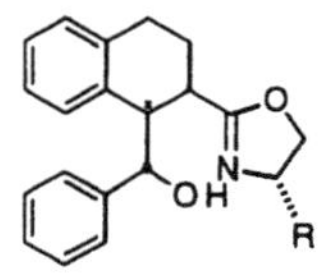 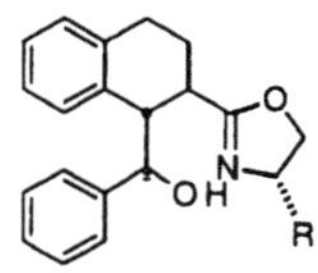 *lk* topicity 25:25	63
7	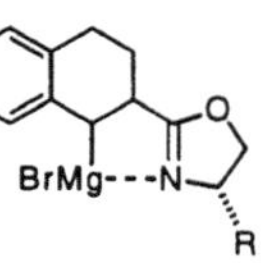 + PhCHO R = *i*-Pr	THF −78 °C	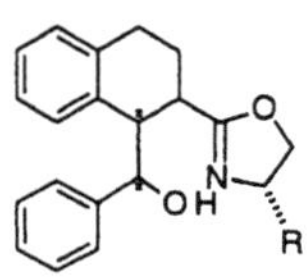 *ul* topicity 66-71	:	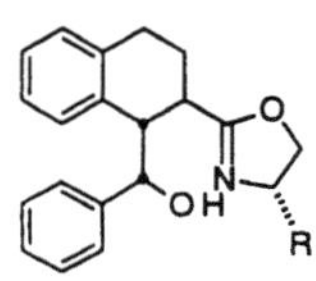 *ul* topicity 34-29	52,63
8	" R = *i*-Pr	THF −70 °C	*ul* topicity 55	:	*ul* topicity 45	52
9	" R = *i*-Pr	THF −60 °C	*ul* topicity 42	:	*ul* topicity 58	52
10	" R = *i*-Pr	THF −45 °C	*ul* topicity 58	:	*ul* topicity 42	52
11	" R = Ph	THF −78 °C	*ul* topicity 71	:	*ul* topicity 29	52
12	" R = Bn	THF −78 °C	*ul* topicity 70	:	*ul* topicity 30	52
13	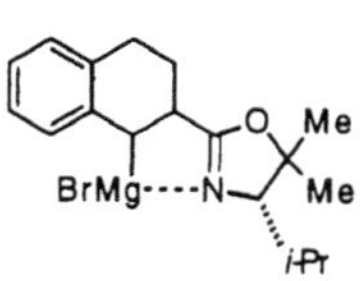 + PhCHO	THF −78 °C	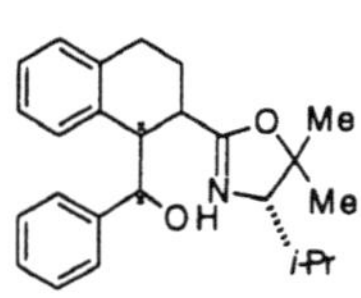 *ul* topicity 80	:	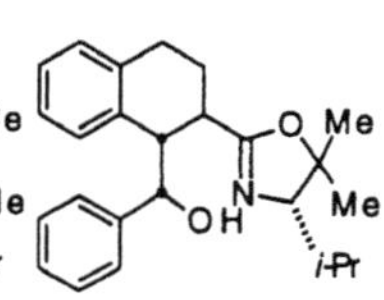 *ul* topicity 20	52

(continued)

Table 4. (Continued)

Entry	Organometallic and Aldehyde	Conditions	Products		Reference
14	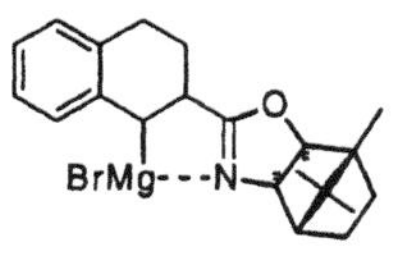 + PhCHO	THF –78 °C	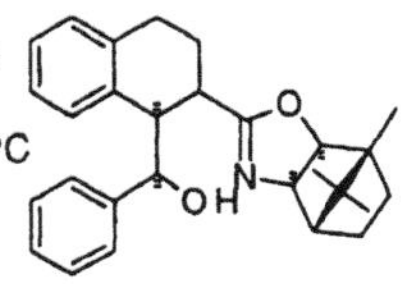 *ul* topicity 33 :	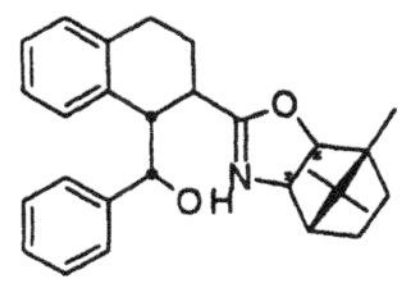 *ul* topicity 67	52
15	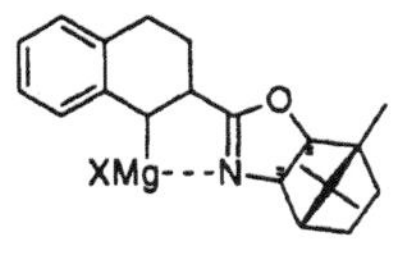 + PhCHO	THF –65 °C	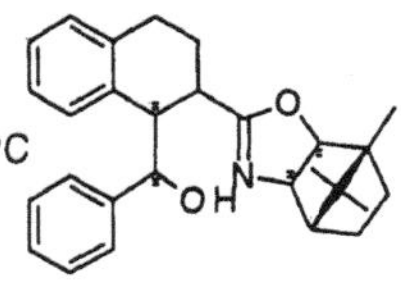 *ul* topicity 20 :	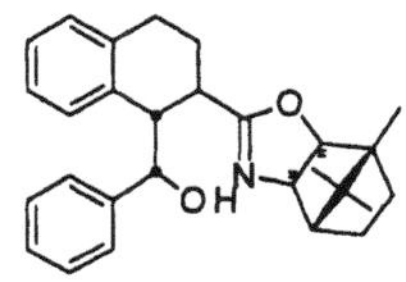 *ul* topicity 80	52

Notes: [a]Of the four possible diastereomers, only those found are illustrated.

VIII. EXAMPLES: CONFIGURATIONALLY LABILE, NONRACEMIC

Two approaches have been taken to capitalize on the diastereoselectivity observed in the addition of isoquinoline Grignards to aldehydes found by the Seebach group (Table 3, entries 2 and 3). The Seebach group used the chiron approach (chiral tetrahydroisoquinoline pivalamide), and Kelly Rein and Pingsheng Zhang in my group developed oxazoline chiral auxiliaries (Table 4). Consistent with the results in the racemic series, in neither approach did lithium reagents afford useful diastereoselection (Table 4, entries 1 and 6).

As shown in entries 2 and 3 of Table 4, Seebach found that a relatively simple Grignard reagent was 100% diastereoselective in its addition to benzaldehyde (entry 2), but a more complex system showed relatively modest selectivity (entry 3). The latter example was used in a synthesis of (+)-corlumine (vide infra), and stands out as a rare exception to the usually exclusive *ul* topicity exhibited by magnesiotetrahydroisoquinolines in addition reactions.[62]

In 1989, Kelly Rein began an investigation into the use of an oxazoline chiral auxiliary to effect diastereoselective additions of Grignard reagents based on the Seebach discovery. The initial effort[63] used an oxazoline auxiliary derived from valine (entry 7), one that we had used previously in studies of asymmetric alkylations of lithiated tetrahydroisoquinolines.[64,65] Although the selectivity provided by this auxiliary was modest, it was used in the asymmetric synthesis of several hydroxybenzyl isoquinoline alkaloids (vide infra), partly motivated by a need to establish the absolute configuration of the addition products. Later, after we became convinced that the approach was worthwhile, Pingsheng Zhang undertook a more systematic investigation of the effect of auxiliary structure and reaction temperature on selectivity,[52,66] as summarized in entries 4,5, and 7–15 of Table 4.[g]

With oxazolines substituted only in the 4-position, the selectivity peaked at about 70% with either R = *i*-Pr, Ph, or Bn (cf. entries 4, 5, 7, 11, and 12).[52] Note that in all these examples, the aldehyde appears to approach the Grignard from the side opposite R (i.e. β-face as illustrated, assuming chelation of the magnesium by the oxazoline nitrogen). This is in contrast to the chirality sense of the reaction of the lithium analogue with alkyl halides, where approach of the electrophile is from the α-face.[65] Although this came as a surprise, we thought we might be able to increase the propensity toward β-approach by making the α-face more crowded. The first approach was to methylate the oxazoline at the 5-position, which forces the isopropyl group to rotate into a conformation that places a methyl group closer to the ligands on the magnesium (Figure 4a, b).[h] The modification achieved the desired effect, boosting the diastereo-selectivity to 80% (entry 11). The second approach used a different auxiliary, this time one derived from camphor quinone. In this instance (Figure 4c), we were also hoping to find a structure that would facilitate separation of the two diastereomeric addition products. The camphor auxiliary proved useful in that regard (entries 14, 15) since it afforded 80% diastereoselectivity, and one recrystallization of the crude product mixture yielded a single diastereomer.[66]

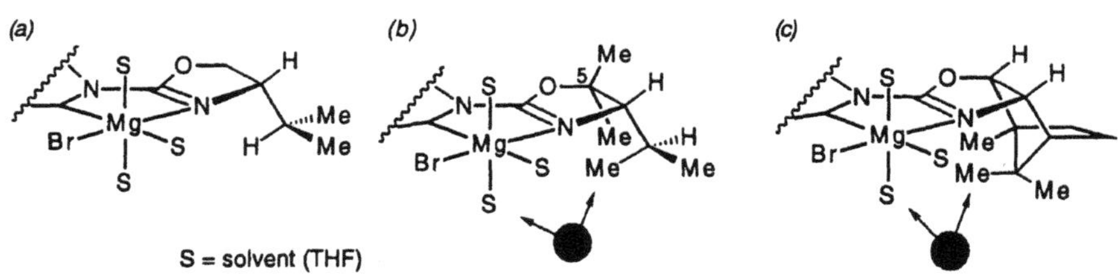

Figure 4. Probable conformation of the magnesiated tetrahydroisoquinolyl oxazoline, based on analogy to the X-ray crystal structure of a magnesiated tetrahyrdoisoquinoline pivalamide.[53] The aldehyde probably coordinates to one of the ligand sites cis to the isoquinolyl carbon prior to reaction. (a) Conformation of isopropyl in 5-unsubstituted auxiliary; (b) Steric crowding produced by rotation of isopropyl in the 5,5-dimethyl derivative. (c) Similar crowding in a camphor-derived auxiliary (enantiomer of that drawn in Table 4, entries 14 and 15).

Being cognizant of the fact that benzylic "carbanions" are not configurationally stable, Pingsheng Zhang had begun an investigation into the dynamics of these systems by ^{13}C NMR. In looking at the resonance of the metal-bearing carbon (Figure 4a), he had noticed a coalescence at –65 °C.[52] Prompted by this observation, he varied the temperature of the addition, with some unusual results. As outlined in Table 4, entries 7–10, the selectivity fell from about 70 to 55% between –78° and –70 °C; at –60 °C the chirality sense was reversed, then reversed again at –45 °C. With the camphor auxiliary, raising the temperature from –78° to –65°C increased the selectivity 67% to 80% (entries 12–13).[52] This bizarre behavior underscores the point made previously (Scheme 5) about seeking mechanistic insight when the carbanion is not configurationally stable. Obviously, I have no explanation for this behavior other than to say that we are probably observing the results of very subtle changes in relative rates of pyramidal inversion at C-1 relative to the rate of addition to benzaldehyde.

Nevertheless, the camphor auxiliary provided a single diastereomer in the synthetically useful yield of 50% (isolated, analytically pure, single diastereomer), which compares favorably to the isolated yield of 58% of the single diastereomer obtained using the chiron approach (entry 2). As described in a subsequent section, the auxiliary approach proved useful for several synthetic applications as well.

IX. STEREOCHEMICAL RATIONALE, PART 2 (OXIDATION POTENTIALS)

At first glance, it may seem that the lack of selectivity exhibited by most of the organolithiums in Tables 3 and 4 are due to similar transition state energies of *ul* and *lk* transition structures such as shown in Figure 3b for the configurationally stable α-aminoorganolithiums. However, Kelly Rein of my group, in collaboration with Zhihong Chen and Luis Echegoyen, had found that the addition of benzophenone to the lithiated pivalamide shown in Table 3, entry 1 yielded a deep blue solution that was shown to contain benzophenone ketyl by ESR,[46] presumably formed by oxidation of the organolithium. The same is true of 1-lithio-tetrahydroisoquinolyloxazoline of Table 4, entry 6.[46] Interestingly, neither of these lithium derivatives give a blue color or a ketyl if transmetallated with magnesium bromide prior to addition of benzophenone. Moreover, the lithiated isoquinoline with the oxazoline auxiliary is totally nonselective in its addition to benzophenone, whereas the Grignard species shows 91% diastereoselectivity.[46] Scheme 6 summarizes this data, and also indicates two other relevant cases: Beak's 2-lithio-N-BOC-pyrrolidine adds to benzophenone selectively[24] while 2-lithio-*N*-methylpyrrolidine and -piperidine (cf. Table 2, entries 1 and 2) add stereorandomly.[45] In the latter cases, Qian Lui Zhang, in collaboration with Y. Zuo and Luis Echegoyen, found that the addition of benzophenone is accompanied by a blue color and the appearance of benzophenone ketyl in the ESR spectrum.[45] Also, recall (Table 4, entry 6) that

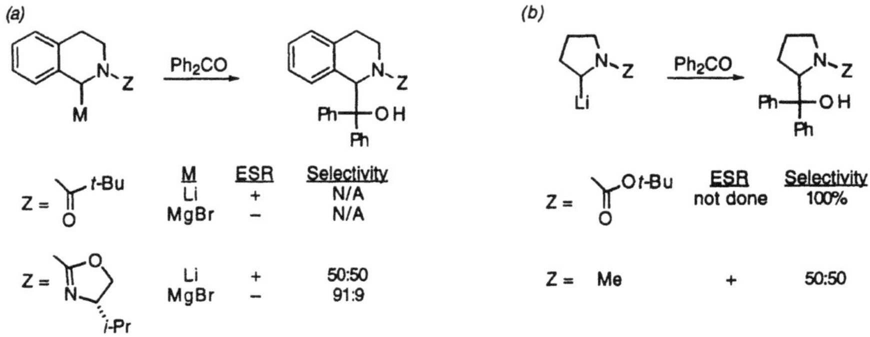

Scheme 6. (a) Detection of benzophenone ketyl by ESR and correlation with stereoselectivity in the addition of lithiated tetrahydroisoquinolines.[46] (b) Selective[24] and nonselective[45] addition of benzophenone to lithiated pyrrolidines.

addition of the lithiated isoquinolyl oxazoline to benzaldehyde afforded all four diastereomers,[46,63] again consistent with an SET mechanism under the assumptions listed in the text accompanying Scheme 5. A nonselective addition that did not involve SET would be expected to show some selectivity at the carbanionic carbon, as was the case in the addition of 2-lithio-*N*-methylpyrrolidine and piperidine to benzaldehyde (cf. Table 2, entries 1 and 2 with entries 3 and 4).

Although the presence of a radical in a reaction mixture does not necessarily place it on the reaction path, the correlation of benzophenone reduction to the steric course of addition of a chiral lithiated oxazoline suggests that SET is operative in these cases.

Why does SET intervene in some of these processes and not others? Eberson has developed a theory that extends Marcus theory to organic processes.[72] The core of this theory is that whether a polar mechanism or an SET mechanism is followed depends on the energy difference between the oxidation potential of a nucleophile and the reduction potential of an electrophile. According to Eberson–Marcus theory, reactions whose free energies of electron transfer, ΔG_{ET}, are more endothermic than 23 kcal/mole (at room temperature) are unlikely to occur by SET because the electron transfer will be too slow. Likewise, if $\Delta G_{ET} < -20$ kcal/mol, then it will be a fast process. In situations between these extremes, a more refined theoretical treatment becomes necessary. This theory has been used by Arnett to show that the aldol addition of lithium pinacolonate to benzaldehyde is a polar process,[73] and by Bordwell to show that anionic substitutions on alkyl halides are polar as well.[74]

The reduction potential ($E_{1/2}$, Pt electrode, THF, tetrabutylammonium perchlorate supporting electrolyte) of benzaldehyde is −2.371V[73] and that of benzophenone is –2.248,[75] meaning that benzophenone is more easily reduced by 0.123V, which corresponds to a difference in free energy of electron transfer of 2.8 kcal/mol. Certain of the organometallics mentioned above oxidize benzaldehyde, and therefore may be presumed to also oxidize benzophenone, but the converse is not necessarily true. Nevertheless, using Eberson's theory and the stereochemical data outlined above, we can predict the relative order of oxidation potential of several of these α-aminoorganometallics, as shown in Figure 5. These trends indicate that lithiated amides and amidines tend to have less positive oxidation potentials than lithiated urethanes, and that Grignard reagents tend to have more positive oxidation potentials than the corresponding organolithiums.

Given these trends, one might ask why certain organometallics that are so closely related structurally behave so differently in their tendency to undergo SET in reactions with electrophiles (e.g. lithiated piperidine that differs in the nitrogen substituent, such as oxazoline, *t*-BOC, and *N*-methyl; lithiated vs. magnesiated piperidinooxazolines, isoquinolyloxazolines, and isoquinoline pivalamides). Certainly, the differences observed between a Li–C bond with a Li–Mg bond depend to a large degree on bond length, bond strength, and degree of ionic character. A chemical probe was developed by Liotta to evaluate tendencies toward SET in

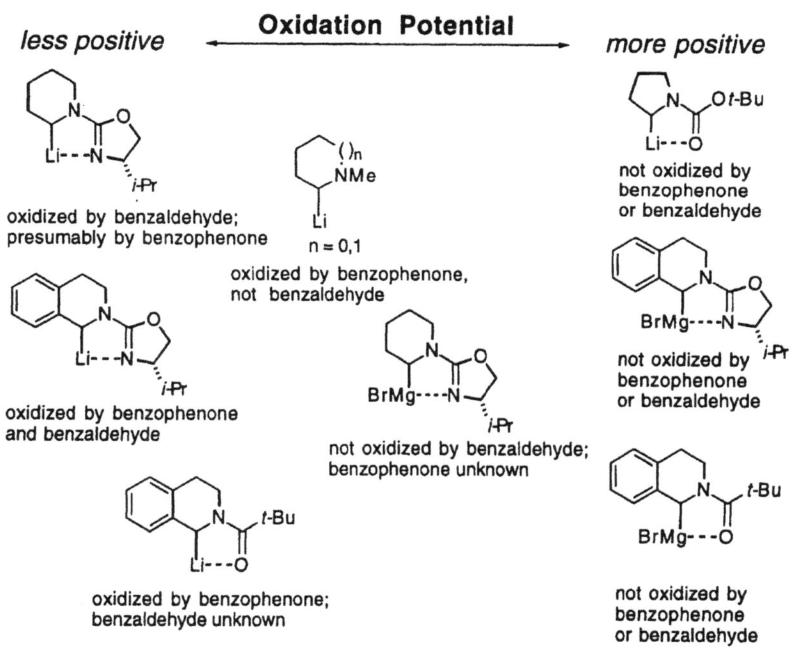

Figure 5. Relative oxidation potentials of α-aminoorganometallics based on tendency toward oxidation by benzophenone and benzaldehyde.

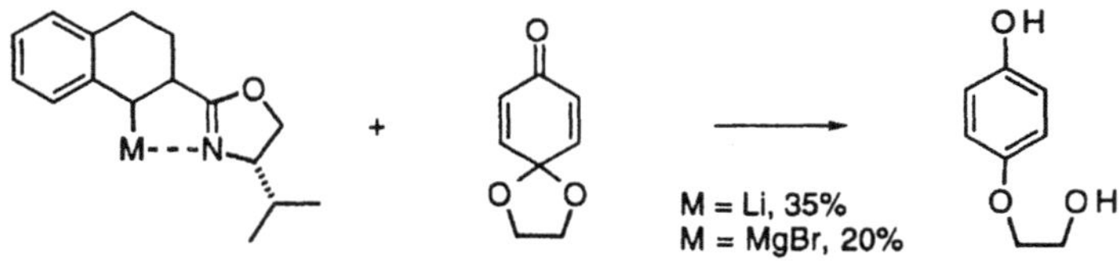

Scheme 7. Probe of single electron transfer in reactions of metalated tetrahydroisoquinolyloxazolines.[46,76]

additions to carbonyls.[76] Thus, SET reduction of the quinone monoacetal shown in Scheme 7 would produce fragmentation of the acetal to a phenol. When lithiated and magnesiated tetrahydroisoquinolyloxazolines were treated with Liotta's reagent, phenol was obtained in 35% yield from the lithium species and 20% from the Grignard,[46] indicating that SET is a possible pathway for both metals, but is more likely with the lithium reagent.

While such differences in behavior are easy to accept when different metals are involved, they are less expected in a series of 2-lithioheterocycles. Pross has suggested that both SET and polar pathways involve an initial single electron shift.[77] According to Pross's theory, if the coupling of the two spins is feasible following a single electron shift, a polar pathway is followed. But *any factor (steric, electronic, or geometric) that operates so as to inhibit or hinder the coupling process will tend to favor a SET pathway over a polar one.*[77] Our findings are consistent with Pross's theory.

X. APPLICATIONS TO ALKALOID SYNTHESIS

Seebach's 1984 discovery of 100% diastereoselectivity in the addition of tetrahydroisoquinoline Grignards (Table 3, entries 2 and 3) to benzaldehyde led to efficient syntheses of several hydroxybenzylisoquinoline alkaloids, all via *u*-hydroxybenzylisoquinolines (Scheme 8). This strategy is similar to that used (at about the same time) by the Meyers group in the synthesis of a number of isoquinoline alkaloids, where the key step was alkylation of a tetrahydroisoquinoline formamidine with an alkyl halide (1 new stereocenter).[27,28] It is worth mentioning that two of the primary tenets of retrosynthetic analysis[78,79] are to make a bond disconnection that results in the greatest simplification, such as between two stereocenters, and to cleave a bond that divides a target into two halves of approximately equal complexity. In this instance (Scheme 8), the addition of the metallated tetrahydroisoquinoline to the aldehyde accomplishes this nicely. The relative configuration of the two stereocenters is initially *u* (*erythro*), but treatment of the *u*-hydroxybenzylisoquinoline pivalamide with trifluoroacetic acid/trifluoroacetic anhydride effects *N*- to

O-acyl migration with inversion of configuration at the carbinol carbon, yielding the *l* (*threo*)-diastereomers stereospecifically.[57,80]

The obvious extension to the strategy outlined in Scheme 8 was to apply it to the synthesis of enantiopure alkaloids. The Seebach group tried the approach outlined in Scheme 9, which began with (*S*)-dopa as the chiral educt.[62] As has already been indicated (Table 4, entry 3), this reaction failed to reproduce the high diastereoselectivity found in simpler examples. Since only two of the four possible addition products were formed, one may surmise that SET was not responsible for the low selectivity. The minor isomer was the *l* addition product. The reasons for the loss of selectivity are not known, but undoubtedly involve very subtle differences in Grignard structure, possibly caused by the lithium carboxylate and the two methoxy substituents, although neither of these components on their own caused a loss of selectivity (cf. Table 3, entry 3 and Table 4, entry 2).

Our approach was to use a chiral auxiliary in a scheme such as this, with the hopes of avoiding some of the problems encountered previously, and to improve on the overall yields. An issue that had to be addressed was the selectivity of addition of oxygenated isoquinolines. As shown in Scheme 10, Pingsheng Zhang found that the methylenedioxytetrahydroisoquinoline was selective in its additions, but the dimethoxy analogue was not. When magnesium bromide was used in the transmetallation, both *u*- and *l*- products were produced with the dimethoxy compound.[52,66]

hydroisoquinolyloxazolines [46,76].

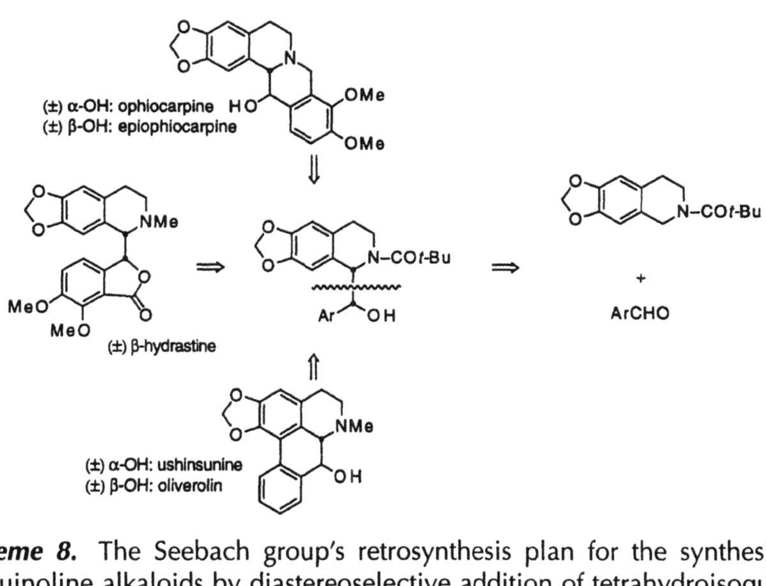

Scheme 8. The Seebach group's retrosynthesis plan for the synthesis of isoquinoline alkaloids by diastereoselective addition of tetrahydroisoquinoline Grignard reagents to aromatic aldehydes.[57,80]

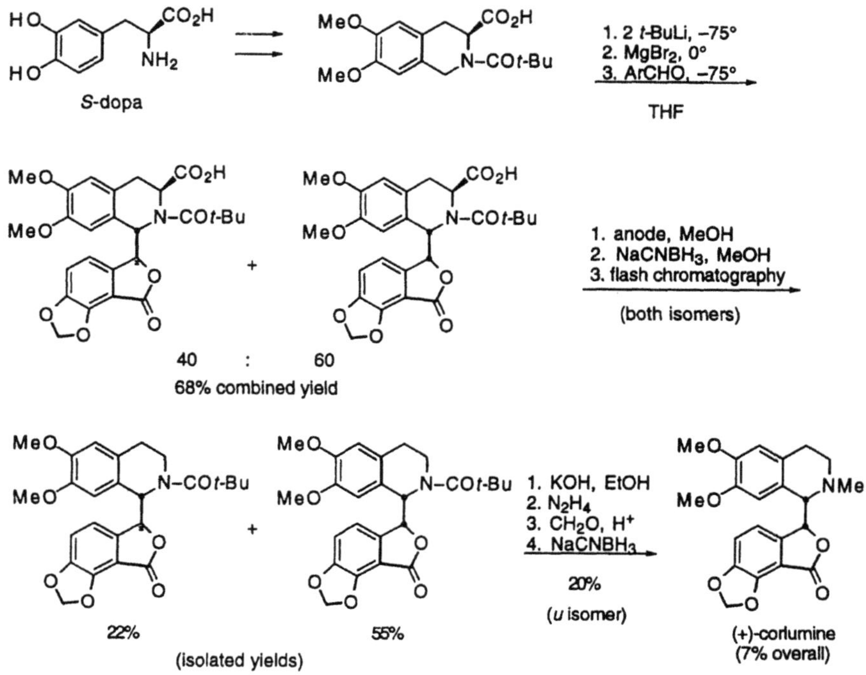

Scheme 9. Seebach's synthesis of (+)-corlumine using a chiral tetrahydroisoquinoline educt.[62]

Transmetallation with magnesium chloride restored the *u* selectivity, and was equally effective with both of the other two examples (unsubstituted and methylenedioxy) as well. Again, the reason for these subtle differences are not known, but may involve an interaction between the halogen and the 7-methoxy substituent that

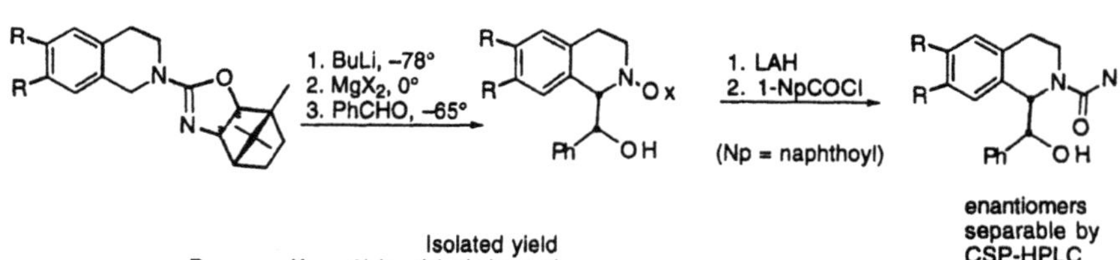

R	X	%ds	Isolated yield (single isomer)
H	Cl or Br	80	50%
-OCH$_2$O-	Cl or Br	78	53%
MeO	Cl	82	58%

Scheme 10. Generality of the camphor-derived oxazoline chiral auxiliary[52,66] and analysis of the enantiomer ratio by chiral stationary phase HPLC.[71]

causes a slight change in geometry of the magnesium complex, that in turn affects the relative energies of competing transition states.[52,66]

For synthesis targets, we chose the phthalide isoquinolinelactones bicuculline and corlumine,[81] and the hemiacetals egenine[82] and corytensine,[83,84] shown in Scheme 11. These targets were chosen for several reasons. First, the relative and absolute configuration of bicuculline (and its μ-diastereomer adlumidine), as well as their diol reduction products are firmly established,[85] so that a synthesis of either (or both) would place the stereochemical assignments on a firm footing. Second, synthesis of egenine and corytensine would confirm their structures. Third, shortly before we began this work, two new alkaloids were reported,[86] decumbensine and epi-α-decumbensine (Figure 6), and we were interested in synthesizing them as well. Fourth, there appeared to be some confusion regarding the difference in structure of egenine[82] and corytensine.[83] As a result of an error in transcribing the X-ray crystal structure from ORTEP to a two-dimensional drawing,[84] it was

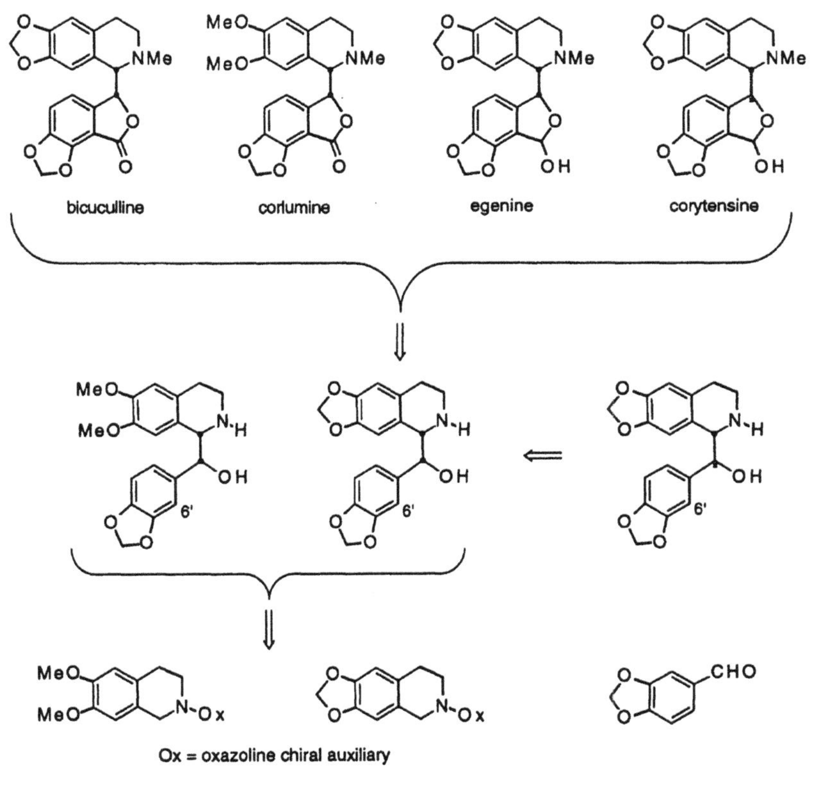

Scheme 11. Synthetic targets and retrosynthetic analysis for phthalide isoquinolines using the camphor-oxazoline chiral auxiliary.[52,63,87]

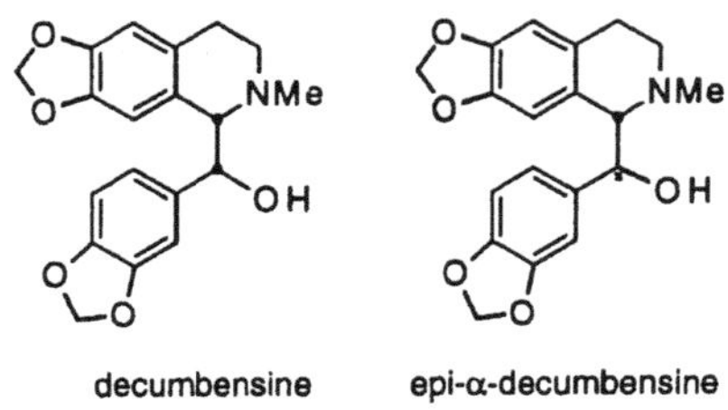

Figure 6. Decumbensine and epi-α-decumbensine.[86]

originally concluded that the difference between the two was the configuration at the hemiacetal carbon.[83] Since both alkaloids had been isolated by chromatography on silica gel, this struck us as unlikely. Fifth, the synthesis of corlumine could be compared directly to the Seebach effort (Scheme 9).[62] Finally, it was presumed that synthesis of these targets would suffice to demonstrate feasibility of this strategy toward the other alkaloid classes outlined in Scheme 8, since phthalides can be converted to such other classes by known routes (see refs. 27,28,57 and references cited therein).

Note that the approach outlined in Scheme 11 uses piperonal as the aldehyde component and requires functionalization at the 6′-position after the key addition step. Kelly Rein tried a more direct approach initially, as shown in Scheme 12, but reduction of the aldehyde was the major pathway.[88] No addition product was found, but the isoquinolyloxazoline could be recovered in 30% yield. Grignard reagents may reduce carbonyls by either β-hydride elimination or electron transfer. Since there are no β-hydrogens in this Grignard, SET is the only possible alternative for the production of the observed lactone.

She then developed a successful approach using the valine-derived oxazoline auxiliary [(*S*)-4-isopropyloxazoline].[63,87] After subsequent optimization as already described,[52,66] Pingsheng Zhang used the auxiliary derived from camphor quinone to synthesize the key hydroxybenzylisoquinoline intermediates shown in Scheme 13.[52] The absolute configuration of the new stereocenters of the addition products from these two auxiliaries are opposite. In the schemes and accompanying discussion below, the camphor-derived auxiliary is illustrated (Scheme 13), but the rest of the synthesis of all the alkaloids except corlumine was actually executed on the enantiomer of that drawn, because it was done with the product of addition using the valine-derived auxiliary. The references in the discussion refer to the relevant papers for each step.

The methylenedioxyisoquinolyloxazoline and the dimethoxyisoquinolyloxazoline were deprotonated with butyllithium, transmetallated with magnesium halide, and added to piperonal with ~80% diastereoselectivity, as shown in Scheme 13a and c. The major isomers were isolated by flash chromatography in the yields indicated. Hydride reduction cleaved the auxiliary and afforded the enantiomeri-

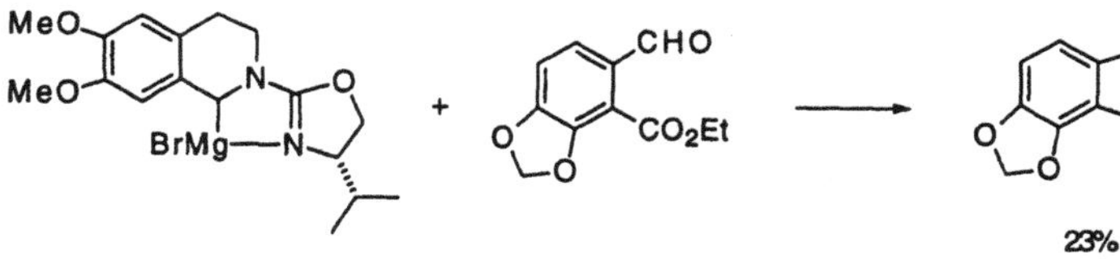

Scheme 12. A convergent synthesis is side-tracked by SET.[88]

cally pure *u*-amino alcohols.[52] Methylation was achieved by cyclization with phosgene and reduction.[52,63,87] To obtain the *l*-compound, the *u*-amino alcohol was inverted by acylation and rearrangement, affording the enantiopure *l*-amino alcohol after reductive cleavage of the pivaloate group (Scheme 13b), and methylation was accomplished as before.[87]

The two *N*-methyl (bis)-methylenedioxy compounds (Scheme 13a, b) corresponded to the proposed structures for decumbensine and epi-α-decumbensine,[86] and we had hoped to confirm the original structural assignments and establish the

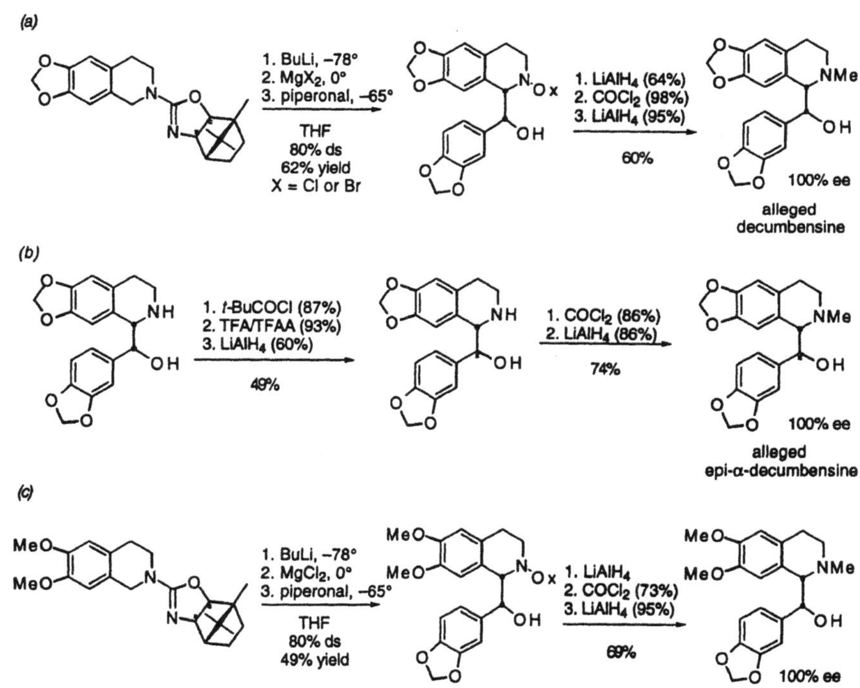

Scheme 13. Synthesis of key intermediates for phthalide isoquinoline alkaloid synthesis.[52,63,87]

configuration of our compounds, but their NMR spectra did not match. Rozwadowska had noticed the same thing, and suggested that epi-α-decumbensine might in fact be corytensine.[89,90] We thought that if epi-α-decumbensine and corytensine were the same, then decumbensine and egenine were also probably identical. So, we decided to make them both to settle the matter. But, it was at about this time that we noticed the mistake about the structure of corytensine,[83,84] along with some misassigned signals in the spectra of both egenine and corytensine (summarized and corrected in refs. 63 and 87), and therefore concluded that we would have to choose different targets for comparison to establish configurations. For this, we decided on bicuculline diol and adlumidine diol,[85] made by reduction of egenine and corytensine.

Scheme 14 shows Kelly Rein's directed metallation strategy for functionalization of the 6′-position. In all three cases, metallation and acylation failed to go to completion. In each case, the product of acylation was accompanied by significant amounts of recovered starting material. Thus the yields are low, but look better if based on unrecovered starting material. Nevertheless, the NMR spectra of the products largely matched literature data (vide infra).[87] Reduction of egenine and corytensine afforded bicuculline diol and adlumidine diol,[87] and comparison of rotation and spectral data matched literature values.[85] These correlations confirmed that the chirality sense of the addition was as indicated, and as independently established by chemical correlation (reductive deoxygenation) and chiral stationary phase HPLC.[71] Pingsheng Zhang later tried a number of approaches to improve on this last step, including use of better directing groups on the benzylic oxygen, but nothing seemed to help. Although the goal of improving this metallation was not achieved, Pingsheng did discover a variant of the Snieckus rearrangement.[91]

The identity of egenine and the decumbensine, and of corytensine with epi-α-decumbensine deserves some comment. In their report, J.-S. Zhang et al. recorded the proton NMR spectrum of both alkaloids, but the carbon spectrum of only decumbensine, which showed 19 carbons (egenine has 20).[86] Our synthetic egenine showed 20 lines at 100 MHz, but only 19 at 20 MHz; signals at 123.99 and 124.11 merged at lower field.[63,87] The carbon spectrum recorded by J.-S. Zhang et al. was obtained at 22.63 MHz. Insufficient epi-α-decumbensine was available for a carbon spectrum. In the proton NMR, the hydrogens at C-7′ of egenine and corytensine come at 6.34 and 6.25, respectively, and were mistaken for aromatic peaks by both Shamma[82] and J.-S. Zhang et al.[86] With our synthetic samples, we used a combination of COSY, HETCOR, off-resonance, and NOE techniques to establish the correct assignments.[63] Under EI conditions, neither egenine nor corytensine showed a molecular ion in the mass spectrum,[82,83,92] although both do under DCI conditions.[87] The mass spectra of decumbensine and epi-α-decumbensine were obtained under CI conditions,[86] but apparently no molecular ion was observed. So between the absence of a molecular ion in the MS, the low-field methine hydrogen in the proton NMR, and the merger of two signals in the carbon NMR, J.-S. Zhang et al. were honestly misled.

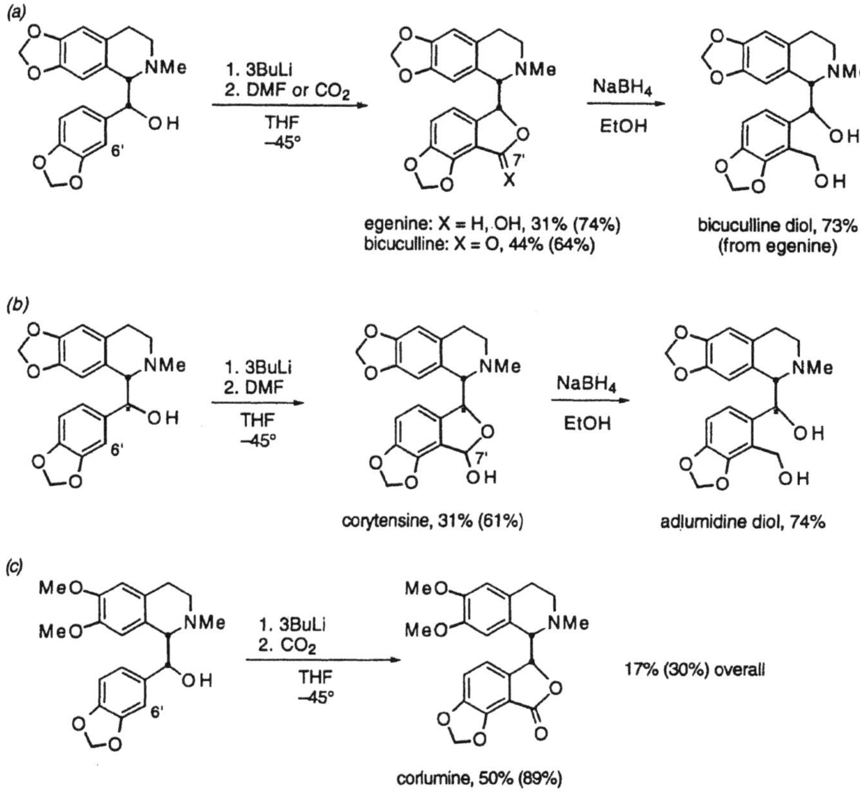

Scheme 14. Directed metalation and acylation of hydroxybenzylisoquinolines for the synthesis of egenine, bicuculline, corytensine, and corlumine.[52,63,87] Yields in parenthesis are based on unrecovered starting material. Reduction of egenine and corytensine yielded bicuculline diol and adlumidine diol, whose rotation and spectral data matched literature values.[85]

The total synthesis of corlumine by the chiron route beginning with *S*-dopa (Scheme 9) and the auxiliary route (Scheme 13c and 14c) can now be compared directly. Beginning with the tetrahydroisoquinoline species that is metallated and added to the aldehyde, the chiron route proceeds in 7% overall yield, while the auxiliary route is 17% (30% based on unreacted starting material in the last step). The comparison neglects the steps necessary for the conversion of (*S*)-dopa into the dimethoxytetrahydroisoquinoline pivalamide and of the synthesis and attachment of the chiral auxiliary, but the steps that are compared cover comparable transformations.

XI. SUMMARY

The addition of stereogenic metal-bearing carbons to prochiral carbonyls is a complex process that may proceed by polar or radical pathways. Determining the steric course of the reaction is only possible with configurationally stable carbanions, but mechanistic understanding is not necessarily a prerequisite to development of useful methods. An auxiliary-based protocol for the synthesis of hydroxybenzylisoquinolines is the most efficient method currently available for the synthesis of enantiopure α-hydroxybenzylisoquinolines such as the phthalideisoquinolines bicuculline, corlumine, egenine, and corytensine, and also of the other alkaloid classes available from them, such as aporphines and protoberberines.

ACKNOWLEDGMENTS

I am grateful to my colleagues, whose names are mentioned in the references and the text, for their experimental skill and fortitude in a demanding field, and to my colleague Luis Echegoyen and his students for help with ESR experiments and many helpful discussions. Our contributions to this area were supported financially by the NIH and the Petroleum Research Fund, administered by the American Chemical Society.

NOTES

[a]For detailed glossaries of stereochemical terminology, see pp. 15–40 of ref. 1 and pp. 1191–1210 of ref. 2.

[b]In some cases, there may be only one product. For example, the addition of methyllithium to acetaldehyde afford the same product by addition to either (heterotopic) carbonyl face. The transition states are nevertheless enantiomeric due to differing bond lengths between the carbonyl carbon and the two methyls.

[c]For a history of the evolution of Cram's rule, from 1952 to 1993, see pp. 121–130 of ref. 1.

[d]For numerous reviews of heteroatom-stabilized carbanions, see volumes 1 and 3 of ref. 13.

[e]Stereoselective radical couplings are currently being developed. For reviews, see refs. 14–16.

[f]This is not unlike the rationale for the *pref/parf* system of assigning relative configuration.[32]

[g]For the analysis of *u/l* selectivity, NMR is the method of choice, since the two diastereomers are easily distinguished by the coupling pattern of the two methine protons: the coupling constant for the *erythro*-isomer is typically 3–5 Hz, whereas the *threo*-pair is typically 7–9 Hz.[57,67–70] To assign absolute configuration, chiral stationary phase HPLC is the method of choice.[71]

[h]Chelation of an octahedral magnesium (with the bromine *trans* to the chelating atom) as shown in this figure is based on analogy to the X-ray crystal structure of a pivalamide Grignard (Table 3, entry 2) obtained by the Seebach group.[53] The rationale also assumes that the aldehyde replaces one of the THF ligands (*cis* to the carbanionic carbon atom on the magnesium) prior to addition.

REFERENCES

1. Gawley, R. E.; Aubé, J. *Principles of Asymmetric Synthesis*; Pergamon (Elsevier Science): Oxford, 1996, Vol. 14 (*Tetrahedron Organic Chemistry Series*).
2. Eliel, E. L.; Wilen, S. H.; Mander, L. N. *Stereochemistry of Organic Compounds*; Wiley-Interscience: New York, 1994.

3. Curtin, D. Y.; Harris, E. E.; Meislich, E. K. *J. Am. Chem. Soc.* **1952**, *74*, 2901.
4. Cram, D. J.; Elhafez, F. A. A. *J. Am. Chem. Soc.* **1952**, *74*, 5828.
5. Prelog, V. *Helv. Chim. Acta* **1953**, *36*, 308.
6. Erdik, E. *Organozinc Reagents in Organic Synthesis*; CRC Press: Boca Raton, FL, 1996.
7. Noyori, R.; Kitamura, M. *Angew. Chem. Int. Ed. Engl.* **1991**, *30*, 49.
8. Duthaler, R. O.; Hafner, A. *Chem. Rev.* **1992**, *92*, 807.
9. Soai, K.; Niwa, S. *Chem. Rev.* **1991**, *92*, 833.
10. Maruoka, K.; Yamamoto, H. In *Catalytic Asymmetric Synthesis*; Ojima, I., Ed.; VCH: New York, 1993, p. 413.
11. Mukaiyama, T. *Aldrichimica Acta* **1996**, *29*, 59.
12. Noyori, R. *Asymmetric Catalysis in Organic Synthesis*; Wiley-Interscience: New York, 1994.
13. *Comprehensive Organic Synthesis. Selectivity, Strategy, and Efficiency in Modern Organic Chemistry*; Trost, B. M.; Fleming, I., Eds.; Pergamon: Oxford, 1991.
14. Miracle, G. S.; Cannizzarro, S. M.; Porter, N. A. *Chemtracts-Org. Chem.* **1993**, *6*, 147.
15. Porter, N. A.; Giese, B.; Curran, D. P. *Acc. Chem. Res.* **1992**, *24*, 296.
16. Curran, D. P.; Porter, N. A.; Giese, B. *Stereochemistry of Radical Reactions: Concepts, Guidelines, and Synthetic Applications*; VCH: New York, 1996.
17. Lepley, A. R.; Khan, W. A. *J. Org. Chem.* **1966**, *31*, 2061.
18. Peterson, D. J.; Hays, H. R. *J. Org. Chem.* **1965**, *30*, 1939.
19. Peterson, D. J. *J. Organometall. Chem.* **1970**, *21*, P63.
20. Peterson, D. J. *J. Am. Chem. Soc.* **1971**, *93*, 4027.
21. Peterson, D. J.; Ward, J. F. *J. Organometall. Chem.* **1974**, *66*, 209.
22. Beak, P.; Zajdel, W. J.; Reitz, D. B. *Chem. Rev.* **1984**, *84*, 471.
23. Beak, P.; Reitz, D. B. *Chem. Rev.* **1978**, *78*, 275.
24. Beak, P.; Kerrick, S. T.; Wu, S.; Chu, J. *J. Am. Chem. Soc.* **1994**, *116*, 3231.
25. Gawley, R. E.; Rein, K. S. In *Comprehensive Organic Synthesis*; Schreiber, S., Ed.; Pergamon: Oxford, 1991; Vol. 1, p. 459.
26. Gawley, R. E.; Rein, K. In *Comprehensive Organic Synthesis. Selectivity, Strategy, and Efficiency in Modern Organic Chemistry*; Trost, B. M., Fleming, I., Eds.; Pergamon: Oxford, 1991, Vol. 3, p. 65.
27. Highsmith, T. K.; Meyers, A. I. In *Advances in Heterocyclic Natural Product Synthesis*; Pearson, W. H., Ed.; JAI: Greenwich, CT, 1991, Vol. 1, p. 95.
28. Meyers, A. I. *Tetrahedron* **1992**, *48*, 2589.
29. Prelog, V.; Helmchen, G. *Angew. Chem. Int. Ed. Engl.* **1982**, *21*, 567.
30. Seebach, D.; Prelog, V. *Angew. Chem. Int. Ed. Engl.* **1982**, *21*, 654.
31. Cahn, R. S.; Ingold, C. K.; Prelog, V. *Angew. Chem. Int. Ed. Engl.* **1966**, *5*, 385.
32. Carey, F. A.; Kuehne, M. E. *J. Org. Chem.* **1982**, *47*, 3811.
33. Hoppe, D.; Carstens, A.; Krämer, T. *Angew. Chem. Int. Ed. Engl.* **1990**, *29*, 1424.
34. Carstens, A.; Hoppe, D. *Tetrahedron* **1994**, *50*, 6097.
35. Gawley, R. E.; Zhang, Q. *Tetrahedron* **1994**, *50*, 6077.
36. Meyers, A. I.; Hoeve, W. T. *J. Am. Chem. Soc.* **1980**, *102*, 7125.
37. Meyers, A. I.; Edwards, P. D.; Rieker, W. F.; Bailey, T. R. *J. Am. Chem. Soc.* **1984**, *106*, 3270.
38. Sanner, M. A. *Tetrahedron Lett.* **1989**, *30*, 1909.
39. Meyers, A. I.; Hellring, S. *Tetrahedron Lett.* **1981**, *22*, 5119.
40. Seebach, D.; Syfrig, M. A. *Angew. Chem., Int. Ed. Engl.* **1984**, *23*, 248.
41. Beak, P.; Lee, W. K. *J. Org. Chem.* **1990**, *55*, 2578.
42. Chong, J. M.; Park, S. B. *J. Org. Chem.* **1992**, *57*, 2220.
43. Gawley, R. E.; Zhang, Q. *J. Am. Chem. Soc.* **1993**, *115*, 7515.
44. Gawley, R. E. *Curr. Org. Chem.* **1997**, *1*, 71.
45. Gawley, R. E.; Zhang, Q. *J. Org. Chem.* **1995**, *60*, 5763.

46. Rein, K. S.; Chen, Z.-H.; Perumal, P. T.; Echegoyen, L.; Gawley, R. E. *Tetrahedron Lett.* **1991**, *32*, 1941.
47. Pearson, W. H.; Lindbeck, A. C.; Kampf, J. W. *J. Am. Chem. Soc.* **1993**, *115*, 2622.
48. Kaufmann, E.; Schleyer, P. v. R.; Houk, K. N.; Wu, Y.-D. *J. Am. Chem. Soc.* **1985**, *107*, 5560.
49. Seebach, D.; Wykpiel, W.; Lubosch, W.; Kalinowski, H. O. *Helv. Chim. Acta* **1978**, *61*, 3100.
50. Rondan, N. G.; Houk, K. N.; Beak, P.; Zajdel, W. J.; Chandrasekhar, J.; Schleyer, P. v. R. *J. Org. Chem.* **1981**, *46*, 4108.
51. Boche, G.; Marsch, M.; Harbach, J.; Harms, K.; Ledig, B.; Schubert, F.; Lohrenz, J. C. W.; Ahlbrecht, H. *Chem. Ber.* **1993**, *126*, 1887.
52. Gawley, R. E.; Zhang, P. *J. Org. Chem.* **1996**, *61*, 8103.
53. Seebach, D.; Hansen, J.; Seiler, P.; Gromek, J. M. *J. Organomet. Chem.* **1985**, *285*, 1.
54. Resek, J. E.; Beak, P. *Tetrahedron Lett.* **1993**, *34*, 3043.
55. Kanazawa, A. M.; Correa, A.; Denis, J.-N.; Luche, M.-J.; Greene, A. E. *J. Org. Chem.* **1993**, *58*, 255.
56. Lohmann, J.-J.; Seebach, D.; Syfrig, M. A.; Yoshifuji, M. *Angew. Chemie Int. Ed. Engl.* **1981**, *20*, 128.
57. Seebach, D.; Huber, I. M. P.; Syfrig, M. A. *Helv. Chim. Acta* **1987**, *70*, 1357.
58. Seebach, D.; Yoshifuju, M. *Helv. Chim. Acta* **1981**, *64*, 643.
59. Meyers, A. I.; Hellring, S.; Hoeve, W. T. *Tetrahedron Lett.* **1981**, *22*, 5115.
60. Beeley, L. J.; Rockell, J. M. *Tetrahedron Lett.* **1990**, *31*, 417.
61. Meyers, A. I.; Hellring, S. *J. Org. Chem.* **1982**, *47*, 2229.
62. Huber, I. M. P.; Seebach, D. *Helv. Chim. Acta* **1987**, *70*, 1944.
63. Rein, K. S.; Gawley, R. E. *Tetrahedron Lett.* **1990**, *31*, 3711.
64. Gawley, R. E.; Smith, G. A. *Tetrahedron Lett.* **1988**, *29*, 301.
65. Rein, K.; Goicoechea-Pappas, M.; Anklekar, T. V.; Hart, G. C.; Smith, G. A.; Gawley, R. E. *J. Am. Chem. Soc.* **1989**, *111*, 2211.
66. Zhang, P.; Gawley, R. E. *Tetrahedron Lett.* **1992**, *33*, 2945.
67. Safe, S.; Moir, R. Y. *Can. J. Chem.* **1964**, *42*, 160.
68. Kametani, T.; Matsumoto, H.; Satch, Y.; Nemoto, M.; Fukumoto, K. *J. Chem. Soc., Perkin Trans. 1* **1977**, 376.
69. Osei-Gyimah, P.; Piascik, M. T.; Fowble, J. W.; Feller, D. R.; Miller, D. D. *J. Med. Chem.* **1978**, *21*, 1173.
70. McMahon, R. M.; Thornber, C. W.; Ruchirawat, S. *J. Chem. Soc., Perkin Trans. 1* **1982**, 2163.
71. Rein, K. S.; Gawley, R. E. *J. Org. Chem.* **1991**, *56*, 839.
72. Eberson, L. *Electron Transfer Reactions in Organic Chemistry*; Springer: Berlin, 1987.
73. Arnett, E. M.; Palmer, C. A. *J. Am. Chem. Soc.* **1990**, *112*, 7354.
74. Bordwell, F. G.; J. A. Harrelson, J. *J. Org. Chem.* **1989**, *54*, 4893.
75. Palmer, C. A.; Ogle, C. A.; Arnett, E. M. *J. Am. Chem. Soc.* **1992**, *114*, 5619.
76. Liotta, D.; Saindane, M.; Waykole, L. *J. Am. Chem. Soc.* **1983**, *105*, 2922.
77. Pross, A. *Acc. Chem. Res.* **1985**, *18*, 212.
78. Corey, E. J.; Ohno, M.; Mitra, R. B.; Vatakencherry, P. A. *J. Am. Chem. Soc.* **1964**, *86*, 478.
79. Corey, E. J.; Cheng, X.-M. *The Logic of Chemical Synthesis*; Wiley: New York, 1989.
80. Seebach, D.; Huber, I. M. P. *Chimia* **1985**, *39*, 233.
81. Blaskó, G.; Gula, D. J.; Shamma, M. *J. Nat. Prod.* **1982**, *45*, 105.
82. Gözler, B.; Gözler, T.; Shamma, M. *Tetrahedron* **1983**, *39*, 577.
83. Wu, T.-S.; Huang, S.-C.; Lu, S.-T.; Wu, T.-C.; McPhail, D. R.; McPhail, A. T.; Lee, K.-H. *Heterocycles* **1988**, *27*, 1565.
84. Wu, T.-S.; Huang, S.-C.; Lu, S.-T.; Wu, T.-C.; McPhail, D. R.; McPhail, A. T.; Lee, K.-H. *Heterocycles* **1990**, *31*, 575.
85. Nonaka, G.; Nishioka, I. *Chem. Pharm. Bull.* **1975**, *23*, 294.
86. Zhang, J.-S.; Xu, R.-S.; Quirion, J. C. *J. Nat. Prod.* **1988**, *51*, 1241.

87. Rein, K. S.; Gawley, R. E. *J. Org. Chem.* **1991**, *56*, 1564.
88. Rein, K. S. Unpublished.
89. Rozwadowska, M. D.; Matecka, D.; Brózda, D. *Tetrahedron Lett.* **1989**, *30*, 6215.
90. Rozwadowska, M. D.; Matecka, D.; Brózda, D. *Liebigs Ann. Chem.* **1991**, 73.
91. Zhang, P.; Gawley, R. E. *J. Org. Chem.* **1993**, *58*, 3223.
92. Rozwadowska, M. D.; Matecka, D. *Liebigs Ann. Chem.* **1991**, 287.

ASYMMETRIC ACCESS TO FUNCTIONAL, STRUCTURALLY DIVERSE MOLECULES EXPLOITING FIVE-MEMBERED HETEROCYCLIC SILYLOXY DIENES

Giovanni Casiraghi, Gloria Rassu,
Franca Zanardi, and Lucia Battistini

Advances in Asymmetric Synthesis
Volume 3, pages 113–189.

ISBN: 0-7623-0106-6

ABSTRACT

Furan, pyrrole, and thiophene silyloxy dienes are invaluable tools with which a myriad of densely functionalized nonracemic (and racemic) molecules are constructed. The aldol-type addition of these nucleophiles to diverse acceptors (e.g carbonyl compounds, imines, nitrones, *N*-acyliminium species) is the main route for the assembly of chiral carbon chains with adjacent oxygen, nitrogen, and sulfur functionalities. After a short discussion of the general aspects of the reaction, the most important achievements, which enable the preparation of the individual target compounds, are presented. Therein, this chapter emphasizes the diastereoselective reactions of 2-silyloxy dienes with chiral nonracemic or racemic precursors, with the discussion being arranged according to the nature of the heterocyclic moiety involved (furan, pyrrole, or thiophene). Also discussed is the exploitation of the silyloxy diene reagents toward preparation of ensembles of small molecules (nucleosides and Annonaceous acetogenins), as well as applications involving cycloadditive reaction behaviors. Finally, a glimpse into the mechanisms and models accounting for the diastereoselective character of the basic carbon–carbon bond-forming stage is presented.

I. INTRODUCTION

In recent years, the fields of drug discovery and lead optimization have greatly benefited from the implementation of potent, extremely sensitive techniques for expeditious assays of a huge number of individual substances and ensembles of structurally diverse molecules.[1a–j] Indeed, as the power of random and target-directed screening programs increases, the quest for more extended repertoires of molecules will grow tremendously.

Nature, the largest and most diverse reservoir of renewable resources available, has spread all over the planet a myriad of plant and animal metabolites, mainly chiral organic molecules, with a limitless range of biofunctional profiles, often exhibiting potential for application as chemotherapeutic agents and drugs. However, while biodiversity constitutes an exciting prospect for access to biological activity, the issue of supplying sufficient quantities of useful compounds and novel variants for detailed testing and drug development remains a considerable problem. Advanced biological technologies, as plant cell cultures and fermentation processes, may provide alternative production systems for both primary and secondary metabolites, but so far this route has been only marginally exploited.

On the other hand, organic synthesis, the art of planning and assembling molecules, strongly supplements Nature's way to access molecular diversity and, at least in principle, allows for practical production of virtually any of the naturally occurring substances as well as disparate artificial mimics and newly designed structures. Nowadays, organic synthesis, including de novo chemical approaches, semisynthesis, and chemo-enzymatic technologies, exploits a seemingly limitless number of construction strategies and techniques wherein stereocontrol and asymmetry are cardinal issues.[2] Among the various asymmetric methodologies with which a given target molecule can be generated, carbon–carbon bond-forming reactions involving chirality transmittal from a given enantio-enriched source, auxiliary or catalyst, occupy a leading position in the organic synthesis scenario, provided sufficient versatility is embodied in the synthetic plan. Ideally, for the effects of shape and stereochemistry on the functional profile to be evaluated, efficient approaches to a given class of active compounds should be amenable to implementation of maximal chemical diversity in the series, while meeting such important criteria as practicality, selectivity, and ergonomics. Nowhere are such requirements more stringent than in the total synthesis of substances of biological interest and naturally occurring products.

A flexible, totally chemical methodology designed to address the previously disclosed issues amenable to construction of both individual chiral structures and collections of diverse bioactive small organic molecules entails the use of furan-, pyrrole-, and thiophene-based silyloxy dienes, an homogeneous triad of functional nucleophilic reactants.[3]

As shown in Scheme 1, the central maneuver of the majority of the synthetic plans herein discussed involves the Lewis acid-assisted, regioselective conjugate addition of a given silyloxy diene **A** to a given electrophilic substrate, for example

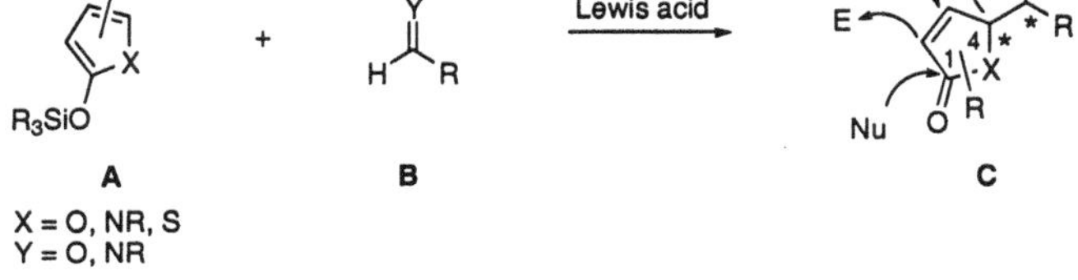

Scheme 1.

an aldehyde or imine of type **B** (a vinylogous variant of the Mukaiyama-aldol cross-addition),[4] to produce intermediary adducts **C**, which can be viewed as ideal templates to be used to prepare densely functionalized molecular frames and target compounds, at will. Accordingly, the diversity of the heteroatom components (X and Y) within **C** coupled with the wide functionalization potential involving the C-1 to C-4 centers, as well as the flexibility with which chirality can be established, emphasize the synthetic effectiveness of this maneuver.

Figure 1 groups representative chiral structures whose syntheses, using the silyloxy dienes chemistry, will be detailed in this review. Here, we will concentrate on the chemistry of oxygen-, nitrogen-, and sulfur-containing silyloxy dienes of general formula **A** (Scheme 1) and their synthetic applications. After a brief section dealing with the preparation and general reactivity of the silyloxy diene reagents (Section II), followed by a summary of earlier work devoted to preparation of

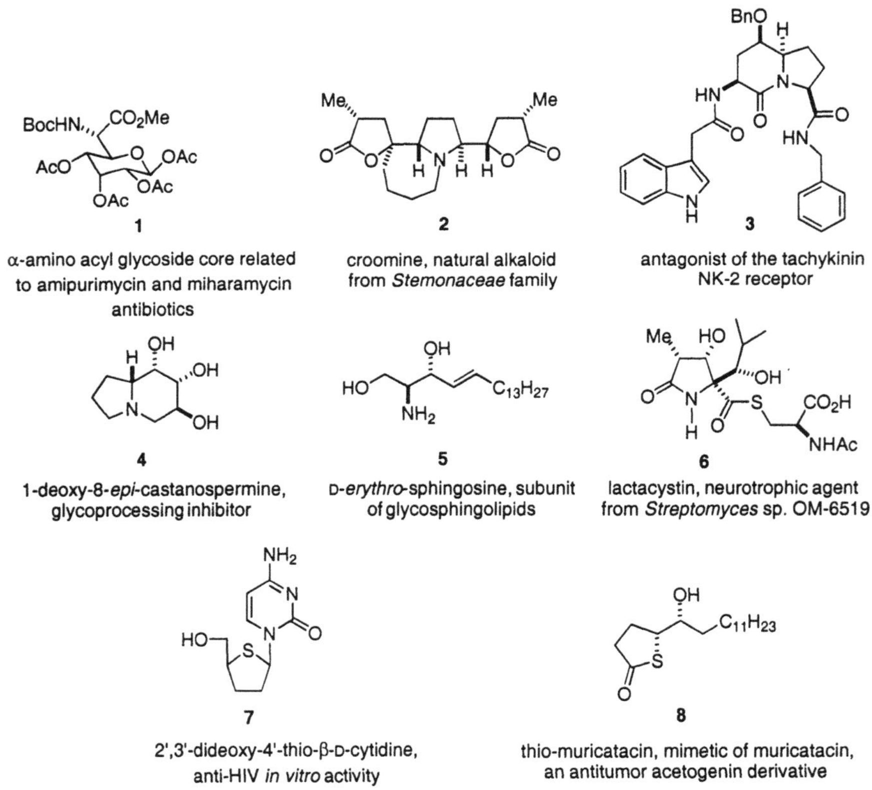

Figure 1.

simple, achiral or racemic compounds (Section III), the chapter will compile work from our and other laboratories mainly discussing total syntheses giving chiral nonracemic compounds of biological interest. Also illustrated will be those syntheses leading to chiral racemic compounds, provided diastereocontrol is exerted during the process. The plan will be to discuss first syntheses involving unsubstituted and substituted 2-silyloxy furans (Section IV), then procedures based on 2-silyloxy pyrroles (Section V), and, finally, examples which make use of 2-silyloxy thiophenes (Section VI). Section IV, which compiles the majority of published achievements on this subject, will be further subdivided according to the homochiral or racemic nature of the targets, while Section V will be organized so as to group together syntheses highlighting a common origin of diastereoselection, i.e. chiron approaches vs. auxiliary-mediated procedures. The final sections will be dedicated to the combined use of the aforementioned silyloxy dienes towards construction of small molecule assemblies (Section VII) as well as miscellaneous applications (Section VIII). As a corollary, Section IX will survey mechanistic hypotheses and models accounting for the diastereoselective character of many of the reported additions involving silyloxy dienes.

This chapter covers published material through end-1996, though a few recent examples from our and other laboratories are briefly condensed in an addendum Section X. Review articles, book chapters, and accounts partially covering this subject matter have previously appeared. These include a chapter by C. W. Jefford et al.[5] discussing the use of 2-(trimethylsilyloxy)furan en route to biologically active lactones (1989), a concise account dealing with the exploitation of furan- and pyrrole-based silyloxy dienes to complex polyol units and carbohydrates (1992),[6] a review mainly compiling work from our laboratory based on the silyloxy diene chemistry (1995),[3] an article discussing applications of (inter alia) heterocyclic silyloxy dienes to asymmetric synthesis of bioactive carbohydrates and alkaloids,[7] and a book chapter comprising a selection of syntheses of carbohydrate mimics and alkaloids exploiting silyloxy dienes.[8]

II. PREPARATION AND GENERAL REACTIVITY OF SILYLOXY DIENE REAGENTS

The five-membered heterocyclic silyloxy dienes of general formula **A**, the key reagents in this review, can be chemically envisioned as silyl derivatives of hydroxylated heterocycles **B**, which represent the enolate forms of the unsaturated butenolide-type compounds **C** (Scheme 2). As a consequence, direct preparation from the respective precursors **C** has been the route of choice. As a general strategy, the reactions simply entail treatment of diversely substituted, readily available unsaturated lactones (X = O), lactams (X = NR), or thiolactones (X = S) with a mix consisting of suitable bases and proper silylating agents, usually silyl halides or trifluoromethane sulfonates.

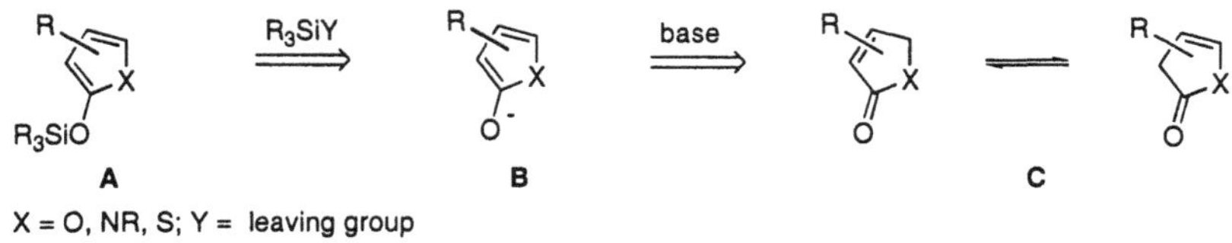

Scheme 2.

A typical procedure leading to 2-(trimethylsilyloxy)furan (**10**, TMSOF) (Scheme 3) involved treatment of precooled (0 °C) 2(3*H*)-furanone **9a** and/or its isomer 2(5*H*)-furanone **9b** (usually a mixture of both) with a slight molar excess of triethylamine and trimethylchlorosilane.[9,10] Distillation under reduced pressure gave **10** in 90% yield, which is storable under an argon atmosphere in the refrigerator for several months. Slightly modified procedures to TMSOF **10** and some substituted congeners advantageously exploited lithium diisopropyl amide[11] or *n*-butyl lithium[12] as enolization reagents. Chromatographically stable 2-(*tert*-butyldimethylsilyloxy)furan (**11**, TBSOF) was conveniently prepared in bulk quantity by reacting a **9a/9b** mixture with Et_3N (1.4 equiv) and TBSOTf (1.1 equiv) in anhydrous CH_2Cl_2 at 0 °C to room temperature.[13]

A nitrogen counterpart, namely *N*-(*tert*-butyloxycarbonyl)-2-(*tert*-butyldimethylsilyloxy)pyrrole (**13**, TBSOP), was readily accessible through reaction of

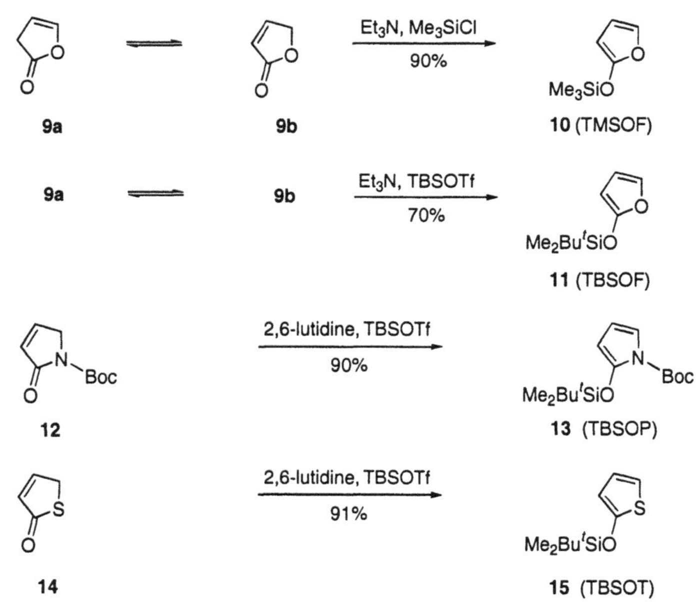

Scheme 3.

N-Boc-pyrrolinone **12**, in turn available in two steps from pyrrole, with 2,6-lutidine (3.0 equiv) and TBSOTf (1.0 equiv).[14] Thiophene-based diene **15** (TBSOT) was similarly produced from 3-thiolen-2-one (**14**) through an enolization/silylation protocol using 2,6-lutidine/TBSOTf mix in CH_2Cl_2.[15] Earlier procedures leading to TMSOF **10**, *N*-methyl-2-(trimethylsilyloxy)pyrrole and 2-(trimethylsilyloxy)thiophene from the respective precursors were developed by Ricci[16] employing (diethylamino)trimethylsilane as the key reagent.

Heterocyclic silyloxy dienes can be viewed both as ambident carbon nucleophiles (a sort of ketene acetals or vinylogous species) and as simple diene reagents; therefore the peculiar reactivity of these substances resides in alkylation or aldol type processes involving the nucleophile C-3 and/or C-5 carbon centers as well as [1,4]-cycloadditive reactions. In practice, the salient nucleophilic reactivity is restricted to Lewis acid-assisted coupling reactions to the C-5 site (Scheme 4, Eq. 1) and this maneuver has been extensively exploited under a variety of synthetic conditions.

Under usual circumstances, reactivity at C-3 has been only scantily observed, though this behavior was prominent with the related borinates (Scheme 4, Eq. 2).[17] Also, silyl enol ethers may undergo Diels–Alder reactions when treated with a variety of activated or inactivated dienophiles, giving rise to the respective cycloadducts (Scheme 4, Eq. 3).

A discussion concerning the detailed reactivity of these diene reagents as well as the mechanisms and models accounting for the diastereoselective behavior of the coupling reactions will be presented at a later stage of this review (vide infra).

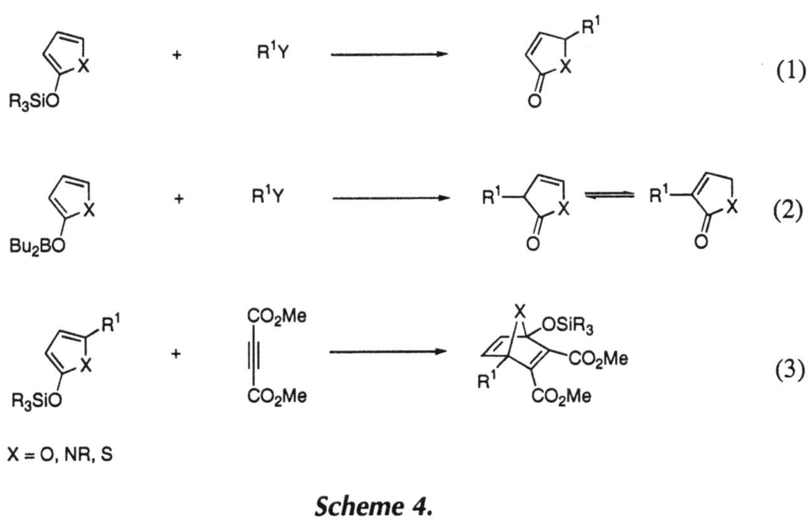

Scheme 4.

III. BACKGROUND

This section briefly summarizes early studies in the achiral or racemic domains dealing with carbon–carbon bond forming reactions involving furan-, pyrrole-, and thiophene-based silyloxy dienes, which appeared during the period 1976 to 1990. Part of this work has been previously reviewed by Jefford and his colleagues in 1989.[5]

In a pioneering study, Asaoka[9] reported the synthesis and reactions of certain silyloxy furans, comprising the parent compound TMSOF **10**, its triethylsilyl congener TESOF, and 5-methyl-, 5-ethyl-, and 3,5-dimethyl-substituted derivatives. Cycloaddition reactions of these dienes with maleic anhydride gave rise to the expected Diels–Alder adducts, which were employed to prepare a variety of 3-hydroxyphthalic anhydrides. As an example (Scheme 5), 5-methyl-2-(trimethylsilyloxy)furan (**16**) reacted with maleic anhydride (**17**) in the absence of solvent giving rise to the expected racemic cycloadduct (±)-**18** which, upon exposure to hydrobromic acid, was transformed into phenolic anhydride **19** in an optimized 67% yield. Yoshii[18] and Asaoka[19,20] also investigated the reactions of TMSOF **10** and certain substituted congeners including 5-methyl derivative **16** with a series of electrophiles such as bromine, aldehydes, ketones, orthocarboxylic esters, acetals, and acylal in the presence of various Lewis acid catalysts, ranging from $SnCl_4$ to $ZnCl_2$, $BF_3{\cdot}OEt_2$, $SbCl_5$, and $TiCl_4$. In several instances, the reaction products derived from regioselective coupling to the C-5 site except for **16**, where a competitive reaction at C-3 occurred. Scheme 6 shows, as an illustrative example, the reaction between **10** and benzaldehyde. In the presence of $SnCl_4$ an unspecified diastereomeric mixture of γ-adducts (±)-**20** formed (95%), which were then transformed to the 1:1 *E/Z* mixture of 4-ylidene butenolides **21** in 91% yield.

Based on these results, Asaoka et al.[21] exploited silyloxy diene **10** to synthesize the racemic macrolide antibiotic (±)-**24** (Scheme 7). Thus, $SnCl_4$-catalyzed reaction of aldehyde (±)-**22** with **10** produced high yield of the expected adduct mixture (±)-**23**, whose multistep transformation allowed the authors to prepare the naturally occurring macrolide (±)-**24** as a racemate.

The reaction between **10** and benzaldehyde was also investigated by Ricci[22] under fluoride anion catalysis (TBAF) in THF at room temperature. Here, the racemic

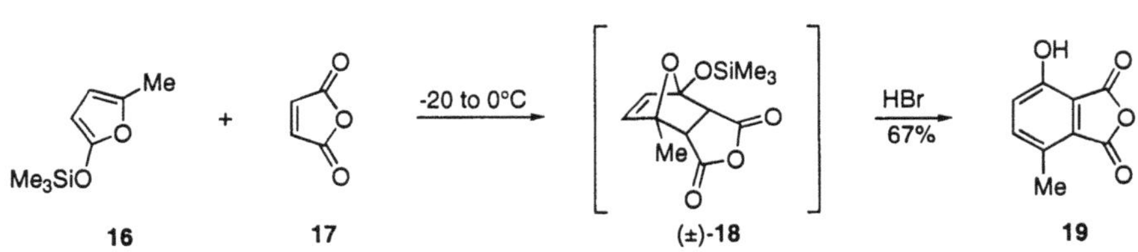

Scheme 5.

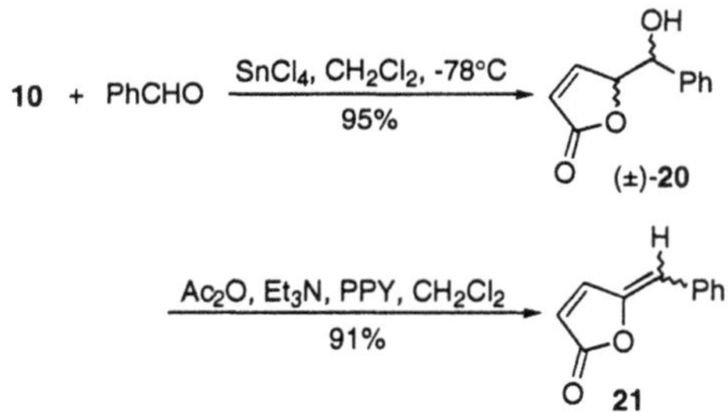

Scheme 6.

adduct mixture (±)-**20** was isolated in 73% yield. As an extension of this chemistry, the same authors[16] succeeded in synthesizing nitrogen and sulfur silyloxy diene relatives **25** and **27**. These compounds smoothly reacted with aldehydes, ketones, lactones, and acyclic and cyclic α,β-unsaturated Michael acceptors giving, in all instances, the expected adducts in high yields. Selected examples are displayed in Scheme 8, showing regioselective Michael additions of silyloxy pyrrole **25** and silyloxy thiophene **27** to acrolein and cyclohexen-2-one, respectively, giving stereochemically undefined products (±)-**26** and (±)-**28**.

The first attempts to rationalize the stereochemical outcome of the vinylogous Mukaiyama aldol addition of TMSOF **10** to simple aldehyde derivatives were independently made by the Brown's[23] and Jefford's groups.[24] As a general phenomenon, high *syn* preference was observed when $SnCl_4$, $ZnBr_2$, $BF_3{\cdot}OEt_2$, TMSOTf or TESOTf were employed as Lewis acid promoters, irrespective of the aldehyde nature. By contrast, the use of fluoride anion sources such as CsF or TBAF resulted in reversal of diastereoselectivity, producing moderate to high *anti* preference. Both authors proposed some mechanistic explanations accounting for the observed behavior, which will be discussed in the appropriate section of this review.

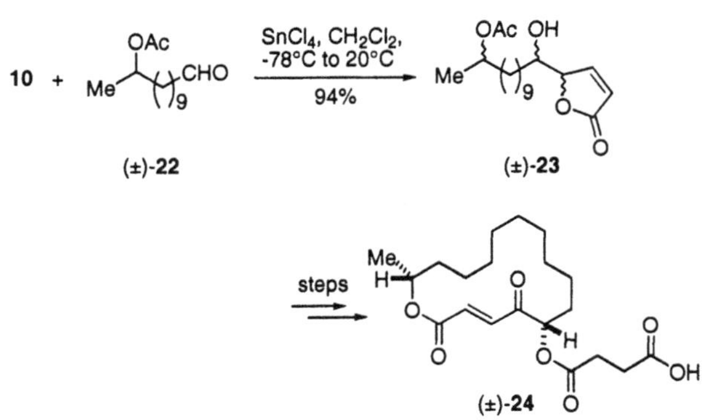

Scheme 7.

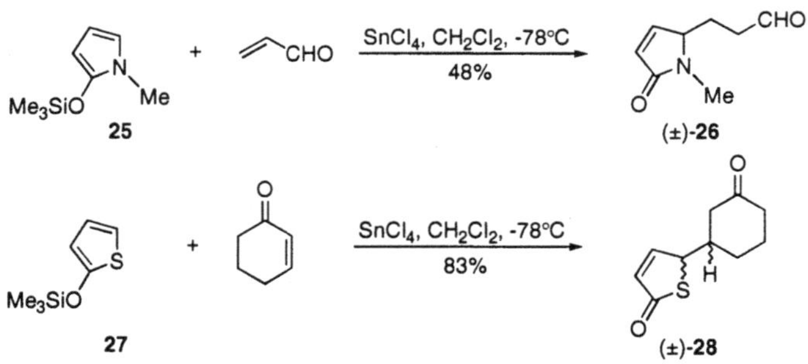

Scheme 8.

The synthetic utility of this reaction was highlighted by the expeditious synthesis of racemic cavernosine derivative (±)-**31**,[25] as well as eldonide (±)-**33**,[26] and freelingnite (±)-**36** (Scheme 9).[27]

In a detailed study directed toward preparation of naturally occurring piperolides, fadyenolides, and relatives, Pelter[11,12] widely utilized (inter alia) 2-(trimethylsilyloxy)-4-methoxyfuran (**37**), quickly obtained from methyl tetronate. It was shown that the Lewis acid-catalyzed reactions of compound **37** with orthocarbonates, orthoesters, acetals, ketones, and aldehydes proceeded with great efficiency, producing the expected adducts as racemic compounds and, when appropriate, diastereomeric mixtures. As an example,[11] the preparation of fadyenolide (±)-**40** from **37** is depicted in Scheme 10. The key reaction was the BF_3 etherate-catalyzed addition of **37** to benzaldehyde dimethyl acetal, giving the racemic butenolide mixture (±)-**38** in 93% yield, which was directly transformed to (±)-**40** via bromide (±)-**39**.

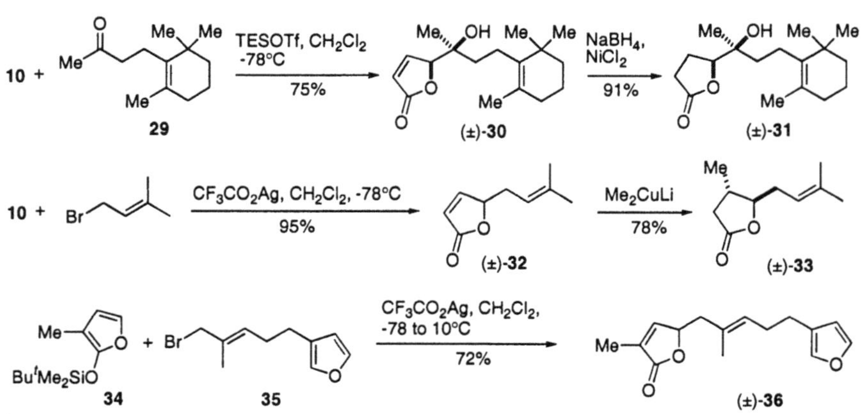

Scheme 9.

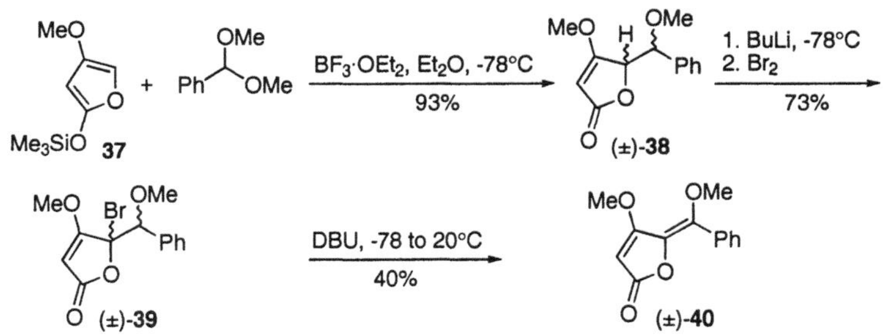

Scheme 10.

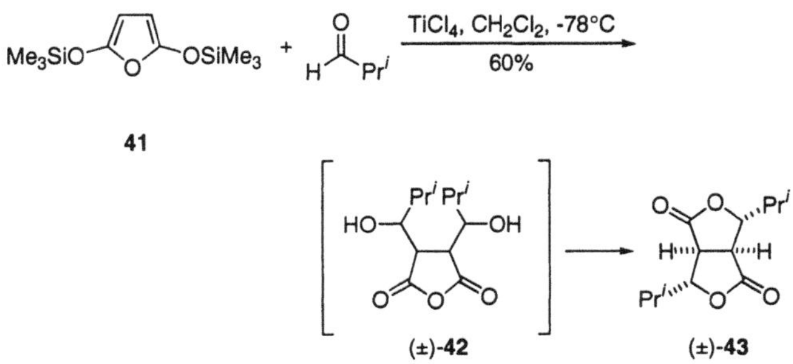

Scheme 11.

In a series of papers,[28–30] Chan investigated the use of 2,5-bis(trimethylsilyloxy)furans of type **41** in reactions involving aldehydes and ketones in the presence of titanium tetrachloride. As shown in Scheme 11, the addition of two molar equivalents of isobutyraldehyde to unsubstituted 2,5-bis(trimethylsilyloxy)furan (**41**) led to diastereoselective formation of racemic all-*cis* 3,7-dioxabicyclo[3.3.0]octanedione (±)-**43** in reasonable yield, through the intermediacy of a stereochemically undefined monocyclic bis-adduct (±)-**42**. Analogous bis(silyloxy)furans also reacted with electron-poor dienophiles, giving *p*-quinones and hydroquinones via uncatalyzed Diels–Alder cycloadditions.

IV. DIASTEREOSELECTIVE REACTIONS USING SILYLOXY FURANS

A. Syntheses of Chiral Nonracemic Compounds

This section is mainly devoted to those approaches which are based on the "chiron concept,"[31] i.e. processes where the chirality of a given target originates from the chirality resident within an enantiopure precursor usually derived from the pool of

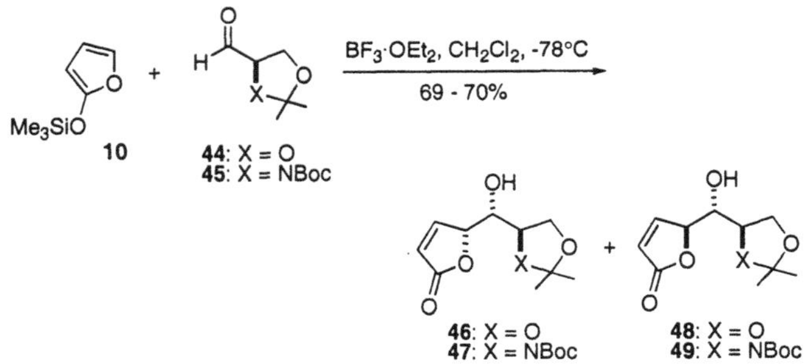

Scheme 12.

chiral available substances. Here, the major concern lies in diastereoselection rather than enantioselection and, as a consequence, the reaction maneuvers involving the chirality transfer from the requisite substrate precursors will be extensively discussed. The examples outlined here make use of the chiral information embodied in the electrophilic species, the silyloxy dienes being envisioned as functional homologative reagents.

The first paper[32,33] which exploited 2-(trimethylsilyloxy)furan (**10**) in a homochiral circumstance described the BF_3 etherate-promoted addition of **10** to 2,3-*O*-isopropylidene-D-glyceraldehyde (**44**), *N*-Boc-2,3-*N*,*O*-isopropylidene-D-serinealdehyde (**45**), and the corresponding imine derivatives. As shown in Scheme 12, the reactions of **10** with aldehydes **44** and **45** produced, as expected, mixtures of D-*arabino*- and D-*ribo*-configured seven carbon butenolides **46–49**, with D-*arabino* compounds **46** and **47** overriding D-*ribo* counterparts **48** and **49** (**46/48** = 83:17; **47/49** = 90:10). The relative and hence absolute configuration of compound **46** derived from an X-ray crystal analysis, while the configuration of the related lactones **47–49** was deduced by analogy considerations, in particular by chiroptical measurements.

Interestingly, α,β-unsaturated lactone **46** was the precursor with which 2,3-dideoxy-3-*C*-methyl-D-*manno*-heptose (**52**) and 2,3-dideoxy-2,3-di-*C*-methyl-D-*glycero*-D-*galacto*-heptose (**53**) were synthesized (Scheme 13).[34] Asymmetric conjugate addition reaction (Me_2CuLi, Et_2O, –80 °C) installed a methyl group at C-3 of intermediate **46**. The presence of the side chain at C-4 precisely governed the stereochemical course (*anti*) of this reaction, affording an 85% yield of **50** as a single diastereoisomer. Methylation via enolate formation (LiHMDS, THF, then MeI) was adopted to convert **50** into **51**. The diastereoselection was moderate (85:15) due to the presence of the 3β,4α-substituents in the furan ring, which act in a mismatched sense. Nonetheless, the major crystalline *trans,trans* diastereoisomer **51** was obtained in 60% yield after chromatographic separation.

At this point, all that remained was to convert the intermediates **50** and **51** into *C*-methylheptoses **52** and **53**, respectively. A common enantioconservative protocol

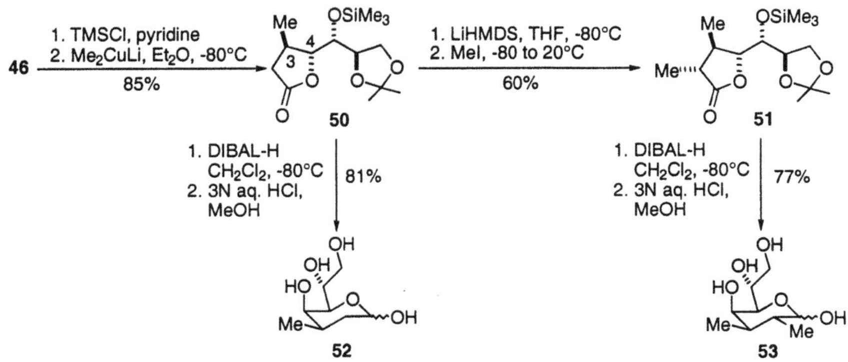

Scheme 13.

of two sequential reactions ensured clean transformations. Lactone-to-lactol reduction occurred under the usual conditions using DIBAL-H in CH_2Cl_2 at –80 °C, while complete cleavage of the silyl and acetonide linkages was quickly accomplished by treatment with 3N aq. HCl in MeOH at room temperature. In this manner, the expected heptoses **52** and **53** were obtained in 81 and 77% yields, respectively, for the two final steps.

This chemistry was also applied to total syntheses of certain octopyranose derivatives by starting with 2,3-*O*-isopropylidene-4-*O*-benzyl-D-threose (**54**) (Scheme 14).[35] Four-carbon elongation of **54** in CH_2Cl_2 with TMSOF **10** (BF_3·OEt_2), followed by protection of the newly formed OH at C-5 as the trimethylsilyl ether, cleanly generated D-*galacto*-configured unsaturated lactone **55** in 69% yield with no other diastereoisomers observed in the reaction mixture. Treatment of **55** in CH_2Cl_2 with solid $KMnO_4$ in the presence of dicyclohexano-18-crown-6-ether at 15–20 °C, followed by silylation, led to lactone **56**, the configuration of which was corroborated by NOE studies. A clean protocol of three mild reactions

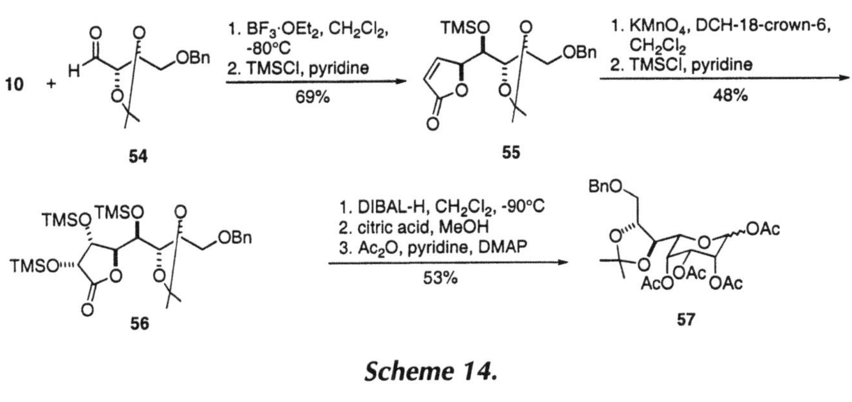

Scheme 14.

then allowed the octopyranose formation. Lactone-to-lactol reduction using DIBAL-H in CH_2Cl_2 at –90 °C, followed by desilylation (citric acid, methanol, 25 °C), afforded a crude pyranose which was converted to tetraacetate **57** by Ac_2O/pyridine/DMAP treatment in 53% overall yield for the three final steps of the sequence.

A reiterative $C_3 + C_4 + C_4$ homologation strategy was chosen for assembly of D-*glycero*-D-*talo*-L-*talo*-undecose pentaacetonide (**63**).[36] The synthesis started with the popular three-carbon chiron 2,3-*O*-isopropylidene-D-glyceraldehyde (**44**) and required two elongation steps utilizing 2-(trimethylsilyloxy)furan (**10**) as a nucleophilic four-carbon homologative reagent. The entire sequence is presented in Scheme 15. Treatment of **44** with **10** in the presence of BF_3 etherate and protection of the crude product as the TMS ether afforded the seven-carbon butenolide **58** as the major component. *Anti*-selective *cis*-dihydroxylation of the double bond was performed using the $KMnO_4$/dicyclohexano-18-crown-6-ether/CH_2Cl_2 system at ambient temperature. There was obtained D-*glycero*-D-*talo*-heptonolactone **59**, with no trace of diastereomeric material. Attention was then directed toward elaborating the lactone framework into an open-chain *aldehydo* sugar. Compound

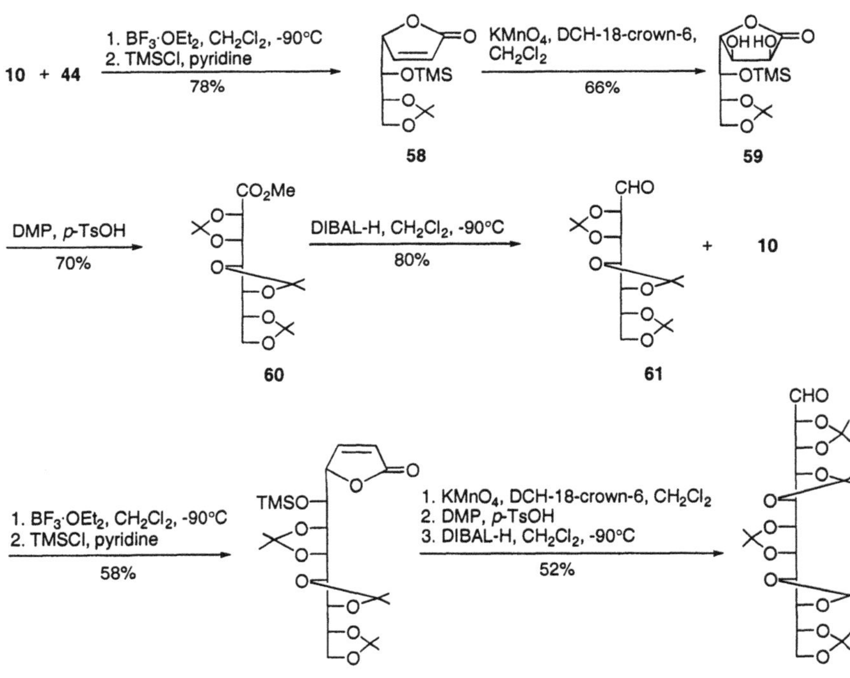

Scheme 15.

59 was directly transformed into methyl ester **60** by treatment with a large excess of dimethoxypropane in the presence of 3 moles equiv of *p*-toluensulfonic acid at room temperature. Controlled reduction of the methyl ester into subtarget aldehyde **61** was achieved, without any epimerization, by careful addition of DIBAL-H in CH_2Cl_2 at −90 °C. The first cycle of the sequence was thus completed and the setting for proper installation of four additional contiguously oxygenated carbon atoms was at hand. Four-carbon elongation of **61** with **10** generated the eleven-carbon unsaturated lactone **62** as the main component, whose absolute configuration was ascertained by an X-ray crystal structure analysis.[37] Reiteration of the reaction sequence of the first cycle allowed clean conversion of lactone **62** to undecose pentaacetonide **63**, with an overall yield of 8.7% for the entire sequence from **44**.

The same stereoselective sequence (one cycle only) was exploited in the synthesis of nonofuranose **67** (Scheme 16).[38] Thus, condensation of TMSOF **10** with dialdose **64**, under usual reaction conditions, gave rise to the butenolide precursor **65** (de > 95%), whose stereochemistry was firmly established via X-ray analysis.[39] This was transformed into crystalline nonofuranuronic acid methyl ester **66** by following exactly the procedure described for compound **60**. Reduction of the ester moiety was achieved by use of DIBAL-H at room temperature furnishing protected furanose **67** in good yield.

Hybrid structures with carbon–carbon-joined amino acid and carbohydrate moieties are often encountered in nature as individual molecules or as the core components of complex nucleoside antibiotics.[40] Cyclic arrays comprise both furanose and pyranose derivatives, bearing either anomeric or terminal substitutions. Furan-based silyloxy diene **10** was utilized for the preparation of some pyranosidic representatives of this progeny.[41,42] A versatile synthetic plan (Scheme 17) called for *C*-glycopyranosyl glycine derivatives **72** and **1** to be generated from the *arabino*-

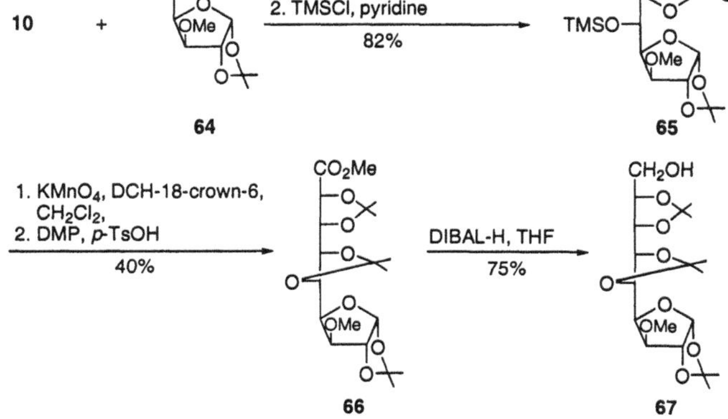

Scheme 16.

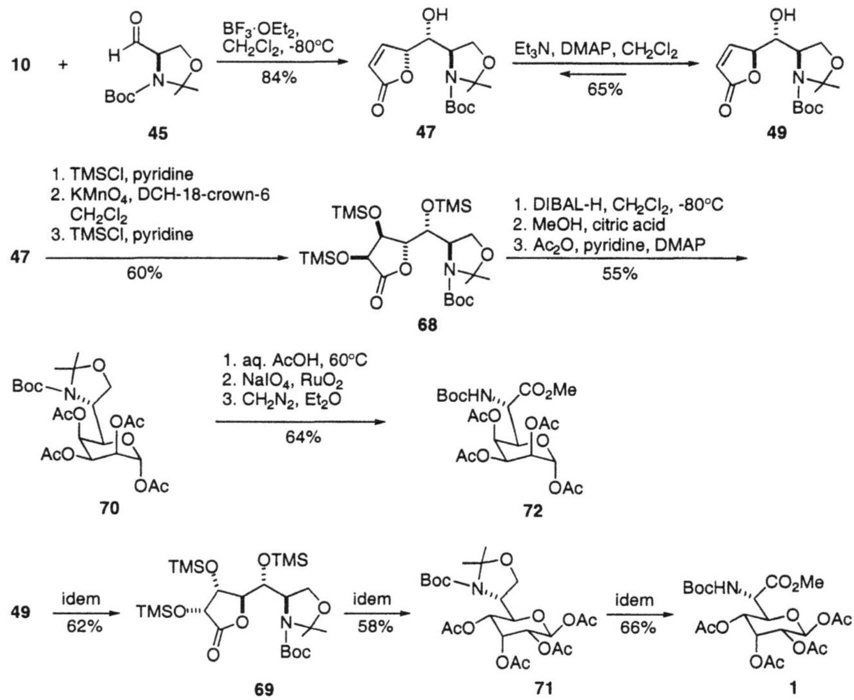

Scheme 17.

and *ribo*-configured butenolides **47** and **49**, respectively. The opening move was the preparation of enantiomerically pure butenolides **47** and **49**, and this was achieved via four-carbon elongation of D-serinal **45** using 2-(trimethylsilyloxy)furan (**10**). This reacted with **45** in CH_2Cl_2 in the presence of BF_3 etherate providing seven-carbon *arabino*-configured butenolide **47** selectively, along with only trace amounts of the *ribo*-counterpart **49**. Base-catalyzed C-4 epimerization of the *arabino*-butenolide using Et_3N provided equilibrium mixtures of *ribo* and *arabino* epimers in a ratio of ca. 65:35 from which the more abundant component **49** was obtained in a pure state. The lactone fragments in **47** and **49** were first elaborated according to a highly stereoselective three-step sequence consisting of protection of the free OH at C-5 as the TMS ether, *anti-cis* dihydroxylation of the butenolide double bond using $KMnO_4$, and persilylation. This provided heptonolactones **68** and **69** in high overall yields.

The ring expansion to pyranoses **70** and **71** required three further operations. DIBAL-H reduction generated γ-lactol intermediates, which, by citric acid–methanol treatment and subsequent peracetylation, were converted to pyranoses **70** and **71**. In the final stages of the synthesis, the remaining carbon to be elaborated was the terminal hydroxymethylene group. Treatment of **70** and **71** with 70% aqueous acetic acid at 60 °C resulted in selective removal of the acetonide groups, giving compounds with unprotected terminal CH_2OH functions. The crude primary alcohols were subjected to oxidation using $NaIO_4$ and catalytic RuO_2, resulting in formation of the expected carboxylic acids, which were finally transformed into the corresponding methyl esters **72** and **1** by CH_2N_2 treatment.

By exploiting the synthetic utility of enantioenriched butenolide matrices, iminoalditol **79** containing five consecutive stereocenters was synthesized from D-glyceraldehyde imine **73** and 2-(trimethylsilyloxy)furan (**10**) as outlined in Scheme 18, utilizing the single chiral element in the precursor imine.[43]

The target called for D-*ribo* butenolide **75** as the chiral matrix. Four-carbon homologation of imine **73** with **10** in CH_2Cl_2 in the presence of BF_3 etherate resulted in formation of butenolide **74**. This was isolated as a 1:1 mixture of two epimers at C-4, which could not be separated owing to rapid equilibration. Mixture **74** was smoothly converted into the *N,N*-diprotected butenolide **75** by reaction with benzyloxycarbonyl chloride under the usual Schotten–Baumann conditions. Although the intermediate **74** was a mixture of isomers, the Cbz-protected butenolide **75** was isolated as the D-*ribo* stereoisomer only. Presumably, the formation of **75** is controlled by the thermodynamics of base-catalyzed lactone equilibration, strongly favoring, in this instance, the D-*ribo* isomer. Based on precedents for related compounds, butenolide **75** was selectively hydroxylated at C-2 and C-3 by using $KMnO_4$ under solid–liquid phase-transfer conditions. There was obtained, after protection (DMP, *p*-TsOH), D-*glycero*-D-*allo*-heptono-1,4-lactone **76** as a homo-

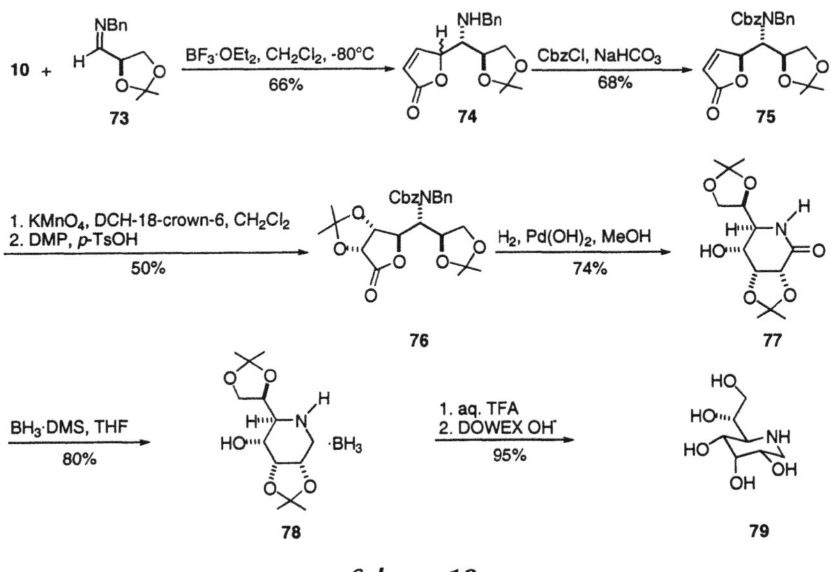

Scheme 18.

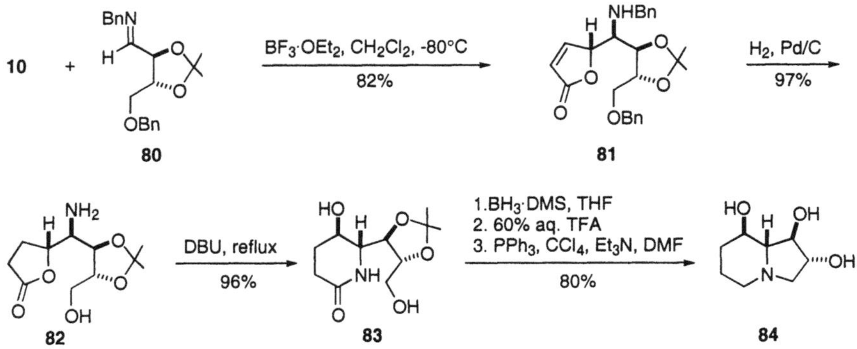

Scheme 19.

geneous material with no trace of other stereoisomers. Hydrogenolytic cleavage of the protective groups at the C-5 nitrogen of **76** using $Pd(OH)_2$ in methanol gave the δ-lactam **77** which was then transformed to **79** in excellent yield by reduction with the BH_3·DMS complex, giving amine–borane complex **78**, followed by deprotection, chromatography on DOWEX OH^-, and lyophilization.

Protected D-threose imine **80** was the enantiomerically pure substrate from which, by reaction with furan-based diene TMSOF **10**, the swainsonine derivative **84** was generated.[44] A short route to **84** is outlined in Scheme 19. The reaction of **80** with **10** in the presence of BF_3 etherate gave butenolide **81** with little if any stereoisomeric contamination. Double bond saturation with concomitant removal of the two benzyl protective groups to provide amine **82** was effected under conventional hydrogenation conditions. Next, upon treatment with DBU, amino lactone **82** underwent clean ring expansion to provide δ-lactam **83** in good yield. Finally, treatment of this compound with an excess of BH_3·DMS complex in THF effected the reduction of the lactam carbonyl to the corresponding amine-borane adduct. Deprotection of this adduct was accomplished by acidic treatment giving a densely oxygenated piperidine intermediate, which was directly cyclized to indolizidine **84** by exposure to PPh_3–CCl_4–Et_3N mix in DMF.

An attempt to extend this simple reaction protocol to preparation of hydroxylated quinolizidines of type **90** resulted in a quite unexpected transformation (Scheme 20).[45] Thus, preparation of the key homologue intermediate **86** was accomplished without incidents by starting with TMSOF **10** and D-arabinose imine **85**, exactly adopting the reaction sequence previously illustrated for the synthesis of intermediate **83** (vide supra). The relative, and hence absolute configuration of **86** was certified by an X-ray analysis,[46] thus confirming the chirality of the five stereocenters within the molecule.

The subsequent transformation to piperidine polyol **87** was effected in the usual manner, by reduction of the lactam carbonyl (BH_3·DMS) and subsequent acidic

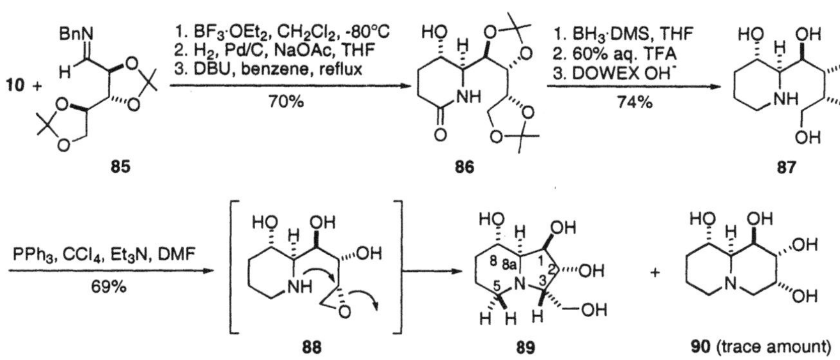

Scheme 20.

deprotection. Surprisingly enough, when **87** was subjected to the Mitsunobu-like annulation protocol with PPh_3–CCl_4–Et_3N mix, indolizidine **89** was mainly obtained, with only trace amounts of the expected quinolizidine **90**. This behavior can be rationalized by assuming preliminary formation of an epoxide intermediate **88**, which underwent preferential 5-*exo*-tetragonal ring closure to **89**, an expanded alexine derivative. The configurational assignment of **89** was mainly based on 2D-NOESY ^{1}H NMR experiments, showing diagnostic NOE contacts between H–C-3 and H–C-5 in the β position, and between H–C-8a and both the protons of the C-3 hydroxymethyl substituent.

Annonaceous acetogenins are derivatives of C_{35} or C_{37} fatty acids, embodying zero to three 2,5-linked tetrahydrofuran core motifs. These compounds (more than 230 characterized representatives) have been isolated from the *Annonaceae*, a widely distributed family of tropical plants, and were shown to possess powerful cytotoxic activities against several cell lines.[47] The heterocyclic silyloxy dienes of this report might admirably serve to access the core units of these interesting compounds as well as a number of nitrogen and sulfur mimics when one realizes that oxygen, nitrogen, and sulfur silyloxy dienes are nucleophilic equivalents of 2-oxo-tetrahydrofuran, pyrrolidine, and thiolane (unsaturated) species.

Indeed, this approach has been successfully put into practice by exploiting, for example, the diastereoselective coupling of TMSOF **10** to five-membered ring oxonium or *N*-acyliminium electrophiles. For syntheses of *trans*-2,5-oligo-tetrahydrofurans such as trinuclear derivative **96**, a clever plan was adopted by Koert[48] utilizing, as a pivotal move, the aldol coupling between **10** and the oxonium species derived from chiral acetoxylactol **91** (Scheme 21).

Although the process proved only partially diastereoselective, a 30% yield of pure *trans,threo*-butenolide **92** was obtained, along with significant amounts of *trans,erythro-, cis,threo-, and cis,erythro*-stereoisomers. Compound **92** was elaborated into cyano derivative **93** by a four-step sequence including double bond

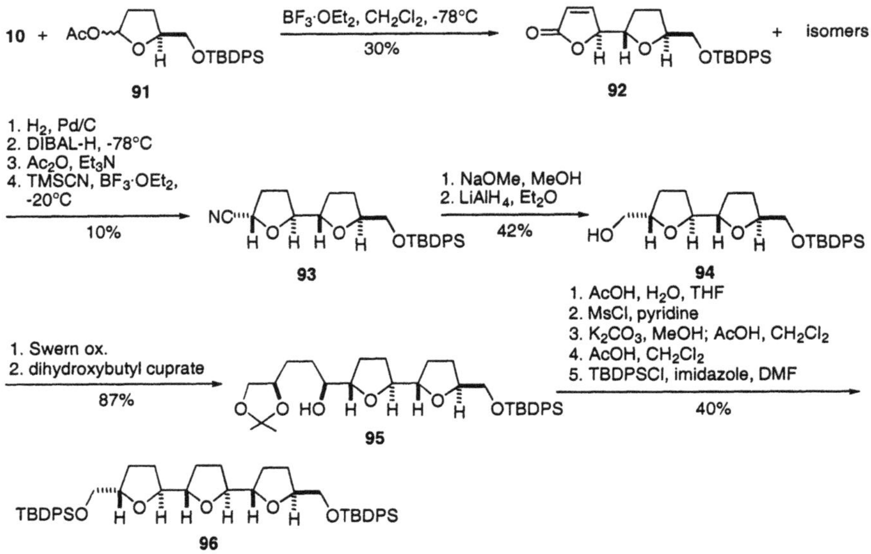

Scheme 21.

saturation, lactone-to-lactol reduction, acetylation, and nucleophilic cyanation (TMSCN). Methanolysis of **93** followed by reduction of the methyl ester so formed gave rise to bicyclic carbinol **94**, which was transformed to **95** by Swern oxidation and stereoselective four-carbon elongation with acetonide-protected (2*R*)-1,2-dihydroxybutyl-mixed cuprate. Finally, a series of sequential reactions, namely, deacetonidation, epoxide formation, and annulation, followed by silylation of the terminal hydroxymethyl, completed the preparation of *trans,threo, trans,threo,trans*-oligomer **96** in a rather low 0.5% overall yield for the entire sequence.

Using a conceptually similar approach, Figadère et al.[49] successfully synthesized long-chain-substituted bis-THF derivative **99**, by starting with L-glutamic acid-derived acetoxylactol **97**. Thus, as shown in Scheme 22, trityl perchlorate-catalyzed coupling of **10** to **97** furnished *trans,threo*-disposed butenolide **98** (67% yield) accompanied by a relevant quantity of the corresponding *trans,erythro*-congener. Double bond hydrogenation and removal of the protective group by TBAF treatment cleanly allowed the formation of lactone **99**, a versatile intermediate en route to dinuclear acetogenin compounds.

The same author[50] also exploited *N*-Boc-iminium species of type **100** to assemble mixed oxygen–nitrogen dinuclear butenolide compounds **101**. Using $BF_3 \cdot OEt_2$, a mixture of diastereomeric products formed with substantial preference for the *threo*-disposed compounds.

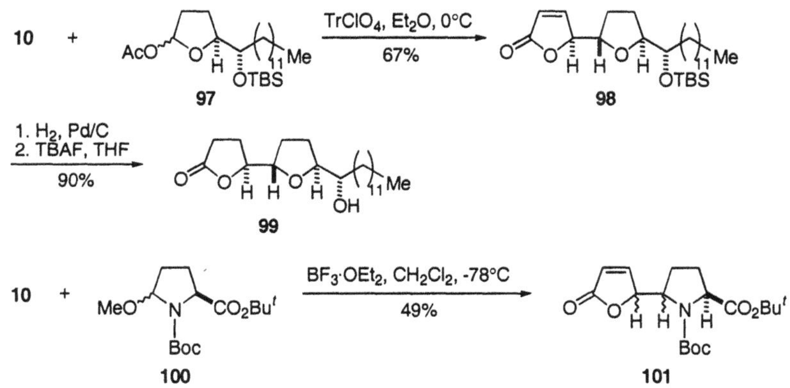

Scheme 22.

N-Carbamoyl iminium species, obtainable from suitably protected heterocyclic hemiaminals, proved extremely useful electrophiles in the Mukaiyama aldol coupling with silyloxy diene nucleophiles. The stereocontrolled maneuver easily generates variously shaped frames to be used as corner stones for assembly of important naturally occurring and synthetic molecules. Along this line, Martin[51] recently reported the asymmetric total synthesis of (+)-croomine (**2**), a medicinally interesting alkaloid isolated from plants belonging to the *Stemonaceae* family. The general plan emphasizes the repetitive use of 3-substituted silyloxyfurans to forge the two lactone ring components (A and D) within the complex croomine structure. The nice procedure commenced with the synthesis of disubstituted silyloxyfuran **104**, which was accomplished via silyloxyfuran **103**, as shown in Scheme 23.

The first vinylogous Mukaiyama reaction of **104** with the acyl iminium ion derived from pyroglutamate-related hemiaminal **105**—a sort of Mannich procedure—afforded, after removal of the *N*-Boc protection, the *trans,threo*-adduct **106**, which arose from the addition of the furan **104** to the less hindered face of the iminium species. Highly stereoselective hydrogenation of **106** with 3% Rh on carbon followed by intramolecular nucleophilic displacement of the terminal side chain bromine by the nitrogen of the proline appendage and acidic hydrolysis produced the salt **107** in 71% yield, thereby setting the stage for the second Mukaiyama addition. Thus, treatment of **107** with $POCl_3$ in DMF resulted in formation of a thermally unstable iminium salt, which was trapped by the silyloxy ether **103**. There was obtained the desired *threo*-configured tetracyclic adduct **108**, accompanied by a minor amount of its *erythro* counterpart (2:1 ratio). Diastereoselective hydrogenation from the less hindered face of **108** finally delivered (+)-croomine (**2**) in 85% yield, which corresponds to a 5% overall yield for the entire nine-step sequence from **102**.

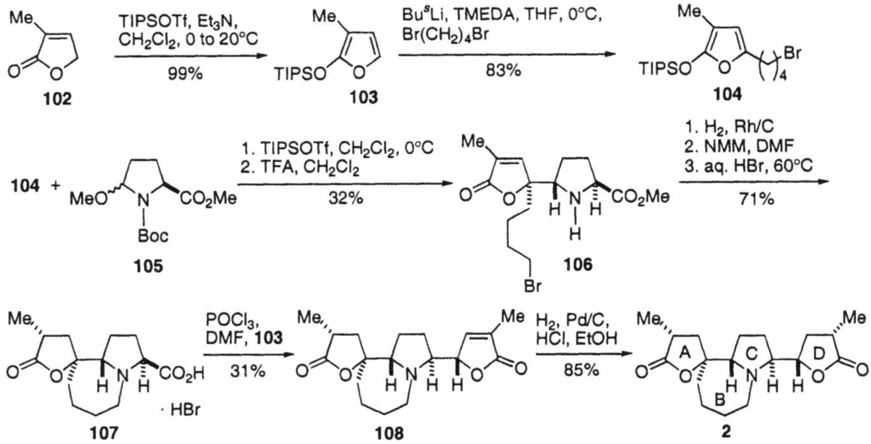

Scheme 23.

A similar stereocontrolled addition of a silyloxyfuran reagent to a cyclic iminium ion intermediate was the synthetic choice with which Hanessian[52] built up a prototypical non-peptidic renin inhibitor model (i.e. structure **114**). As illustrated in Scheme 24, the synthesis of the requisite iminium precursor **110** commenced with (*S*)-mandelate **109**, and involved an efficient sequence of 10 conventional steps (Wittig two-carbon elongation, Michael addition, OH to NH_2 Mitsunobu intervention).

The key event was the diastereoselective addition of **10** to **110** (BF_3·OEt_2), producing the *trans,threo*-dinuclear adduct **111**, predominantly (6:1 *threo*/*erythro*

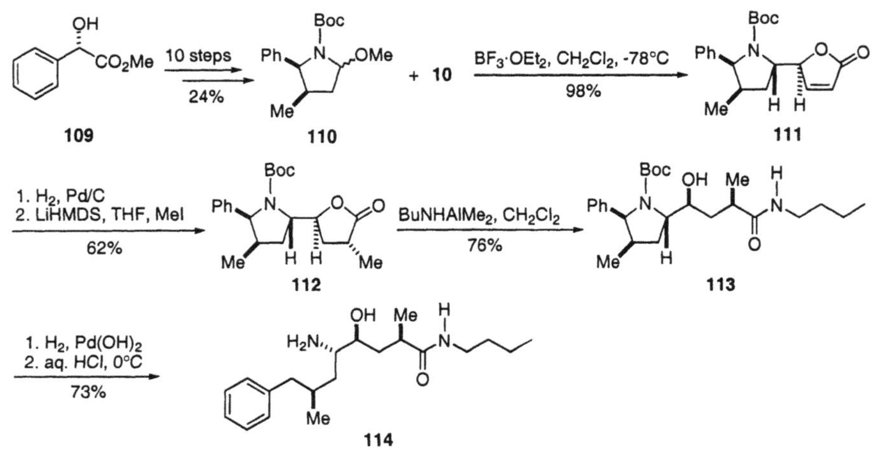

Scheme 24.

ratio). Hydrogenation of the double bond followed by stereoselective enolate methylation (LiHMDS, MeI) gave the desired product **112**, whose absolute configuration was certified by an X-ray analysis. Weinreb transamidation (BuNHAlMe$_2$) directly afforded conformationally constrained model structure **113**, which was finally converted to target hydroxyethylene isostere **114** by hydrogenolytic pyrrolidine ring opening and acidic deprotection in a 8% overall yield from mandelate **109**. Noticeably, the cyclic structure **113** showed no activity in the renin binding affinity assay, while the acyclic amino derivative **114** proved weakly active.

The same author[53] exploited 2-(trimethylsilyloxy)furan (**10**) and its γ-methyl substituted analogue **16** as synthetic equivalents of the tetrahydrofuran or γ-lactone units of clinically promising tricyclic β-lactams incorporating a carbacephem structure. As shown in Scheme 25, the key reaction involved $ZnCl_2$-catalyzed diastereoselective coupling of silyloxy diene nucleophiles **10** and **16** to chiral, commercially available azetidinone **115**. Thus, treatment of **115** with **10** in the presence of $ZnCl_2$ gave the *trans,threo*-configured butenolide adduct **116** preferentially (98:2 dr), which was easily transformed to vinyl lactone **117** by Michael addition (CH_2:CHMgBr·CuI) and azetidinone protection. Carbonyl reduction and change of O-TMS for O-TBS group allowed preparation of **118** in 32% yield from **117**. Transformation to **120** was ensured via acylation to **119**, followed by ring annulation involving ozonolysis and trimethylphosphite treatment. Having the synthesis of the desired tricyclic structure **120** secured, there remained to remove both the TMS and benzyl protective groups within the molecule. After a number of unsuccessful attempts, a rather sophisticated route was carried out, comprising temporary masking of the double bond, benzyl-to-allyl ester interconversion, and double bond unmasking. This led to **121**, the immediate precursor of tricyclic carbacephem target **122**, which was isolated as the stable amidinium salt. During the study, an expeditious synthesis of a lactone congener **123** bearing a quaternary stereocenter was also accomplished, utilizing 5-methyl-2-(trimethylsilyloxy)furan (**16**) and taking advantage of the previously explored chemistry. Despite their promising structures, compounds **122** and **123** proved inactive against a number of bacterial strains.

As a further example emphasizing how silyloxy diene chemistry admirably serves to assemble densely functionalized, multichiral architectures of biological relevance, the stereocontrolled synthesis of a bicyclic β-turn peptidomimetic, **3**, was reported by the Hanessian group (Scheme 26).[54] Once more, the cardinal operation features a highly selective Lewis acid catalyzed addition of TMSOF **10** to *N*-Boc iminium ion obtained from **124**, which, in turn, derives from L-pyroglutamate. The addition of a catalytic amount of BF_3 etherate to hemiaminal **124** generated a cyclic iminium species which was trapped by **10**, regioselectively, affording the *trans,threo*-binuclear adduct **125** with a 10:1 *threo/erythro* diastereomeric ratio. Hydrogenation of **125** to a saturated lactone, followed by selective

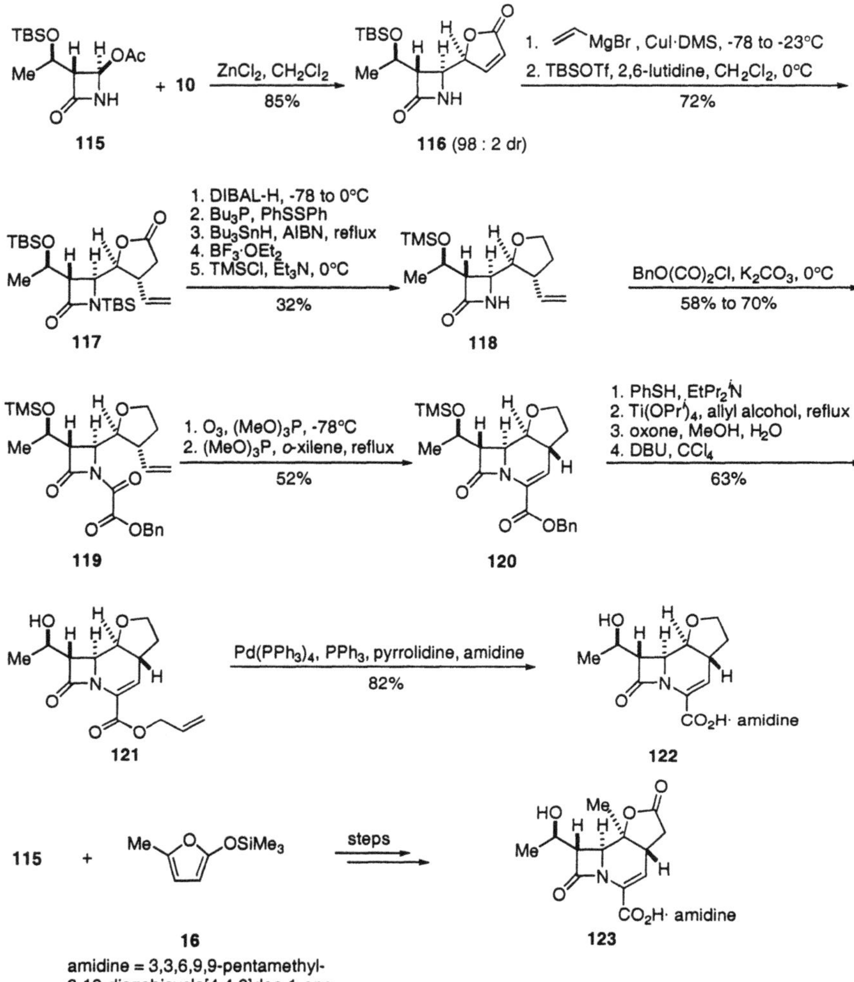

Scheme 25.

removal of the *N*-Boc protection gave **126**, which underwent clean annulation to indolizidinone derivative **127** by exposure to NaOMe in methanol.

Transformation of **127** to the requisite stereoisomer **128** called for inversion of the C-8 and C-8a stereocenters, and this was effected by a clever three-step protocol involving Dess–Martin periodinane oxidation, DBU-promoted epimerization of the juncture carbon atom α to the carbonyl, and selective reduction of the ketone function. To access the key peptidomimetic core **131**, the installation of properly protected 3,6-*cis* disposed carboxylic and amino functionalities was required, and

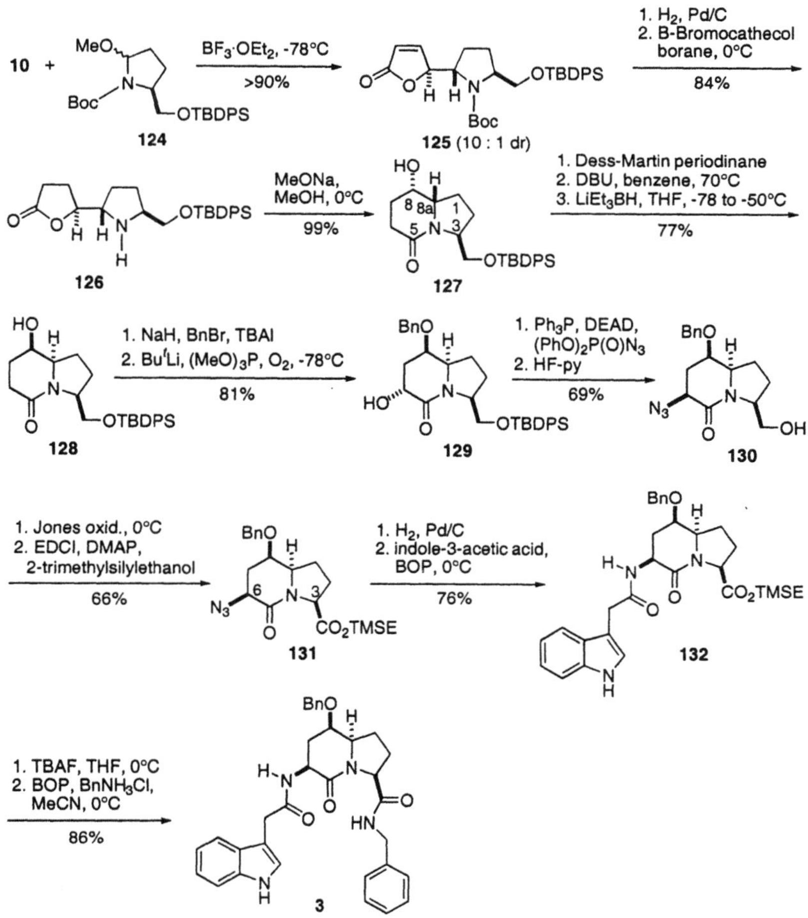

Scheme 26.

this goal was gained by first introducing a hydroxyl group at C-6 with high α-facial selectivity, followed by a Mitsunobu-type azidation maneuver. Thus, benzylation of **128** furnished a protected intermediate, which was enolized (Bu^tLi) and treated with trimethylphosphite in the presence of a flow of oxygen, an efficient source of hydroxyl group. There was obtained compound **129** in an optimized 81% yield, with the expected (6*R*) absolute configuration. Introduction of the azide group (diphenylphosphoryl azide) within **129** occurred with complete inversion, affording, after removal of the silyl protection, a 69% yield of the advanced intermediate **130**. Hydroxymethyl-to-carboxyl oxidation (Jones reagent) and 2-trimethylsilylethyl esterification cleanly afforded **131**, which incorporated the proper functional group diversification required for ready implementation to the target peptidomimetic **3**. The azide **131** was selectively reduced to a free amine, which was coupled with indole-3-acetic acid (standard BOP conditions) to generate peptide **132**. Finally, deprotection of the carboxyl moiety (TBAF) and conventional peptide chemistry successfully provided the target peptidomimetic model **3**, with a nice ~15% overall yield from hemiaminal **124**. Receptor binding assays of compound **3** revealed selective, though weak, antagonist activity for the NK-2 receptor. Noteworthy, the utility of indolizidine unit **131** was extended to preparation of various amide derivatives; also, constrained dipeptide frames such as **131** constitute promising scaffolds for the construction of solid-supportable peptidomimetic libraries.

In a study directed toward exploitation of norephedrine-derived 2-methoxyoxazolidines as masked formyl cation equivalents, Scolastico[55,56] reported the use of TMSOF **10** to assemble enantiomerically enriched polyhydroxylated molecular frameworks. As shown in Scheme 27, addition of **10** to chiral orthoamide **133** proceeded with high yield and selectivity at −100 °C using $BF_3{\cdot}OEt_2$ (or TBSOTf) giving a major butenolide adduct **134** (de > 98%), whose C-4 and C-5 oxygens possess relative *erythro* stereodisposition (*N,O-threo*). Butenolide **134** was sequentially submitted to double bond dihydroxylation ($KMnO_4$) and lactone reductive opening ($NaBH_4$) to generate polyol **135**. As the final step, auxiliary removal by means of 1,3-propanedithiol/$BF_3{\cdot}OEt_2$ furnished enantiopure D-ribose dithioacetal **136**.

Similar chiralized cyclic acyl cation equivalents, available from homochiral diols and amino alcohols, were exploited by Pelter and colleagues[57] in Lewis acid-catalyzed additions with TMSOF **10** or 4-methoxy-2-(trimethylsilyloxy)furan (**37**). After a number of poorly diastereoselective coupling attempts between **10** and cyclic ortho esters derived from L-diisopropyl tartrate, the authors turned their attention to oxazolidine-based acyl cation equivalents, readily available from (*R*)-phenylglycinol or (1*R*,2*R*)-norpseudoephedrine.

As an example, Scheme 28 illustrates that the BF_3 etherate catalyzed reaction of silyl enol ether **10** with oxazolidine **137** proceeded diastereoselectively, producing

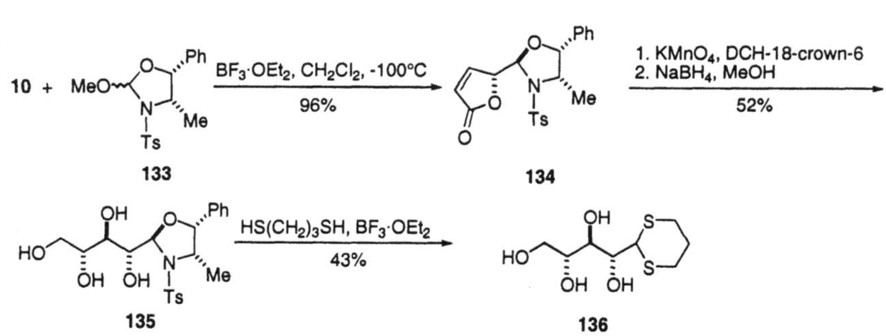

Scheme 27.

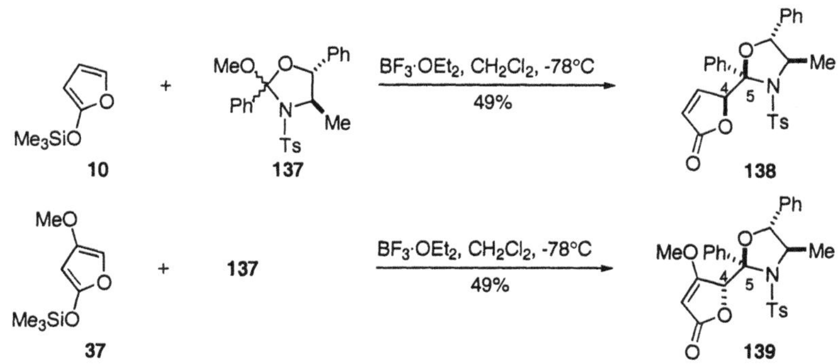

Scheme 28.

the sole (4*S*,5*R*)-adduct **138** in 49% yield. Remarkably, however, when the same oxazolidine **137** was reacted with tetronic acid-related silyloxy diene **37**, reverted facial selectivity of the diene pro-stereogenic carbon was observed, resulting in exclusive formation of the (4*R*,5*R*)-product **139**. To assess the synthetic potential of the butenolide adducts of type **138** and **139**, conjugate addition reactions were performed on these substrates, involving organocuprate reagents or carbanions derived from arylaldehyde dithioacetals and, as expected, high *trans* selectivity relative to the C-4 stereocenter was observed.

In a study directed toward the synthesis of naturally occurring goniobutenolides A and B, which were thought to possess a *threo* relationship between the two vicinal hydroxyls, Sharpless and coworkers[58] utilized the well-experienced asymmetric dihydroxylation of olefins (AD) procedure in order to install the hydroxyl functions as well as the proper chirality of the targets. Mukaiyama coupling of cinnamyl aldehyde dimethyl acetal **140** with 2-(trimethylsilyloxy)furan (**10**), followed by refluxing of the aldol adducts in glacial acetic acid, furnished a 1:1 mixture of (*E*,*Z*)-configured triene **141** and (*E*,*E*)-configured triene **142** in 80% combined yield (Scheme 29). Exposure of alkenes **141** and **142** to AD-β mix resulted in efficient and highly enantioselective (99% ee) formation of *threo-diols* **143** and **144**, respectively. Since the spectral and chirooptical data of **143** and **144** did not match those of goniobutenolides A and B, it was assumed that the C-6 and C-7 hydroxyls within the natural compounds possess an *erythro* relationship.

To validate this hypothesis, the same research group proceeded in synthesizing the *erythro* goniobutenolides **147** and **148**. This time, installation of chirality was effected prior to the condensation reaction. The required *erythro* diol functionality dictated the use of the (*Z*)-ethyl cinnamate olefin **145**, which was transformed to enantioenriched diol **146** upon treatment with AD/DHQ-IND mix (63% ee), followed by acetonidation and ester-to-aldehyde reduction (DIBAL-H). Additive-eliminative coupling of **146** with **10** allowed the synthesis of a mixture of separable (*Z*)- and (*E*)-olefins, which were finally deprotected to furnish butenolides **147** and

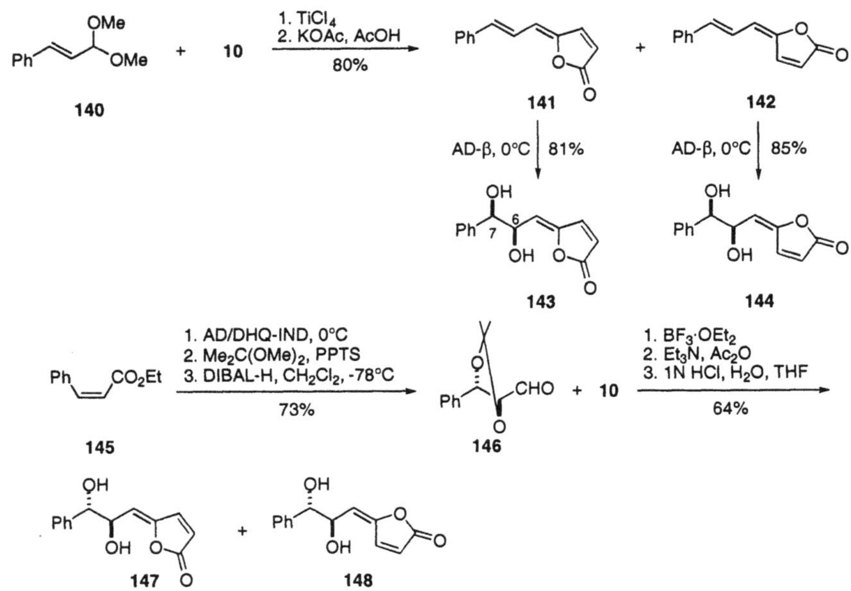

Scheme 29.

148 (the enantiomers of goniobutenolide A and B, respectively), whose structure firmly established the relative and absolute configuration of the natural products.

Similar chemistry was developed by Ko and Lerpiniere[59] in order to obtain highly enantioenriched goniobutenolide A (**153**) and goniobutenolide B (**154**). Starting from *threo*-diol **149** (Scheme 30), in turn obtained in 95% ee from protected (*E*)-cinnamyl alcohol by Sharpless AD procedure, the cyclic sulfate intermediate **150** formed, which underwent clean rearrangement-opening reaction to afford *erythro*-configured diol **151** in 79% yield. This sulfide was then transformed to

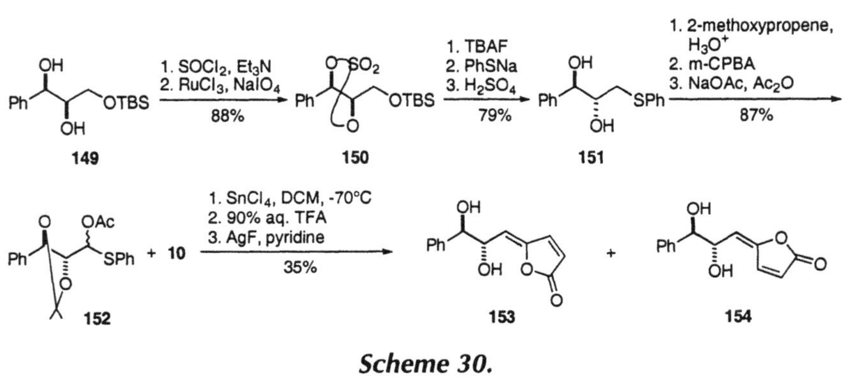

Scheme 30.

acetoxysulfide **152** by conventional chemistry. $SnCl_4$-catalyzed reaction between TMSOF **10** and **152** gave a mixture of protected and unprotected diastereomeric adducts, which were all treated with AgF in pyridine to afford a 1.6:1 separable mixture of goniobutenolide A (**153**) and goniobutenolide B (**154**) in a 35% combined overall yield from **152**.

Of note, unlike the majority of the processes discussed in this and the other sections, both plans to goniobutenolides A and B detailed in Schemes 29 and 30 adopted addition–elimination protocols where the two new stereocenters created during the key carbon–carbon addition (i.e., C-4 and C-5 in **141**, **142**, **153**, and in **154**) were lost during the subsequent (or concomitant) elimination stage. This renders the issue of the diastereocontrol of the TMSOF-to-aldehyde coupling ininfluent, since the chirality of the products is related either to the chirality resident within the aldehyde substrates, or is implemented by asymmetric catalysis at a later stage of the synthesis.

A highly stereoselective and efficient total synthesis of (+)-goniofufurone (**157**), a cytotoxic styryl-lactone isolated from the stem bark of *Goniothalamus giganteus*, was recently accomplished by Hanaoka,[60] exploiting, as a key operation, the diastereoselective coupling of TMSOF **10** with a suitable homochiral aldehyde precursor (Scheme 31).

Thus, aldehyde **155** was condensed with **10** under chelation-controlled conditions [$Ti(OPr^i)_2Cl_2$, CH_2Cl_2] giving 4,5-*anti*-5,6-*syn*-configured butenolide **156** as the dominant adduct. Successive desilylation and debenzylation, followed by spontaneous ring closure and isomerization of the C-4 stereocenter (TBAF) afforded (+)-goniofufurone (**157**).

Enantiomerically pure nitrones, easily accessible from the corresponding aldehyde derivatives, represent a useful class of electrophilic precursors to access different kinds of nitrogen-containing chiral units. Very recently,[61] the *N*-benzyl nitrone of (*S*)-lactaldehyde **158** was exploited as a functional three-carbon synthon en route to highly oxygenated δ-amino lactone blocks (Scheme 32). After extensive investigation, optimal conditions were found to couple 2-(trimethylsilyloxy)furan

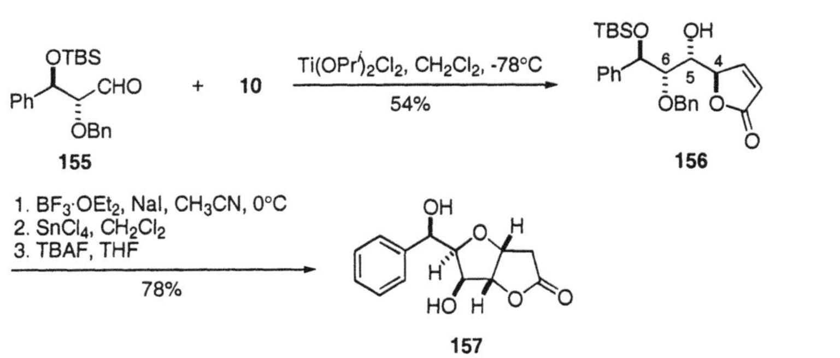

Scheme 31.

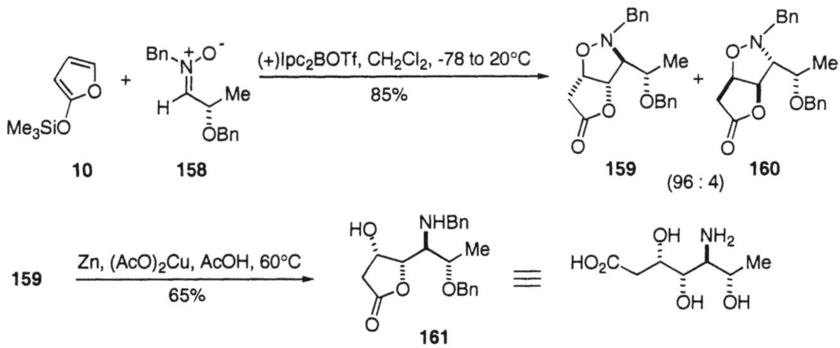

Scheme 32.

(**10**) to nitrone **158**, consisting of use of stoichiometric amounts of (+)-Ipc_2BOTf or $TiCl_4$ in CH_2Cl_2. There, the product consisted of a 96:4 mixture of 4,5-*anti*-5,6-*anti*-isomer **159** and 4,5-*anti*-5,6-*syn*-derivative **160**. Remarkably, the study highlighted that, in all instances, 4,5-*anti*-configured stereoisomers dominated, the 5,6-*syn* stereodisposition being preferred when catalytic TMSOTf was employed, and the 5,6-*anti* relationship favored with catalytic ZnI_2 or stoichiometric Lewis acid promoters. To assess the utility of these isoxazolones, **159** was subjected to reductive cleavage of the *N,O* linkage using the Zn/Cu couple, resulting in formation of lactone **161**, a synthetic equivalent of L-*altro*-β,γ,ε,-trihydroxy-δ-aminoheptanoic acid.

In a study directed to develop the chemical potential of a new chiralized derivative of glyoxylic acid, namely the Oppolzer's sultam-derivatized aldehyde **162**, Bauer[62] synthesized quite interesting six carbon butenolides **163** and **164**, as shown in Scheme 33. Thus, reaction of **162** with **10** in the presence of 3 mol% of $Eu(fod)_3$ in a 99:1 methylene chloride/methanol solvent mixture resulted in formation of products **163** and **164** in 90% combined yield, with *anti*-isomer **163** predominating (95:5 dr). Noteworthy is the high diastereoselection displayed by this aldol-type addition which, contrary to what was observed for the majority of related additions, shows marked *anti* preference. Likely, this unusual behavior can be attributed to

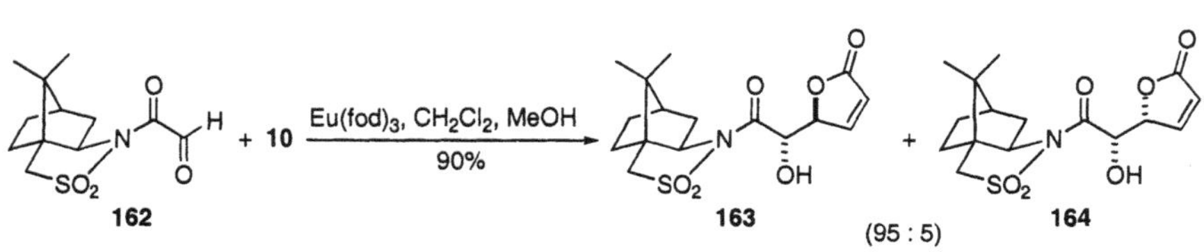

Scheme 33.

the powerful directing effect of the sultam moiety within **162**, which overrides the intrinsic bias of these systems towards *syn*-disposed derivatives.

B. Syntheses of Racemic or Achiral Compounds

Herein compiled are those examples emphasizing diastereoselective nucleophilic additions of silyloxy furan reagents to various racemic or achiral electrophilic species, giving rise to racemic compounds or achiral derivatives, where the newly formed stereocenters are lost during the synthesis.

The reaction between TMSOF **10** and acyclic *N*-acyl-*N,O*-acetals to give chemically interesting aminobutenolides was investigated by Harding.[63] As an example (Scheme 34), $BF_3 \cdot OEt_2$-catalyzed addition of **10** to representative hemiaminal **165** produced racemic *syn*-butenolide (±)-**166** in 57% yield, accompanied by a minor amount (~10%) of the corresponding *anti*-configured isomer. The reaction was easily extended to a variety of related *N,O*-acetals, resulting in preparation of a wide number of adducts, where the 4,5-*syn*-isomers invariably dominated.

In situ generated cyclic five-membered iminium electrophiles have been exploited by different research groups to assemble bicyclic pyrrolidine–furanone subunits typically found in the *Stemonaceae* alkaloids. In a preparatory study directed toward total syntheses of representative members from the *Stemona* family (e.g. croomine (**2**), vide supra), Martin[64] explored the Mannich-type reactions of 2-(trimethylsilyloxy)furan (**10**) and 3-methyl-2-(*tert*-butyldimethylsilyloxy)furan (**34**) with the *N*-acyliminium ion generated from ethoxypyrrolidine **167**. As shown in Scheme 35, the coupling reactions proved rather stereoselective, furnishing

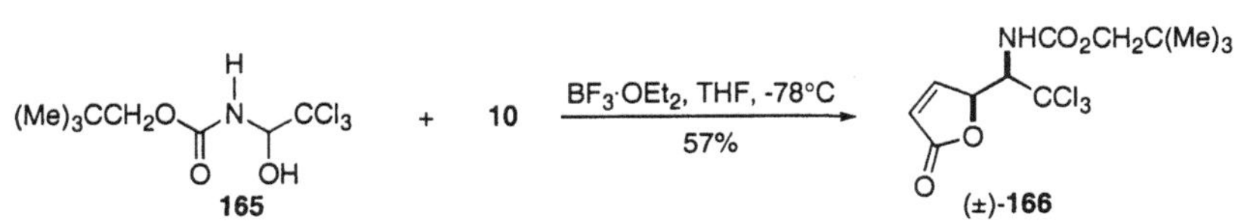

Scheme 34.

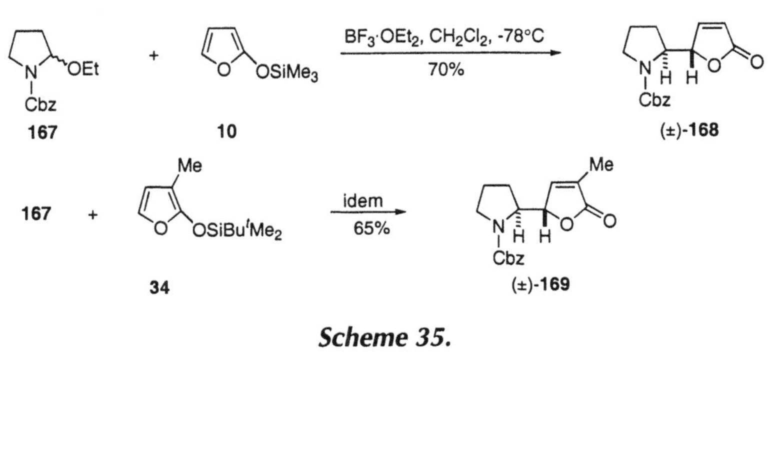

Scheme 35.

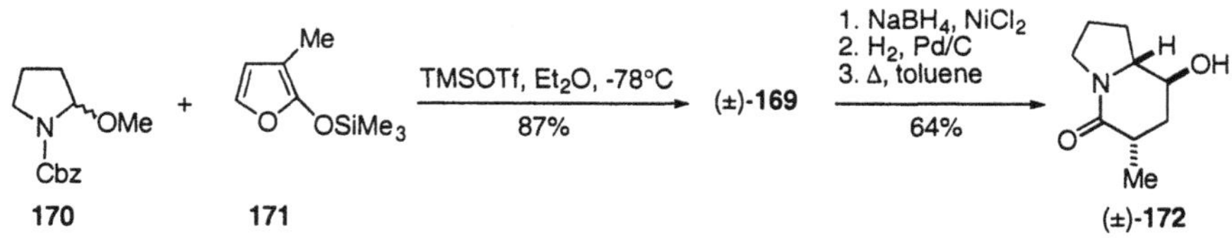

Scheme 36.

binuclear adducts (±)-**168** and (±)-**169**, respectively. The observed preference for *threo*-disposed derivatives (5:1 to 8:1) is consistent with the previously discussed results involving chiral iminium species.

The same approach was also independently investigated by Morimoto[65] by coupling silyloxyfuran **171** to **170** under a range of Lewis-acid catalysts (Scheme 36). Preparatively useful diastereoselection favoring *threo*-adducts (90:10 dr) was observed with TMSOTf, $Sn(OTf)_2$, and $Ph_3P{=}O{\cdot}Tf_2O$ mix in diethyl ether. The major isomer (±)-**169** was transformed, by simple chemistry, to racemic indolizidinone (±)-**172** via double bond transfer hydrogenation, removal of the Cbz-protection, and thermal γ-lactone to δ-lactam ring expansion.

The same author[66] developed an efficient synthetic method to racemic γ-lactone substituted 1-azabicyclo[5.3.0]decane unit (±)-**178**, which represents the tricyclic core system of various representatives of the *Stemona* alkaloids (Scheme 37). As in the above discussed studies, the fundamental move was the highly stereocon-

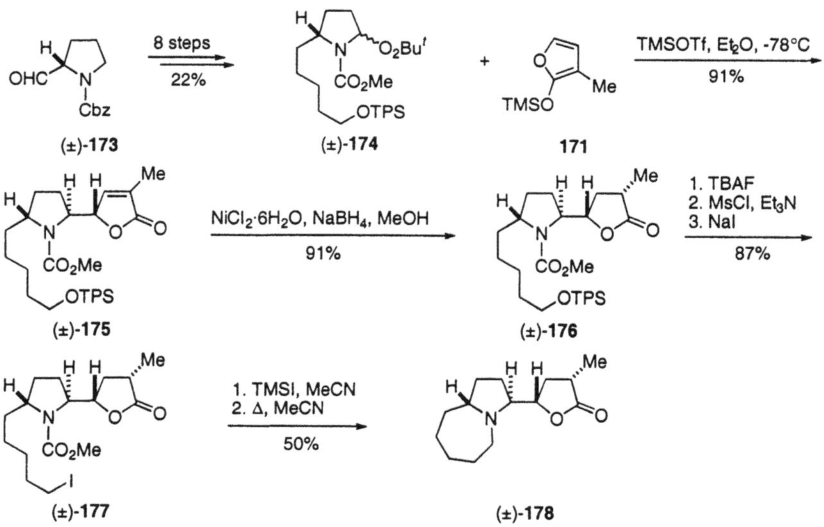

Scheme 37.

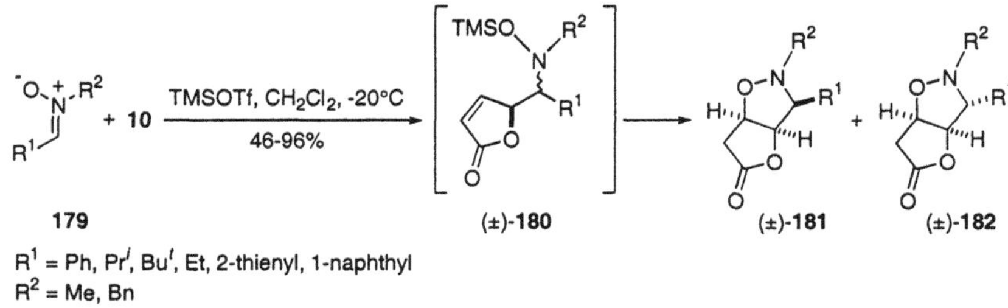

Scheme 38.

trolled addition of 3-methyl-2-(trimethylsilyloxy)furan (**171**) to a long-chain substituted iminium species generated from *N*-carboxymethyl pyrrolidine peroxide (±)-**174**, which was prepared in eight steps (22% overall yield) from racemic prolinal (±)-**173**. In the event, the wanted *threo*-adduct (±)-**175** was obtained in high yield (TMSOTf as catalyst), which was stereoselectively reduced to saturated lactone (±)-**176** by $NaBH_4/NiCl_2$ treatment. The creation of the seven-membered ring within the target compound required intramolecular nucleophilic displacement of the terminal hydroxyl by the pyrrolidine nitrogen, and this was effected by changing iodine for OH to (±)-**177**, deprotection of the *N*-methoxycarbonyl group, and thermal annulation to (±)-**178**.

Remarkable chemistry was developed by Trombini[67] utilizing aldonitrones of type **179** as electrophilic sources in addition processes with silyloxyfurans. As depicted in Scheme 38, a variety of nitrones **179** were found to react with TMSOF **10** in the presence of a catalytic amount of TMSOTf giving *syn*-configured bicyclic isoxazolidines (±)-**181**, accompanied by variable amounts of the corresponding *anti*-isomers (±)-**182**. Mechanistically, it was assumed that labile butenolide adducts of type (±)-**180** formed first, which underwent oxa-Michael ring closure to bicyclic species. Indeed, lactones (±)-**180** were detected by IR and ^1H NMR analyses, but any attempt to isolate them failed, owing to rapid conversion to annulated adducts. The stereochemical course of these reactions is noteworthy. While *C*-aryl-*N*-methyl nitrones preferentially led to the *syn*-products (±)-**181** (55–92% de), *C*-alkyl-*N*-benzyl derivatives showed an inverted preference for the *anti* products (±)-**182** (40–76% de), and *C*-alkyl-*N*-methyl congeners didn't exhibit any selectivity.

An efficient and scalable total synthesis of racemic mitomycin C, a potent agent extensively used for cancer chemotherapy, was executed by Fukuyama[68] taking advantage of the silyloxydiene-based chemistry. The optimized synthesis, illustrated in Scheme 39, follows a precedent on the same subject, adopting a similar plan.[69] The first maneuver was the $SnCl_4$-assisted coupling of 5-(ethylthio)-2-(trimethylsilyloxy)furan (**184**) to readily available chalcone **183** giving, after addition of pyridine, the desired enol ether (±)-**185**. Upon heating, intramolecular

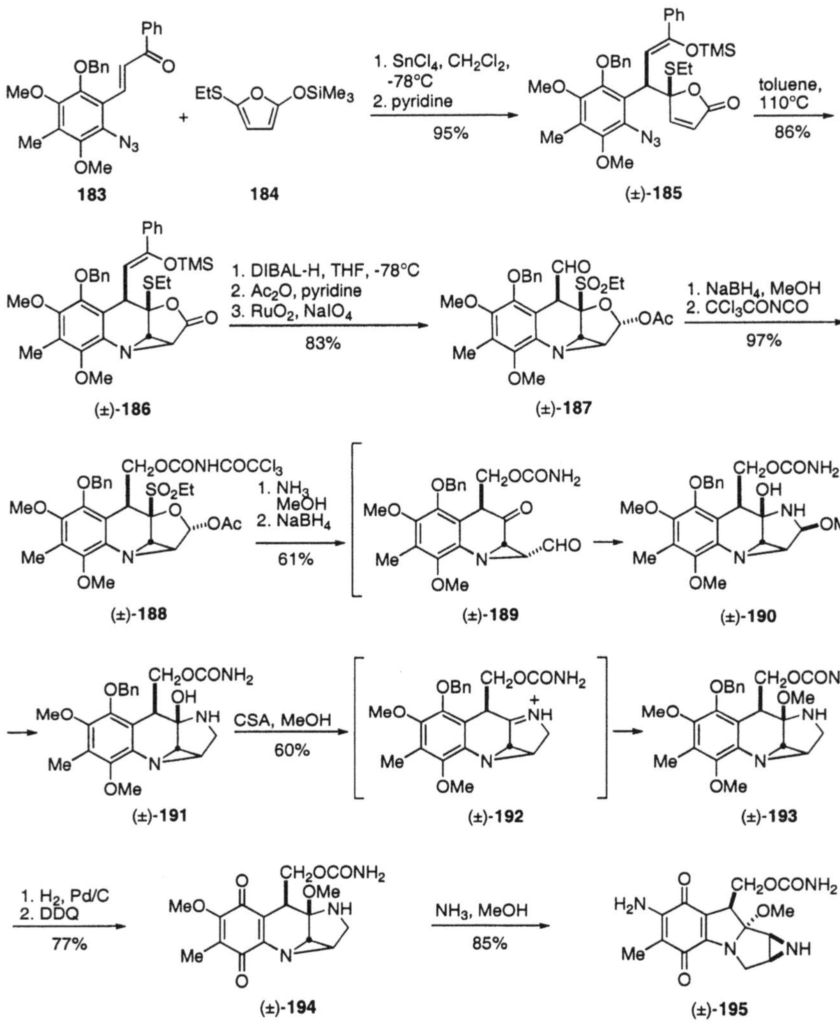

Scheme 39.

azide–olefin cycloaddition occurred to give aziridine (±)-**186**, exclusively. Reduction of the lactone moiety, followed by acetylation of the lactol so formed, and RuO_2-catalyzed periodate oxidation furnished aldehyde (±)-**187** in 83% yield. The formyl group was then reduced with $NaBH_4$, and the resulting alcohol was protected with trichloroacetyl isocyanate to give carbamate (±)-**188**.

When treated with methanolic ammonia, compound (±)–**188** underwent ammonolysis to give intermediate (±)-**190** via keto aldehyde (±)-**189**. Addition of

$NaBH_4$ then produced the requisite hemiaminal (±)-**191** in 61% overall yield from (±)-**188**. The required angular methoxy group of (±)-**193** was introduced via iminium ion (±)-**192** by acidic treatment in methanol. Hydrogenolytic removal of the benzyl ether followed by DDQ cleanly afforded racemic isomitomycin A [(±)-**194**], which was rearranged into racemic mitomycin C [(±)-**195**] by treatment with saturated methanolic ammonia. This rearrangement, typical in the mitomycin family, simply involves the nucleophilic attack of the pyrrolidine nitrogen to the *N*-linked quinone carbon with concomitant breakdown of the carbon–aziridine linkage.

During an extended research program aimed at developing short syntheses of pyranonaphthoquinone antibiotics, Brimble advantageously exploited conjugate addition of 2-(trimethylsilyloxy)furan (**10**) to a variety of quinone compounds, leading to furobenzo-furan and furonaphtho-furan templates. As an example of this remarkable chemistry,[70–73] an expedient synthesis of deoxyfrenolicin (±)-**201** is reported in Scheme 40.

The synthesis commenced with uncatalyzed conjugate addition of **10** to naphthoquinone **196**, leading to adduct (±)-**198** in 60% yield, through the intermediacy of butenolide (±)-**197**. Subsequent rearrangement of (±)-**198** to furo[3,2-*b*]naphtho[2,3-*d*]pyran (±)-**199** was then effected by exposure to ceric ammonium nitrate in aqueous acetonitrile. Hydrogenation of the hemiacetal (±)-**199** over palladium on carbon, followed by methylation (CH_2N_2), afforded methyl ester (±)-**200** in 71% yield, by a procedure involving concomitant reduction of the hemiacetal moiety and hydrogenolysis of the γ-lactone ring. It now remained to

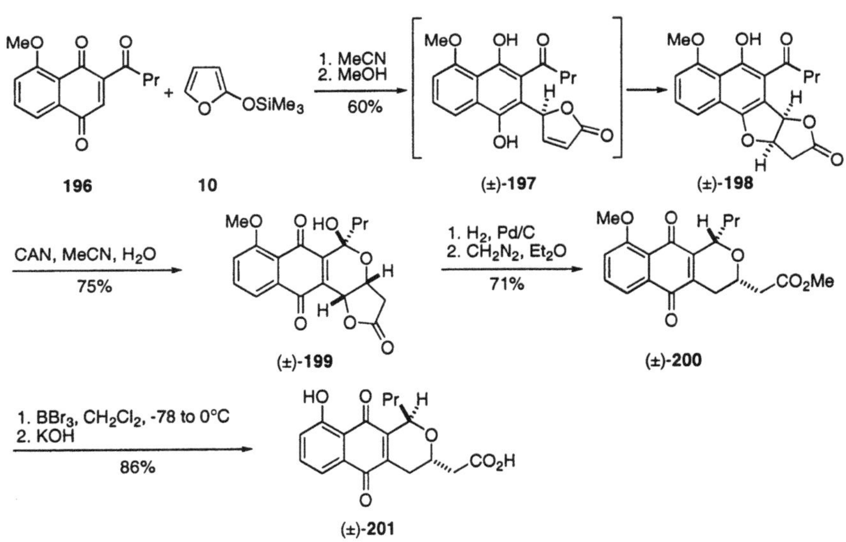

Scheme 40.

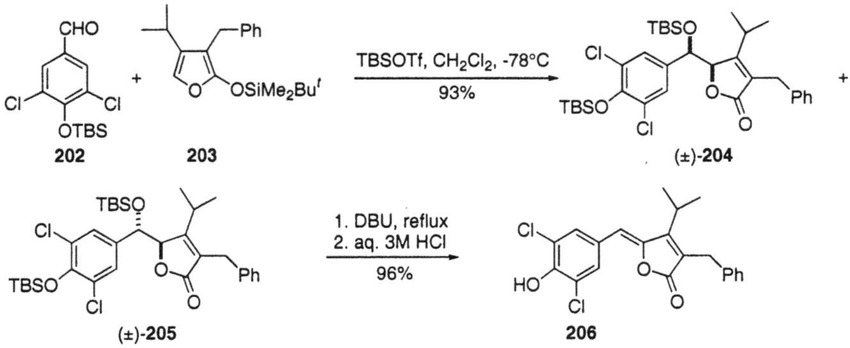

Scheme 41.

convert methyl ester (±)-**200** into natural deoxyfrenolicin (±)-**201**. Noteworthy, BBr_3 treatment allowed deprotection of the methyl ester with the concurrent reaction being the epimerization at C-1. Hydrolysis of the methyl ester within (±)-**200** using KOH finally afforded the target compound (±)-**201**. Similar chemistry allowed the same author to prepare a series of interesting quinone heterocycles, comprising 5-*epi*-7-deoxykalafungin and 5-*epi*-7-*O*-methylkalafungin.[72]

In a series of papers, Boukouvalas and colleagues[74–76] investigated, as a key reaction, the γ-homologation between 4-substituted or 3,4-disubstituted silyl enol ethers and achiral prochiral aldehydes. In all cases, the addition maneuvers resulted in preferential formation of racemic *syn*-adducts accompanied by minor quantities of their *anti* counterparts, with *syn/anti* ratios ranging from 2.7:1 to 13.0:1. The subsequent eliminative procedures effected on the mixture of *syn*- and *anti*-isomers led to the elimination of the stereocenters within the products, affording achiral compounds which, indeed, represent interesting, biologically active natural targets.

Thus, for example,[74] TBSOTf-promoted addition of 3-benzyl-4-isopropyl-2-(*tert*-butyldimethylsilyloxy)furan (**203**) to aromatic aldehyde **202** (Scheme 41) directly produced OTBS-protected *syn*- and *anti*-lactones (±)-**204** and (±)-**205** (2.7:1 ratio) in 93% yield. DBU-promoted *cis*-selective elimination, followed by acidic quenching, then afforded achiral nostoclide I (**206**), a naturally occurring cytotoxic lactone.

By adopting a conceptually similar approach, the same authors[75,76] succeeded in synthesizing the potent antibiotic patulin (**212**) and its biogenetic precursor, neopatulin (**216**) (Scheme 42). Aldol reaction of silyl enol ether **208** with benzyloxyacetaldehyde (**207**) under $BF_3 \cdot OEt_2$ catalysis provided α,β-unsaturated lactone (±)-**209** as a mixture of diastereoisomers (4.5:1 *syn/anti* ratio). Protection of the secondary hydroxyl as a pivaloyl ester and desilylation furnished (±)-**210**, which was conveniently cyclized to racemic (±)-**211** by oxidation of the primary alcohol to an aldehyde (TEMPO), followed by annulation. Treatment of (±)-**211** with DBU

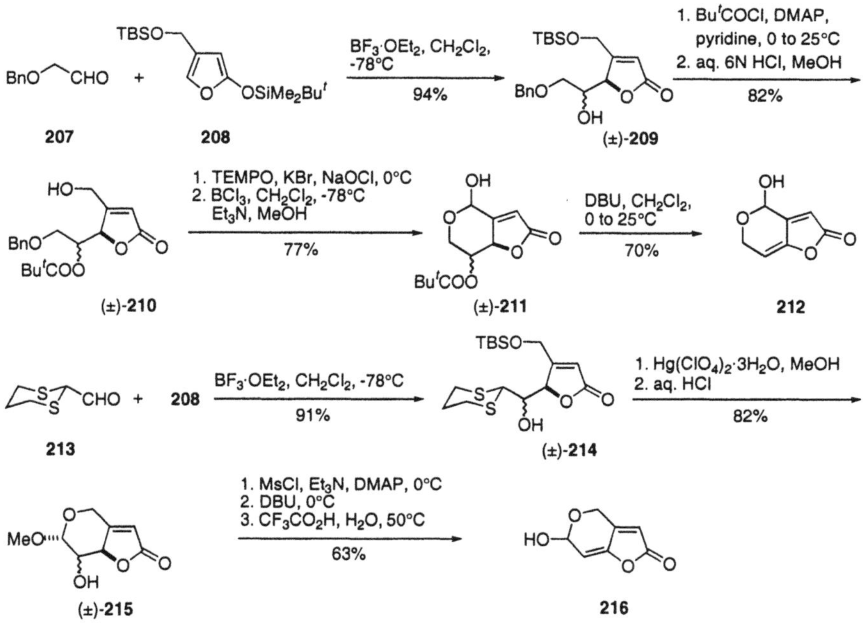

Scheme 42.

finally furnished patulin (**212**) in a good 42% overall yield for the entire six-step sequence from **207**.

The same enol ether **208** reacted with 2-formyl-1,3-dithiane (**213**) (BF_3 etherate) to give a 13:1 mixture of *syn/anti* alcohols (±)-**214**, which were readily transformed to neopatulin (**216**) by a five-step sequence involving dethioacetalization and desilylation to afford intermediate mixture (±)-**215**, followed by dehydration (MsCl; then DBU) and acidic hydrolysis (aq. CF_3CO_2H).

Unstable, highly electrophilic alkoxycarbonyl quaternary salts of triazine regioselectively reacted with silyl enol ethers, including TMSOF **10**, to give intermediary adducts which, in some instances, could be oxidatively transformed into 5-substituted 1,2,3-triazines in good yields.[77] As an example, Scheme 43 displays

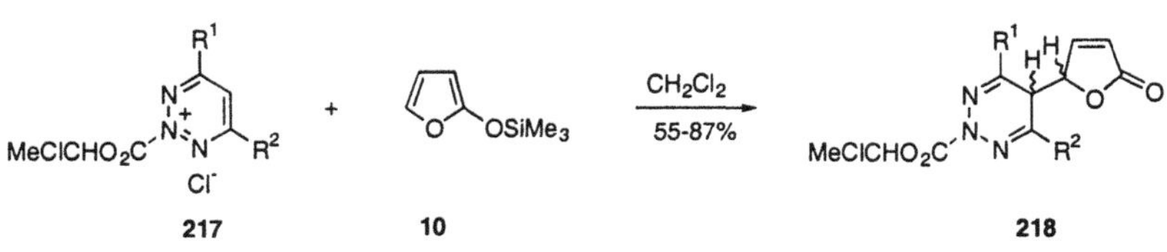

Scheme 43.

how 2,5-dihydrotriazines **218** were produced, by starting with **10** and triazine salts **217**. Disappointingly, this reaction proved unselective, affording a mixture of four diastereoisomers, which also failed to give the expected triazines, due to rapid decomposition of the adducts **218** under the oxidative work-up.

V. DIASTEREOSELECTIVE REACTIONS USING SILYLOXY PYRROLES

The 2-silyloxy pyrrole scaffold, unlike its oxygen and sulfur counterparts, embodies, as premium feature, an heteroatom–the nitrogen–amenable to further substitution. This special characteristic, while expanding the structural diversity of the silyloxy pyrrole family with an additional substitution site, renders these reagents prone to incorporate a range of suitable chiral inductors by anchoring them to the free pyrrole nitrogen. As a consequence, two basic approaches exploiting these reagents can be envisioned to propagate chirality into a given nonracemic target, the conventional induction from intrinsically chiral precursors (the so-called chiron approach), and the induction from temporarily chiralized precursors (the auxiliary driven technique). The bulk literature achievements in this domain fall into two distinct categories: syntheses using inherently chiral electrophilic sources (Section V. A.), and syntheses using temporarily chiralized silyloxy pyrroles (Section V. B.). In addition, a few reports investigated diastereoselective reactions involving achiral–prochiral electrophilic components and achiral silyloxy pyrrole compounds, and pertinent examples will be discussed separately (Section V. C).

A. Induction from the Chiral Electrophilic Precursor

Densely functionalized aminated organic compounds, including hydroxylated amino acids, azasugars, amino sugars, and a variety of monocyclic and bicyclic alkaloidal systems form a wide family of important molecules whose role in biology and medicinal fields is well documented. 2-Silyloxy pyrroles, such as TBSOP **13**, can be envisioned as ideal scaffolds with which a variety of aminated derivatives of biological interest can be generated by paralleling the previously disclosed chemistry involving the furan cousins.

The synthetic potential of silyloxy pyrrole **13** was first explored by our research group aimed to synthesize highly oxygenated pyrrolidinones, precious intermediates for the assembly of polyhydroxylated pyrrolizidine-type alkaloids.[14,78]

Treatment of 2,3-*O*-isopropylidene-D-glyceraldehyde (**44**) in anhydrous Et_2O with TBSOP **13** at –85 °C in the presence of $SnCl_4$ gave crystalline D-*arabino*-configured α,β-unsaturated γ-lactam **219** as the predominant reaction product in a gratifying 80% isolated yield (Scheme 44). Doubling the scope of the synthesis, when **44** was allowed to react at the same temperature with **13** in Et_2O in the presence of $BF_3{\cdot}OEt_2$, reversal of stereochemistry occurred, resulting in predominant formation of crystalline D-*ribo*-configured epimer **220** (70% yield), along with

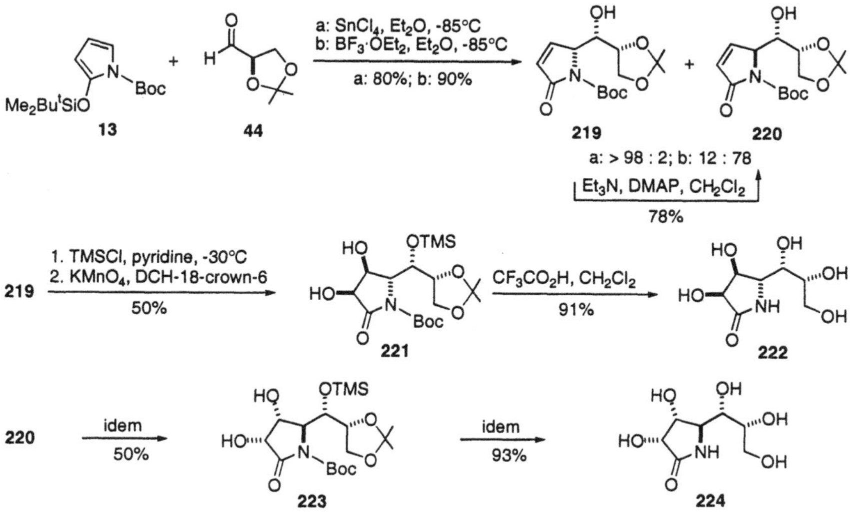

Scheme 44.

less than 20% of **219**. Also, clean and almost quantitative epimerization at C-4 was observed when lactam **219** was subjected to treatment with Et_3N in CH_2Cl_2 at room temperature in the presence of DMAP, and this provided a good alternative preparation of thermodynamically more stable D-*ribo* lactam **220**. At this point, the stereochemistry of both the adducts was investigated by single crystal X-ray analysis, thus confirming the relative disposition of the two newly formed stereocenters.[78] Next, after protection of the free OH group as trimethylsilyl ether, the double bond of both epimers was selectively dihydroxylated ($KMnO_4$) producing diastereomeric pyrrolidinones **221** and **223** in 50% isolated yields. In the event, the compounds were obtained as 2,3-*cis*-3,4-*anti*-diastereoisomers only, since the functionalization of the double bond is strictly governed by the presence of a bulky substituent at C-4 which hinders the *syn* face of the lactam ring. Finally, the acetonide, TMS, and Boc protections in **221** and **223** were cleanly removed by treatment with 0.2 M trifluoroacetic acid in CH_2Cl_2 giving, after silica-gel chromatography, the free lactams **222** and **224** in 91 and 93% yields, respectively.

Synthetically, unsubstituted 2-silyloxy pyrroles of type **13** can be viewed as prochiral γ-anion equivalents of γ-amino butyric acid (GABA). In this prospect, a series of chirally defined seven-to-nine carbon alditolyl GABAs (e.g. compounds **226**, **229**, and **232**) was straightforwardly prepared,[79] via homologation of suitable aldehydo sugar precursors with *N*-(*tert*-butoxycarbonyl)-2-(*tert*-butyldimethylsilyloxy)pyrrole (**13**).

As shown in Scheme 45, saturated lactam adducts **225**, **228**, and **231** were first prepared by coupling of **13** with the respective aldehydes **44**, **227**, and **230**, followed

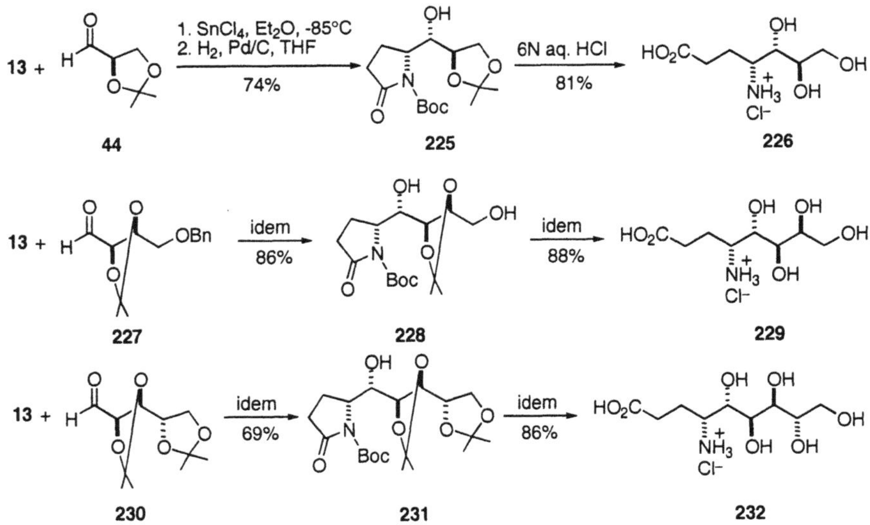

Scheme 45.

by saturation of the double bonds within the intermediary adducts. Noticeably, irrespective of the nature of the chirons involved, all homologative processes proved extremely regio- and stereoselective, with 4,5-*syn*-5,6-*anti*-isomers predominating and, as a rule, chirality transmittal from the aldehyde precursor was driven from the stereocenter at the α-carbon with negligible participation, if any, of the remainder chiral sites. Finally, acidic treatment (6N aq. HCl) provided hydrolytic ring opening and concomitant deprotection of the acid-labile acetonides, furnishing the expected 4-amino-2,3,4-trideoxyaldonic acids **226**, **229**, and **232** as hydrochloride salts.

The chemistry and biological implications of hydroxylated indolizidine and pyrrolizidine alkaloids, both natural and synthetic, have attracted enormous interest in recent years, and this is due, at least in part, to the unique value of these alkaloids as inhibitors of glycoprocessing enzymes and therapeutic agents.[80] In searching for a flexible protocol suitable to preparation of castanospermine-related compounds, our group envisioned a strategy based on the realization that the bicyclic core of the targets could be built up using a pyrrole reagent such as **13** to forge the five-membered ring component and a hydroxylated four-carbon aldehyde of type **227** to furnish the six-membered ring structure (Scheme 46).[81]

Optimally, threose **227** was treated with **13** in diethyl ether at −80 °C in the presence of $SnCl_4$. The addition occurred regio- and stereoselectively at the C-5 carbon of **13** to form crystalline α,β-unsaturated lactam **233** exclusively in 80% isolated yield. Treatment of **233** with TMSOTf in CH_2Cl_2 in the presence of thiophenol cleanly afforded lactam **234** (75%), which was hydrogenated and deprotected to compound **235** (Pd/C, NaOAc, THF) in 95% isolated yield. Lactam

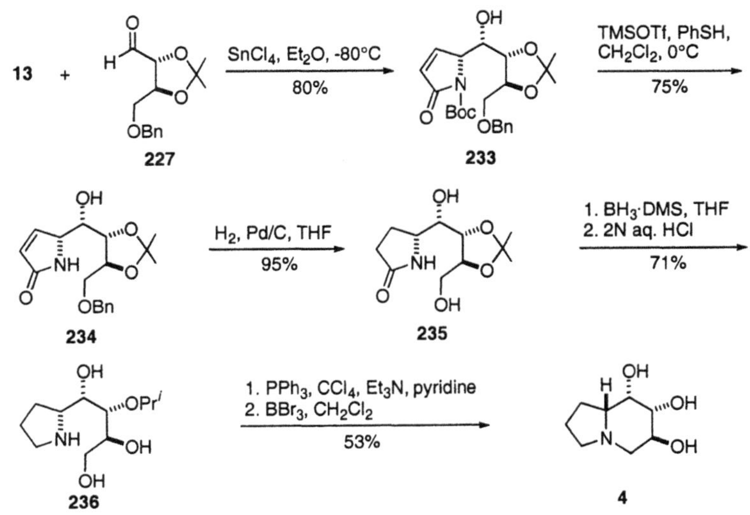

Scheme 46.

235 was directly exposed to an excess of BH_3·DMS in THF at room temperature and the crude amine–borane adduct thus formed was subjected to acidic treatment (HCl) at room temperature. Surprisingly, this treatment afforded isopropyl ether **236** (71%) likely to have arisen from reduction of the lactam with concomitant regioselective opening of the acetonide at the O-7–C(Me)$_2$ linkage and over-reduction. Amino alcohol **236** was ready for the final and crucial cyclization step. This was achieved by subjecting compound **236** to PPh_3–CCl_4–Et_3N in pyridine at room temperature. There was obtained, after ion-exchange resin purification and BBr_3-promoted dealkylation, the expected (6*S*,7*R*,8*S*,8a*R*)-6,7,8-trihydroxyindolizidine (1-deoxy-8-*epi*-castanospermine, **4**) in 53% isolated yield.

Silyloxy pyrrole **13** also served admirably to construct the skeleton of important hydroxylated pyrrolizidines, by starting with three-carbon aldehyde units. As an example, the divergent synthesis of two diastereomeric *cis*-1,2-dihydroxypyrrolizidines **240** and **241** is detailed in Scheme 47.[82]

The common matrix, the unsaturated lactam **219**, was first prepared, as previously described, by starting with **13** and **44**. The double bond within **219** was then hydrogenated and the protective groups removed by acidic treatment. This afforded triol **237** in 65% isolated yield. Exposure of **237** to methanesulfonyl chloride in pyridine gave rise to trimesyl derivative **238**, which underwent annulation to **239** by carbonyl reduction (BH_3·DMS) followed by DBU treatment. For **240** to be prepared, a simple enantioconservative demesylation reaction was adopted (6% sodium amalgam in isopropanol) whereas for diastereoisomer **241**, a double nucleophilic displacement with benzoate anion was selected, resulting in inversion

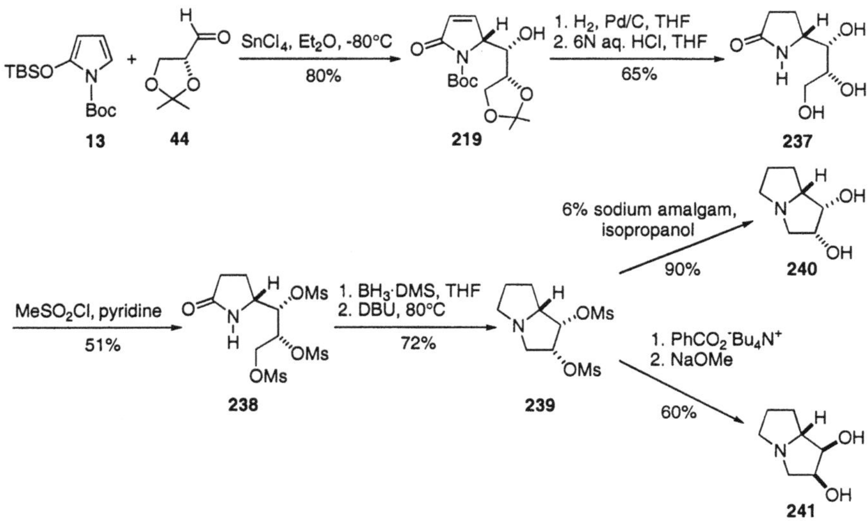

Scheme 47.

of configuration at the C-1 and C-2 stereocenters. In a strictly analogous manner, the enantiomeric counterparts of **240** and **241** were accessible by employing protected L-glyceraldehyde as a three-carbon chiral source.

As illustrated in the previous sections of this account, TBSOP **13** served as a versatile four-carbon building block to synthesize variously substituted nitrogen-containing compounds, through the intermediacy of structurally defined α,β-unsaturated lactam templates. Aimed at further exploring the potential of **13** en route to biologically important aminated compounds, it was reasoned that oxidative extrusion of the C-1 and C-2 carbon atoms in the unsaturated lactam precursors of type **242** (Scheme 48), via fission of the C-2–C-3 linkage, would provide a straightforward entry to hydroxylated α-amino aldehyde and α-amino acid derivatives. Accordingly, TBSOP **13** can be envisioned as a glycine anion equivalent.[83,84]

Successful implementation of this strategy to chiral syntheses of polyhydroxy-α-amino acids, a family of compounds strictly related to naturally occurring (+)-polyoxamic acid and the polyoxin complex, was attained starting from enantiopure sugar aldehydes. As an example, the synthesis of 4-*epi*-polyoxamic acid (**246**) is illustrated in Scheme 48. According to an optimal protocol, crystalline lactam **242**, easily available from **44** and **13** with > 95% de, was first subjected to double bond dihydroxylation with $KMnO_4$/18-crown-6 ether/CH_2Cl_2 reagent system. This gave lactam **243** as the sole stereoisomer in 65% isolated yield. Hydrolytic lactam opening (LiOH, THF) and subsequent oxidative diol fission at the C-2–C-3 linkage ($NaIO_4$) provided protected 2-amino-2-deoxy-D-arabinose **244** in 88% yield for the two consecutive reactions. Exposure of **244** to $NaIO_4$/catalytic RuO_2

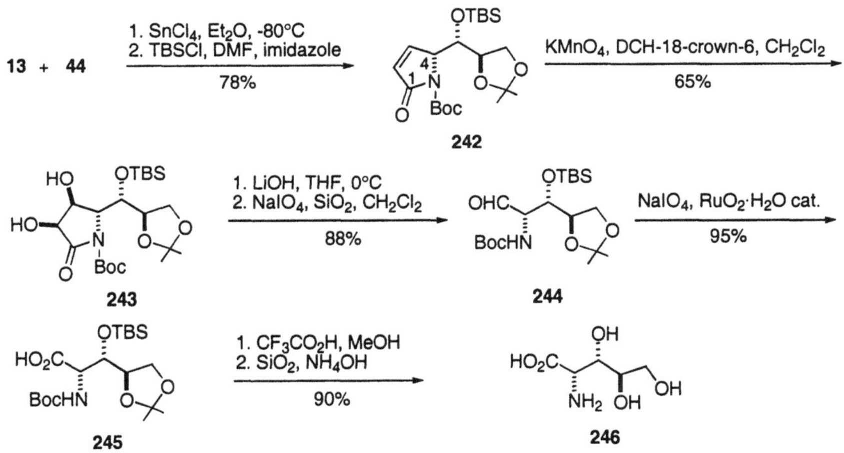

Scheme 48.

in CH_3CN-CCl_4-water-acetone solvent mixture furnished protected amino acid **245** almost quantitatively (> 95%), which was fully deprotected by 1:1 trifluoroacetic acid/methanol treatment to amino acid **246**. The same reaction protocol was successfully extended to other aldehydo-sugar derivatives easily obtainable from common precursors. Regardless of the aldehyde chirality and substitution, a wide variety of hydroxylated amino acids **247–251** (listed in Scheme 49) was obtained via the corresponding aminosugar intermediates in preparatively useful yields ranging from 25 to 34% for the complete sequences.

As a further extension of this technique,[85] arabinofuranosylglycine **254** was synthesized, in a direct manner, from benzylated *O*-acetylarabinofuranose **252**, utilizing **13** as a masked glycine anion equivalent (Scheme 50). Thus, reaction of **13** with **252** in Et_2O in the presence of $TrClO_4$ at 0 °C afforded unsaturated lactam **253** almost exclusively, which was directly transformed to the protected amino acid **254** by following the reaction sequence outlined for the acyclic amino acid **246**.

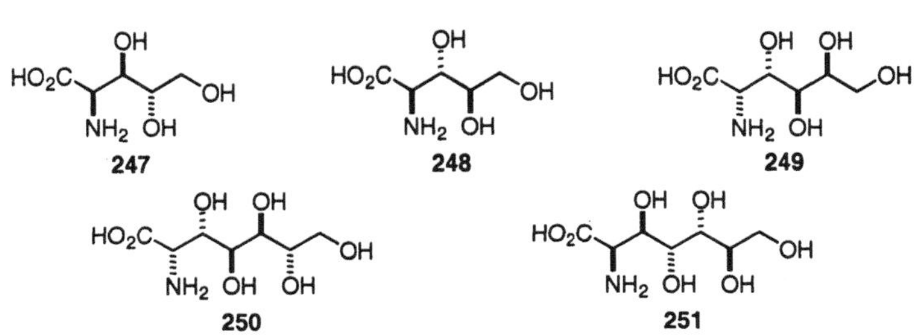

Scheme 49.

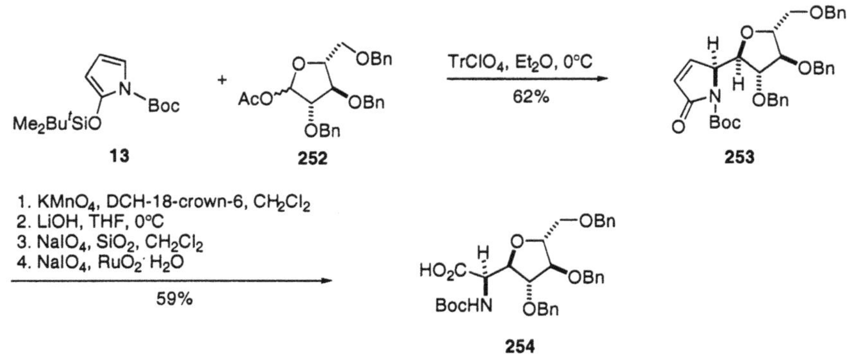

Scheme 50.

A closely related sacrificial synthetic plan led us to prepare the quite rare diaminoarabinose **259**, the sugar component of the naturally occurring antifungal antibiotic prumycin.[86] According to Scheme 51, the starting maneuver was the homologation of protected L-serinal **255** with TBSOP **13**. Remarkably, the BF_3 etherate-assisted Mukaiyama aldol reaction proved regioselective, furnishing, after aqueous $NaHCO_3$ quenching, an 80% yield of two 5-*O*-silylated adducts wherein 4,5-*syn*-lactam **256** predominated (75% de). At first, the formyl function of the five-carbon sugar **258** was installed by oxidative sacrifice of the C-1 and C-2 atoms within the seven-carbon precursor **256**.

Thus, selective dihydroxylation of the lone carbon–carbon double bond within **256** furnished diol **257**, which was sequentially subjected to hydrolytic ring opening and diol fission. There was obtained protected diaminopentose **258** in high yield, which was used as such during the next stages of the synthesis. The crucial annulation reaction was cleanly performed through selective deacetonidation by the mild reagent citric acid–methanol system, which also ensured concomitant

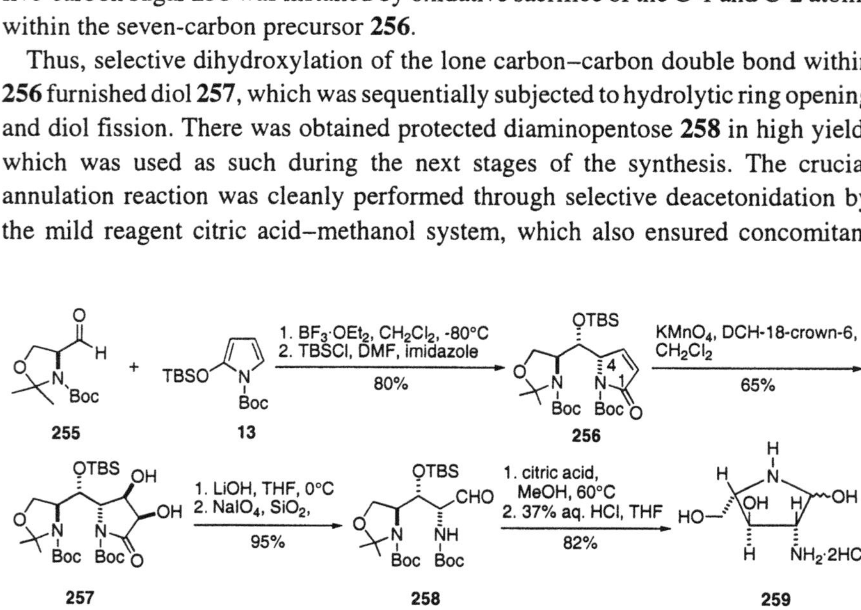

Scheme 51.

pyrrolidinose formation. The final unmasking of the *N*-Boc and *O*-TBS protective groups was effected by HCl in THF, affording the bis-hydrochloride salt **259**, which was isolated as an inseparable mixture of α- and β-anomers.

The synthetic potential of pyrrolic enol ether **13** was also demonstrated during the asymmetric synthesis of *trans*-2,3-*cis*-3,4-dihydroxy-L-proline (**265**), an unusual constituent of the adhesion protein of byssus, the silky linkage of the sea bivalve *Mytilus edulis* (mussel).[87] The route commenced with the preparation of adduct **260** from **13** and protected D-glyceraldehyde **44** under BF_3 etherate catalysis. Thus, as depicted in Scheme 52, treatment of D-*ribo*-configured lactam **260** with $KMnO_4$, as usual, followed by exposure of the diol so formed to dimethoxypropane in the presence of *p*-toluensulfonic acid, directly afforded lactam **261** in 60% yield.

The transformation of **261** into formyl pyrrolidinone **262** called for selective deprotection of the terminal acetonide (citric acid in methanol) and subsequent oxidative excision of the C-6 and C-7 carbon atoms (aq. $NaIO_4$). The formyl function within **262** was selectively reduced to a carbinol by $NaBH_4$, which was protected as TBS-ether **263**. For optimal conversion, the reduction of the lactam carbonyl to methylene was performed according to a two-step protocol, consisting of partial reduction to an hemiaminal ($LiEt_3BH$) and subsequent deoxygenation (Et_3SiH, $BF_3{\cdot}OEt_2$). There was obtained protected L-prolinol **264**, which was cleanly elaborated into the target amino acid **265** by conventional chemistry. As a whole, the plan emphasizes the use of the popular three carbon synthon **44** as a chiral CO_2H equivalent and the exploitation of **13** as a pyrrolidine ring surrogate.

Seven-carbon lactam adduct **219**, easily accessible from silyloxy pyrrole **13** and glyceraldehyde **44** (vide supra), was also employed to assemble interesting 3-

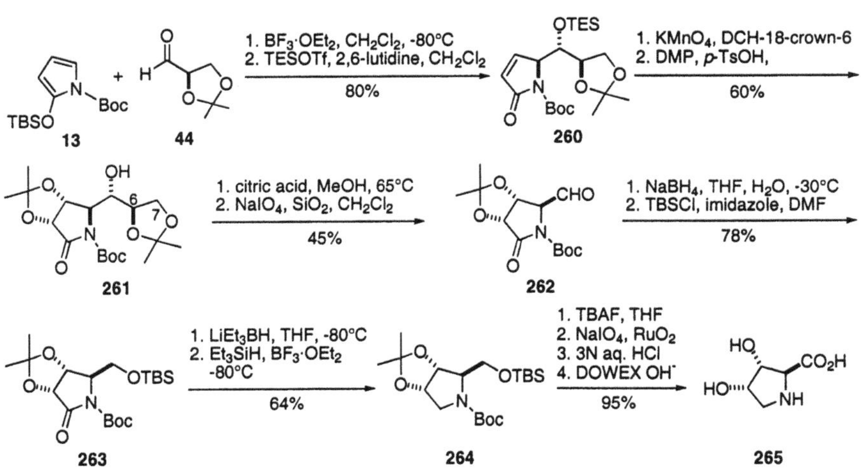

Scheme 52.

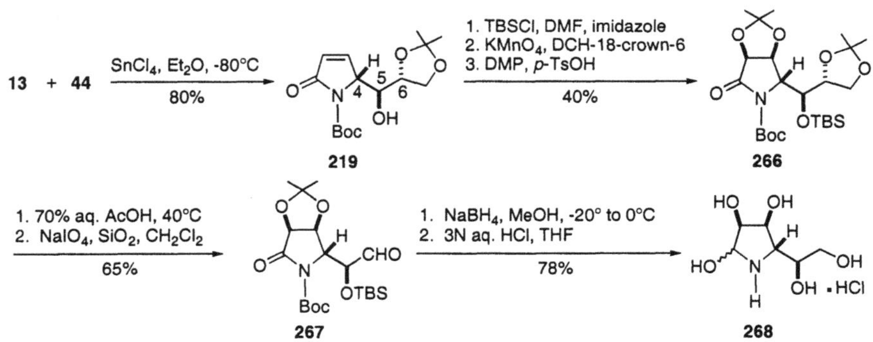

Scheme 53.

amino-3-deoxy- and 4-amino-4-deoxyhexoses, according to a divergent plan (Schemes 53 and 54).[88] To access 4-amino-4-deoxy-D-talose **268**, for example (Scheme 53), the opening reaction was the diastereoselective $SnCl_4$-promoted coupling between isopropylidene protected D-glyceraldehyde **44** and TBSOP **13**, furnishing crystalline 4,5-*syn*-5,6-*anti*-configured α,β-unsaturated lactam **219** in high yield and diastereoselectivity (de > 90%).

Subsequent protection of the free C-5 hydroxy group as TBS-ether, followed by stereoselective *cis-anti* dihydroxylation of the double bond and protection of the formed diol, then produced heptonolactam **266** in 40% yield for the three steps. One-carbon shortening at the right side of the molecule to aldehyde **267** was performed by first removing the terminal isopropylidene moiety within **266** (70% aq. AcOH) and then cleaving the C6–C-7 bond by periodate treatment under solid–liquid phase transfer conditions employing $NaIO_4$-impregnated wet silica gel. At this stage, all that remained was the reduction of both the formyl group and the lactam carbonyl, and this was concomitantly effected by $NaBH_4$ treatment in

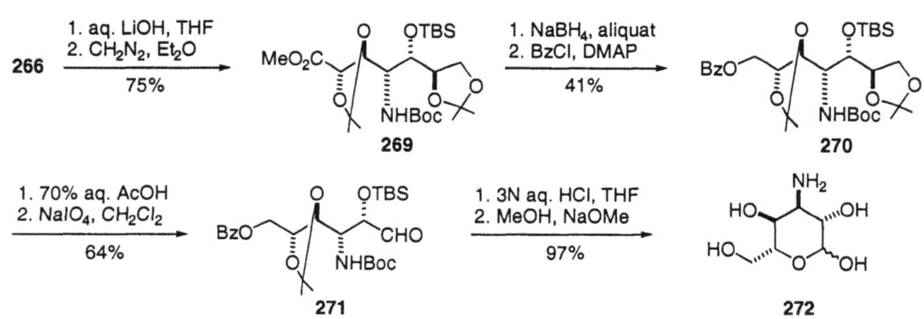

Scheme 54.

methanol. Protected 4-amino-4-deoxy-D-talose formed, which was fully deprotected to the hydrochloride **268** by HCl treatment.

To enlarge the scope of this chemistry, advanced intermediate **266** was also elaborated into D-altrose derivative **272**, as shown in Scheme 54. Here, the task was to transform the C-1 lactam carbonyl of **266** into the terminal CH_2OH of the target aminosugar **272**, and this was performed by a four-step protocol comprising hydrolytic ring opening (LiOH, THF) and methyl ester formation (CH_2N_2) to give the open-chain γ-amino ester **269**, followed by ester-to-carbinol reduction ($NaBH_4$, aliquat) and benzoylation of the hydroxymethyl function so formed. In the event, fully protected aminoheptitol **270** was obtained in 31% yield from **266**. In the final stages of the sequence, the isopropylidene protection at the right-hand terminus of the chain of **270** was selectively cleaved by acidic treatment (AcOH), and the resulting diol was oxidatively shortened by one carbon ($NaIO_4$) to generate the differently protected aldehydo-hexose **271** in 64% yield from **270**. Removal of all the acid-sensitive protective groups of **271** by aqueous HCl in THF, followed by MeONa-catalyzed debenzoylation, finally furnished 3-amino-3-deoxy-D-altrose (**272**), here depicted in its pyranose form, in almost quantitative yield.

A recent achievement of our group[89] was the diastereoselective synthesis of D-*erythro*-sphingosine, an essential component of glycosphingolipids, by adopting the well-experienced silyloxy pyrrole-based homologative technique. The chiral synthesis of **5** (Scheme 55) commenced with the preparation of D-*arabino*-configured lactam **219**, which was accomplished in 80% isolated yield and 95% de by $SnCl_4$-assisted coupling of D-glyceraldehyde **44** to **13**, as previously described. The protection of the free hydroxyl group within **219** as TBS-ether was then effected as usual by exposure to TBSCl in DMF in the presence of imidazole. Addition of the silylating mixture to **219** after 10 h at room temperature afforded the expected lactam **242** with only minor (less than 12%) loss of the chiral integrity. However, when the reaction mixture was left for longer (4 days), only a minor amount of **242** was isolated, with the concomitant reaction being the epimerization of the stereocenter at C-4 giving rise to the D-*ribo*-configured isomer **273**.

The preparation of **5** called for 4,5-*anti*-heptenonolactam **273** as the requisite precursor. According to our plan, removal of the C-1 and C-2 carbons at the left side of the molecule was required, and this was accomplished by a two-step maneuver comprising double bond dihydroxylation and oxidative fission of the C-2–C-3 diol linkage. Thus, selective *anti-cis* dihydroxylation of the double bond within **273** was accomplished by exposure to $KMnO_4$/dicyclohexano-18-crown-6 ether to give diol **274** in 68% yield, which was subjected to hydrolytic ring opening (LiOH, THF) and oxidative shortening by two carbon atoms ($NaIO_4$) to generate protected aminoribose **275** in 87% yield. Aldehyde-to-carbinol reduction ($NaBH_4$) and protection of the newly formed primary hydroxyl as TBDPS-ether furnished alditol **276** in 70% yield. For the advanced erythrose intermediate **277** to be prepared, additional sacrifice of the right-hand terminal carbon of **276** had to be performed via selective deacetonidation (70% aq. AcOH) and $NaIO_4$-promoted diol

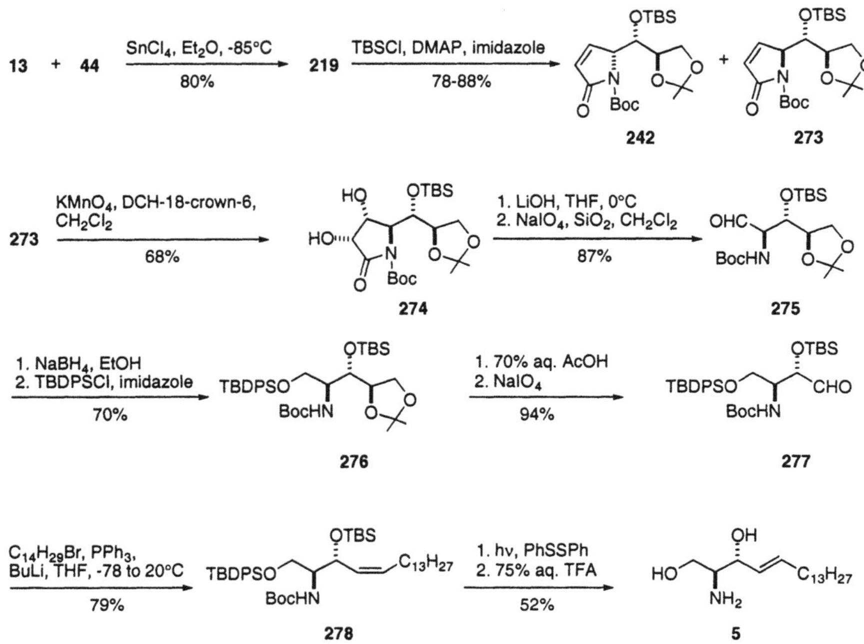

Scheme 55.

cleavage. In the event, the key aldehyde **277** was obtained in 94% yield for the two steps. All that remained was the creation of the suitable unsaturated hydrocarbon appendage by a Wittig homologation of **277** using tetradecylidene triphenylphosphorane ($C_{14}H_{29}Br$/BuLi). The reaction proved stereoselective, furnishing adduct **278** as an inseparable 98:2 *Z/E* isomeric mixture (79% yield). Photoinduced double bond isomerization in the presence of diphenyl sulfide produced the expected *E*-disposed olefin, which was finally deprotected to pure crystalline D-*erythro*-sphingosine (**5**) in 52% yield, which corresponds to a nice 11% overall yield for the 12-steps sequence from **44**.

In a study directed towards preparation of scantily studied azafuranose nucleosides, an ingenious approach was envisaged by starting with TBSOP **13**—the source of the heterocyclic core—and using D-glyceraldehyde **44** as a surrogate of the hydroxymethyl cation.[90] As a remarkable example, the synthesis of 4′-azauridine **286** is illustrated in Scheme 56.

At first, *anti-anti*-configured lactam adduct **220** was prepared, which was quickly transformed to pyrrolinone **280** by a sequence of five steps, comprising double bond saturation and acidic deprotection to **279**, followed by two-carbon shortening of the side chain and protection. As usual, unsaturated matrix **281** was obtained from **280** via enolate phenyl selenation–oxidation protocol. This material was subjected

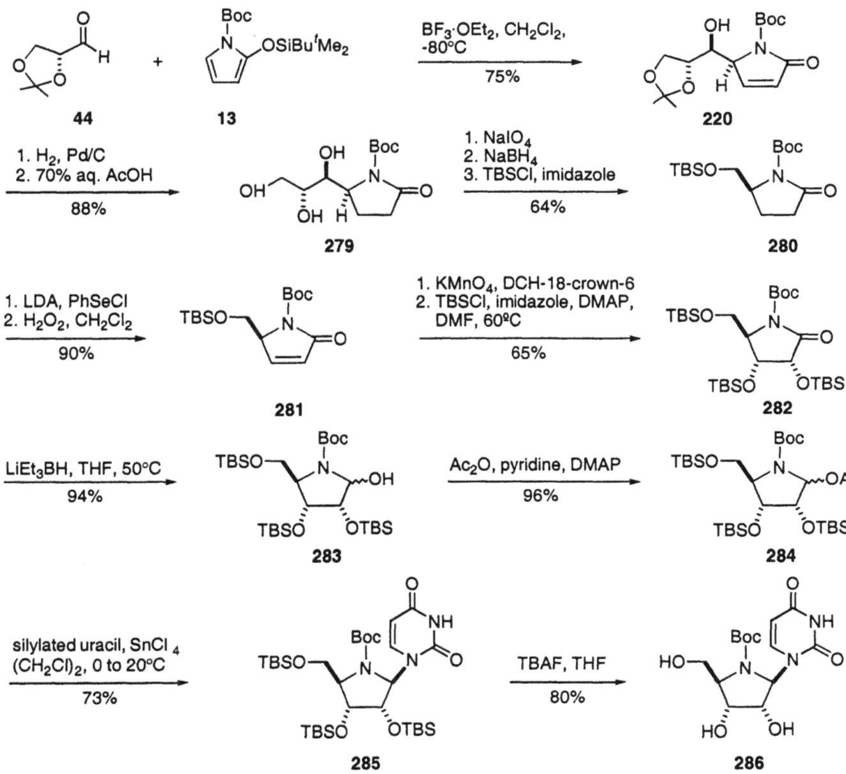

Scheme 56.

to stereospecific dihydroxylation ($KMnO_4$) and the crude diol so formed was protected to **282**, which was then converted to azaribose **283** by superhydride treatment. Exposure of **283** to Ac_2O/pyridine/DMAP provided **284** (96%), which was coupled with silylated uracil. This treatment provided protected 4′-azauridine **285** as planned in a gratifying 73% yield (92:8 β/α isomeric ratio). TBAF-promoted desilylation and final silica gel chromatographic purification provided pure β-*N*-Boc-4′-azauridine **286** in 80% isolated yield. Of note, this simple approach was extensively exploited to prepare a variety of azanucleosides, and also adapted to assemble collections of many congeners (vide infra).

B. Induction from Temporarily Chiralized Silyloxy Pyrroles

As briefly stated at the beginning of this section, pyrrole silyloxy dienes have the unique advantage, vis-à-vis the corresponding oxygen and sulfur relatives, to embody nitrogen, an anchoring point for useful chiral auxiliary ligands.

Baldwin[91] first adopted this concept during a skillful synthesis of (+)-lactacystin (**6**), a neurotrophic substance discovered in *Streptomyces* sp. OM 6519. There, the pivotal *N*-chiralized methyl-substituted silyloxy diene **291** was prepared according to the Seebach "chirality self-regeneration" philosophy[92] by starting with (*R*)-glutamic acid (**287**) (Scheme 57). Thus, to access **291**, (*R*)-glutamic acid (**287**) was first converted to hydroxymethyl pyrrolidinone **288** by known procedures, and this compound was then protected to chiral oxazolidine **289** using benzaldehyde. Accordingly, the native chiral information is transmitted to and stored in the protective moiety of **289**, so memorizing chirality when the native sp^3 chiral center at C-5 will disappear during the crucial silyl enol ether formation. α-Methylation of **289** via enolization and subsequent double bond formation according to the phenyl selenation–oxidation maneuver ensured clean preparation of pyrrolinone **290**, which was elaborated into the requisite enol ether **291** following the usual methodology (TBSOTf, lutidine).

The key reaction of this clever synthesis involved stereoselective Mukaiyama aldol reaction of **291** with isobutyraldehyde, thereby ensuring the construction of the C-5 quaternary center and the C-1′ secondary hydroxyl function of the target

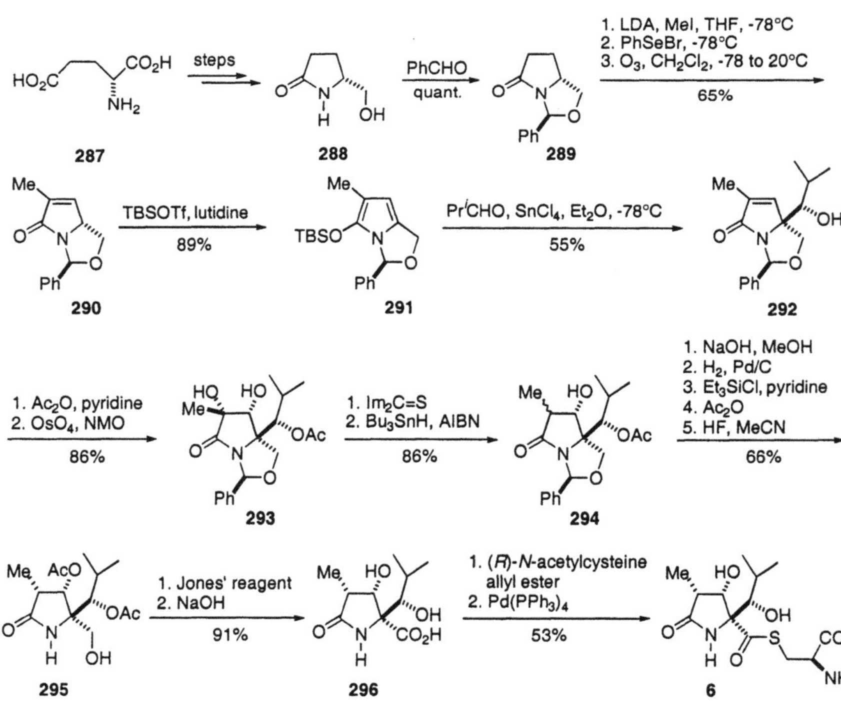

Scheme 57.

in the correct stereochemical form. The aldol reaction of **291** with isobutyraldehyde was achieved at –78 °C in diethyl ether to afford γ,γ-disubstituted lactam **292** as a crystalline solid. The 4,5-*threo* disposition of the *N,O*-substituents in **292** and the β-location of both the phenyl and the hydroxybutyl moiety are noteworthy (see Section IX). After acetylation, **292** was converted to diol **293** by osmilation, and this product was subjected to α-deoxygenation to afford the 1:1 epimeric mixture **294**. Base-catalyzed epimerization of this mixture fortunately allowed enrichment into the desired *syn*-isomer, with concomitant, if unwanted, removal of the *O*-acetyl protecting group. Hydrogenolytic cleavage of the benzylidene auxiliary followed by suitable protecting group manipulation gave lactam **295**, which was then oxidized and deprotected to the key acid precursor **296**. Final transformation of **296** into (+)-lactacystin (**6**) was carried out by coupling with (*R*)-*N*-acetylcysteine allyl ester followed by deallylation.

Enantiopure 2-silyloxy pyrroles bearing *N*-chiral substituents, easily available from 2,5-dimethoxy-2,5-dihydrofuran (**297**) and suitable chiral amines, have been independently exploited by Poli[93,94] and Royer[95,96] en route to variously substituted homochiral pyrrolidinones and pyrrolidines. As shown in Scheme 58,[93,94] treatment of dihydrofuran **297** with (*R*)-α-methylbenzyl amine in refluxing acetic acid gave rise to pyrrolinone **298**, which was directly transformed to silyloxy pyrrole **299** using the TBSCl–DIPEA reagent system. Addition of **299** to methyl orthoformate

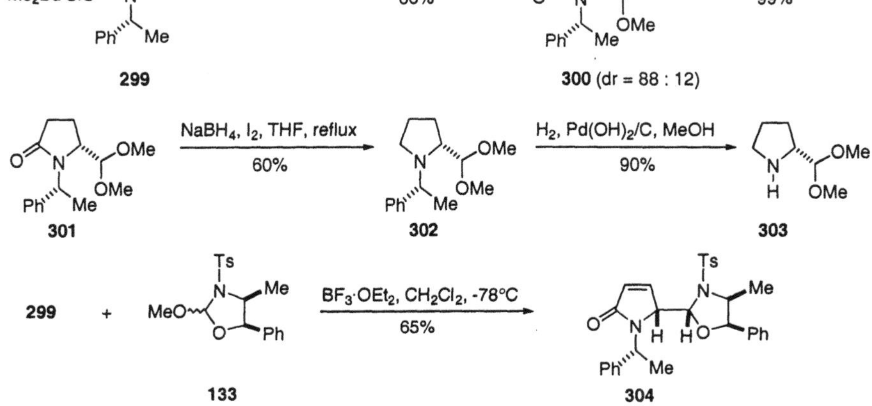

Scheme 58.

under $BF_3 \cdot OEt_2$ assistance produced (5*R*)-substituted pyrrolinone **300** preferentially, accompanied by a minor amount (12%) of the corresponding (5*S*)-epimer. The double bond within **300** was then saturated to compound **301**, which was then transformed to pyrrolidine **302** by $NaBH_4/I_2$ treatment. The auxiliary moiety within **302** was cleaved hydrogenolytically, allowing preparation of protected (*R*)-prolinal **303**.

A more useful procedure to a protected proline aldehyde derivative was also envisaged by starting with **299** and the previously experienced chiral oxazolidine **133**. According to a matched-sense addition, adduct **304** formed exclusively, whose configuration was firmly secured by X-ray analysis. Structural correlation between **304** and **300** confirmed the absolute configuration of the latter material. The diastereoselective coupling of **299** with cinnamaldehyde was also investigated during a preliminary study directed toward preparation of goniofufurone mimics. In the event, a 79:21 mixture of two unsaturated adducts formed whose stereochemistry was not ascertained at that point.

According to the same conceptual approach, Royer[95] investigated in detail the diastereoselective coupling of a variety of prochiral aldehydes to phenyl alaninol-derived silyloxy pyrrole **306**, in turn prepared from chiralized pyrrolinone **305**. As an example (Scheme 59), the Lewis acid-assisted Mukaiyama aldol addition of **306** with acetaldehyde resulted in preferential formation of the *syn*-adducts **307** and **308**, the best selectivity [(*RR/SS*) = 74:18; *syn/anti* = 92:8] being observed with $TiCl_4$. The scope of the reaction was further enlarged to α,β-unsaturated carbonyl compounds highlighting high chemical efficiency, flanked by unsatisfactory regio- and stereocontrol.

Very recently, the same author[96] exploited chiral pyrrole **306** to access both enantiomers of aza-muricatacin (vide infra, compounds **331** and *ent*-**331**) via $BF_3 \cdot OEt_2$-promoted diastereoselective coupling with tridecanal.

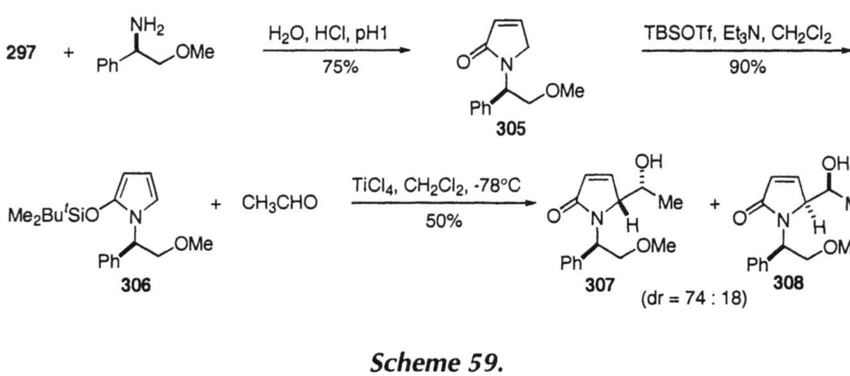

Scheme 59.

C. Diastereoselective Reactions Leading to Racemic Compounds

Research in our laboratory have recently focused on the use of 2-silyloxy pyrroles for syntheses of a variety of β-hydroxylated α-amino acids.[84,97] Indeed, it was envisaged (Scheme 60) that α-amino acid structures of type **A** and **A′** could derive from unsaturated lactams **B** and **B′**, respectively, via oxidative sacrifice of the C-2 and C-3 pyrroline carbons and subsequent implementation of the carboxylic function at C-4. Compounds **B** and **B′** in turn, can be generated by alkylation of silyl enol ethers **C** and **C′** at the γ-position, with substituted enol ether **C′** being easily derived from its unsubstituted parent **C** by means of suitable γ-substitution. Overall, this plan emphasizes the use of dienes **C** and **C′** as chemical surrogates of the α-glycine anion and α-aminoacyl anion, respectively.

Based on this strategy, the synthesis of racemic *threo*-3-hydroxyleucine [±-(**311**)] started with unsubstituted silyloxy pyrrole **13**, which was coupled with isobutyraldehyde in the presence of $SnCl_4$ (Scheme 61). There was obtained, after proper silylation, *syn*-configured adduct (±)-**309**, as expected, in 70% isolated yield. Implementation of the CO_2H moiety α to the amino function was then performed by a multistep procedure consisting of double bond dihydroxylation, lactam ring opening, and periodate diol fission to form α-amino aldehyde (±)-**310**, followed by formyl-to-carboxyl oxidation. Final acidic deprotection furnished racemic *threo*-3-hydroxyleucine [(±)-**311**] in 59% yield from (±)-**309**. Conversely, the preparation of racemic α-methylthreonine [(±)-**315**] called for γ-methyl-substituted silyloxy pyrrole **312** as the starting scaffold, and this material was readily prepared from **13** via γ-methylation (MeI, CF_3CO_2Ag) and silyl enol ether formation (TBSOTf, lutidine).

By paralleling the above procedure and using acetaldehyde as the electrophilic partner, *syn*-adduct (±)-**313** was first synthesized, which was then converted to the target α-amino acid (±)-**315** through the intermediacy of amino aldehyde (±)-**314**.

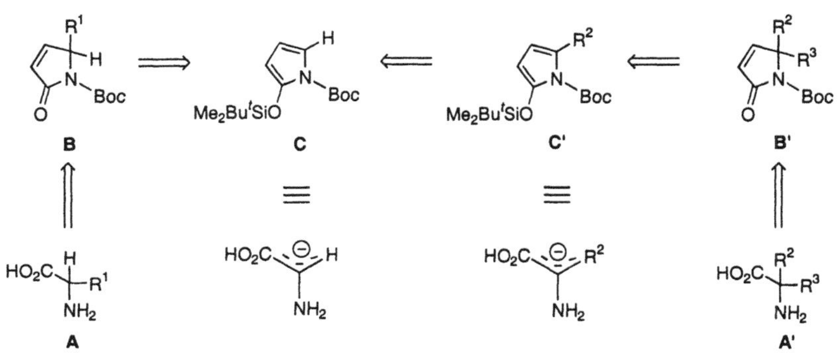

Scheme 60.

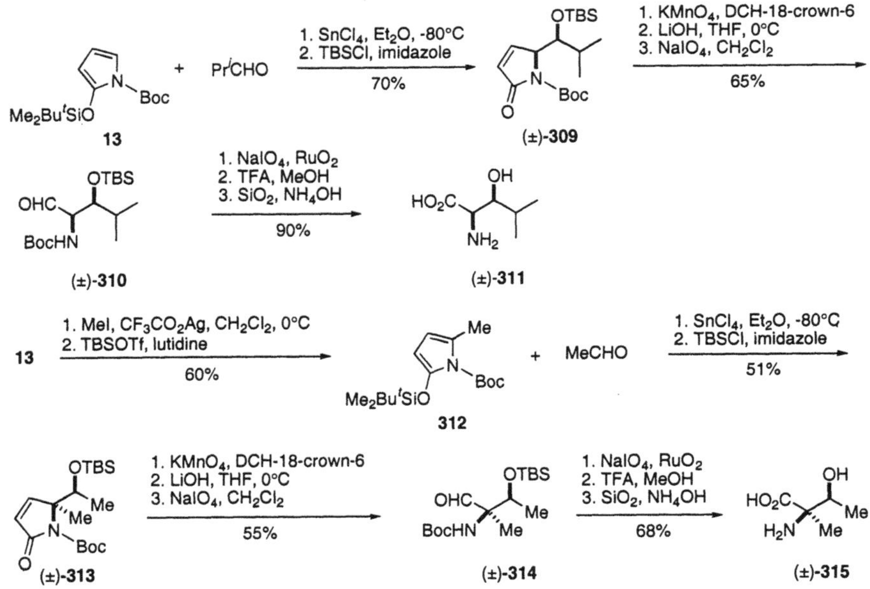

Scheme 61.

According to this plan, racemic threonine as well as racemic allothreonine were also prepared with comparable efficiency.

In his continuing effort to prepare oligonuclear representatives of the Annonaceous acetogenin family, Figadère[50] attempted direct coupling between TBSOP **13** and *N*-Boc-protected 2-methoxypyrrolidine **316** under Lewis acid assistance (Scheme 62).

The results showed that the coupling was moderately efficient with either BF_3 etherate (38% yield) or $TrClO_4$ (40% yield), whereas the diastereomeric ratio of the formed racemic adducts (±)-**317** ranged from 75:25 to 88:12. Of note, no attempts to elucidate the relative stereochemistries of the formed binuclear adducts were carried out.

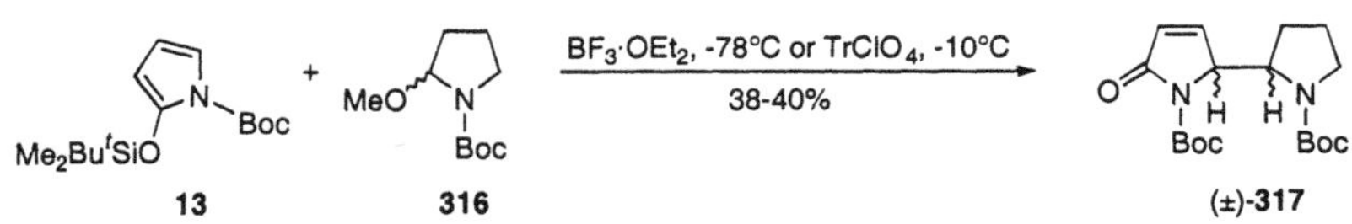

Scheme 62.

VI. DIASTEREOSELECTIVE REACTIONS USING SILYLOXY THIOPHENES

As compared to the extensive investigations and synthetic applications of the furan- and pyrrole-based 2-silyloxy dienes summarized in the previous sections of this article, the chemistry of their sulfur counterparts, 2-silyloxythiophenes, have received scarce attention, till now. This probably reflects the moderate occurrence of sulfur-carrying biomolecules vis-à-vis the more abundant and challenging congeners bearing oxygen and nitrogen functionalities. The first asymmetric synthesis exploiting 2-(*tert*-butyldimethylsilyloxy)thiophene (**15**) was reported by our own research group in 1995,[15,98] about 10 years after the pioneering work by Ricci et al.,[16] and involved preparation of enantiopure 2′,3′-dideoxy-4′-thiocytidines, representatives of a scantily investigated progeny of nucleoside mimics.

According to Scheme 63, a short synthesis of anti-HIV active 4′-thiocytidine **7** called for thiosugar **322** as the immediate precursor. Diastereoselective addition of 2-(*tert*-butyldimethylsilyloxy)thiophene (**15**) to L-glyceraldehyde **318** in the presence of BF_3 etherate in CH_2Cl_2 resulted in preferential formation of 4,5-*syn*-adduct **319**, accompanied by less than 10% of its 4,5-*anti*-diastereoisomer. Hydrogenation of the major compound furnished crystalline thiolactone **320**, which was readily transformed to aldehyde **321** by cleavage of the acetonide protection and periodate C-5–C-6 fission. The key thiosugar **322** was obtained from **321** by simple chemistry involving mild reduction of the aldehyde carbonyl to hydroxymethyl ($NaBH_4$), protection (TBDPSCl, imidazole), and thiolactone-to-thiolactol conversion ($LiAlH_4$), followed by acetylation (Ac_2O, pyridine, DMAP). The final cytosine

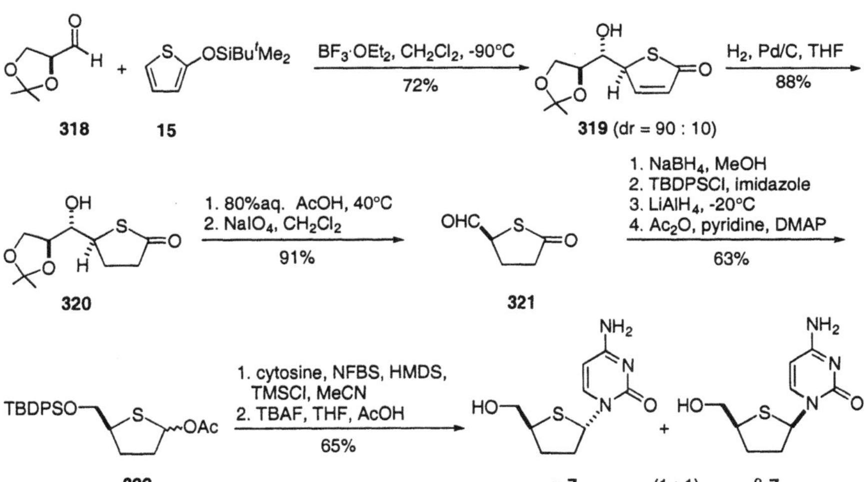

Scheme 63.

coupling was conducted in the presence of nonafluorobutanesulfonate and HMDS in acetonitrile and furnished, after proper deprotection, a separable 1:1 mixture of α- and β-D-cytidine derivatives α-**7** and β-**7**. The same chemistry was applied to preparation of the analogous nucleosides of the L-series, comprising 2′,3′-dideoxy-4′-thio-α- and β-L-cytidine, whose absolute configuration was established by X-ray analysis of a synthesis intermediate.[98]

VII. CONSTRUCTION OF SMALL MOLECULE ENSEMBLES

The whole chemistry discussed in the previous sections of this review and the variety of the target molecules available from 2-silyloxy furan, pyrrole, and thiophene reagents highlight the merits of this versatile triad as a chemically uniform scaffold series with which molecular diversity can be planned and realized. In principle, a given synthetic protocol adopting uniform chemistry and tolerating the diversity of the silyloxy diene reagents may be applied according to a parallel execution or a combined logic allowing generation of collections of oxygen-, nitrogen-, and sulfur-containing functional molecules, either as individual compounds or as mixtures. As an emblematic example, a unified synthesis of 2′,3′-dideoxynucleosides and their sulfur and nitrogen mimetics was planned by exploiting the reagent triad TBSOF **11**, TBSOT **15**, and TBSOP **13**.[13] Accordingly, the parallel preparation of a 30-component collection of individual, pure pyrimidine nucleosides of both the D- and L-series was executed according to a unified chemical procedure.

Scheme 64 illustrates, as an example, the preparation of five cytidine-related derivatives of the D-series, $\mathbf{C^1}$–$\mathbf{C^3}$, from the three silyloxy diene precursors **11**, **15**, and **13** and isopropylidene protected L-glyceraldehyde **318**. The overall synthetic maneuver comprises three parallel executions, adopting a rather uniform set of reactions. The opening move was the preparation of the unsaturated templates $\mathbf{A^1}$, $\mathbf{A^2}$, and $\mathbf{A^3}$ via stereoselective coupling between **318** and the individual silyloxy dienes. In the next stage of the sequence, the heterocyclic core unit within $\mathbf{A^1}$–$\mathbf{A^3}$ was hydrogenated, while the γ-triol appendage was shortened by two carbon atoms by the usual oxidative cleavage technique, followed by $NaBH_4$ reduction and protection. After suitable manipulation of the carbonyl function to a hydroxyl, this transformation allowed preparation of the key sugar units $\mathbf{B^1}$–$\mathbf{B^3}$, ready for the final nucleobase coupling. A modified Vorbrüggen procedure was adopted to couple the silylated 4-*N*-acetylcytosine to the corresponding sugar components. To ensure kinetic convergence, an excess of nucleobase and a mixed Lewis acid catalytic system was employed, consisting of a 1:1 mixture of $SnCl_4$ and TMSOTf in 1,2-dichloroethane. The coupling furnished, after appropriate deprotection, the expected cytidine nucleosides $\mathbf{C^1}$–$\mathbf{C^3}$ in good yields. It should be noticed that, while in the oxygen and thio-series the nucleosides emerge as 1:1 α/β anomeric mixtures, the aza-sugar reaction ($\mathbf{B^3}$ to $\mathbf{C^3}$) proved highly stereoselective, affording the

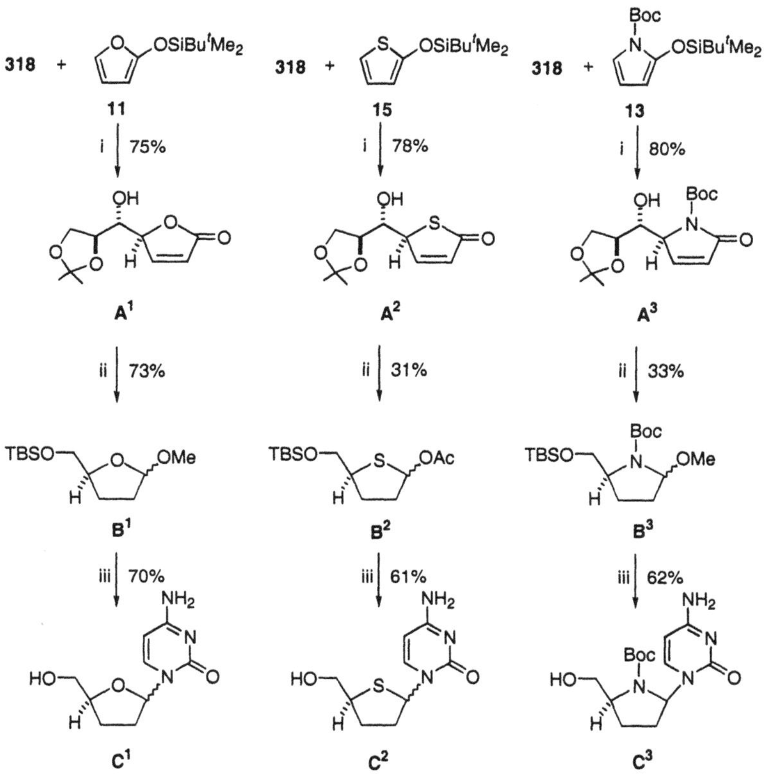

Key protocols: (i) $BF_3{\cdot}OEt_2$, or $SnCl_4$, CH_2Cl_2 or Et_2O, -80°C. (ii) 1. H_2, Pd/C, THF. 2. AcOH, 50°C; then $NaIO_4$, CH_2Cl_2. 3. $NaBH_4$; then TBSCl, imidazole. 4. DIBAL-H or $LiEt_3BH$; then $CH(OMe)_3$, $BF_3{\cdot}OEt_2$ or Ac_2O, pyridine. (iii) 1. silylated 4-*N*-acetylcytosine, $SnCl_4$/TMSOTf (1:1), 1,2-dichloroethane. 2. TBAF; then K_2CO_3, MeOH.

Scheme 64.

β-anomer quite predominantly. Along this path, successful preparation of the corresponding five cytidine nucleosides of the L-series, as well as 10 uridines and 10 thymidines was straightforwardly ensured.

As an exciting prospect of this strategy, a much more rapid preparation of a small library of the same 30 nucleosides was planned and executed. The first step of our plan was the preliminary synthesis of the appropriate sugar precursors **B^1**, **B^2**, and **B^3** in a racemic form. Next, as shown in Scheme 65, a "combine-split" in-solution technical approach was adopted using the three oxygen, sulfur, and nitrogen sugars **B^1**–**B^3** and the three pyrimidine bases, uracil (Ur), thymine (Th), and cytosine (Cy).

Thus, the carbohydrate precursors were mixed in equimolar quantity by dissolving them in anhydrous 1,2-dichloroethane and the resulting solution was subdivided

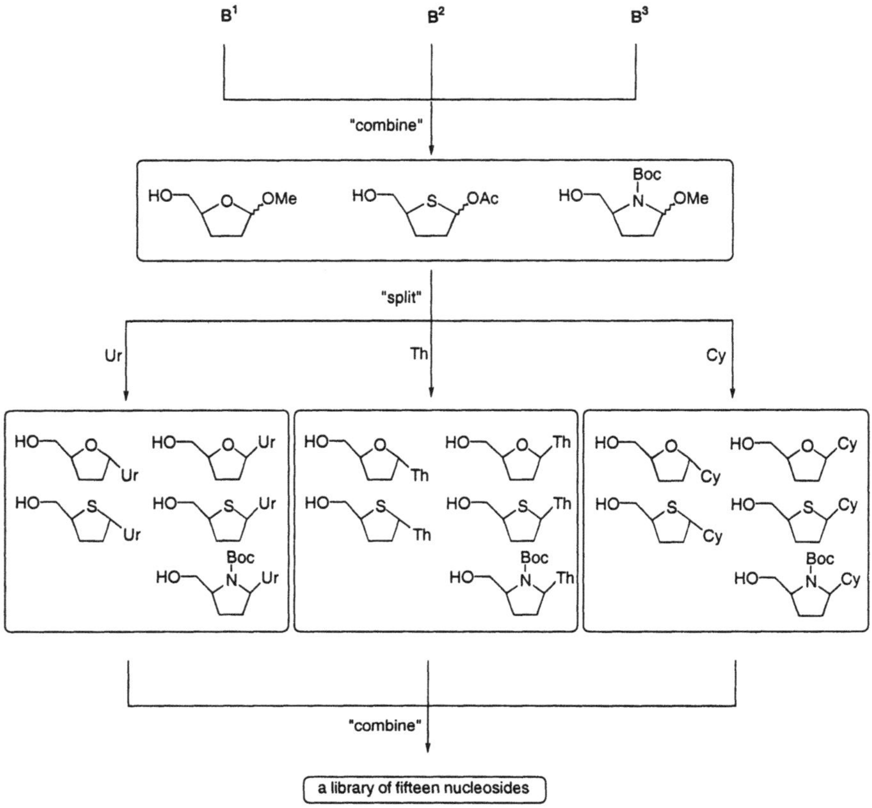

Scheme 65.

into three equal portions. Each portion was independently reacted with one individual activated base—Ur, Th, Cy under $SnCl_4$/TMSOTf assistance—to give, after complete deprotection and separation of the nucleoside fraction, three diverse sublibraries which were finally combined to furnish a full library of nucleosides comprising the expected 15 racemic compounds with a 64% combined yield based on the average molecular weight of the nucleoside mixture. In principle a truly random synthesis would furnish an ensemble of 18 racemates (36 different nucleosides), i.e. $3^2 \times 2^i$, where i = 2 is the number of stereocenters in the resulting nucleosides. In practice, however, due to the stereoselective character of the coupling reaction involving the azafuranose template, strongly favoring the β-anomeric compounds (vide supra), the total number of individuals in the full library was reduced to 15 major racemates out of 18. The three nucleoside sublibraries were analyzed by reverse-phase HPLC. Having all the pure components in hand strongly facilitated the analysis of the mixtures by simply comparing the experimental HPLC traces with standard profiles of artificial mixtures of pure nucleosides. Remarkably, the HPLC analyses indicated the presence of the expected nucleosides with no detectable by-products. In addition, the analyses indicated that the nucleoside mixture was rather close to equimolar.

The above disclosed concepts have been recently applied by our own research group to preparation of small collections of potentially bioactive monocyclic and oligocyclic systems related to the acetogenins of the *Annonaceae*.[47] It is worth noticing, at this point, that having a triad of equally reacting oxygen, nitrogen, and sulfur templates not only ensures preparation of naturally occurring tetrahydrofuran units, but also provides access to many structural variants carrying pyrrolidine or tetrahydrothiophene (thiolane) moieties. In the oligonuclear series, the exciting perspective of preparing core units incorporating either homogeneous or heterogeneous arrays can be put to practice by timely intervention of the three different heterocyclic scaffolds according to a step-growth oligomerization rationale (vide infra).

A remarkable goal of our laboratory[99] was the parallel assembly of six muricatacin compounds, comprising the natural (*R,R*)- and (*S,S*)-muricatacin pair (**330** and *ent*-**330**) and the corresponding sulfur- and nitrogen-containing relatives (**8** and **331**, *ent*-**8** and *ent*-**331**), by starting with both enantiomers of glyceraldehyde acetonide and exploiting the triad of silyloxy dienes **11**, **15**, and **13**. As illustrated in Scheme 66, for (*R,R*)-muricatacin series, the three parallel syntheses commenced with (+)-(*R*)-glyceraldehyde acetonide (**44**), which was coupled to **11**, **15**, and **13** under $BF_3{\cdot}OEt_2$ or $SnCl_4$ assistance. Under optimal conditions, 4,5-*syn*-configured adducts **46**, **323**, and **219** preferentially formed, contaminated by only marginal amounts of the corresponding 4,5-*anti*-diastereoisomers which were easily separated by flash chromatography on silica gel. The de values ranged from 88 to 89%. At this point of the synthesis, it was necessary to saturate the double bond within

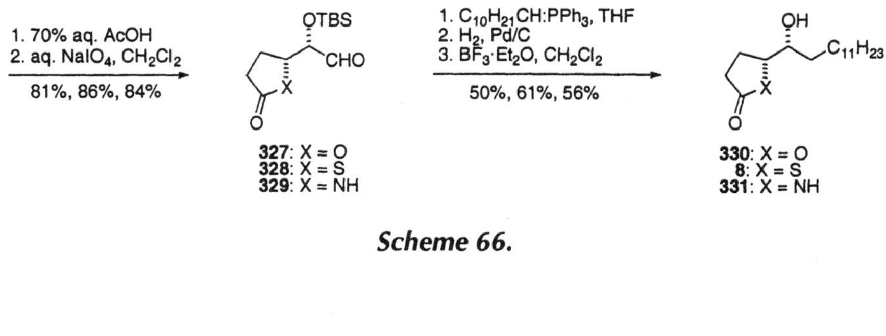

Scheme 66.

the above unsaturated adducts and protect the free hydroxyl at C-5 before fission of the C-6–C-7 carbon–carbon bond.

Thus, compounds **46**, **323**, and **219** were subjected to sequential catalytic hydrogenation (H_2, Pd on carbon) and protection of the OH function as TBS ether (TBSCl, imidazole, DMF) to afford seven-carbon intermediates **324–326** in excellent yields. Oxidative removal of the C-7 carbon atom within **324–326** to generate the six-carbon aldehydes **327–329** was effected by cleaving of the terminal acetonide (70% aq. AcOH, 50 °C) followed by fission of the diol so formed by $NaIO_4$-impregnated wet silica in CH_2Cl_2. The sequences afforded configurationally stable compounds **327–329** in high yields, which were used as such in the next stages of the syntheses.

There remained to carry out suitable elongations on these aldehydes in order to complete the elaboration of the muricatacin side chains. Indeed, Wittig olefination of **327–329** in THF with the appropriate C_{11} ylide gave rise to the corresponding unsaturated compounds, which were quickly transformed to muricatacin derivatives **330**, **8**, and **331** by catalytic hydrogenation (Pd on carbon, THF), followed by BF_3 etherate-promoted desilylation. The overall yields from **44** were 25, 24, and 32%, respectively, over eight steps. The synthetic paths to (+)-muricatacin enantiomers from L-glyceraldehyde were similar to those of their (4*R*,5*R*)-configured

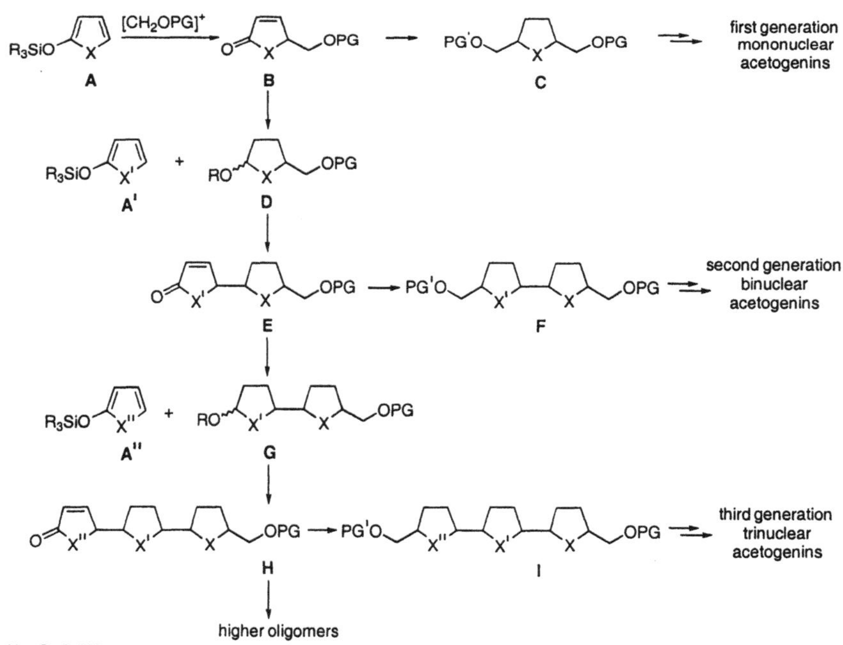

Scheme 67.

counterparts, and proceeded uneventfully to afford (+)-muricatacin *ent*-**330** and its sulfur and nitrogen relatives, *ent*-**8** and *ent*-**331** in 24, 22, and 32% overall yields, respectively, from L-glyceraldehyde.

An important group of plant natural products comprises a wide series of non-terpenoid C_{35} or C_{37} long-chain fatty acids, usually named Annonaceous acetogenins, that are derived from the fatty acid metabolism.[47] The extended differentiation of the *Annonaceae* realm (over 2000 different plants) is the basis for the exceedingly large structural and stereochemical diversity of the acetogenins which have been isolated (more than 230 different compounds). The dominant structural motif of this class of compounds is represented by one or more adjacent or nonadjacent THF-rings, usually spanning the inner region of the carbon backbone. Owing to the common tetrahydrofuran nature of the acetogenin core, synthetic access to these structures through homologative protocols involving silyloxy furan reagents (or their sulfur and nitrogen relatives) seems to be quite advantageous and straightforward. Indeed, along this line, some preliminary attempts to create certain intermediary fragments have been performed,[48,49] culminating in the preparation of the oligo-THF units **94, 96**, and **99** (vide supra).

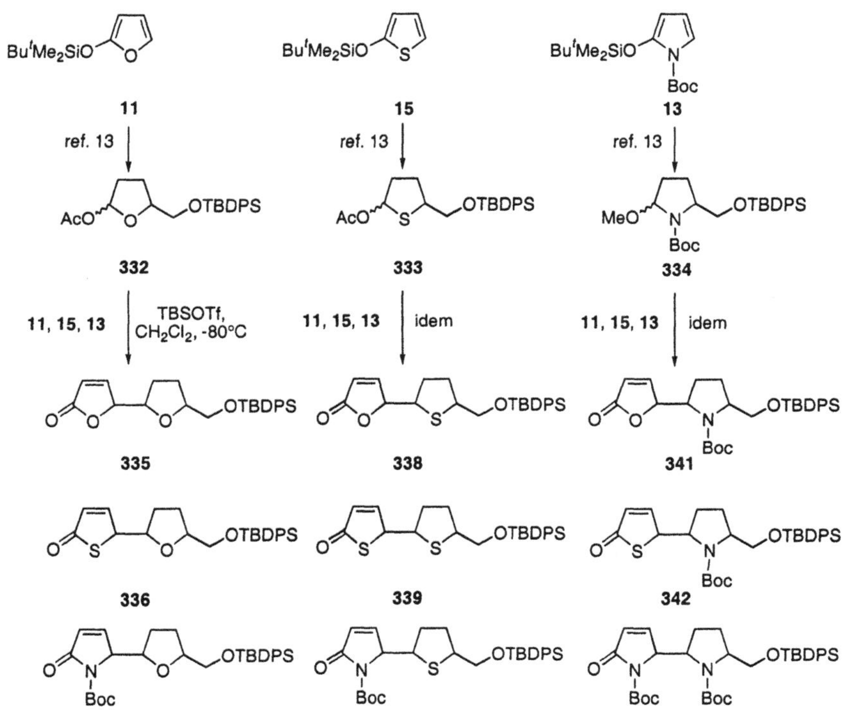

Scheme 68.

A more comprehensive, tunable design aimed at the construction of an expanded library of Annonaceous acetogenins and variants thereof exploiting combined intervention of oxygen, sulfur, and nitrogen silyloxy diene templates has been recently launched by our laboratories.[100] A synopsis of the plan is displayed in Scheme 67, featuring the cardinal role played by the triad of the heterocyclic silyloxy dienes **A** which furnish both the nucleophilic and the electrophilic modules (e.g. **D** and **G**). Thus, the execution of the plan consists of a step-growth oligomerization wherein the intermediates of each cycle (e.g. **B**, **E**, and **H**) serve to prepare the corresponding acetogenin generation (e.g. mono-, bi-, and trinuclear species) through intermediates **C**, **F**, and **I**. The first panel of the work, culminating in the construction of a collection of 16 structurally defined binuclear acetogenin core motifs of type **335–343**, is shown in Scheme 68.

The electrophilic scaffolds **332**, **333**, and **334** were first prepared from the corresponding silyloxy dienes **11**, **15**, and **13** via the previously disclosed protocols (vide supra). Then, each heteroatom-containing module was separately homologated with the individual silyloxy diene nucleophiles **11**, **15**, and **13**, giving rise to the expected 3 × 3 compound matrix, comprising all the possible combinations of heteroatoms (compounds **335–343**). Of notice, the actual stereochemical outcome of the homologative reactions, using catalytic TBSOTf, proved partially diastereoselective, producing in all instances *threo,trans*-configured compounds, often accompanied by substantial quantities of the corresponding C-4 epimeric *erythro,trans*-derivatives and/or *threo,cis*-compounds.

VIII. MISCELLANEOUS APPLICATIONS

The peculiar reactivity of 2-silyloxy reagents of this report mainly derives from their structure, these substances being both carbon nucleophilic species and electron-rich dienes. While the former reactivity has been widely explored in the furan, pyrrole, and thiophene series (vide supra), the latter behavior has been only marginally exploited, and the major achievements are restricted to the furan domain.

A study focusing on asymmetric Diels–Alder reactions of chiral 5-substituted 2-silyloxy furans appeared in 1995 by Sha and colleagues.[101] Remarkably, it was found (Scheme 69) that when dienes **344** were reacted with acetylene dicarboxylate in the presence of (+)-Eu(hfc)$_3$, the expected adducts **345** and **346** formed with high diastereofacial selectivity (7.5:1 to 9.3:1). On the other hand, without any catalyst or with an achiral Eu(fod)$_3$ catalyst, the reactions gave much lower facial selectivities (2:1 to 4:1). Interestingly, cycloadduct **346b** served as precious intermediate to prepare an enantiopure desmethyl C,D-ring derivative of Taxol.[102]

A more extended project aimed at the total synthesis of complex, naturally occurring metabolites was recently proposed by the Schlessinger's group that is centered upon the application of intrinsically chiral 2-silyloxy furan reagents in Diels–Alder cycloadditive processes.[103–106]

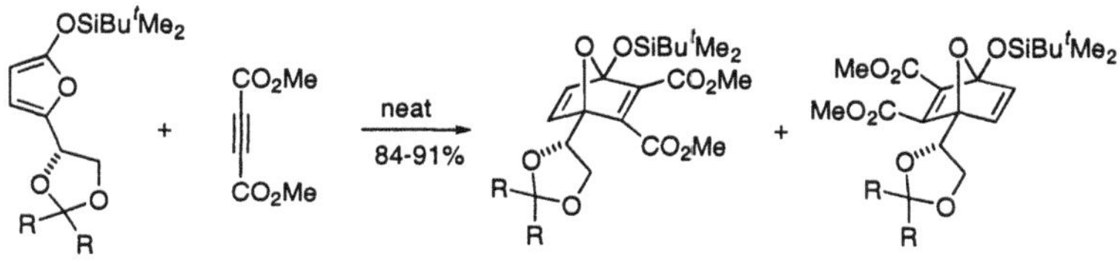

Scheme 69.

A first work[103] deals with diastereoselective Diels–Alder reactions of variously substituted silyloxy furans of type **347** and **350** with different dienophiles such as α,β-unsaturated esters, nitriles, sulfones, and imides. As leading examples, cycloadditions involving methyl acrylate and dimethyl fumarate giving oxobicyclo *endo*-adducts of type **348** and **351** are displayed in Scheme 70.

Thus, reaction of **347** with methyl acrylate in CH_2Cl_2 in the presence of chlorotitanium triisopropoxide produced the *endo*-adduct **348** with a significant 15:1 diastereomeric ratio. The oxobicyclic adduct **348** was further subjected to several transformations; among them, PPTS treatment to produce cyclohexenone **349** has a remarkable value for natural product synthesis. In the same manner, addition of **350** to dimethyl fumarate produced *endo*-adduct **351** with excellent diastereoselectivity, which was simply transformed to the vinylogous amide **352** by PPTS treatment.

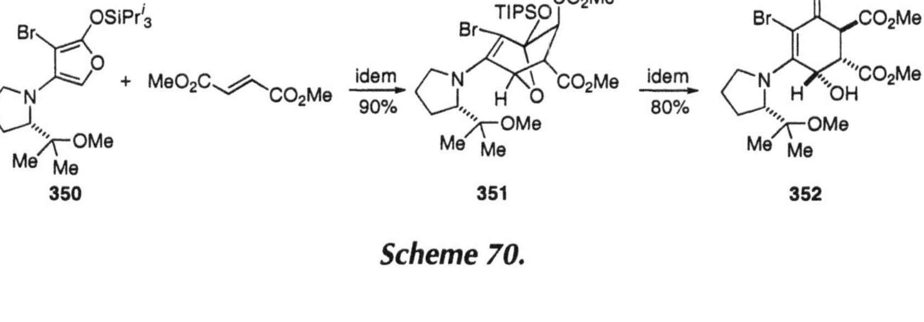

Scheme 70.

The dienophile domain was further explored by the same research group[104] with reactions involving diene **350** and a series of electron-poor dienophiles as sulfone acrylate esters, phenyl vinyl sulfone, *N*-phenyl maleimide, methyl vinyl ketone, and acrolein. With the exception of the acrolein adduct, all of these reactions proved highly *endo*-selective (12:1 to 50:1) and facial-selective (99:1) affording synthetically useful oxabicycloheptanone adducts.

The extensive investigation by the Schlessinger's group in this area has culminated in the clever total synthesis of (+)-cyclophellitol (**363**),[105] a bioactive natural product isolated from the culture filtration of mushroom *Phellinus* sp. (Scheme 71).

The crucial move was the cycloaddition of chiral furan **353** to racemic dimethyl 2,3-pentadienedioate (±)-**354** in THF at −100 °C producing marginally stable [2:2:1] oxabicyclic enamine **355** in 91% yield. Of note, adduct **355** emerged from a double diastereofacially selective reaction with concomitant, remarkable kinetic resolution of racemic allenic dimethyl ester (±)-**354**. Acidic removal of the chiral amine appendage within **355** (2 M aq. HCl, MeCN, 0 °C) afforded a bromo ketone

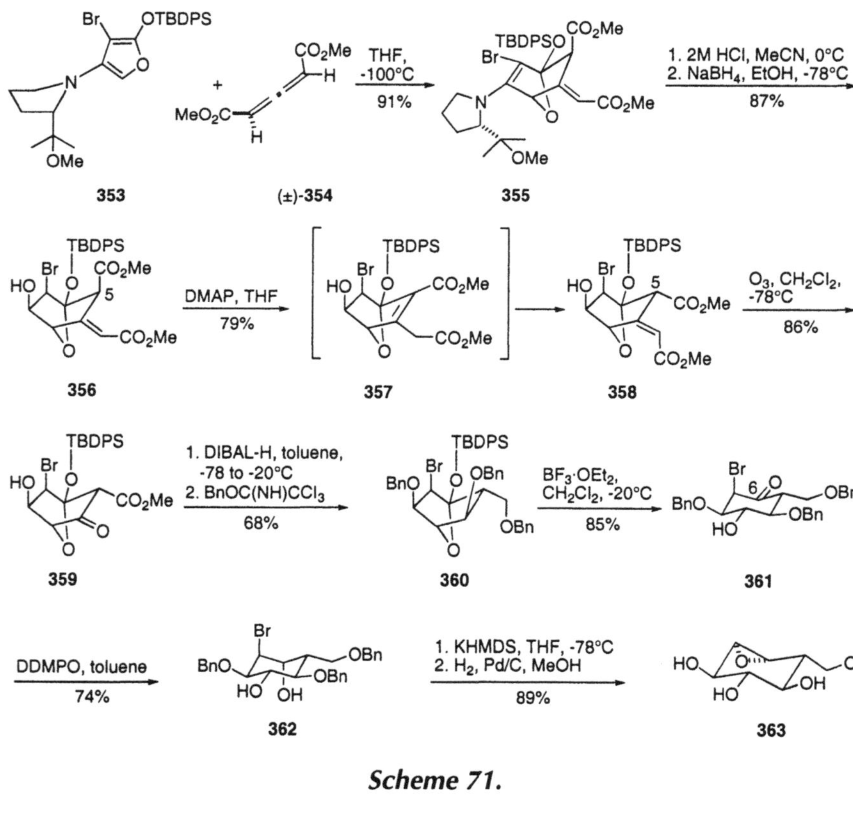

Scheme 71.

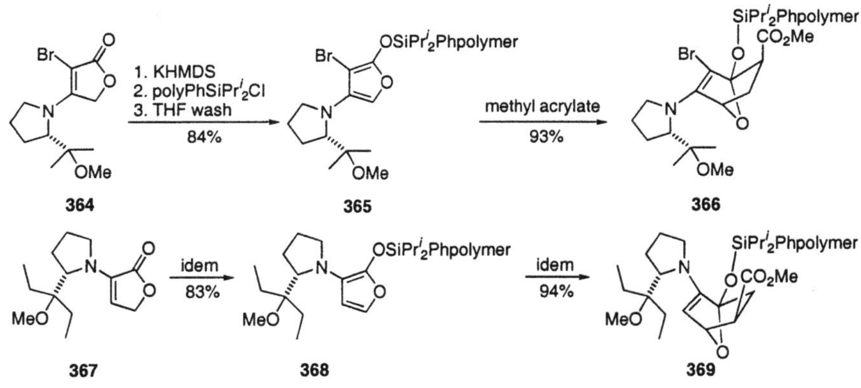

Scheme 71.

intermediate, which was stereoselectively reduced ($NaBH_4$) to the *syn*-bromohydrin **356**, in 87% yield for the two steps from **355**. The stereochemistry of the axial ester residue at C-5 was then adjusted by exposure of **356** to DMAP in THF; in the event, epimerization at C-5 occurred uneventfully, producing compound **358** via isomeric intermediate **357**.

Ozonolysis of **358** afforded unstable β-keto ester **359**, which was transformed into oxabicyclic ketal **360** by simultaneous reduction of the keto and ester moieties, followed by benzyl ether protection of the free hydroxyl groups. Careful cleavage of the silyl ketene acetal functionality ($BF_3{\cdot}OEt_2$) cleanly transformed bicyclic compound **360** into cyclohexanone **361**, which was ready for the last significant reduction maneuver, namely stereoselective ketone-to-hydroxyl reduction to afford C-6 axial alcohol derivative **362**. Thus, **361** was treated with diisobutylaluminium 2,6-di-*tert*-butyl-4-methylphenoxide (DDMPO), furnishing **362** in 74% yield, which was finally converted to (+)-cyclophellitol (**363**) by epoxide ring formation (KHMDS) and hydrogenolytic deprotection.

An exciting development[106] of the above disclosed chemistry involved adaptation of the silyloxy furan-based methodology to the solid-phase technology. As shown in Scheme 72, the requisite polymer-bound diene reagents **365** and **368** were directly created by trapping the suitable enolate precursors, derived from furanones **364** and **367**, using a styrene-divinylbenzene polymer-linked chloride. Polymer-linked homochiral furans **365** and **368** were then reacted with methyl acrylate to form polymer-bound adducts **366** and **369**, respectively, with excellent *endo* and facial selectivities. TBAF-mediated cleavage of the adducts from the polymer matrix and subsequent manipulations allowed the authors to prepare a variety of useful multifunctional intermediates in an optically pure state.

IX. MECHANISMS AND MODELS: ADDITIONS TO PROCHIRAL C=X BONDS

A faithful reader of the various sections of this article would have certainly noticed how important was the diastereocontrol exerted during the central C–C bond-forming maneuvers of the many reviewed synthetic examples.[107] The crucial chiral induction event mainly originates from the Lewis acid-assisted aldol type addition of a silyloxy reagent to a given achiral–prochiral or chiral–prochiral C=X bond. Here, the basic reaction can be rationalized assuming that silyloxy dienes—a sort of vinylic silyl ketene acetals (thioacetals, aminals)—are ambident nucleophiles with both the α- and γ-carbons prone to react. As a rule, the product of this maneuver formally originates, in all instances, from an attack of the γ-nucleophilic carbon of the five-membered heterocycle to the electrophilic C=X bond with the formation of two novel stereocenters. When aldehydes are involved, the product is a vicinal diol (or an hydroxy amine or an hydroxy thiol), while the carbon–carbon bond-forming maneuver represents the vinylogous version of the Mukaiyama aldol reaction, a scantily exploited operation of the organic chemistry repertoire.[108] Alternatively, one can suppose a sort of hetero Diels–Alder-like approach of the two reaction partners, though appropriate cycloadducts have never been isolated, or even detected, in spite of a careful inspection of the reaction products. In order for the various reaction mechanisms to be rationalized, this section will be subdivided based on the nature of the prochiral electrophilic species exploited. Also, a brief insight concerning the auxiliary-driven induction will be presented.

A. Additions to Prochiral C=O Bonds

This approach groups the majority of known procedures involving oxygen, nitrogen, and sulfur 2-silyloxy dienes. Apart from a few exceptions, the reactions between prochiral-substituted or unsubstituted dienes with prochiral carbonyl compounds invariably displayed *syn*-selective behaviors (simple diastereoselection) with good Felkin-type inductions (facial diastereoselection) when a chiral carbon adjacent to the carbonyl is present. An emblematic example is represented by the well-exploited reaction between silyloxy dienes of type **I** and homochiral α-alkoxyaldehydes **II**, giving rise to 4,5-*syn*-5,6-*anti*-configured adducts **III**, preferentially, accompanied by minor amounts of 4,5-*anti*-5,6-*anti*-disposed diastereoisomers **IV** and no detection of the remainder two isomers arising from an anti-Felkin attack (4,5-*syn*-5,6-*syn*- and 4,5-*anti*-5,6-*syn*-isomers). Two major transition state models can be suggested accounting for this remarkable stereochemical outcome, namely the "open-chain" models of type **TS2** or **TS3**, and the "Diels–Alder-like" models of type **TS1** and **TS4**.

As shown in Figure 2, the preferred product **III** can derive from either transition-state model **TS1** or **TS2**, while the minor isomer **IV** can emerge from **TS3** or **TS4**. Open-chain model **TS3** seems to be favored over **TS2** due to stereoelectronic

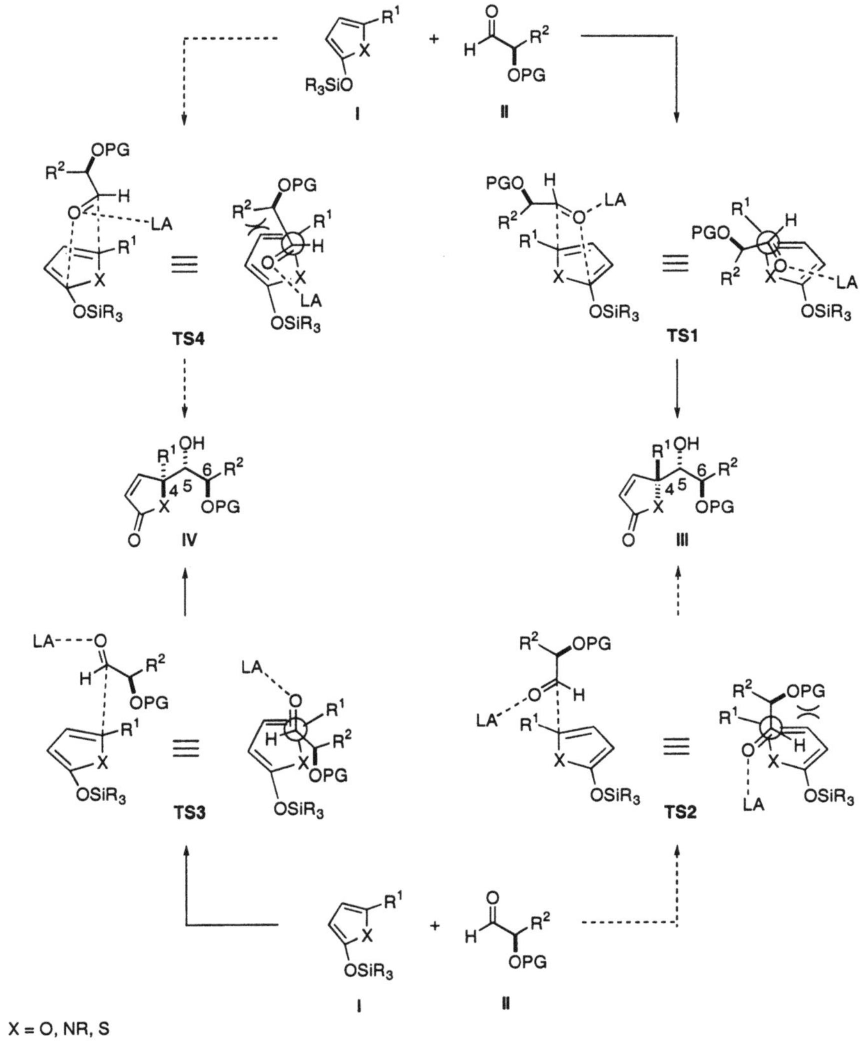

Figure 2.

implications (minor hindrance and favorable charge dispersion), whereas Diels–Alder model **TS1** appears to be favored over **TS4** owing to favorable orbital overlap and less severe congestion. On the other hand, facing the experimental results, a Diels–Alder like approach according to **TS1** (*si–si* trajectory) seems to be the route of choice due to the favorable orbital overlap contribution. We should remember that 4,5-*anti* preference favoring adducts of type **IV** has been shown in reactions involving fluorine anion agents (TBAF),[23–25] while 5,6-*syn* preference, arising from α-chelation stereocontrol, has only scantly been observed since this behavior is limited to the use of strongly coordinating Lewis acids such as $Ti(OPr^i)_2Cl_2$.[60]

Also, worth of notice is the remarkable 4,5-*syn*-5,6-*anti* preference displayed by α-amino substituted aldehydes (e.g. serinal derivatives), as well as the scarce influence on the usual reaction behavior exerted by some substituents located in the various positions of the silyloxy diene partners.

B. Additions to Cyclic Iminium and Oxonium Species

A series of syntheses described in this article took advantage of Lewis acid-assisted coupling of 2-silyloxy dienes of type **II** (X = O, NR, S) to in situ-generated cyclic oxonium (Y = O) and iminium (Y = NR) species of type **I**, generating useful binuclear templates **III** and **IV** (Figure 3).

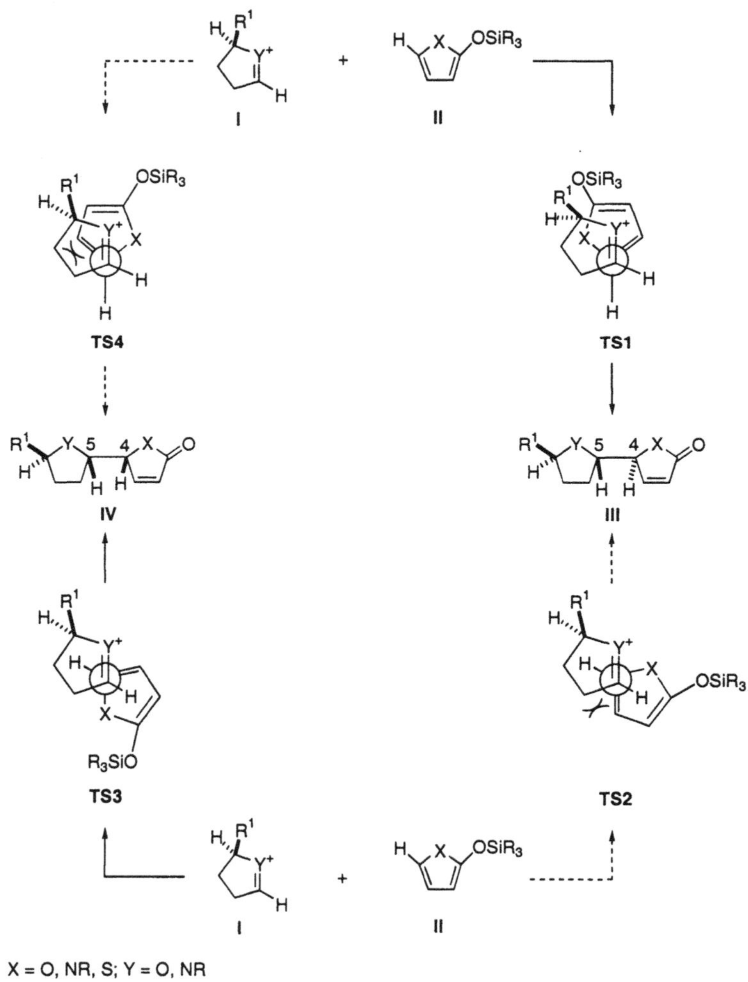

Figure 3.

As generally observed, irrespective of the nature of the electrophilic substrates, all reactions showed moderate to high 4,5-*threo* selectivity (simple diastereoselectivity) producing adducts **III**, predominantly. The formation of *threo*-products **III** can occur from either transitions state **TS1** or **TS2**, while the minor *erythro*-products **IV** would be generated via **TS3** or **TS4**. Based upon the experimental results and purely empirical conjectures, Diels–Alder arrangement **TS1** should be seemingly preferred, even though a contribution from an open-chain approach **TS2** cannot be excluded. Noticeably, when certain substituents on the cyclic electrophilic reagent **I** are present, their α or β location governs the trajectory of the incoming nucleophile, strongly favoring a *trans* stereo-disposition of the groups in the resulting adducts (facial diastereoselectivity).

C. Additions to Imines, Nitrones, and Acyclic Iminium Derivatives

The potential of the addition of heterosubstituted carbon nucleophiles to imines, nitrones, and iminium derivatives allowing the assembly of vicinal aminoalcohol, diamine, and thioamine segments was also explored in the silyloxy diene chemistry field. However, in spite of the synthetic relevance of this maneuver, a uniform rationale covering the stereochemical results herein discussed is quite hard to formulate, owing to the unpredictable behaviors displayed by the different reactions. When imines or acyclic iminium ions were employed, kinetic preference for 4,5-*syn*-configured adducts was observed, though examples leading to reverted diastereoselection (4,5-*anti*) were also reported.[109] The same stereochemical variability also featured a set of reactions involving nitrone-based reagents, wherein the 4,5-*syn* vs. 4,5-*anti* preference was severely dictated by the nature of the nitrone substituents.

D. Auxiliary-Driven Induction

The majority of the addition reactions grouped herein, leading to homochiral intermediates, were based on chirality transfer from a given intrinsically chiral electrophilic substrate, while the silyloxy diene partner was achiral. However, few, highly interesting examples dealt with chirality transmittal from the silyloxy diene in which a chiral auxiliary appendage was the stereocontrol element. Two successful applications of this asymmetric tactic have been proposed by the Baldwin's[91] and Royer's groups,[95,96] exploiting chiral inductors anchored to suitable 2-silyloxy pyrroles.

As shown in Figure 4, in both instances, the predominant coupling adducts **292** and **307** possess a *N,O-syn* configuration, which can be rationalized by the two transition states **TS1** and **TS2**, respectively. Of note, while the *syn* preference is dictated, as usual, by stereoelectronic implications (vide infra), the facial selectivity (i.e. the selection of the trajectory with which the incoming electrophile attacks a given face of the nucleophile) is governed by the auxiliary component. Quite surprisingly, the absolute configuration of the reaction product **292** claims the

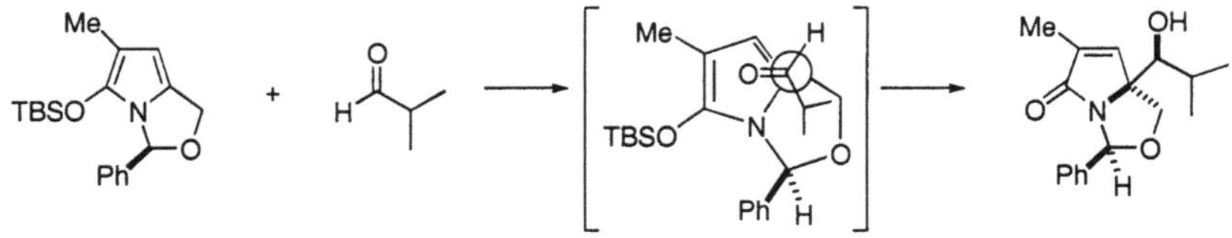

291 TS1 292

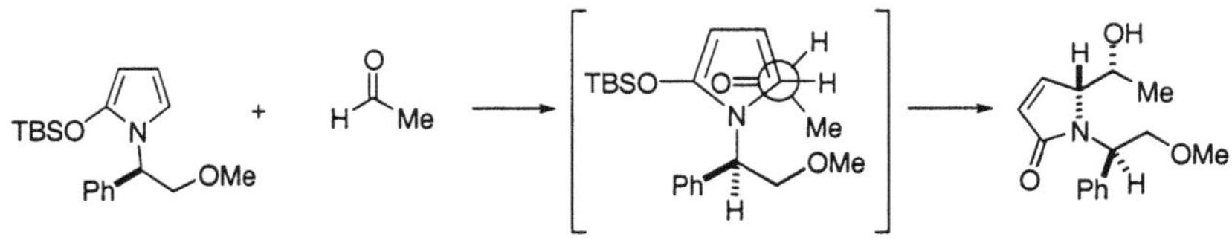

306 TS2 307

Figure 4.

addition of the aldehyde to the same face as the bulky phenyl substituent, possibly due to a sort of favorable interaction between the lipophilic moieties within the reacting molecules. On the other hand, the reaction of **306** with acetaldehyde occurs with a reverted behavior (*si–si* trajectory), as expected, according to a conventional open-chain model. The superior performance of the reaction leading to **292** (90:10 dr) as compared to that of the **306**-to-**307** conversion (74:18 dr) may be confidentially attributed to the more favorable steric constraint of the bicyclic model **TS1** vis-à-vis the more flexible monocyclic counterpart **TS2**.

X. ADDENDUM

After submission of the original manuscript, few relevant research papers appeared concerning certain sections of this chapter,[110–114] while a review report listing synthetic methodologies to γ-alkylidene butenolides exploiting, *inter alia*, 2-oxyfuran derivatives was compiled.[115] To update the literature information on the entitled topic through the end 1997, short commentaries follow.

Section IV.A.

Continuing his work on the assembly of natural oligo-tetrahydrofuranic compounds, Figadère[110] reported a replicative synthetic strategy toward a number of THF building blocks which can be used as key intermediates in the total synthesis of Annonaceous acetogenins.

As depicted in Scheme 73, replicative *C*-glycosylation of muricatacin-related acetoxytetrahydrofuran **370** with trimethylsilyloxyfuran **10** followed by double bond saturation, cleanly afforded homologated binuclear THF scaffolds **371** and

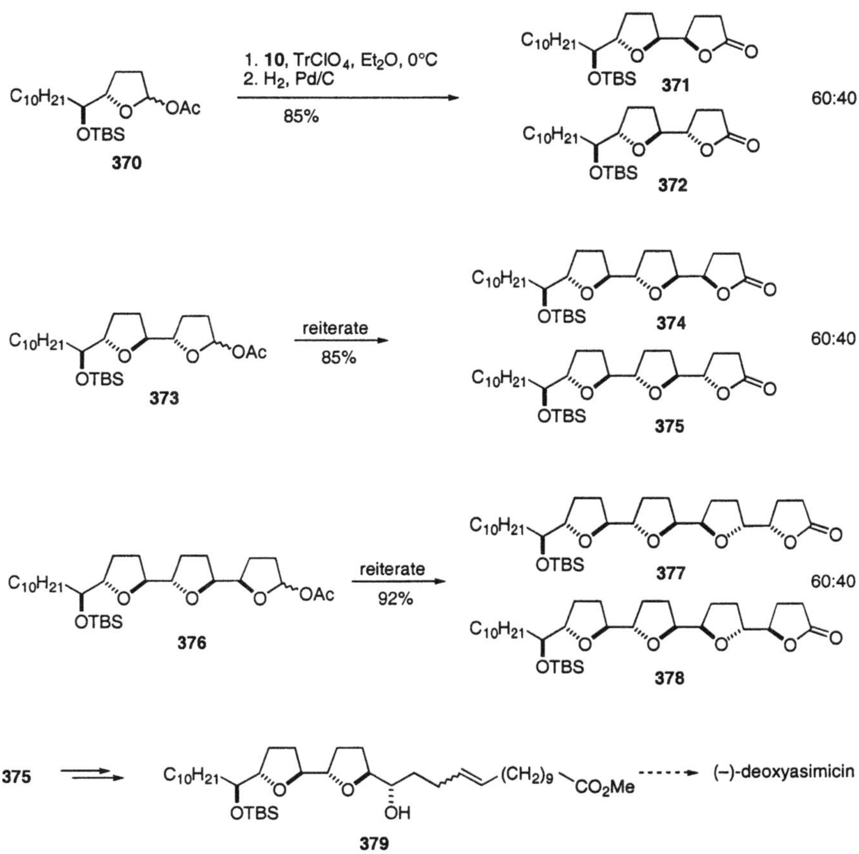

Scheme 73.

372, which were then used to forge a series of structurally defined higher oligomers **374**, **375**, **377**, and **378**.

As an example, trinuclear lactone **375** was elaborated into elongated fragment **379**, whose conversion into naturally occurring (–)-deoxyasimicin is precendented.

Parallel, concise syntheses of both enantiomers of *trans*-β-hydroxypipecolic acid, conformationally restricted serine analogues, were executed by our research team[111] exploiting, as the introductory move, the diastereoselective vinylogous aldol coupling between TBSOF **11** and both enantiomers of glyceraldehyde *N*-benzyl imine (Scheme 74).

To access L-pipecolic acid **384**, the major aldol coupling adduct **380** was sequentially subjected to catalytic hydrogenation and base-promoted ring expansion to piperidinone **381**, which was cleanly reduced to crystalline pyrrolidine **382**

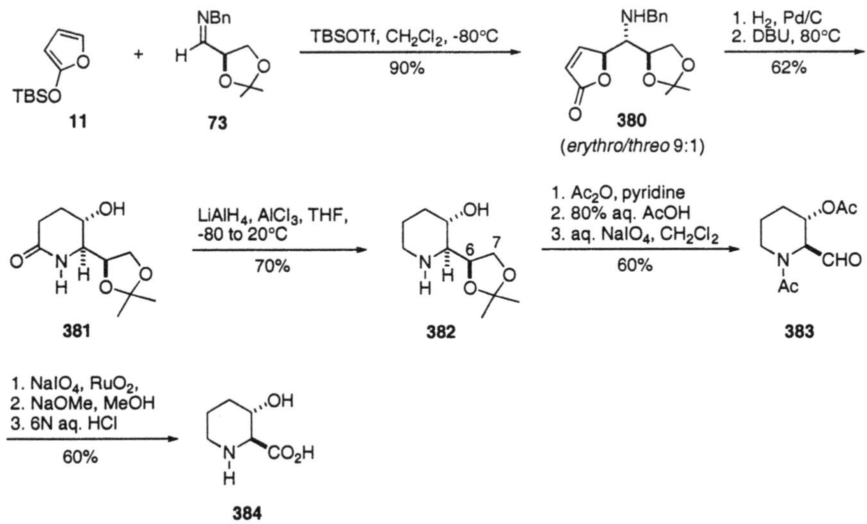

Scheme 74.

in good yield (X-ray analysis). Protective group adaptation and oxidative fission of the C-6–C-7 diol to a carboxyl function finally completed the total synthesis of **384**, through the intermediacy of amino aldehyde **383**.

Section IV.B.

In a wide study focused on the addition reactions of silyloxy dienes to 2-substituted 1,4-benzoquinones and 1,4-naphthoquinones, Brimble and Elliott[112] reported a series of cycloadditive coupling reactions involving TMSOF **10**.

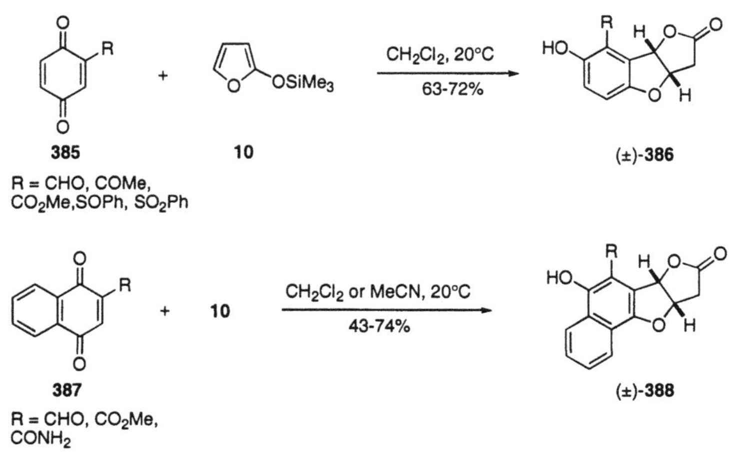

Scheme 75.

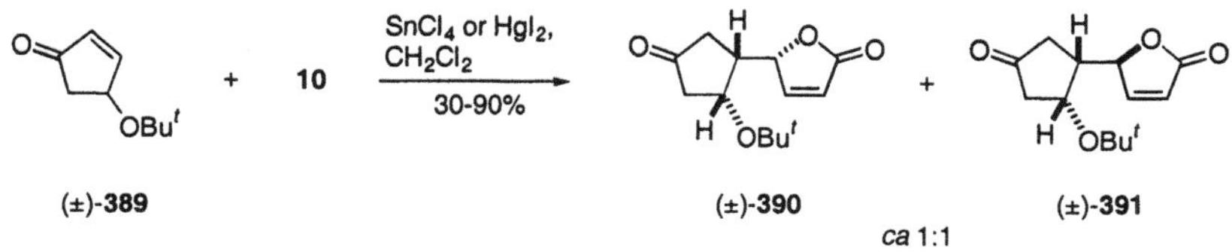

Scheme 76.

As shown in Scheme 75, all quinone substrates of type **385** behave similarly, producing the corresponding racemic adducts (±)-**386**, while naphthoquinones **387** gave rise to the expected racemic tetracyclic derivatives (±)-**388**. It is the opinion of the authors that the mechanism of these additions may lie at the borderline between a concerted (Diels–Alder) and a stepwise addition–ring closure process.

A partially diastereoselective conjugate addition of TMSOF **10** to racemic enone (±)-**389** leading to a 1:1 mixture of *cis-anti-* and *cis-syn-*adducts (±)-**390** and (±)-**391** was recently reported by Haynes et al. (Scheme 76) as part of a wider study aimed at the total synthesis of biologically active brefeldin A.[113]

Section VIII.

A remarkable example of enantioselective Michael addition reactions of 2-(trimethylsilyloxy)furan (**10**) and its 3-methyl derivative to 3-[(*E*)-2-butenoyl)]-1,3-oxazolidin-2-one (**392**) was reported by Kitajima and Katsuki[114] using suitable chiral scandium or copper complexes (Scheme 77).

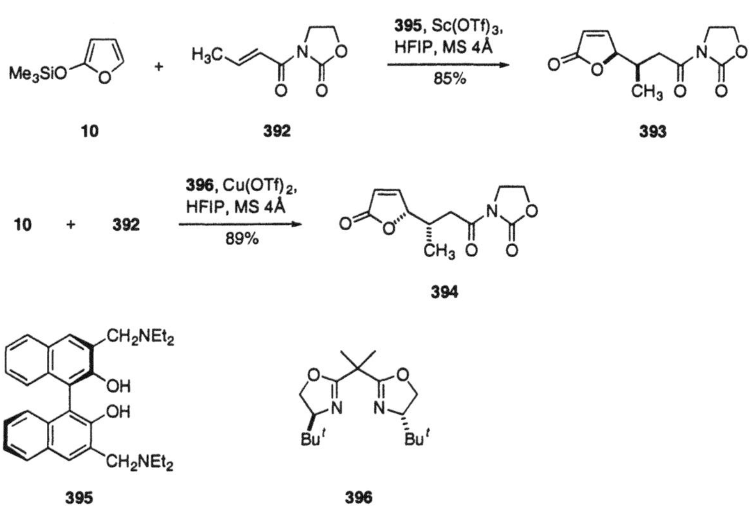

Scheme 77.

Thus, for example, when **10** was reacted with **392** in the presence of 10 mol% of $Sc(OTf)_3$, 12 mol% of chiral ligand **395**, and hexafluoroisopropanol (HFIP) as an additive, (*R,R*)-configured *anti*-adduct **393** was obtained in 85% yield, 68% ee, and 50:1 *anti:syn* ratio. On the other hand, utilizing $Cu(OTf)_2$-bis(oxazoline) **396** complex as a catalyst, (*S,S*)-*anti*-adduct **394** formed in 89% yield, 95% ee, and 8.5:1 *anti:syn* ratio. To our knowledge, the disclosed reactions represent the first example of catalytic asymmetric procedures involving the heterocyclic silyloxy dienes of this compilation.

ACKNOWLEDGMENTS

We would like to express our sincere thanks to the effective members of our research group, Dr. Luigi Pinna and Dr. Mara Cornia, for their invaluable contributions throughout the work. We would also like to acknowledge the many colleagues, fellows, and students whose names are in our papers. Research in the authors' laboratories has been generously supported by the Consiglio Nazionale delle Ricerche (CNR), Ministero dell'Università e della Ricerca Scientifica e Tecnologica (MURST), and Regione Autonoma Sardegna.

REFERENCES AND NOTES

1. (a) Moos, W. H.; Green, G. D.; Pavia, M. R. *Annu. Rep. Med. Chem.* **1993**, *28*, 315–324; (b) Gallop, M. A.; Barrett, R. W.; Dower, W. J.; Fodor, S. P. A.; Gordon, E. M. *J. Med. Chem.* **1994**, *37*, 1233–1251; (c) Gordon, E. M.; Barrett, R. W.; Dower, W. J.; Fodor, S. P. A.; Gallop, M. A. *J. Med. Chem.* **1994**, *37*, 1385–1401; (d) Madden, D.; Krchnák, V.; Lebl, M. *Perspec. Drug Dis. Design* **1995**, *2*, 269–285; (e) Ellman, J. A. *Chemtracts-Org. Chem.* **1995**, *8*, 1–4; (f) Pirrung, M. C. *ibid.* **1995**, *8*, 5–12; (g) Czarnik, A. W. *ibid* **1995**, *8*, 13–18; (h) Mitscher, L. A. *ibid* **1995**, *8*, 19–25; (i) Thompson, L. A.; Ellman, J. A. *Chem. Rev.* **1996**, *96*, 555–600; (j) Balkenhohl, F.; von dem Bussche-Hünnefeld, C.; Lansky, A.; Zechel, C. *Angew. Chem, Int. Ed. Engl.* **1996**, *35*, 2288–2337.
2. Hudlicky, T. *Chem. Rev.* **1996**, *96*, 3–30.
3. Casiraghi, G.; Rassu, G. *Synthesis* **1995**, 607–626.
4. Mukaiyama, T.; Narasaka, K.; Banno, K. *Chem. Lett.* **1973**, 1011–1014. Mukaiyama, T.; Banno, K.; Narasaka, K. *J. Am. Chem. Soc.* **1974**, *96*, 7503–7509. Costa, A. L.; Tagliavini, E. In *Seminars in Organic Synthesis*; Società Chimica Italiana, Ed.; Editor: Milano, 1996, pp. 185–210.
5. Jefford, C. W.; Jaggi, D.; Sledeski, A. W.; Boukouvalas, J. In *Studies in Natural Products Chemistry*; Atta-ur-Rahman, Ed.; Elsevier: Amsterdam, 1989, Vol. 3, pp. 157–171.
6. Casiraghi, G.; Rassu, G. In *Studies in Natural Products Chemistry*; Atta-ur-Rahman, Ed.; Elsevier: Amsterdam, 1992, Vol. 11, pp. 429–480.
7. Casiraghi, G.; Zanardi, F.; Rassu, G.; Spanu, P. *Chem. Rev.* **1995**, *95*, 1677–1716.
8. Casiraghi, G.; Rassu, G.; Zanardi, F. In *Carbohydrate Mimics: Concepts and Methods*; Chapleur, Y., Ed.; VCH: Weinheim, 1998, Chap. 17, pp. 327–348.
9. Asaoka, M.; Miyake, K.; Takei, H. *Chem. Lett.* **1977**, 167–170.
10. 2-(Trimethylsilyloxy)furan (**10**) is sold by Aldrich.
11. Pelter, A.; Al-Bayati, R. I. H.; Ayoub, M. T.; Lewis, W.; Pardasani, P.; Hansel, R. *J. Chem. Soc., Perkin Trans. 1* **1987**, 717–742.
12. Pelter, A.; Al-Bayati, R. I. H.; Lewis, W. *Tetrahedron Lett.* **1982**, *23*, 353–356.
13. Rassu, G.; Zanardi, F.; Battistini, L.; Gaetani, E.; Casiraghi, G. *J. Med. Chem.* **1997**, *40*, 168–180.
14. Casiraghi, G.; Rassu, G.; Spanu, P.; Pinna, L. *J. Org. Chem.* **1992**, *57*, 3760–3763.
15. Rassu, G.; Spanu, P.; Pinna, L.; Zanardi, F.; Casiraghi, G. *Tetrahedron Lett.* **1995**, *36*, 1941–1944.

16. Fiorenza, M.; Reginato, G.; Ricci, A.; Taddei, M.; Dembech, P. *J. Org. Chem.* **1984**, *49*, 551–553.
17. Dibutyl borinates, quickly available from the corresponding α,β-unsaturated heterocycles by treatment with dibutylboron triflate (Pr^i_2EtN, CH_2Cl_2, –78 °C), reacted with a variety of aliphatic and aromatic aldehydes, giving the expected 3-substituted adducts in high yields. Casiraghi, G.; Zanardi, F.; Battistini, L. Unpublished results.
18. Yoshii, E.; Koizumi, T.; Kitatsuji, E.; Kawazoe, T.; Kaneko, T. *Heterocycles* **1976**, *4*, 1663–1668.
19. Asaoka, M.; Sugimura, N.; Takei, H. *Bull. Chem. Soc. Jpn.* **1979**, *52*, 1953–1956.
20. Asaoka, M.; Yanagida, N.; Ishibashi, K.; Takei, H. *Tetrahedron Lett.* **1981**, *22*, 4269–4270.
21. Asaoka, M.; Yanagida, N.; Takei, H. *Tetrahedron Lett.* **1980**, *21*, 4611–4614.
22. Fiorenza, M.; Ricci, A.; Romanelli, M. N.; Taddei, M.; Dembech, P.; Seconi, G. *Heterocycles* **1982**, *19*, 2327–2329.
23. Brown, D. W.; Campbell, M. M.; Taylor, A. P.; Zhang, X.-A. *Tetrahedron Lett.* **1987**, *28*, 985–988.
24. Jefford, C. W.; Jaggi, D.; Boukouvalas, J. *Tetrahedron Lett.* **1987**, *28*, 4037–4040.
25. Jefford, C. W.; Jaggi, D.; Bernardinelli, G.; Boukouvalas, J. *Tetrahedron Lett.* **1987**, *28*, 4041–4044.
26. Jefford, C. W.; Sledeski, A. W.; Boukouvalas, J. *Tetrahedron Lett.* **1987**, *28*, 949–950.
27. Jefford, C. W.; Sledeski, A. W.; Rossier, J.-C.; Boukouvalas, J. *Tetrahedron Lett.* **1990**, *31*, 5741–5744.
28. Brownbridge, P.; Chan, T.-H. *Tetrahedron Lett.* **1980**, *21*, 3423–3426.
29. Brownbridge, P.; Chan, T.-H. *Tetrahedron Lett.* **1980**, *21*, 3427–3430.
30. Brownbridge, P.; Chan, T.-H. *Tetrahedron Lett.* **1980**, *21*, 3431–3434.
31. Hanessian, S. *The Total Synthesis of Natural Compounds: the Chiron Approach*; Pergamon Press: Oxford, 1983.
32. Casiraghi, G.; Colombo, L.; Rassu, G.; Spanu, P. *Tetrahedron Lett.* **1989**, *30*, 5325–5328.
33. Casiraghi, G.; Colombo, L.; Rassu, G.; Spanu, P.; Gasparri Fava, G.; Ferrari Belicchi, M. *Tetrahedron* **1990**, *46*, 5807–5824.
34. Casiraghi, G.; Pinna, L.; Rassu, G.; Spanu, P.; Ulgheri, F. *Tetrahedron: Asymmetry* **1993**, *4*, 681–686.
35. Rassu, G.; Spanu, P.; Casiraghi, G.; Pinna, L. *Tetrahedron* **1991**, *47*, 8025–8030.
36. Casiraghi, G.; Colombo, L.; Rassu, G.; Spanu, P. *J. Org. Chem.* **1991**, *56*, 2135–2139.
37. Gasparri Fava, G.; Belicchi Ferrari, M.; Casiraghi, G.; Rassu, G.; Spanu, P. *J. Cryst. Spectrosc. Res.* **1991**, *21*, 629–633.
38. Casiraghi, G.; Colombo, L.; Rassu, G.; Spanu, P. *J. Org. Chem.* **1990**, *55*, 2565–2567.
39. Gasparri Fava, G.; Ferrari Belicchi, M.; Belletti, D.; Casiraghi, G.; Rassu, G. *J. Cryst. Spectrosc. Res.* **1991**, *21*, 261–264.
40. Knapp, S. *Chem. Rev.* **1995**, *95*, 1859–1876.
41. Casiraghi, G.; Colombo, L.; Rassu, G.; Spanu, P. *J. Chem. Soc., Chem. Commun.* **1991**, 603–604.
42. Casiraghi, G.; Colombo, L.; Rassu, G.; Spanu, P. *J. Org. Chem.* **1991**, *56*, 6523–6527.
43. Rassu, G.; Pinna, L.; Spanu, P.; Culeddu, N.; Casiraghi, G.; Gasparri Fava, G.; Belicchi Ferrari, M.; Pelosi G. *Tetrahedron* **1992**, *48*, 727–742.
44. Casiraghi, G.; Rassu, G.; Spanu, P.; Pinna, L.; Ulgheri, F. *J. Org. Chem.* **1993**, *58*, 3397–3400.
45. Rassu, G.; Casiraghi, G.; Pinna, L.; Spanu, P.; Ulgheri, F.; Cornia, M.; Zanardi, F. *Tetrahedron* **1993**, *49*, 6627–6636.
46. Gasparri Fava, G.; Belicchi Ferrari, M.; Pelosi, G.; Casiraghi, G. Unpublished results.
47. Gu, Z.-M.; Zhao, G. X.; Oberlies, N. H.; Zeng, L.; McLaughlin, J. L. In *Recent Advances in Phytochemistry; Arnason*, J. T., Mata, R., Romeo, J. T., Eds.; Plenum Press: New York, 1995, Vol. 29, pp. 249–310. Zeng, L.; Ye, Q.; Oberlies, N. H.; Shi, G.; Gu, Z.-M.; He, K.; McLaughlin, J. L. *Nat. Prod. Rep.* **1996**, 275–306. Koert, U. *Synthesis* **1995**, 115–132. Figadère, B. *Acc. Chem. Res.* **1995**, *28*, 359–365. Cavé, A.; Figadère, B., Laurens, A.; Cortes, D. In *Progress in the Chemistry of Organic Natural Products*; Hertz, W., Ed.; Springer-Verlag: Wien, New York, 1997, Vol. 70, pp. 81–288.

48. Koert, U.; Stein, M.; Harms, K. *Tetrahedron Lett.* **1993**, *34*, 2299–2302.
49. Figadère, B.; Chaboche, C.; Peyrat, J.-F.; Cavé, A. *Tetrahedron Lett.* **1993**, *34*, 8093–8096.
50. Pichon, M.; Figadère, B.; Cavé, A. *Tetrahedron Lett.* **1996**, *37*, 7963–7966.
51. Martin, S. F.; Barr, K. J. *J. Am. Chem. Soc.* **1996**, *118*, 3299–3300.
52. Hanessian, S.; Raghavan, S. *Bioorg. Med. Chem. Lett.* **1994**, *4*, 1697–1702.
53. Hanessian, S.; Reddy, G. B. *Bioorg. Med. Chem. Lett.* **1994**, *4*, 2285–2290.
54. Hanessian, S.; McNaughton-Smith, G. *Bioorg. Med. Chem. Lett.* **1996**, *6*, 1567–1572.
55. Bernardi, A.; Cardani, S.; Carugo, O.; Colombo, L.; Scolastico, C.; Villa, R. *Tetrahedron Lett.* **1990**, *31*, 2779–2782.
56. Bernardi, A.; Piarulli, U.; Poli, G.; Scolastico, C.; Villa, R. *Bull. Soc. Chim. Fr.* **1990**, *127*, 751–757.
57. Pelter, A.; Ward, R. S.; Sirit, A. *Tetrahedron: Asymmetry* **1994**, *5*, 1745–1762.
58. Xu, D.; Sharpless, K. B. *Tetrahedron Lett.* **1994**, *35*, 4685–4688.
59. Ko, S. Y.; Lerpiniere, J. *Tetrahedron Lett.* **1995**, *36*, 2101–2104.
60. Mukai, C.; Kim, I. J.; Hanaoka, M. *Tetrahedron Lett.* **1993**, *34*, 6081–6082.
61. Castellari, C.; Lombardo, M.; Pietropaolo, G.; Trombini, C. *Tetrahedron: Asymmetry* **1996**, *7*, 1059–1068.
62. Bauer, T. *Tetrahedron: Asymmetry* **1996**, *7*, 981–984.
63. Harding, K. E.; Coleman, M. T.; Liu, L. T. *Tetrahedron Lett.* **1991**, *32*, 3795–3798.
64. Martin, S. F.; Corbett, J. W. *Synthesis* **1992**, 55–57.
65. Morimoto, Y.; Nishida, K.; Hayashi, Y.; Shirahama, H. *Tetrahedron Lett.* **1993**, *34*, 5773–5776.
66. Morimoto, Y.; Iwahashi, M. *Synlett* **1995**, 1221–1222.
67. Camiletti, C.; Poletti, L.; Trombini, C. *J. Org. Chem.* **1994**, *59*, 6843–6846.
68. Fukuyama, T.; Yang, L. *J. Am. Chem. Soc.* **1989**, *111*, 8303–8304.
69. Fukuyama, T.; Yang, L. *J. Am. Chem. Soc.* **1987**, *109*, 7881–7882.
70. Brimble, M. A.; Hodges, R.; Stuart, S. J. *Tetrahedron Lett.* **1988**, *29*, 5987–5990.
71. Brimble, M. A.; Brimble, M. T.; Gibson, J. J. *J. Chem. Soc., Perkin Trans. 1* **1989**, 179–184.
72. Brimble, M. A.; Stuart, S. J. *J. Chem. Soc., Perkin Trans. 1* **1990**, 881–885.
73. Brimble, M. A.; Lynds, S. M. *J. Chem. Soc. Perkin Trans. 1* **1994**, 493–496.
74. Boukouvalas, J.; Maltais, F.; Lachance, N. *Tetrahedron Lett.* **1994**, *35*, 7897–7900.
75. Boukouvalas, J.; Maltais, F. *Tetrahedron Lett.* **1995**, *36*, 7175–7176.
76. Boukouvalas, J.; Maltais, F. *Tetrahedron Lett.* **1994**, *35*, 5769–5770.
77. Itoh, T.; Matsuya, Y.; Hasegawa, H.; Nagata, K.; Okada, M.; Ohsawa, A. *J. Chem. Soc., Perkin Trans. 1* **1996**, 2511–2515.
78. Rassu, G.; Casiraghi, G.; Spanu, P.; Pinna, L.; Gasparri Fava, G.; Belicchi Ferrari, M.; Pelosi, G. *Tetrahedron: Asymmetry* **1992**, *3*, 1035–1048.
79. Rassu, G.; Pinna, L.; Spanu, P.; Ulgheri, F.; Cornia, M.; Zanardi, F.; Casiraghi, G. *Tetrahedron* **1993**, *49*, 6489–6496.
80. Liddell, J. R. *Nat. Prod. Rep.* **1996**, *13*, 187–193. Casiraghi, G.; Zanardi, F.; Rassu, G.; Pinna, L. *Org. Prep. Proced. Int.* **1996**, *28*, 641–682. Kibayashi, C. In *Studies in Natural Products Chemistry*; Atta-ur-Rahman, Ed.; Elsevier: Amsterdam, 1992, Vol. 11, pp. 229–275. Michael, J. P. *Nat. Prod. Rep.* **1997**, *14*, 21–41.
81. Casiraghi, G.; Ulgheri F.; Spanu, P.; Rassu, G.; Pinna, L.; Gasparri Fava, G.; Belicchi Ferrari, M.; Pelosi, G. *J. Chem. Soc., Perkin Trans. 1* **1993**, 2991–2997.
82. Casiraghi, G.; Spanu, P.; Rassu, G.; Pinna, L.; Ulgheri, F. *J. Org. Chem.* **1994**, *59*, 2906–2909.
83. Casiraghi, G.; Rassu, G.; Spanu, P.; Pinna, L. *Tetrahedron Lett.* **1994**, *35*, 2423–2426.
84. Rassu, G.; Zanardi, F.; Cornia, M.; Casiraghi, G. *J. Chem. Soc., Perkin Trans. 1* **1994**, 2431–2437.
85. Rassu, G.; Zanardi, F.; Battistini, L.; Casiraghi, G. *Tetrahedron: Asymmetry* **1995**, *6*, 371–374.
86. Soro, P.; Rassu, G.; Spanu, P.; Pinna, L.; Zanardi, F.; Casiraghi, G. *J. Org. Chem.* **1996**, *61*, 5172–5174.
87. Zanardi, F.; Battistini, L.; Nespi, M.; Rassu, G.; Spanu, P.; Cornia, M.; Casiraghi, G. *Tetrahedron: Asymmetry* **1996**, *7*, 1167–1180.

88. Spanu, P.; Rassu, G.; Ulgheri, F.; Zanardi, F.; Battistini, L.; Casiraghi, G. *Tetrahedron* **1996**, *52*, 4829–4838.
89. Spanu, P.; Rassu, G.; Pinna, L.; Battistini, L.; Casiraghi, G. *Tetrahedron: Asymmetry* **1997**, *8*, 3237.
90. Rassu, G.; Pinna, L.; Spanu, P.; Ulgheri, F.; Casiraghi, G. *Tetrahedron Lett.* **1994**, *35*, 4019–4022.
91. Uno, H.; Baldwin, J. E.; Russell, A. T. *J. Am. Chem. Soc.* **1994**, *116*, 2139–2140.
92. Seebach, D.; Sting, A. R.; Hoffmann, M. *Angew. Chem., Int. Ed. Engl.* **1996**, *35*, 2708–2748.
93. Poli, G.; Ciofi Baffoni, S.; Maccagni, E.; Sardone, N. *Tetrahedron Lett.* **1995**, *36*, 8669–8672.
94. Ciofi Baffoni, S. Laurea Thesis, University of Firenze, 1996.
95. Baussanne, I.; Royer, J. *Tetrahedron Lett.* **1996**, *37*, 1213–1216.
96. Baussanne, I.; Schwardt, O.; Royer, J.; Pichon, M.; Figadère, B.; Cavé, A. *Tetrahedron Lett.* **1997**, *38*, 2259–2262.
97. Zanardi, F.; Battistini, L.; Rassu, G.; Cornia, M.; Casiraghi, G. *J. Chem. Soc., Perkin Trans. 1* **1995**, 2471–2475.
98. Gasparri Fava, G.; Belicchi Ferrari, M.; Pelosi, G.; Zanardi, F.; Casiraghi, G.; Rassu, G. *J. Chem. Cryst.* **1996**, *26*, 509–513.
99. Rassu, G.; Pinna, L.; Spanu, P.; Zanardi, F.; Battistini, L.; Casiraghi, G. *J. Org. Chem.* **1997**, *62*, 4513–4517.
100. Zanardi, F.; Battistini, L.; Rassu, G.; Pinna, L.; Mor, M.; Culeddu, N.; Casiraghi, G. *J. Org. Chem.* **1998**, *63*, 1368–1369.
101. Sha, C.-K.; Shen, C.-Y.; Lee, R.-S.; Lee, S.-R.; Wang, S.-L. *Tetrahedron Lett.* **1995**, *36*, 1283–1286.
102. Sha, C.-K.; Lee, S.-J.; Tseng, W.-H. *Tetrahedron Lett.* **1997**, *38*, 2725–2728.
103. Schlessinger, R. H.; Pettus, T. R. R.; Springer, J. P.; Hoogsteen, K. *J. Org. Chem.* **1994**, *59*, 3246–3247.
104. Schlessinger, R. H.; Wu, X.-H.; Pettus, T. R. R. *Synlett* **1995**, 536–538.
105. Schlessinger, R. H.; Bergstrom, C. P. *J. Org. Chem.* **1995**, *60*, 16–17.
106. Schlessinger, R. H.; Bergstrom, C. P. *Tetrahedron Lett.* **1996**, *37*, 2133–2136.
107. For a general, modern treatise dealing with the various aspects of the asymmetric carbon–carbon bond-forming reactions, including conditions and mechanisms, see: *Houben-Weil, Methods of Organic Chemistry*; Helmchen, G.; Hoffman, R. W.; Mulzer, J.; Shaumann, E., Eds.; George Thieme Verlag: Stuttgart, 1995, Vol. E21a–e.
108. Singer, R. A.; Carreira, E. M. *J. Am. Chem. Soc.* **1995**, *117*, 12360–12361.
109. Quite recently, the coupling reaction between 2-(*tert*-butyldimethylsilyloxy)furan (**11**) and glyceraldehyde imine **73** was re-investigated, during a study directed toward preparation of enantiopure *trans*-3-hydroxypipecolic acids (Scheme 74). When the reaction was carried out in CH_2Cl_2 at –80 °C in the presence of 0.5–0.6 molar equivalents of TBSOTf, 4,5-*anti*-5,6-*anti* adduct **380** was preferentially obtained (90% yield, 90% de), accompanied by only trace amounts of the corresponding 4,5-*anti*-5,6-*syn* isomer. See ref. 111.
110. Figadère, B.; Peyrat, J.-F.; Cavé, A. *J. Org. Chem.* **1997**, *62*, 3428–3429.
111. Battistini, L.; Zanardi, F.; Rassu, G.; Spanu, P.; Pelosi, G.; Gasparri Fava, G.; Belicchi Ferrari, M.; Casiraghi, G. *Tetrahedron: Asymmetry* **1997**, *8*, 2975–2987.
112. Brimble, M. A.; Elliott, R. J. R. *Tetrahedron* **1997**, *53*, 7715–7730.
113. Haynes, R. K.; Lam, W. W.-L.; Yeung, L.-L.; Williams, I. D.; Ridley, A. C.; Starling, S. M.; Vonwiller, S. C.; Hambley, T. W.; Lelandais, P. *J. Org. Chem.* **1997**, *62*, 4552–4553.
114. Kitajima, H.; Katsuki, T. *Synlett* **1997**, 568–570.
115. Negishi, E.; Kotora, M. *Tetrahedron* **1997**, *53*, 6707–6738.

ASYMMETRIC CATALYSIS USING HETEROBIMETALLIC COMPOUNDS

Masakatsu Shibasaki and Hiroaki Sasai

Advances in Asymmetric Synthesis
Volume 3, pages 191–233.

ISBN: 0-7623-0106-6

I. INTRODUCTION

The development of catalytic asymmetric reactions is one of the major lines of research in the field of organic chemistry. So far, a number of chiral catalysts have been reported, and some of them have exhibited a much higher catalytic efficiency than enzymes, which are natural catalysts.[1] Most of the synthetic asymmetric catalysts, however, show limited activity in terms of either enantioselectivity or chemical yields. The major difference between synthetic asymmetric catalysts and enzymes is that the former activate only one side of the substrate in an intermolecular reaction, whereas the latter cannot only activate both sides of the substrate but can also control the orientation of the substrate. If this kind of synergistic cooperation can be realized in synthetic asymmetric catalysis, the concept will open up a new field in asymmetric synthesis, and a wide range of applications may well ensue.

In this chapter we will introduce the newly developed heterobimetallic asymmetric catalysis, in which the catalyst functions, just like an enzyme, as both a Brϕnsted base and a Lewis acid.[2] Furthermore, we have also succeeded in developing oligomeric homometallic complexes, in which the same sort of amphoteric metals act cooperatively to perform efficient catalytic asymmetric reactions.

II. HETEROBIMETALLIC ASYMMETRIC CATALYSTS WITH RARE EARTH ELEMENTS

A. Preparation and Structure of Rare Earth–Alkali Metal–Tris(1,1′-bi-2-naphthoxide) Complexes (LnMB)

Our preliminary attempts to obtain a basic chiral rare earth complex of type **I** (Figure 1) have led us to create several new chiral heterobimetallic complexes which catalyze various types of asymmetric reactions. We will discuss the alkali metal-free, rare earth complexes such as **1** in Section IV. The rare earth-alkali metal-tris(1,1′-bi-2-naphthoxide) complexes (LnMB, where Ln = rare earth, M = alkali metal, and B = 1,1′-bi-2-naphthoxide) have been efficiently synthesized from the corresponding metal chloride and/or alkoxide,[3–6] and the structures of the LnMB complexes have been unequivocally determined by a combination of X-ray crys-

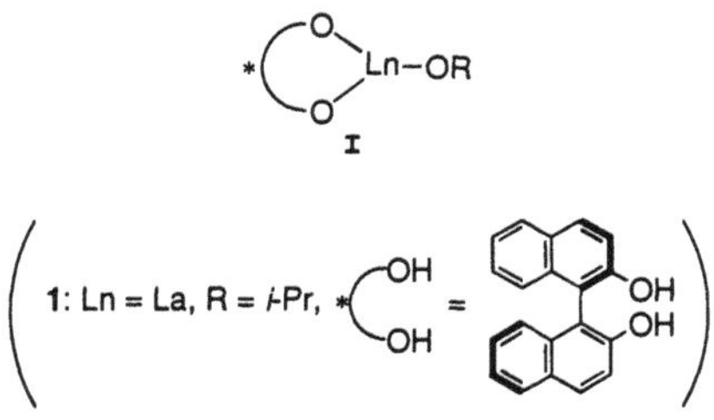

Figure 1. Structure of our first target directed towards chiral basic catalysts.

tallography and LDI-TOF-mass spectroscopy (Figure 2).[7–9] For example, the effective procedure for the synthesis of LLB (where L = lanthanum and lithium, respectively) is the treatment of $LaCl_3 \cdot 7H_2O$ with 2.7 mol equiv of BINOL dilithium salt (**3**), and NaO-*t*-Bu (0.3 mol equiv) in THF at 50 °C for 50 h, as shown in Scheme 1.[6] Alternatively, also shown in Scheme 1, we established another efficient procedure for the preparation of LLB, this time starting from La(O-*i*-Pr)$_3$,[10] the exposure of which to 3 mol equiv of BINOL (**2**) in THF is followed by the addition of butyllithium (3 mol equiv) at 0 °C. It is noteworthy that these heterobimetallic asymmetric complexes including LLB are stable in organic solvents such as THF, CH_2Cl_2, and toluene, which contain small amounts of water, and are also insensitive to oxygen.

B. Catalytic Asymmetric Nitroaldol Reaction

The nitroaldol (Henry) reaction has been recognized as a powerful synthetic tool and has been utilized in the construction of numerous natural products and other useful compounds.[11,12] As shown in Scheme 2, We succeeded in realizing the first example of a catalytic asymmetric nitroaldol reaction by use of a catalytic amount

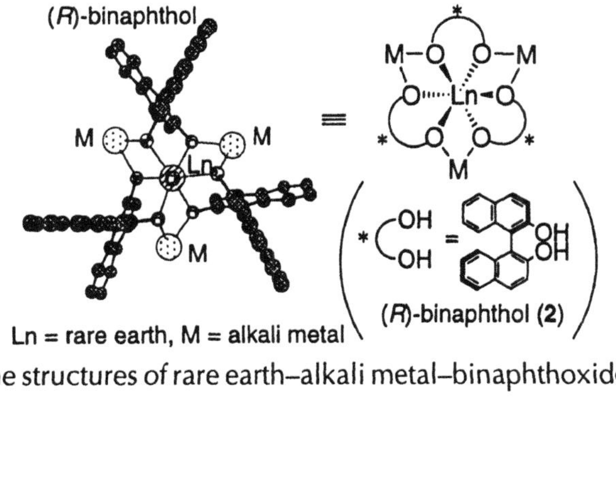

Figure 2. The structures of rare earth–alkali metal–binaphthoxide complexes (LnMB).

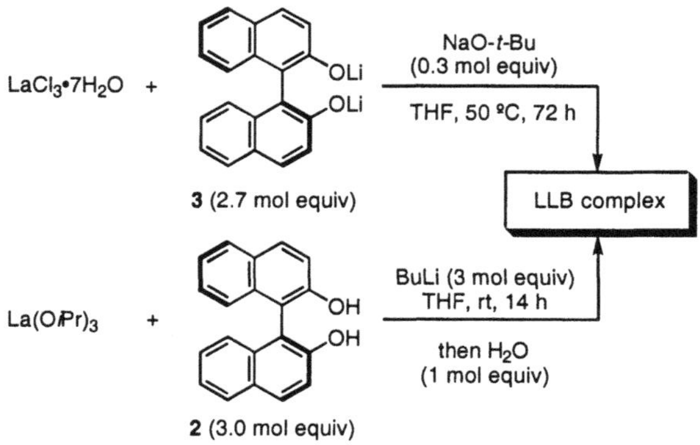

Scheme 1. Best preparative procedures for LLB.

of LLB.[3] THF appeared to give the best results as far as solvent effects are concerned.

The rare earth metals are generally regarded as a group of 17 elements with similar properties, especially with respect to their chemical reactivity. However, in the case of the above-mentioned catalytic asymmetric nitroaldol reaction, we observed pronounced differences both in the reactivity and enantioselectivity of the various rare earth metals used.[5] The unique relationship between the ionic radii of rare earth metals and the optical purities of the nitroaldols is depicted in Figure 3. For example, when benzaldehyde (**11**) and nitromethane (**5**) were used as starting materials, the corresponding Eu complex gave **12** in 72% ee (91%) in contrast to 37% ee (81%) in the case of the LLB (–40 °C, 40 h). These results suggest that small changes in the structure of the catalyst (ca. 0.1 Å in the ionic radius of the rare earth cation) cause a drastic change in the optical purity of the nitroaldols produced. Although in general nitroaldol reactions are regarded as equilibrium processes, no detectable retro-nitroaldol reactions were observed in the Ln-BINOL complex-catalyzed asymmetric nitroaldol reactions.

Having succeeded in obtaining the first results from a catalytic asymmetric nitroaldol reaction, we then attempted to apply the method to a catalytic asymmetric synthesis of biologically important compounds. The product nitroaldols were

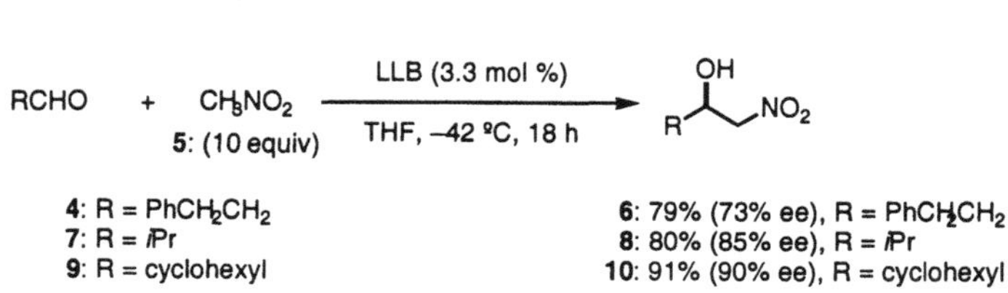

Scheme 2. Catalytic asymmetric nitroaldol reactions promoted by LLB.

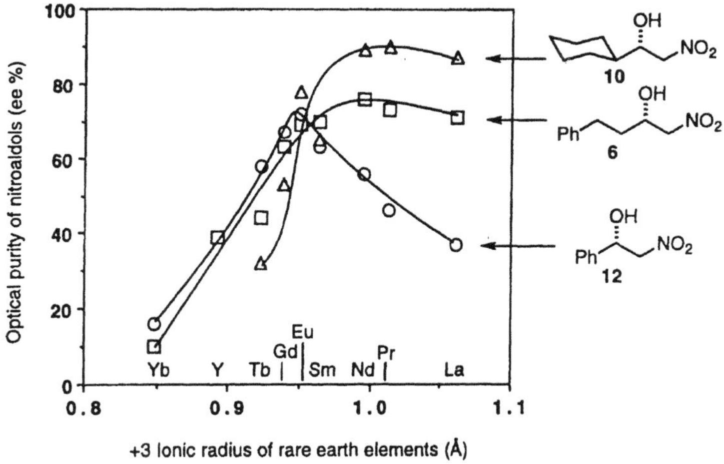

Figure 3. Effects of ionic radii of rare earth elements on the optical purities of nitroaldol derivatives.

readily converted into β-amino alcohols and/or α-hydroxy carbonyl compounds. Thus, convenient syntheses of three kinds of optically active β-blockers are presented in Scheme 3.[5,13–15] For example, using 10 mol equiv of nitromethane (**5**) and **13** at –50 °C in the presence of 3.3 mol% of (*R*)-LLB catalyst, a 90% yield of nitroaldol **14** in 94% ee was obtained. Reductive alkylation of the nitroaldol **14** to **15** was accomplished in 80% yield by a PtO_2-catalyzed hydrogenation in the presence of 5 mol equiv of acetone in methanol. (*S*)-Metoprolol (**15**) was thus

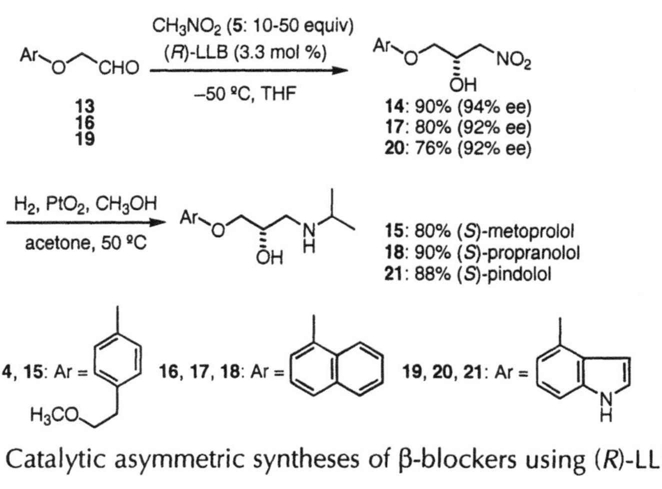

Scheme 3. Catalytic asymmetric syntheses of β-blockers using (*R*)-LLB as a catalyst.

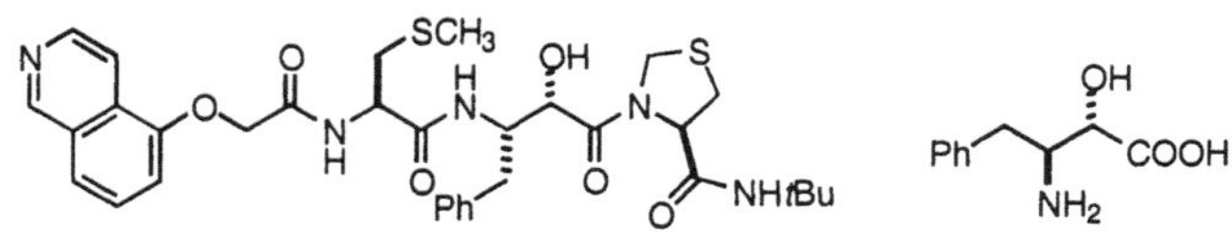

Figure 4. Structures of HIV protease inhibitor KNI-272 and its component (2*S*, 3*S*)-3-amino-2-hydroxy-4-phenylbutanoic acid (*erythro*-AHPA, **23**).

synthesized in only two steps from pro-chiral aldehyde **13**.[5,7,15] Interestingly, the nitroaldols **14**, **17**, and **20** were found to have the (*S*)-absolute configuration when (*R*)-LLB was used. The nitronates thus appear to react preferentially with the *si* face of the aldehydes, in contrast to the enantiofacial selectivity which might have been expected on the basis of previous results (cf. Scheme 2). These results suggest that the presence of an oxygen atom at the β-position greatly influences the enantiofacial selectivity.

Furthermore, diastereoselective catalytic nitroaldol reactions of optically active α-amino aldehydes with nitromethane using LLB were found to proceed in a highly diastereoselective manner.[16] The adducts (3-amino-2-hydroxy-1-nitro derivatives) are attractive intermediates for the synthesis of unnatural *erythro*-amino-2-hydroxy acids, which are important components of several biologically active compounds. As an example, the promising HIV-protease inhibitor KNI-272 (**22**)[17] contains (2*S*,3*S*)-3-amino-2-hydroxy-4-phenylbutanoic acid (*erythro*-AHPA, **23**) as a subunit (Figure 4).

As shown in Table 1, we found that treatment of *N*-phthaloyl-L-phenylalanal (**24**) with nitromethane (**5**) at –40 °C in the presence of (*R*)-LLB (3.3 mol%) gave practically a single stereoisomer of (2*R*,3*S*)-2-hydroxy-4-phenyl-3-phthaloylamino-1-nitrobutane (**25**) in 92% yield (>99:1 *erythro* selectivity). The enantiofacial selectivity for the C-2 hydroxyl group of **25** agreed with results previously observed in enantioselective nitro aldol reactions for non-β-oxa-aldehydes using LLB. Interestingly, the reaction of the (*S*)-aldehyde **24** with **5**, using the (*S*)-LLB

Table 1. Diasteroselective Nitroaldol Reaction Catalyzed by LLB

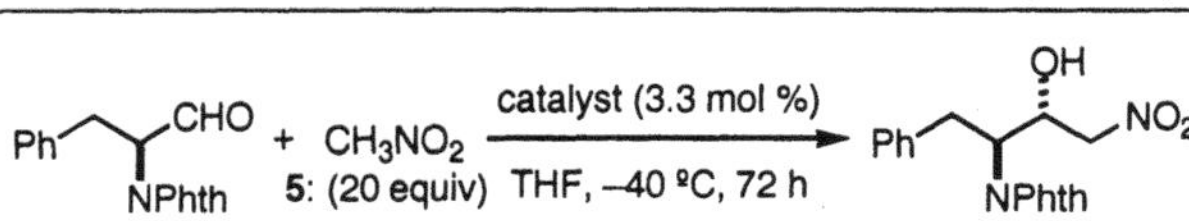

Entry	*Catalyst*	*Yield (%)*	*erythro (% ee)*	*:*	*threo*
1	(*R*)-LLB	92	99 (96)	:	1
2	(*S*)-LLB	96	74 (90)	:	26

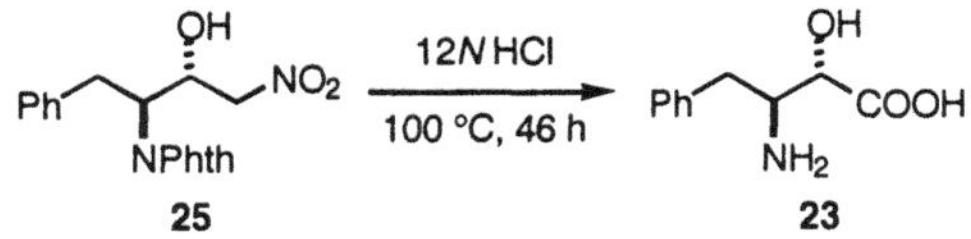

Scheme 4. Synthesis of *erythro*-AHPA (**23**).

complex as a catalyst, led to a reduced diastereoselectivity. Conversion of the nitroaldol adduct **25** into **23** was achieved in one pot (80% yield) (Scheme 4).

LLB-type catalysts were also able to promote diastereoselective and enantioselective nitroaldol reactions starting from prochiral materials. However, limited enantioselectivities (<78% ee) and diastereoselectivities (ca. 2:1–3:1) were obtained using LLB.[13] In order to obtain both high enantio- and diastereoselectivity, we focused our attention on the preparation of a novel asymmetric catalyst. We began our investigations by preparing a number of complexes from La(O-*i*-Pr)$_3$, butyllithium (3 mol equiv), and (*R*)-BINOL derivatives (3 mol equiv), in which alkyl, alkenyl, alkynyl, aryl, and/or cyano groups were introduced at certain positions. First, the utility of the complex as an asymmetric catalyst was assessed using a nitroaldol reaction of **5** with hydrocinnamaldehyde (**4**). Although 3,3′-dimethyl-BINOL[18]- and 3,3′-bis(trimethylsilyl)BINOL[19]-derived complexes gave racemic **6**, and the BIPOL[20]-derived catalyst gave **6** in only 39% ee, surprisingly, substitution at the 6,6′-position of BINOL proved effective in obtaining superior asymmetric catalysts.[21] The structures of these rare earth–lithium-6,6′-disubstituted-BINOL complexes (LnLB*; where B* = 6,6′-disubstituted-BINOL) are similar to that of the (*R*)-LLB complex depicted in Figure 5 (as elucidated by ^{1}H- and ^{13}C-NMR). The results obtained for the nitroaldol reaction at –40 °C for 91 h in THF using 3.3 mol% of the various catalysts and 10 equiv of nitromethane are shown in Table 2. Using catalysts **26–34**, nitroaldols were routinely obtained with optical purity *superior* to that of the adducts obtained using LLB.

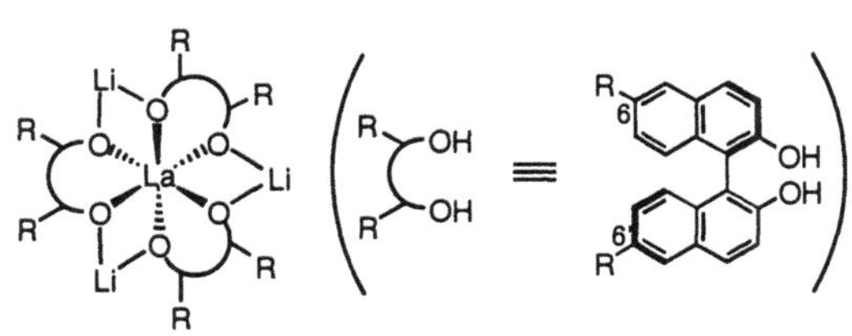

LLB: R = H
26: R = Br
27: R = CH_3
28: R = C≡N
29: R = C≡CH
30: R = C≡CPh
31: R = C≡CSi(CH_3)$_3$
32: R = C≡CSiEt$_3$
33: R = C≡CTBS
34: R = C≡CSi(CH_3)$_2$Ph

Figure 5. Structural modification of LLB.

Table 2. Catalytic Asymmetric Nitroaldol Reactions Promoted by Various LLB-Type Complexes (LLB*)

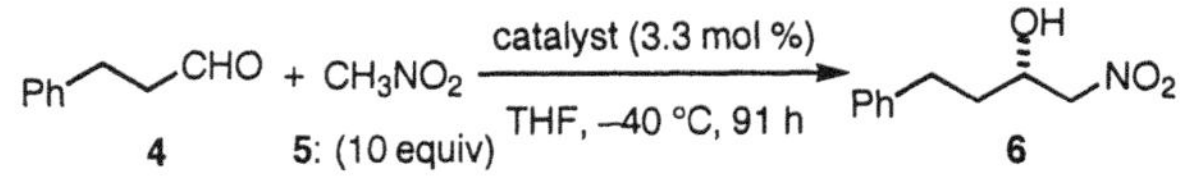

Entry	*Catalyst*	*Yield (%)*	*ee (%)*
1	LLB	79	73
2	**26**	80	67
3	**27**	84	63
4	**28**	67	55[a]
5	**29**	69	71
6	**30**	74	79
7	**31**	85	88
8	**32**	84	85
9	**33**	59	85
10	**34**	54	86

Note: [a] 6,6′-Dicyano-BINOL of 93% ee was used.

Another advantage was conferred by introducing 6,6′-substituents to BINOL. In general, catalytic asymmetric syntheses of fluorine-containing compounds are rather difficult.[22] In collaboration with Dr. Iseki and Dr. Oishi at Daikin Industries, Ltd. (Tsukuba, Japan) we found that a catalytic asymmetric nitroaldol reaction of the rather unreactive α,α-difluoro aldehydes proceeded satisfactorily when using the heterobimetallic asymmetric catalysts generated from 6,6′-bis((triethylsilyl)ethynyl)BINOL, as shown in Table 3.[23]

Table 3. Catalytic Asymmetric Nitroaldol Reactions of α,α-Difluoro Aldehydes

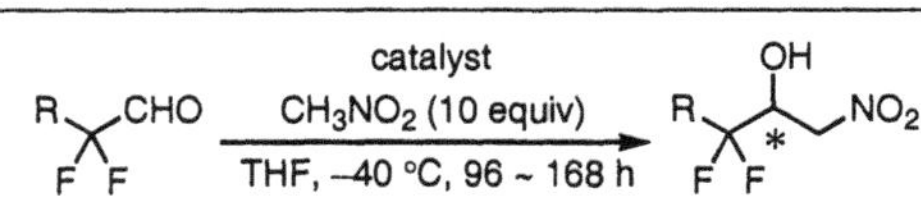

35: R = $PhCH_2CH_2CH_2$
37: R = c-C_6H_{11}

36: R = $PhCH_2CH_2CH_2$
38: R = c-C_6H_{11}

Entry	*Aldehyde*	*Catalyst*[a] *(mol%)*	*Product*	*Yield (%)*	*ee (%)*[b]
1	**35**	**A** (20	**36**	74	55
2	**35**	**B** (5)	**36**	67	74
3	**35**	**C** (5)	**36**	55	92
4	**37**	**C** (5)	**38**	58	95

Notes: [a] Catalyst: **A** = (*R*)-LLB; **B** = (*R*)-SmLB; **C** = $SmLi_3$tris((*R*)-6,6′-bis((triethylsilyl)ethynyl)binaphthoxide)
[b] Absolute configuration of **36** was determined to be (*S*) by X-ray crystallography of a derivative.

Table 4. *syn*-Selective Catalytic Asymmetric Nitroaldol Reactions

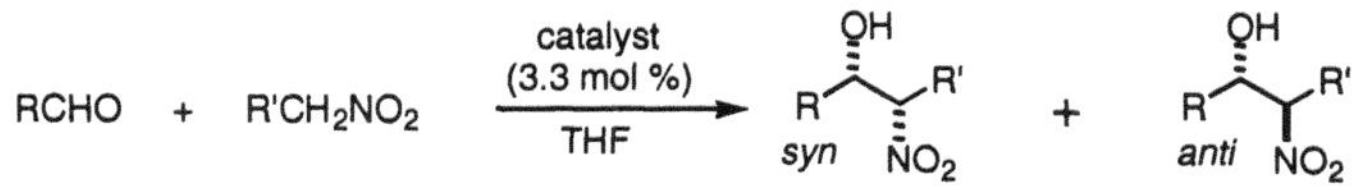

4: R = $PhCH_2CH_2$ **39**: R' = CH_3
48: R = $CH_3(CH_2)_4$ **42**: R' = Et
45: R' = CH_2OH

40 (*syn*), **41** (*anti*): R = $PhCH_2CH_2$, R' = CH_3
43 (*syn*), **44** (*anti*): R = $PhCH_2CH_2$, R' = Et
46 (*syn*), **47** (*anti*): R = $PhCH_2CH_2$, R' = CH_2OH
49 (*syn*), **50** (*anti*): R = $CH_3(CH_2)_4$, R' = CH_2OH

Entry	*Aldehyde*	*Nitroalkane*	*Catalyst*	*Time (h)*	*Temp (°C)*	*Nitro-aldols*	*Yield (%)*	*syn/anti*	*ee of syn (%)*
1	**4**	39	**LLB**	75	−20	**40 + 41**	79	74:26	66
2	1	**39**	**31**	75	−20	**40 + 41**	72	85:15	92
3	**4**	**39**	**32**	75	−20	**40 + 41**	70	89:11	93
4	**4**	**42**	**LLB**	138	−40	**43 + 44**	89	85:15 93:7	87
5	**4**	**42**	**32**	138	−40	**43 + 44**	85	84:16 92:8	95
6	**4**	**45**	**LLB**	111	−40	**46 + 47**	62	87:13 92:8	66
7	**4**	**45**	**32**	111	−40	**46 + 47**	97		97
8	**48**	**45**	**LLB**	93	−40	**49 + 50**	79		78
9	**48**	**45**	**32**	93	−40	**49 + 50**	96		95

C. Diastereoselective and Enantioselective Nitroaldol Reactions

With more effective asymmetric catalysts in hand, we next applied the catalysts **31** and/or **32** to diastereoselective nitroaldol reactions. We were very pleased to find that, in all cases, high *syn* selectivity and enantioselectivity were obtained using 3.3 mol% of the new catalysts.[24,25] Representative results are shown in Table 4. It is noteworthy that structurally simple aldehydes such as hexanal (**48**) (which has no neighboring group to assist with asymmetric induction) gave nitroaldols in high optical purity. It appears that the *syn* selectivity in the nitroaldol reaction can best be explained as arising from steric hindrance in the bicyclic transition state—as can readily be seen in the Newman projection. This suggests that the transition state leading to the *syn*-adduct is the more favorable (Figure 6).[26] In the favored transition state, the catalyst is believed to simultaneously act as a Lewis acid and base at different sites. In contrast the nonchelation controlled transition state appears to afford *anti*-adducts with lower optical purities.

The *syn*-selective asymmetric nitroaldol reaction was successfully applied to the catalytic asymmetric synthesis of *threo*-dihydrosphingosine (**53**), which elicits a variety of cellular responses by inhibiting protein kinase C (Scheme 5).[25] The nitroaldol reaction of hexadecanal (**51**) with 3 equiv of nitroethanol gave the corresponding nitroaldol adducts in high *syn* selectivity (91:9) in 78% yield, with the *syn*-adduct **52** being obtained with up to 97% ee. In this case, the LLB-catalyzed

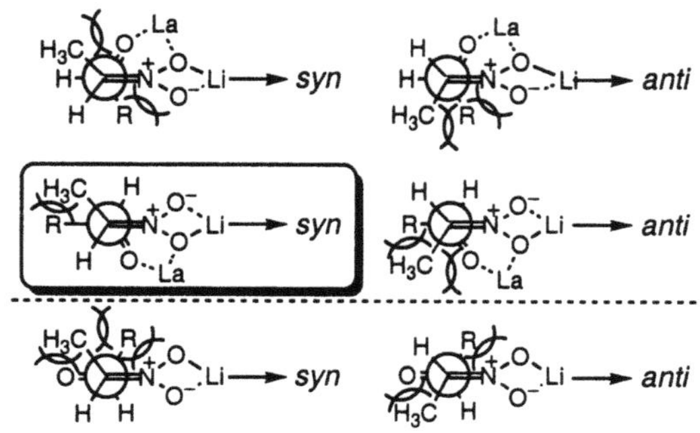

Figure 6. Proposed transition states of diastereoselective and enantioselective nitroaldol reactions.

reaction under similar conditions proceeded slowly to give a 86:14 ratio of the *syn*-and *anti*-adducts in 31% yield with a lower optical purity (83% ee). Hydrogenation of **52** in the presence of 10% Pd on charcoal afforded *threo*-dihydrosphingosine (**53**) in 71% yield.

D. Second-Generation LLB Catalyst (LLB-II)

Catalytic asymmetric nitroaldol reactions promoted by LLB or its derivatives require at least 3.3 mol% of asymmetric catalysts for efficient conversion. Moreover, even in the case of 3.3 mol% of catalysts, the reactions are rather slow. In order to enhance the activity of the catalyst, a consideration of the possible mechanism for catalytic asymmetric nitroaldol reactions is clearly a necessary prerequisite to formulation of an effective strategy. One possible mechanism for the catalytic asymmetric nitroaldol reactions is shown at the top of Scheme 6. Although the lithium nitronate is generated first, there appears to be a significant possibility

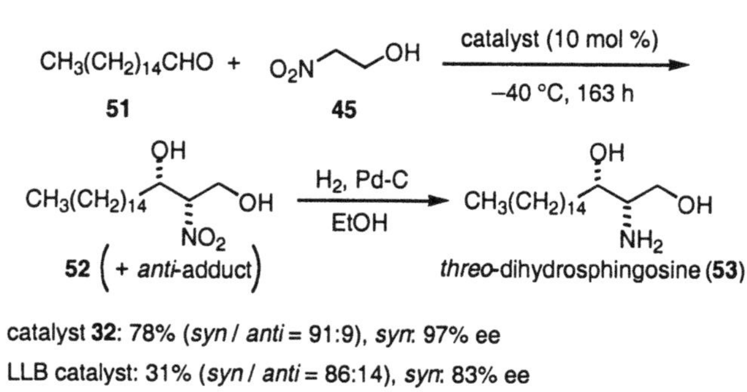

Scheme 5. Catalytic asymmetric synthesis of *threo*-dihydrosphingosine.

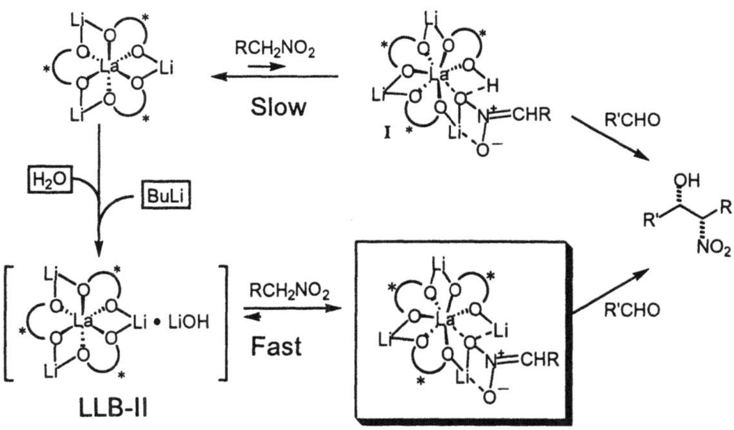

Scheme 6. Proposed mechanism for the catalytic asymmetric nitroaldol reaction promoted by LLB and/or LLB-II.

that the aldehyde first coordinates to La. We strove to detect the postulated intermediate **I** using various methods.

These attempts, however, proved to be unsuccessful, probably owing to the low concentrations of the intermediate, which we thought might be ascribable to the presence of an acidic OH group in close proximity. In order to remove a proton from **I**, we added almost 1 equiv of base to the LLB catalyst. After many attempts, we were finally pleased to find that 1 mol% of second-generation LLB (LLB-II), prepared from LLB, 1 mol equiv of H_2O, and 0.9 mol equiv of butyllithium efficiently promoted the catalytic asymmetric nitroaldol reactions. Moreover, we also found that the use of LLB-II (3.3 mol%) accelerated these reactions. The use of other bases such as NaO-*t*-Bu, KO-*t*-Bu and Ca(O-*i*-Pr)$_2$ gave less satisfactory results. The results are shown in Table 5.[27]

The structure of LLB-II has not yet been unequivocally determined. We propose here, however, that it is a complex of LLB and LiOH; a proposed reaction course for its use in an improved catalytic asymmetric nitroaldol reaction is shown at the bottom of Scheme 6. Although a molecular ion peak was not observed, a fragment ion peak for the complex of LLB and lithium nitronate was observed at 816 by FAB mass spectral analysis. The above-mentioned results suggest quite an interesting chemical phenomenon, namely the tight complexation of LLB and LiOH and the high rate of aggregation between LLB and the lithium nitronates.

Using the second-generation LnLB catalyst consisting of 6,6′-bis((trimethylsilyl)ethynyl)BINOL and Sm, as shown in Scheme 7, we succeeded in achieving an efficient catalytic asymmetric synthesis of arbutamine (**56**), a useful β-agonist.[9,28]

Table 5. Comparisons of Catalytic Activity between Either LLB and Second-Generation LLB (LLB-II) or **32** and **32**-II

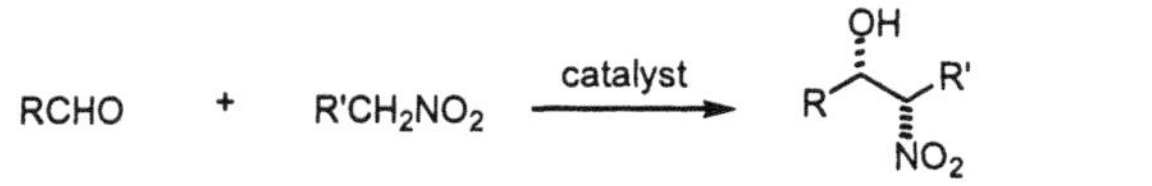

9: R = C_6H_{11} 5: R' = H 10: R = C_6H_{11}, R' = H
4: R = $PhCH_2CH_2$ 39: R' = CH_3 40: R = $PhCH_2CH_2$, R' = CH_3
42: R' = Et 43: R = $PhCH_2CH_2$, R' = Et
45: R' = CH_2OH 46: R = $PhCH_2CH_2$, R' = CH_2OH

Entry	*Substrate*	*Catalyst*[a] *(mol %)*	*Time (h)*	*Temp (°C)*	*Product*	*Yield (%) (Syn/anti)*	*ee (%) of syn*
1	**9 + 5**	LLB (1)	24	–50	**10**	5.6	88
2	**9 + 5**	LLB-II (1)	24	–50	**10**	73	89
3	**9 + 5**	LLB-II (3.3)	4	–50	**10**	70	90
4	**4 + 39**	**32** (1)	113	–30	**40**	25 (70/30)	62
5	**4 + 39**	**32**-II (1)	113	–30	**40**	83 (89/11)	94
6	**4 + 42**	**32** (1)	166	–40	**43**	trace	—
7	**4 + 42**	**32**-II (1)	166	–40	**43**	84 (95/5)	95
8	**4 + 45**	**32** (1)	154	–50	**46**	trace	—
9	**4 + 45**	**32**-II (1)	154	–50	**46**	76 (94/6)	96

Note: [a]LLB-II: LLB + H_2O (1 mol equiv) + BuLi (0.9 mol equiv); **32**-II: **32** + H_2O (1 mol equiv) + BuLi (0.9 mol equiv).

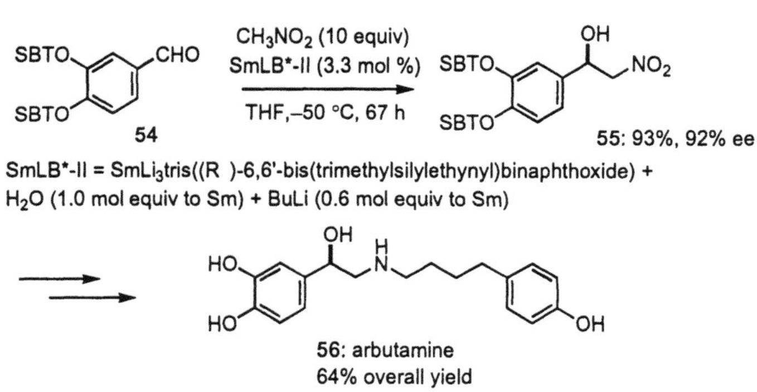

Scheme 7. A catalytic asymmetric synthesis of arbutamine.

E. Tandem Inter- and Intramolecular Catalytic Asymmetric Nitroaldol Reaction

Tandem catalytic asymmetric syntheses are especially useful because optically active compounds with several chiral centers can be constructed from simple achiral compounds in one pot, using a small amount of an asymmetric catalyst. Therefore, we examined a tandem inter-intramolecular catalytic asymmetric nitroaldol reaction of **57** with **5**, using LnLB complexes as catalysts.[29] Optimized conditions for the synthesis of **59b** are shown in Scheme 8. That is, after the complete disappearance of **57** at –40 °C in the presence of 5 mol% of PrLB, the reaction mixture was warmed to room temperature and stirred for an additional 24 h at the same temperature, allowing for the complete conversion of **59a** into another diastereoisomer, **59b**. The structure of **59b** was determined by X-ray structural analysis also shown in Scheme 8. Although the purification of **59b** by silica gel column chromatography was quite difficult, direct crystallization of the crude mixture from CH_2Cl_2 produced **59b** with 79% ee, which was isolated in 41% yield. Almost optically pure **59b** was obtained after recrystallization.

The proposed mechanism by which **59b** was produced is shown in Scheme 9. PrLB is a multifunctional heterobimetallic asymmetric catalyst displaying both Lewis acidity and Brønsted basicity; the Pr atom works as a Lewis acid, and the Li–naphthoxide portion functions as a Brønsted base. Thus, **57** is activated by the Pr atom, and nitromethane (**5**) is deprotonated by the Li–naphthoxide portion, resulting in the formation of **I**. Both **57** and **5** then react intermolecularly to give **II**. In addition, a lithium nitronate, again formed from **II**, reacts kinetically with an internal ketone to give **III**, followed by the generation of **59a** and a regeneration of PrLB. At room temperature, however, an equilibrium appears to exist between **III**,

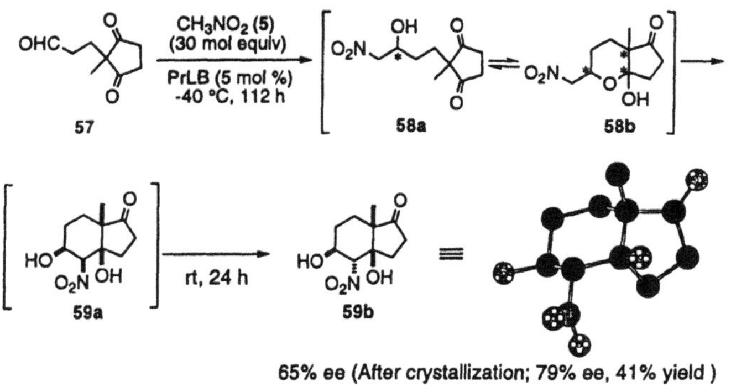

Scheme 8. The optimized conditions and the structure of **59b** obtained by the tandem inter- and intramolecular catalytic asymmetric nitroaldol reaction of **57** with **5**.

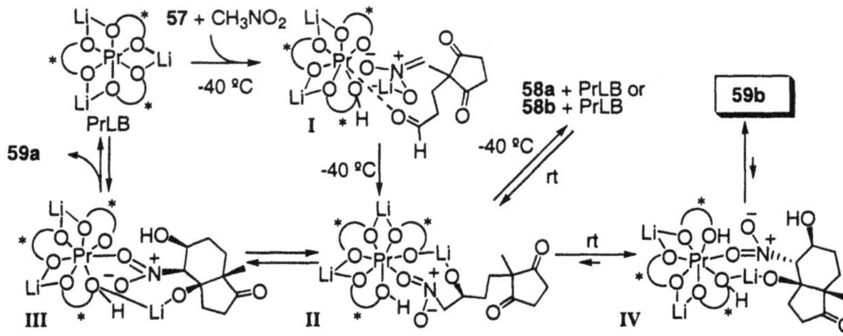

Scheme 9. Proposed mechanism for the tandem inter- and intramolecular catalytic asymmetric nitroaldol reaction.

II, and **IV**, and the reaction proceeds towards the generation of the thermodynamically more stable **59b**.

The reaction of a mixture of **58a** and **58b**, isolated from interaction of **57** and **5** in the presence of PrLB at –40 °C, with **5** in the presence of PrLB at –40 °C for 59 h and at room temperature for 28 h gave **59b** with 65% ee. Similarly, the reaction of a mixture of **58a** and **58b** obtained under similar conditions, in the presence of NaO-*t*-Bu instead of PrLB, generated **59b** with 66% ee. These results suggest that the optical purity of **59b** depends only on the enantioselection of the intermolecular nitroaldol reaction and that, in an intramolecular reaction, there is no kinetic resolution of **58**, with the cyclization of **58** to **59b** being controlled by the absolute configuration of the **58a** hydroxy group.

Because **57** was easily prepared from 2-methyl-1,3-cyclopentanedione and acrolein in one step, and because after recrystallization the adduct **59b** can be obtained in optically highly pure form, this tandem inter- and intramolecular catalytic asymmetric reaction could serve a practical purpose in preparing a chiral synthon for natural products and/or chiral ligands.

F. Catalytic Asymmetric Michael Reactions Promoted by LSB

Catalytic asymmetric Michael reactions are one of the most important synthetic methods for obtaining asymmetric centers.[30] 1,3-Dicarbonyl compounds in particular are highly promising Michael donors for the enantioselective construction of carbon–carbon bonds. Although LSB (L = lanthanum, S = sodium) was ineffective as an asymmetric catalyst for nitroaldol reactions,[7] we surprisingly found that LSB was effective in the catalytic asymmetric Michael reaction of various enones with malonates, giving Michael adducts in up to 92% ee and almost quantitative yield.[8] Some typical results are summarized in Table 6. In general, the use of THF as a solvent gave the best results, whereas in the case of the LSB-catalyzed reaction of *trans*-chalcone (**71**) with **65**, the use of toluene was essential in giving **72** in good

Table 6. Catalytic Asymmetric Michael Reactions Promoted by (*R*)-LnMB (10 mol%)

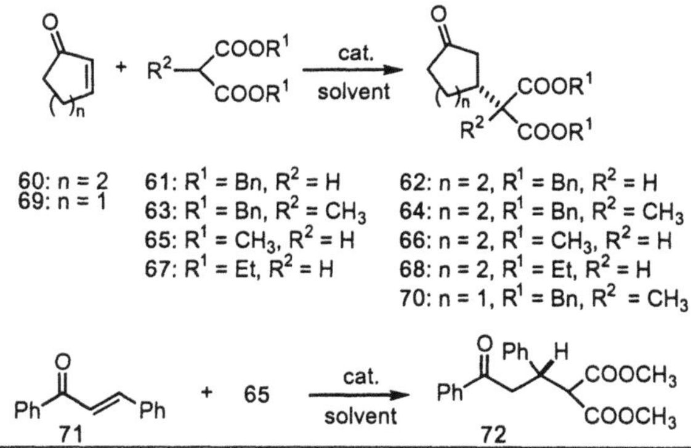

Entry	*Enone*	*Michael Donor*	*Product*	*Cat.*	*Solvent*	*Temp (°C)*	*Time (h)*	*Yield (%)*	*ee (%)*
1	**60**	**61**	**62**	LSB	THF	0	24	97	88
2	**60**	**61**	**62**	LSB	THF	rt	12	98	85
3	**60**	**61**	**62**	LSB	toluene	rt	12	96	82
4	**60**	**61**	**62**	LLB	THF	rt	12	78	2
5	**60**	**61**	**62**	LPB	THF	rt	12	99	48
6	**60**	**63**	**64**	LSB	THF	0	24	91	92
7	**60**	**63**	**64**	LSB	THF	rt	12	96	90
8	**60**	**65**	**66**	LSB	THF	rt	12	98	83
9	**60**	**67**	**68**	LSB	THF	rt	12	97	81
10	**69**	**63**	**70**	LSB	THF	–40	36	89	72
11	**71**	**65**	**72**	LSB	THF	–50	36	62	0
12	**71**	**65**	**72**	LSB	toluene	–50	24	93	77
13	**71**	**65**	**72**	PrSB	toluene	–50	24	96	56
14	**71**	**65**	**72**	GdSB	toluene	–50	24	54	6

enantiomeric excess. Center-metal effects were also investigated for the Michael reaction of **71** with **65**, indicating that LSB was the best catalyst for catalytic asymmetric Michael reactions.

What is the origin of the catalytic activity and mode of enantioselectivity in the LSB-catalyzed Michael reactions? In order to clarify the nature of the interaction between the enone and the asymmetric catalyst, the complexation was studied by ^{1}H NMR after mixing cyclohexenone (**60**) and the asymmetric bimetallic complexes and observing the chemical shift of the α-proton of the enone **60** (Figure 7). It is well known that, in general, praseodymium complexes induce an upfield shift, while europium complexes induce a downfield shift.[31] In addition, ordinary Lewis acids such as $La(OTf)_3$ and Et_2AlCl also induce a downfield shift. We were pleased

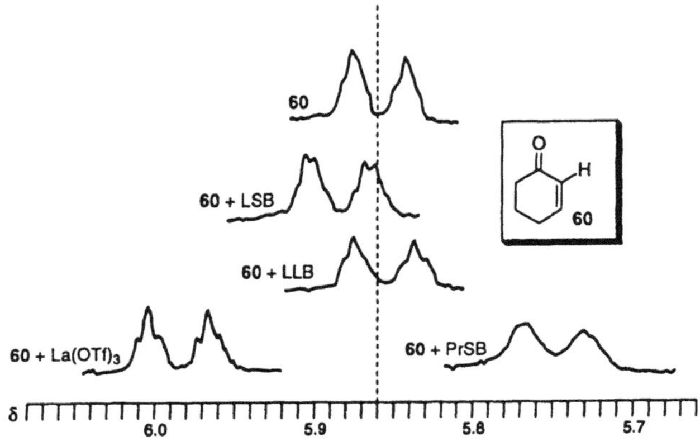

Figure 7. Chemical shift of the α-proton on cyclohexenone (**60**).

to find that complexation with LSB induced a small downfield shift on the α-proton of **60**, and PrSB, a moderately effective asymmetric catalyst for Michael reactions, induced a large upfield shift. Interestingly, in the case of either EuSB or LLB, which gave only near-racemic Michael adducts, the ^{1}H NMR spectra showed no changes in the chemical shift of the α-proton of **60**. These NMR studies indicated that the carbonyl group of the enone coordinated to the lanthanum and/or praseodymium metal in the LnSB molecule, while the double bond of the enone did not coordinate to either LLB and/or EuSB. These changes in chemical shift were observed even in the presence of **65**. The chemical phenomena described above might be understood by considering the differing dihedral angles with which the BINOL moiety binds to the center metal in each case.

Using the X-ray structure of LnSB as a reference point, computational simulations of the enantioselection process were carried out using a Rappé's universal force field (UFF).[32] As shown in Scheme 10, with enone **60** coordinated to the lanthanum metal cation, the plane of the cyclohexenone ring should be almost parallel to the nearest of the naphthyl ring systems, facilitating an attack by the coordinated sodium enolate of **65** to give the Michael adduct **66**. UFF calculations and conformational searches for models of the pro-(*R*)- and pro-(*S*)-adduct systems clearly indicated that the (*R*)-LSB complex complexes better as a pro-(*R*)-adduct than as a pro-(*S*)-adduct ($\Delta E = 4.9$ kcal/mol). LSB can thus generate an intermediate such as that shown at the top of Scheme 10, giving high ee. The resultant sodium enolates of the optically active Michael adducts appear to abstract a proton from an acidic OH so as to regenerate the LSB catalyst. The proposed catalytic cycle is shown in Scheme 11. Thus, the basic LSB complex also acts as a Lewis acid, controlling the orientation of the carbonyl function and so activating the enone for attack. It appears that the multifunctional nature of the LSB catalyst makes possible the formation of Michael adducts with high ees even at room temperature.

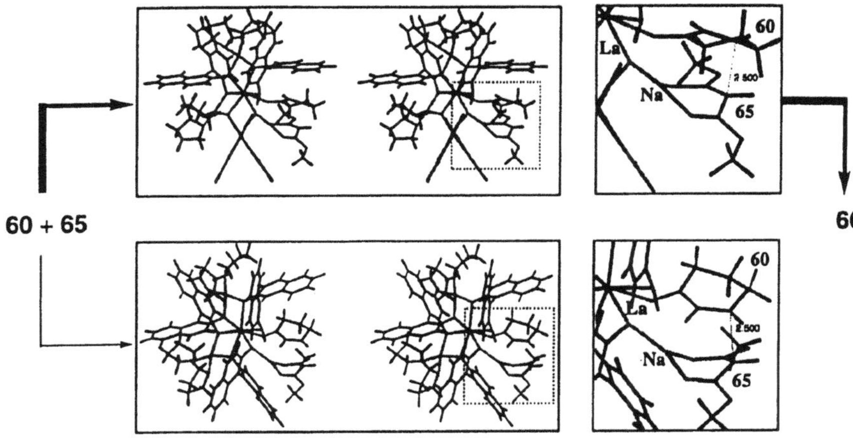

Scheme 10. UFF computational simulations of the Michael reaction of **60** with **65** catalyzed by (*R*)-LSB. Top: Pro-(*R*)-model (favorable model). Bottom: Pro-(*S*)-model (unfavorable model). Enlarged pictures of the reaction sites are drawn on the right side of each of the stereoviews.

LSB was next applied to another type of catalytic asymmetric Michael reaction, in which the asymmetric center is induced on the side of the adduct originating from the Michael donor. Although the optical purities of the Michael adducts obtained in THF in the presence of 5 mol% of LSB are modest, we were very pleased to find that in CH_2Cl_2 the asymmetric Michael reaction gave the desired adducts in

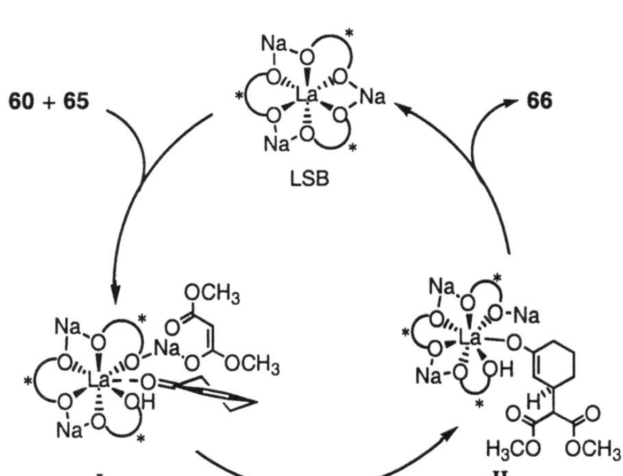

Scheme 11. Proposed catalytic cycle for the asymmetric Michael reaction promoted by LSB.

Table 7. Catalytic Asymmetric Michael Reactions Promoted by LSB in CH_2Cl_2

Michael Donor	*Michael Acceptor*	*Product*	*Catalyst Amount (mol %)*	*Temp (°C)*	*Time (h)*	*Yield (%)*	*ee (%)*
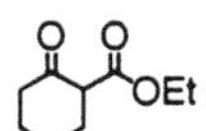 **73**	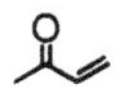 **74**	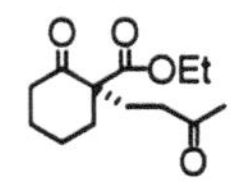 **75**	10	-50	19	85	93
			5	-50	19	89	91
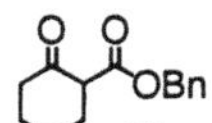 **76**	**74**	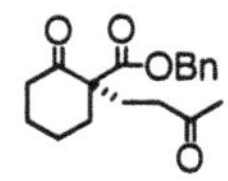 **77**	10	-50	12	73	91
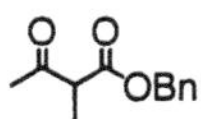 **78**	**74**	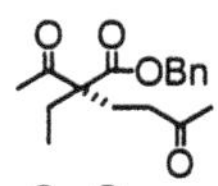 **79**	5	-50	16	98	89
73	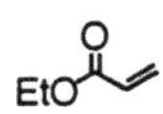 **80**	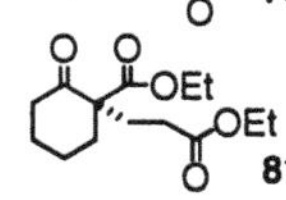 **81**	10	0→rt	17	60	76

up to 93% ee. Representative results are shown in Table 7. Moreover, in this solvent it was found that the catalytic asymmetric Michael reaction for **75** was not significantly affected by the choice of rare earth metal.[33]

In both types of catalytic asymmetric Michael reactions, the use of either second-generation LSB or 6,6′-substituted BINOL-derived LSB-type catalysts did not result in significantly improved results.

G. Catalytic Asymmetric Hydrophosphonylation of Imines Promoted by the Rare Earth–Potassium–BINOL Catalyst (LnPB)

α-Aminophosphonic acids are interesting compounds for use in the design of enzyme inhibitors. The concept of mimicking the tetrahedral transition states of enzyme-mediated peptide bond hydrolyses has led to the successful design and synthesis of phosphonamide-containing peptides as a promising new class of proteinase inhibitors.[34] We have succeeded in developing the first catalytic asymmetric hydrophosphonylation of imines using LnK_3tris(binaphthoxide) complexes (LnPB), which give optically active α-aminophosphonates in modest to high enantiomeric excesses.[35]

As shown in Table 8, the reaction of imine **82** with 1.5 equiv of dimethyl phosphite and 10 mol% of LPB in THF-toluene (1:7) at room temperature for 96 h gave α-aminophosphonate **83** with 96% ee in 70% yield. Moreover, in the case of imine **84**, the use of 5 mol% of LPB gave **85** with 92% ee in 82% yield.

The proposed mechanism for this catalytic asymmetric hydrophosphonylation is shown in Scheme 12. The first step of this reaction is the deprotonation of dimethyl

Table 8. Catalytic Asymmetric Hydrophosphonylation of Acyclic Imines[a]

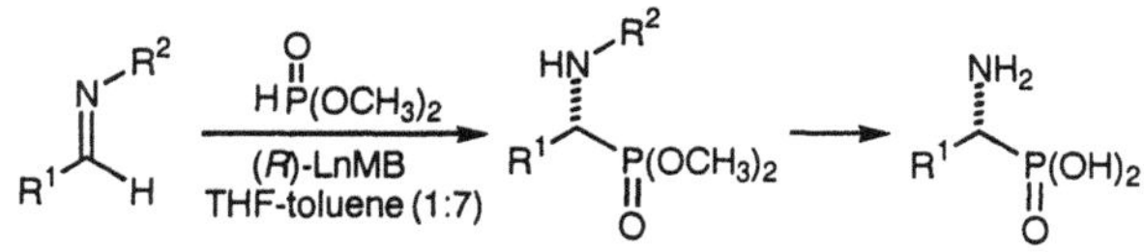

82: R^1 = *i*-Pr, R^2 = $CH(p\text{-}CH_3OC_6H_4)_2$ **83**: R^1 = *i*-Pr, R^2 = $CH(p\text{-}CH_3OC_6H_4)_2$
84: R^1 = *i*-Pr, R^2 = $CHPh_2$ **85**: R^1 = *i*-Pr, R^2 = $CHPh_2$
86: R^1 = Et, R^2 = $CHPh_2$ **87**: R^1 = Et, R^2 = $CHPh_2$
88: R^1 = C_5H_{11}, R^2 = $CHPh_2$ **89**: R^1 = C_5H_{11}, R^2 = $CHPh_2$
90: R^1 = (*E*)-PhCHCH, R^2 = $CHPh_2$ **91**: R^1 = (*E*)-PhCHCH, R^2 = $CHPh_2$
92: R^1 = $C_{12}H_{25}$, R^2 = $CHPh_2$ **93**: R^1 = $C_{12}H_{25}$, R^2 = $CHPh_2$

Entry	*Imine*	*Cat. (mol %)*	*Temp.*	*Time (h)*	*Yield (%)*	*ee (%)*
1	**82**	LPB (20)	rt	21	**83**:62	91
2	**82**	LSB (20	rt	21	**83**:38	49
3	**82**	LLB (20)	rt	21	**83**:46	38
4	**82**	LPB (10)	rt	96	**83**:70	96
5	**84**	LPB (20)	rt	63	**85**:97	97
6	**84**	LPB (5)	rt	143	**85**:82	92
7	**86**	LPB (20)	rt	63	**87**:88	94
8	**88**	LPB (20)	rt	63	**89**:57	92
9	**90**	GdPB (20)	50 °C	40	**91**:86	66
10	**92**	LPB (20)	rt	84	**93**:50	89

Note: [a]All reactions were performed in the presence of 5 equiv of dimethyl phosphite, except entries 4 and 5 (1.5 equiv).

phosphite by LPB to generate the potassium dimethyl phosphite. This potassium phosphite immediately coordinates to a rare earth to give **I** due to the strong oxophilicity of rare earth metals.[36] **I** then reacts with an imine to give an optically active potassium salt of the α-aminophosphonate, which leads via a proton-exchange reaction to an α-aminophosphonate and LPB, thereby completing the catalytic cycle and giving the desired asymmetric hydrophosphonylation.

We also examined the hydrophosphonylation of cyclic imines such as **94**. In collaboration with Professor Martens and Dr. Gröger at the University of Oldenburg, we succeeded in developing an efficient *catalytic* asymmetric hydrophosphonylation of cyclic imines. Some representative results are shown in Table 9.[37]

By the use of 20 mol% of LPB in THF-toluene (1:7) at room temperature, only a modest enantioselectivity of 61% ee accompanied by a modest chemical yield of 53% was observed for the formation of the desired **95** after 144 h. The efficiency of the reaction was improved by increasing the reaction temperature to 50 °C. Therefore, additional efforts to increase the efficiency of the reaction by investigating the influence of further lanthanoid metal components in the catalyst structure

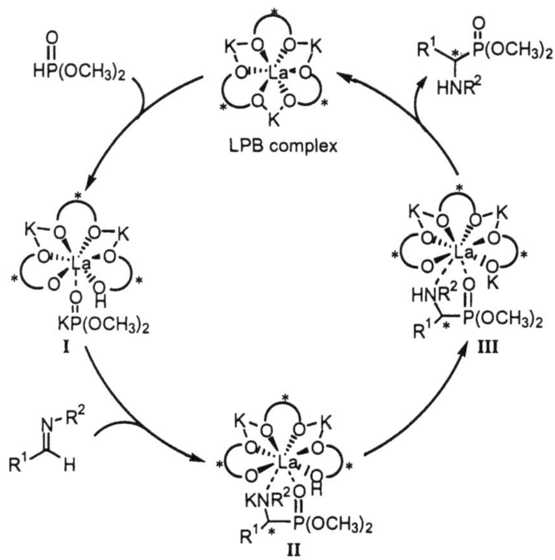

Scheme 12. Proposed catalytic cycle for the hydrophosphonylation of imines.

Table 9. LnPB-Catalyzed Asymmetric Hydrophosphonylation of **94**

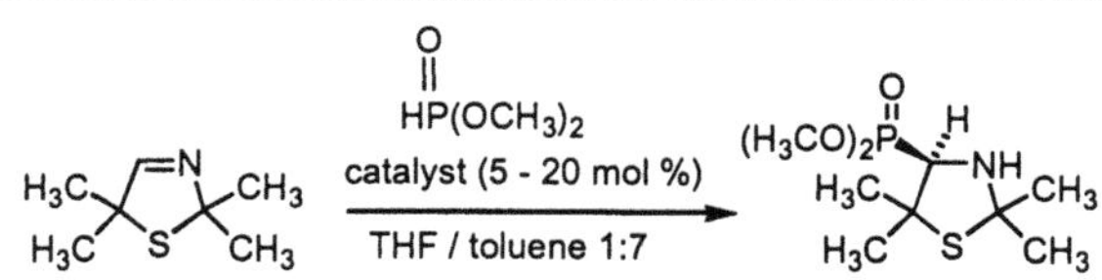

Entry	*Catalyst (mol %)*	*Temp*	*Time (h)*	*Yield (%)*	*ee (%)*
1	LPB (20)	rt	144	53	61
2	LPB (20)	50 °C	50	55	64
3	PrPB (20)	50 °C	50	51	84
4	SmPB (20)	50 °C	40	97	93
5	GdPB (20)	50 °C	50	77	95
6	DyPB (20)	50 °C	50	76	97
7	YbPB (20)	50 °C	50	90	96
8	YbPB (20)	rt	50	86	98
9	YbPB (5)	50 °C	40	63	95

were carried out at 50 °C. A great increase in the ee values was obtained by using a Sm, Gd, Dy, or Yb catalyst with ee values of up to 97% ee and good chemical yields. By carrying out the phosphite addition with YbPB catalyst at room temperature, the product **95** was obtained with a similarly high 86% chemical yield and in an excellent 98% ee. This is the highest enantioselectivity so far observed for catalytic asymmetric hydrophosphonylation.

III. HETEROBIMETALLIC ASYMMETRIC CATALYSTS OTHER THAN RARE EARTH COMPLEXES AND THEIR USE IN CATALYTIC ASYMMETRIC REACTIONS

A. Aluminum–Alkali Metal–BINOL Complex (ALB)

As discussed in previous sections, mechanistic studies of LnMB-catalyzed reactions, in particular LSB-catalyzed Michael reactions, have revealed that LnMB acts as a base catalyst (via the OM moiety) while at the same time showing Lewis acid character (via the Ln center) making possible highly enantioselective reactions. These results suggest that extension of the heterobimetallic concept to center-metals other than rare earths could lead to new heterobimetallic asymmetric catalysts with novel functionality.

We began with the development of an amphoteric asymmetric catalyst assembled from aluminum and an alkali metal.[38] We were pleased to find that the new asymmetric catalyst could be prepared efficiently from $LiAlH_4$ and 2 equiv of (*R*)-BINOL (Scheme 13) and the aluminum–lithium–BINOL complex (ALB) was highly effective in the Michael reaction of cyclohexenone (**60**) with dibenzyl malonate (**61**), giving **62** with 99% ee and 88% yield at room temperature. Representative results are summarized in Table 10.[39] This is the highest ee so far reported for catalytic asymmetric Michael reactions.[8,30,33,39] Although LLB and LSB complement each other in their ability to catalyze asymmetric nitroaldol and Michael reactions,[7,8] aluminum–M–(*R*)-BINOL complexes (M = Li, Na, K, and Ba)[40] are more useful for the catalytic asymmetric Michael reaction (entries 3–6).

With these excellent results in hand, we set out to unequivocally determine the structure of ALB. The structure determined by X-ray crystallographic analysis is

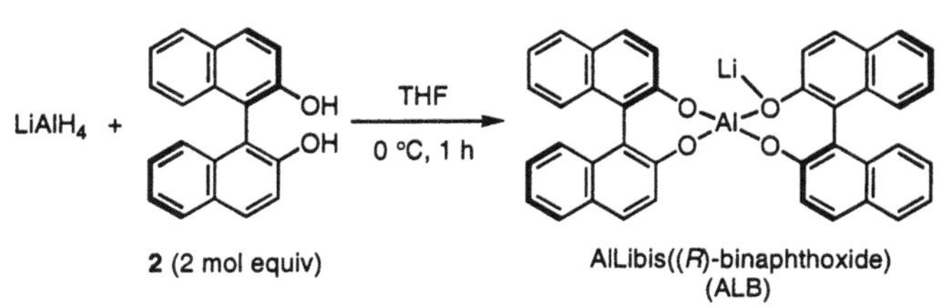

Scheme 13. Preparation of AlLibis ((*R*)-binaphthoxide) (ALB).

Table 10. Catalytic Asymmetric Michael Reactions Promoted by the AlMbis ((*R*)-binaphthoxide) Complex (AMB)

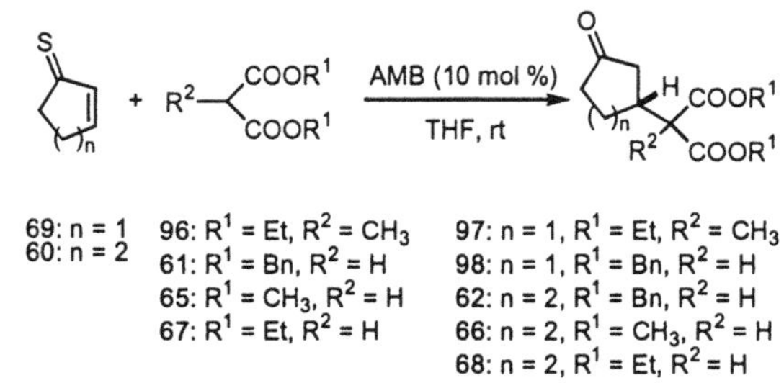

Entry	*Enone*	*Michael Donor*	*Product*	*M*	*Time (h)*	*Yield (%)*	*ee (%)*
1	**69**	**96**	**97**	Li	72	84	91
2	**69**	**61**	**98**	Li	60	93	91
3	**60**	**61**	**62**	Li	72	88	99
4	**60**	**61**	**62**	Na	72	50	98
5	**60**	**61**	**62**	K	72	43	87
6	**60**	**61**	**62**	Ba	6	100	84
7	**60**	**65**	**66**	Li	72	90	93
8	**60**	**67**	**68**	Li	72	87	95

shown in Figure 8.[27,39] The complex has a tetrahedral geometry around aluminum with an average Al–O distances of 1.75 Å. The long Li–O(1) distance of 2.00 Å is indicative of the ionic character between Li^+ and Al-2BINOL$^-$.

The small electronegativity value of lithium (1.0) compared to that of aluminum (1.5) suggests that a lithium enolate should be generated preferentially from a malonate derivative. However, the role of aluminum in this reaction is not well understood. In order to study the interaction between the enone and aluminum, an

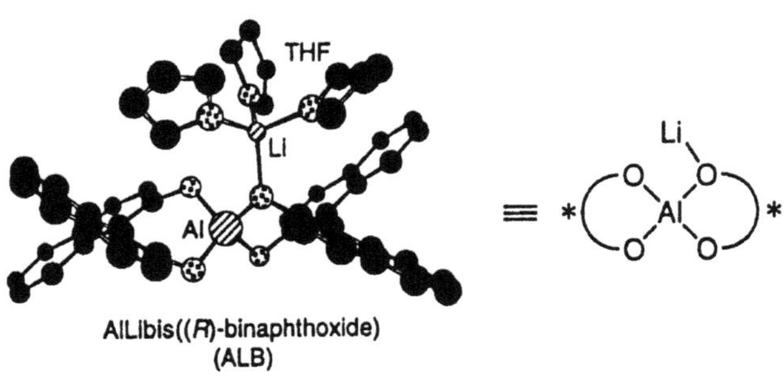

Figure 8. X-ray structure of ALB.

^{27}Al NMR study was performed.[39] One broad signal was observed for the ALB at δ 75 (S_0). Upon addition of 3 equiv of cyclohexenone (**60**) to the complex, two additional signals were obtained at δ 40 (S_1) and δ 23 (S_2). The latter signal (S_2) became stronger and was a major peak under conditions for the Michael reaction. This is strong evidence for an octahedral arrangement of six magnetically equivalent ligands around the aluminum cation in THF solution, since all shifts are upfield relative to S_0.[41] These NMR studies clearly indicate that the carbonyl groups of the enones coordinate to the aluminum. Consequently, this AlLibis(binaphthoxide) complex (ALB) is also believed to act as a heterobimetallic multifunctional asymmetric catalyst,[8] making possible efficient Michael reactions even at room temperature.

The above-mentioned mechanistic considerations suggest that the reaction of a lithium enolate derived from a malonate derivative with an enone should lead to an intermediary aluminum enolate. Further studies were carried out to obtain direct evidence of such an aluminum enolate in an effort to find a way to enhance its influence on the chemical fate of the Michael adducts. On the basis of the electronegativity of aluminum, the protonation of the aluminum enolate should be slower than of the corresponding lithium, sodium and/or lanthanoid enolates. Is it possible that such an Al enolate could then be trapped by an electrophile such as an aldehyde?

As was anticipated, the reaction of cyclopentenone (**69**), diethyl methylmalonate (**96**), and hydrocinnamaldehyde (**4**) in the presence of 10 mol% of ALB gave the three-component coupling product **99** as a single isomer in 91% ee (64% yield) (Table 11). Quite interestingly, the use of LLB, LSB, and/or a lithium-free lanthanum-BINOL complex[42] gave very unsatisfactory results for the same three-component coupling reaction. This is probably due to the lanthanum enolates being more reactive towards the acidic proton. Moreover, the use of benzaldehyde (**11**)

Table 11. Tandem Michael-Aldol Reactions

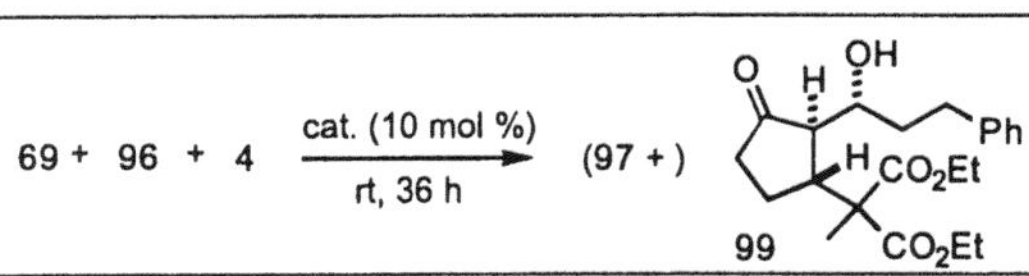

	97		99	
Cat.	*Yield*	*(%) ee (%)*	*Yield (%)*	*ee (%)*
(*R*)-ALB	7	90	64%	91
(*R*)-LLB	46	3	30%[a]	—
(*R*)-LSB	73	86	trace	—
Li-free-La-(*R*)-BINOL	57	83	trace	—

Note: [a]Inseparable mixture.

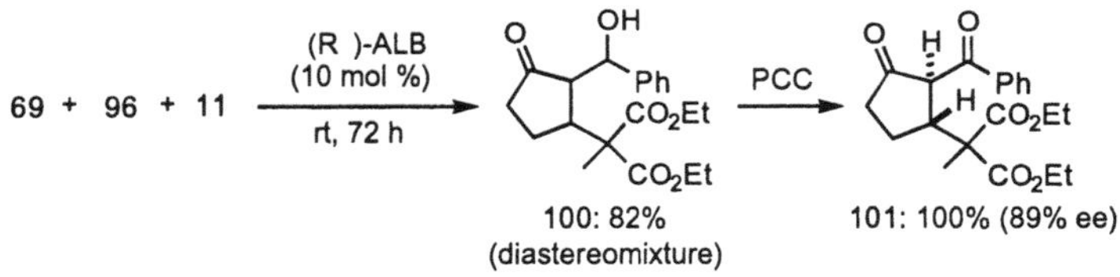

Scheme 14. Three-component coupling reaction using benzaldehyde (**13**) as an electrophile.

instead of hydrocinnamaldehyde (**4**) gave the three-component coupling product **100** in 82% yield (Scheme 14). Although this product **100** was a mixture of diastereomers, the oxidation of **100** gave the corresponding diketone derivative **101** as a single isomer in 89% ee. This is the first example of a catalytic asymmetric tandem Michael-aldol reaction.

We can therefore describe the reaction pathway in these three-component coupling reactions as follows (Scheme 15). The reaction of diethyl methylmalonate (**96**) with the ALB gives the corresponding lithium enolate (**I**). Enolate **I** then reacts with cyclopentenone (**69**), which precoordinated to the aluminum, to give the aluminum enolate **II** in an enantioselective manner. A further reaction of enolate **II**

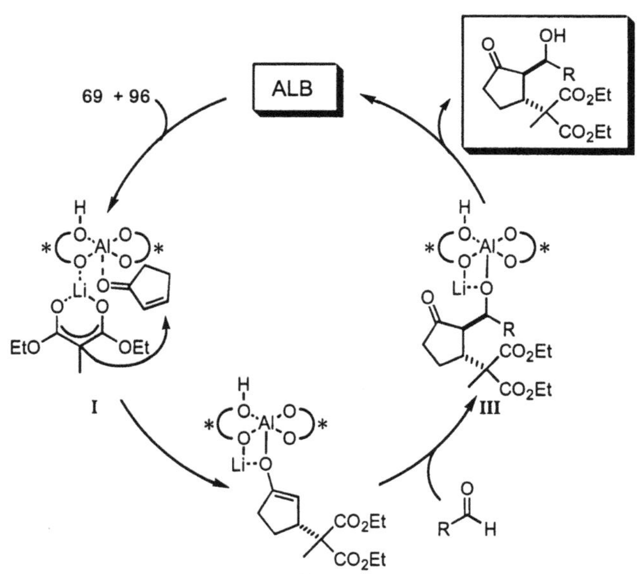

Scheme 15. Possible mechanism for the catalytic asymmetric tandem Michael-aldol reaction promoted by ALB.

with aldehyde would lead to an alkoxide (**III**). Although it is unclear whether the Al or the Li alkoxide is generated, the resulting alkoxide may then abstract a proton from an acidic OH to give the three-component coupling product and regenerate the ALB, thereby completing the catalytic cycle.

B. Catalytic Asymmetric Hydrophosphonylation of Aldehydes Using LLB and/or ALB

In recent years, α-hydroxy phosphonates have attracted much attention due to their wide-ranging biological activity and their usefulness as synthetic intermediates for other biologically important α-substituted phosphoryl compounds. Although the biological activities of α-substituted phosphoryl compounds depend on their absolute configuration, syntheses of optically active phosphoryl compounds have only recently begun to be studied in detail. So far, Shibuya et al. and Spilling et al. have independently reported[43,44] enantioselective hydrophosphonylations of aldehydes using LLB. However, the purity of the LLB catalysts utilized by their groups appeared to be rather low. Thus, we reinvestigated catalytic asymmetric hydrophosphonylation using LLB. As a result, we found that, in the presence of 10 mol% of LLB, hydrophosphonylation of benzaldehyde (**11**) and cinnamaldehyde (**102**) with 1.3 equiv of dimethyl phosphite in THF at –40 °C gave the corresponding α-hydroxy phosphonates **103** and **104** in 76% ee (79% yield) and in 72% ee (78% yield), respectively.[45] Shibuya et al. and Spilling et al. had reported that hydrophosphonylation of **11** using LLB prepared in their groups gave the corresponding α-hydroxy phosphonates in less than 30% ee under similar conditions to our own.[43,44]

It is noteworthy that with slow addition of the aldehydes, the enantiomeric excesses of both **103** and **104** at –40 °C were increased to 83% ee (73% yield) and 79% ee (88% yield), respectively. As solvent, THF again gave the best results. Representative results obtained at –78 °C are summarized in Table 12.[45]

In the catalytic asymmetric hydrophosphonylations, the use of neither 6,6′-bis((triethylsilyl)ethynyl)BINOL nor the second-generation LLB catalyst significantly improved in the results. The effects of slow addition of the aldehydes on the enantioselection can best be explained as follows. Heterobimetallic catalysts such as LLB are believed to activate both nucleophiles and electrophiles. For the hydrophosphonylation of comparatively unreactive aldehydes, the activated phosphite can react only with aldehydes which are precoordinated to lanthanum. However, in the case of reactive aldehydes such as **11** and **102**, the Li-activated phosphite may be able to undergo a competing reaction with the unactivated aldehyde. If such aldehydes are added in one portion, the ee of the product will thus be reduced. Slow addition of the aldehyde, in contrast, has the effect of maximizing the ratio of activated to unactivated aldehyde present in solution, by allowing time for the completion of the catalytic cycle and a regeneration of the catalyst, thereby facilitating aldehyde activation. Reactive aldehydes should, therefore, be added

Table 12. Catalytic Asymmetric Hydrophosphonylation of Aldehydes

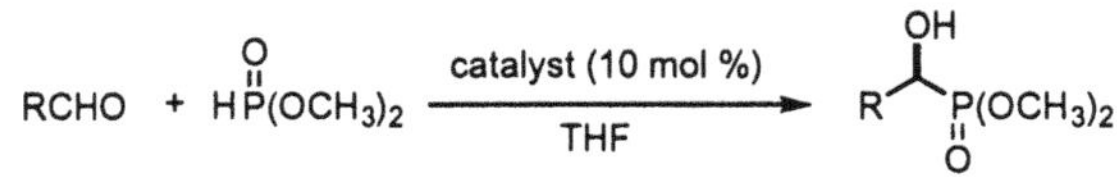

11: R = Ph	103: R = Ph
105: R = p -NO_2-Ph	106: R = p -NO_2-Ph
107: R = p -Cl-Ph	108: R = p -Cl-Ph
109: R = p -Me-Ph	110: R = p -Me-Ph
111: R = p -MeO-Ph	112: R = p -MeO-Ph
113: R = p -Me_2N-Ph	114: R = p -Me_2N-Ph
102: R = (E)-PhCH=CH	104: R = (E)-PhCH=CH
115: R = (E)-PhCH=C(CH_3)	116: R = (E)-PhCH=C(CH_3)
117: R = (E)-$CH_3(CH_2)_2$CH=CH	118: R = (E)-$CH_3(CH_2)_2$CH=CH
119: R = $CH_3(CH_2)_4CH_2$	120: R = $CH_3(CH_2)_4CH_2$

			LLB (–78°C)		*ALB (–40°C)*	
Entry	*Aldehyde*	*Product*	*Time (h)*	*Yield (%) (ee (%))*	*Time (h)*	*Yield (%) (ee (%))*
1	**11**	**103**	8	88 (79)	51	95 (90)
2	**105**	**106**	12	85 (36)	40	85 (71)
3	**107**	**108**	8	80 (63)	38	80 (83)
4	**109**	**110**	7	93 (78)	92	82 (86)
5	**111**	**112**	9	83 (88)[a]	115	88 (78)
6	**111**	**112**	8	87 (93)		
7	**113**	**114**	12	88 (95)[a]	no reaction	
8	**113**	**114**	12	80 (95)		
9	**102**	**104**	8	90 (84)	83	85 (82)
10	**115**	**116**	8	94 (92)	61	47 (56)
11	**117**	**118**	8	63 (75)	39	53 (55)
12	**119**	**120**	8	88 (61)	41	95 (16)

Note: [a]Aldehyde was added at one portion.

slowly in order to avoid the side reaction which proceeds without activation of the aldehyde by LLB (Scheme 16).

Furthermore, we have examined the hydrophosphonylation of aldehydes using ALB as a catalyst, just as described for LLB.[46] As shown in Table 12, ALB is also effective at generating α-hydroxyphosphonates from aldehydes, especially ones with an electron-withdrawing substituent. ALB and LLB can thus be used in a complementary manner for an asymmetric catalysis of the hydrophosphonylation of a wide variety of aldehydes.

C. Gallium–Alkali Metal–BINOL Complex (GaMB)

In an attempt to develop other asymmetric heterobimetallic catalysts with advantageous functions, we then attempted to prepare heterobimetallic asymmetric

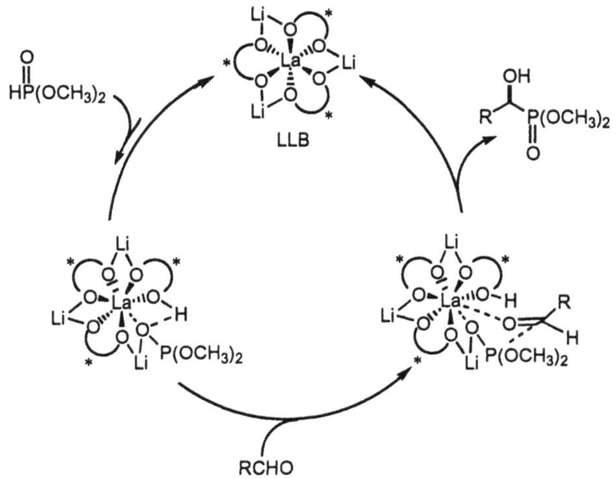

Scheme 16. A proposed catalytic cycle for asymmetric hydrophosphonylation.

catalysts with either B, Ga, or In as the central cation. The boron–alkali metal–BINOL complex did not promote the reaction of **60** with **61** at room temperature. However, the gallium–sodium–BINOL complex (GaSB) and the indium–potassium–BINOL complex (InPB) were both effective catalysts for asymmetric Michael reactions, with GaSB being enantioselectively superior to InPB (Table 13). The GaSB

Table 13. Catalytic Asymmetric Michael Reaction of **60** with **61** in the Presence of Heterobimetallic Asymmetric Catalyst (10 mol%)

Entry	M^1	M^2	*Time (h)*	*Yield (%)*	*ee (%)*
1	B	Li	17	0	—
2	B	Na	21	0	—
3	B	K	15	0	—
4	Ga	Li	43	71	49
5	Ga	Na	143	45	98
6	Ga	K	44	50	86
7	In	Li	24	77	2
8	In	Na	95	25	10
9	In	K	168	61	84

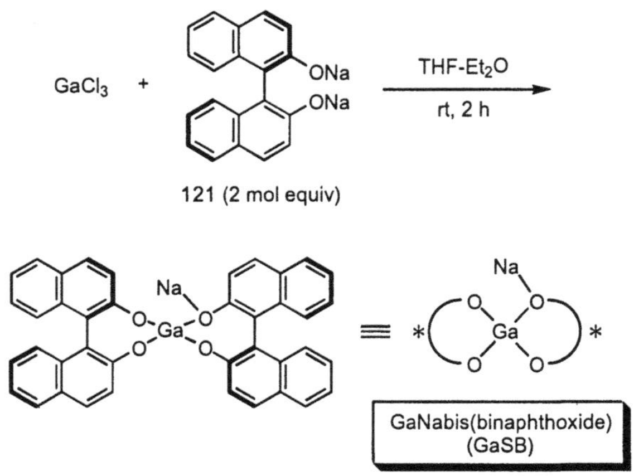

Scheme 17. Preparation of the GaNabis ((*R*)-binaphthoxide) complex (GaSB) and its structure.

catalyst was prepared from $GaCl_3$, NaO-*t*-Bu (4 mol equiv to $GaCl_3$), and (*R*)-BINOL (2 mol equiv to $GaCl_3$) in THF–ether, and the structure, consisting of one Ga atom, one Na atom, and two molecules of BINOL, was determined by ^{13}C NMR and LDI-TOF mass spectral analysis (Scheme 17). While determining the optimal amounts of starting materials to construct the GaSB catalyst, we observed, as expected, that using almost one mole excess of NaO-*t*-Bu did not reduce the optical purity of the Michael adducts but did enhance the reactivity of the catalyst. As shown in Table 14, whereas the asymmetric Michael reaction of cyclohexenone (**60**) with dibenzyl malonate (**61**) catalyzed by 10 mol% GaSB required 143 h at room temperature to give **62** with 98% ee in 45% yield, treatment of **60** and **61** with 10 mol% GaSB and 9 mol% NaO-*t*-Bu at room temperature for only 21 h gave **62** with 98% ee in 87% yield. Addition of the sodium salt of dibenzyl malonate (9 mol%) instead of NaO-*t*-Bu gave an almost identical result (96% ee, quant. yield). The asymmetric Michael reaction of cyclopentenone (**69**) with **61** in the presence of 10 mol% GaSB and 9 mol% NaO-*t*-Bu also proceeded smoothly to give **98** with 98% ee in 96% yield (room temperature, 22 h). Moreover, an effective catalytic asymmetric synthesis of **123** was also realized for the first time (99% ee, 79% yield).

Typical time curves for the Michael reaction of **60** with **61** in the presence of GaSB, GaSB+NaO-*t*-Bu, or NaO-*t*-Bu are shown in Figure 9. The reaction was found to be kinetically controlled, as treatment of racemic **62** with GaSB or GaSB+NaO-*t*-Bu resulted in only racemic **62**. GaSB- and GaSB+NaO-*t*-Bu-catalyzed reactions were second-order for up to ca. 70% of the reaction with a linear correlation coefficient of $r > 0.998$. The calculated rate constants for the GaSB-catalyzed reaction ($k_{obs \cdot GaSB}$) and the GaSB+NaO-*t*-Bu-catalyzed reaction ($k_{obs \cdot GaSB+NaO\text{-}t\text{-}Bu}$) were 0.031 $M^{-1}h^{-1}$ and 1.78 $M^{-1}h^{-1}$, respectively

Table 14. Enhancement of Catalyst Efficiency for Asymmetric Michael Reactions Using Ga or Al Complex

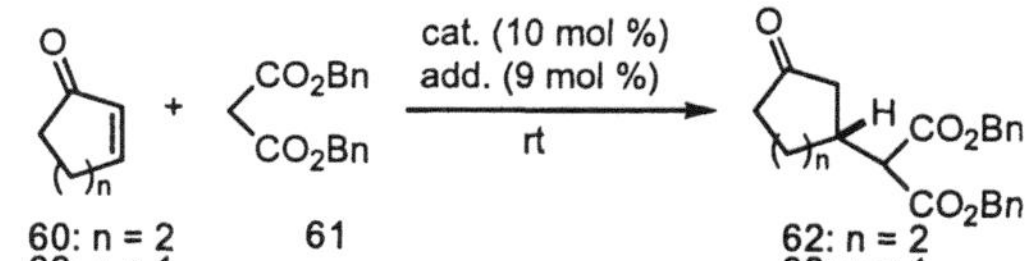

60: n = 2
69: n = 1
122: n = 3

61

62: n = 2
98: n = 1
123: n = 3

Entry	*Enone*	*Catalyst*	*Additive*[a]	*Time (h)*	*Yield (%)*	*ee (%)*
1	**60**	GaSB	—	143	45	98
2	**60**	GaSB	A	21	87	98
3	**60**	GaSB	A[b]	6	81	84
4	**60**	GaSB	A[c]	6	91	60
5	**60**	GaSB	C	21	quant	96
6	**69**	GaSB	—	72	32	89
7	**69**	GaSB	A	22	96	98
8	**122**	GaSB	—	73	trace	—
9	**122**	GaSB	A	73	79	99
10	**60**	ALB	—	72	88	99
11	**60**	ALB	B	12	quant	97
12	**60**	ALB	C	24	99	98
13	**122**	ALB	—	72	34	97
14	**122**	ALB	C	72	83	96

Notes: [a]A: NaO-*t*-Bu, B: Li-hexamethyldisilazide, C: Na-malonate.
[b]2 mol equiv of NaO-*t*-Bu was used to GaSB.
[c]3 mol equiv of NaO-*t*-Bu was used to GaSB.

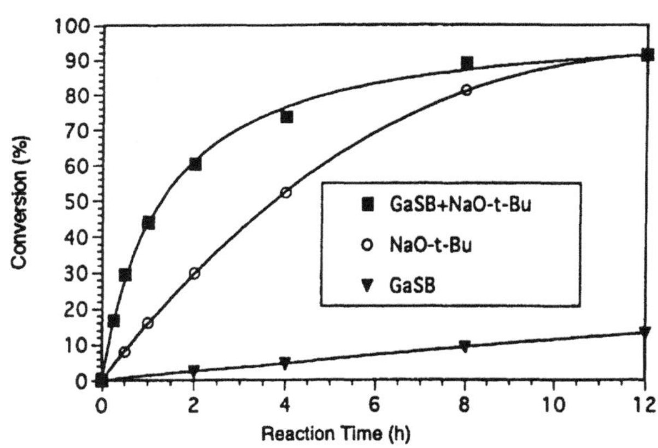

Figure 9. Typical time curves of the asymmetric Michael reaction of cyclohexenone (**60**) (392 mM) with dibenzyl malonate (**61**) (392 mM) in THF in the presence of 10 mol% catalysts at 24 °C.

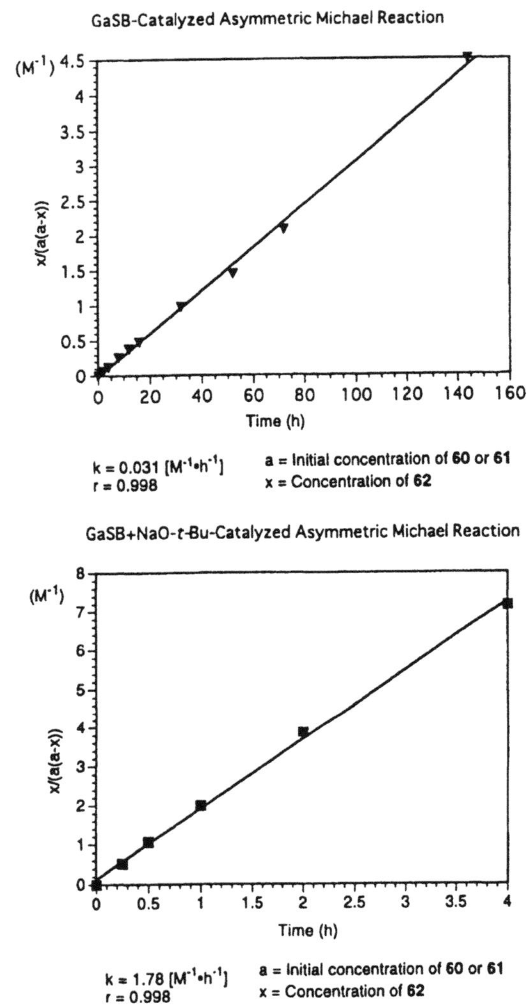

Figure 10. Kinetic analysis of GaSB- and/or GaSB+NaO-*t*-Bu-catalyzed asymmetric Michael reaction of **60** with **61**. Top: GaSB-catalyzed Michael reaction. Bottom: GaSB+NaO-*t*-Bu-catalyzed Michael reaction.

(Figure 10). As expected, the reaction of **60** with **61** in the presence of 10 mol% of the sodium salt of dibenzyl malonate at room temperature for 12 h gave the racemic Michael adduct **62** in 91% yield. These results clearly suggest that the sodium salt of dibenzyl malonate binds to GaSB much more quickly than it reacts with the enone, as observed in the case of LLB, thereby producing Michael adducts with high enantiomeric excess. No free sodium dibenzyl malonate should therefore be

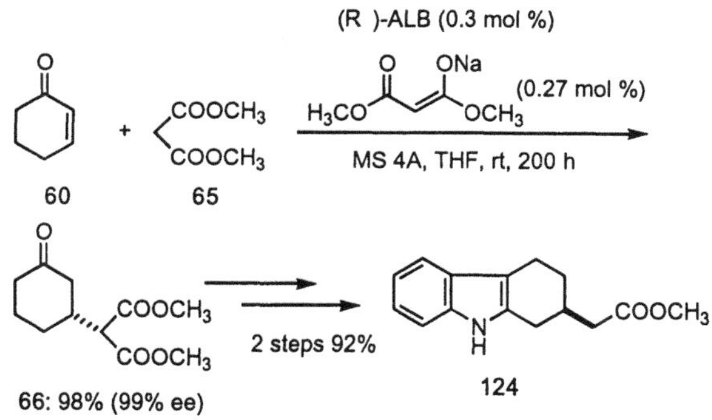

Scheme 18. Practical synthesis of the indole derivative **124** by an ALB (0.3 mol%) and sodium malonate-(0.27 mol%) catalyzed Michael reaction as a key step.

allowed to be present in the reaction medium. From the above assembly, the malonate anion can be transferred to cyclohexenone with an enantiomeric excess similar to that in the case of GaSB itself. This result appears to suggest that cyclohexenone also coordinates to the Ga metal center under the conditions described above.[39,47]

The usefulness of this activation strategy has also been demonstrated for catalytic asymmetric Michael reactions promoted by ALB. Representative results are summarized in Table 14 (entries 10–14). ALB- and ALB+Na–malonate-catalyzed reactions were also second-order with a linear correlation coefficient of $r > 0.995$. The calculated rate constants of the ALB-catalyzed reaction ($k_{obs \cdot ALB}$) and ALB+Na–malonate-catalyzed reaction ($k_{obs \cdot ALB+Na\text{-}malonate}$) were 0.273 $M^{-1}h^{-1}$ and 1.66 $M^{-1}h^{-1}$, respectively.

This strategy has also made possible the use of 0.3 mol% ALB for efficient catalytic asymmetric Michael reactions (Scheme 18). The Michael adduct **66** (99% ee) was converted to the tricyclic compound **124** in 92% overall yield using a Fischer indole synthesis as the key step. Such tricyclic compounds should prove to be attractive intermediates for the catalytic asymmetric synthesis of various indole alkaloids.[48]

D. Catalytic Asymmetric Epoxide Openings Promoted by GaMB

The enantioselective ring opening of symmetrical epoxides is an attractive and quite powerful method in asymmetric synthesis.[49] Since a pioneering report by Whitesell,[50] various types of stoichiometric or catalytic asymmetric epoxide openings have been reported.[51] Only a few practical methods, however, have been reported so far, which in general require the use of silylated compounds as nucleophiles. We became very interested in the development of catalytic asymmetric epoxide ring openings using non-silylated nucleophiles such as RSH, HCN, and

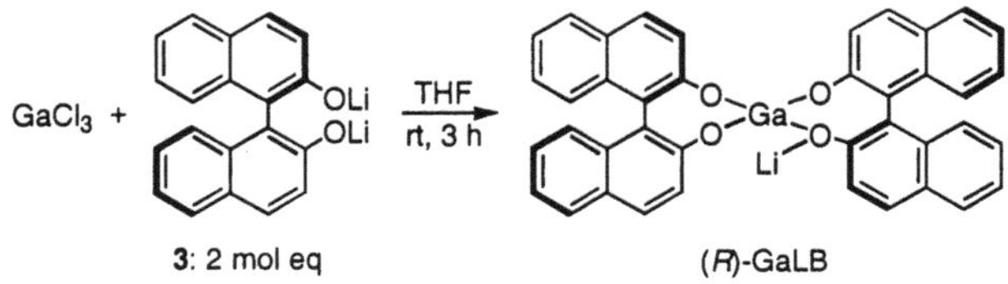

Scheme 19. Preparation of (*R*)-GaLibis(binaphthoxide) ((*R*)-GaLB)

HN_3.[52] In 1985, Yamashita and Mukaiyama reported a catalytic asymmetric epoxide opening by thiols using a Zn–tartrate catalyst.[51b] To the best of our knowledge, this is the only useful example to obtain synthetically versatile β-hydroxy sulfides in an optically active form by way of an epoxide opening.

Mukaiyama's catalyst, however, is still unsatisfactory in terms of broad usefulness, selectivity, and reactivity. We therefore examined a catalytic asymmetric ring opening of symmetrical epoxides with thiols using the heterobimetallic complexes. We envisioned that these complexes would prove to be useful for the catalytic asymmetric opening of **125** with a nucleophile such as $PhCH_2SH$ (**126**). However, LaM_3tris(binaphthoxide) (M = Li or Na) or AlLibis(binaphthoxide) (ALB) showed only low catalytic activity, giving 2-(benzylthio)cyclohexanol (**127**)[51b] in 1–10% yields, although modest to high ees (27–86% ee) were observed. We then examined the new heterobimetallic asymmetric complexes with group 13 elements (B, Ga, In) other than Al. Of these, the GaLibis((*R*)-binaphthoxide) complex ((*R*)-GaLB), which was readily prepared from $GaCl_3$, (*R*)-binaphthol (2 mol equiv to $GaCl_3$), and BuLi (4 mol equiv to $GaCl_3$) in THF, showed a high catalytic activity for the present reaction.[53] The proposed structure of GaLB is shown in Scheme 19. Although racemic **127** was obtained when THF was used as a solvent, the use of toluene gave a 40% ee of (1*R*, 2*R*)-**127** (87% yield) in the presence of 10 mol% of GaLB at rt (Scheme 20).[54] Interestingly, the ee of **127** gradually increased as the reaction proceeded, presumably due to the resultant decrease in the concentration of the remaining thiol. In addition, by using a stoichiometric amount of GaLB, compound **127** could be obtained with 88% ee in 87% yield. These results appear to indicate that an undesired ligand exchange of binaphthol for the thiol **126**,[55] or perhaps just a Lewis acid-catalyzed reaction without participation of a lithium binaphthoxide moiety, was taking place.

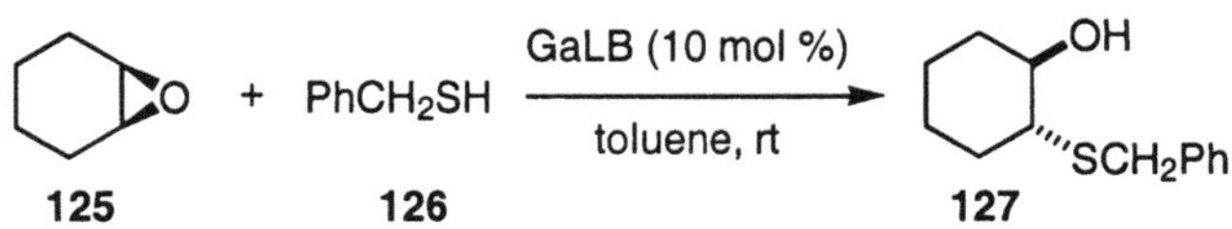

Scheme 20. Enantioselective epoxide opening by $PhCH_2SH$.

Table 15. Catalytic Asymmetric Ring Openings of Symmetrical Epoxides with *t*-BuSH (128) Catalyzed by (*R*)-GaLB with MS 4A

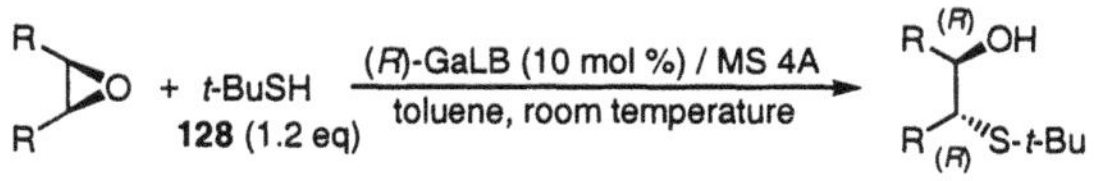

Entry	*Epoxide*	*MS 4A*[a] *(g)*	*Time (h)*	*Product*	*Yield (%)*	*ee (%)*
1	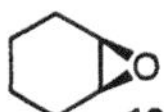 **125**	none	65	129	35	98
2	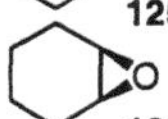 **125**	0.2	9	129	80	97
3	 **130**	0.2	36	131	74	95
4	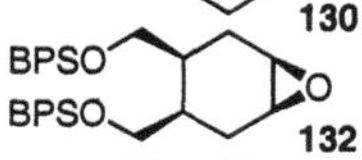 **132**	0.2	12	133	83	86
5	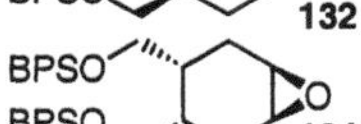 **134**	0.2	137	135	64	91
6	 **136**	0.2	24	137	89	91
7[b]	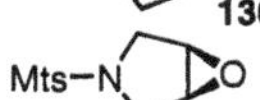 **138**	0.2	72	139	89	89
8	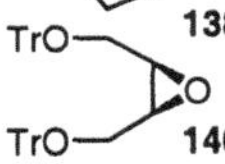 **140**	2.0	48	141	89	82

Notes: [a]Weight per 0.1 mmol of GaLB.
[b]Carried out at 50 °C in the presence of 30 mol% of GaLB. Mts = 2,4,6-trimethylbenzene-sulfonyl.

We hypothesized that using more sterically hindered thiols such as *t*-BuSH (**128**) might prevent such undesired side reactions. As hoped, the reaction of **125** with **128** in the presence of GaLB (10 mol%) afforded **129** with 98% ee even at rt, albeit only in 35% yield. After many attempts, we were pleased to find that the addition of molecular sieve (MS) 4A (0.2 g/mmol of **125**) was extremely effective in enhancing the reaction rate,[56] with **129** being obtained in 80% yield without reduction of ee (Table 15, entries 1–2).[53] GaLB catalysis was found to be applicable to a wide range of symmetrical epoxides; epoxides with functional groups (**130, 132**, and **134**) and epoxides fused to five-membered rings (**136** and **138**) underwent ring opening with high ee (89–96% ee) in good yields (entries 3–7). On the other hand, the reaction of acyclic **140** proceeded very slowly, giving **141** in a low yield. This problem was overcome by an increase in the amount of MS 4A (2.0 g/mmol of **140**), with the reaction reaching completion in 48 h and affording **141** with 82% ee in 89% yield (entry 8). The role of MS 4A (sodium aluminosilicate) is interesting, with ^{13}C NMR spectra revealing that GaLB was transformed into GaSB in the presence of large

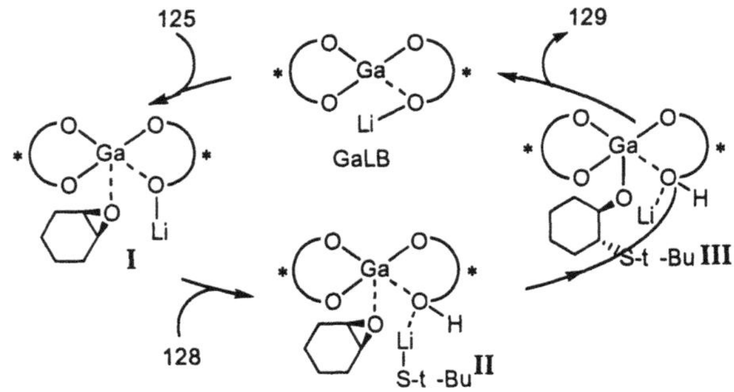

Scheme 21. Working model for the ring-opening of cyclohexene oxide (**125**) with *t*-BuSH (**128**) catalyzed by GaLB.

amounts of MS 4A, and in fact the use of GaSB (10 mol%) instead of GaLB gave **141** with 78% ee in 93% yield (rt, 12 h). In striking contrast to these results, the use of GaSB with the other epoxides gave very low ees, suggesting that the transformation of GaLB to GaSB did not take place to a significant extent under the standard reaction conditions (entries 2–6) using smaller quantities of MS 4A. Another possible role of MS 4A may be to assist the decomplexation of the product from the catalysts, resulting in an enhanced overall reaction rate.

A working model for the catalytic asymmetric epoxide-opening is shown in Scheme 21. GaLB appears to act as an asymmetric multifunctional catalyst,[2,8] with a lithium binaphthoxide moiety functioning as a Brϕnsted base, activating **128** and controlling the orientation of the resulting lithium thiolate by chelation. In addition, gallium metal appears to function as a Lewis acid, activating and also controlling the orientation of **125**, presumably due to coordination of an axial lone pair, and allowing for the cleavage of the C–O bond by a backside attack (**II**). On this basis, the likely absolute configuration of **129** can be predicted to be (1*R*, 2*R*), and in fact (*R*)-GaLB did indeed give (1*R*, 2*R*)-products in every case. Because six-membered S_N2-like transition states are not usually seen for epoxide-opening reactions, due to problems with the bond angle, this geometry (**II**) is suggested as a working model that provides a way to correlate these results.[57] The alkoxide intermediate (**III**) thus

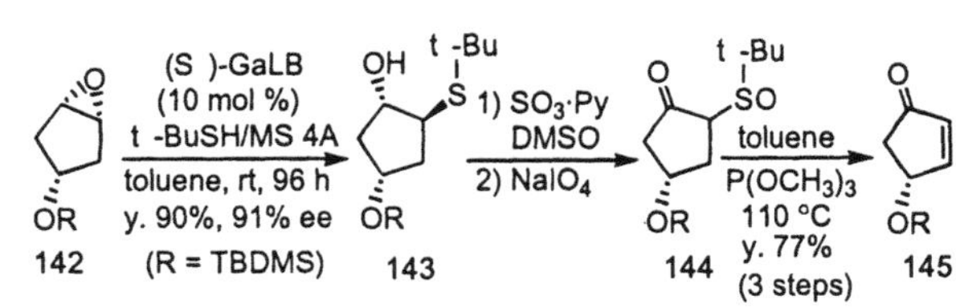

Scheme 22. Catalytic asymmetric synthesis of **145**.

produced may then abstract a proton from an acidic OH group to afford **129** and regenerate GaLB, thereby completing the catalytic cycle.

We have applied this methodology to the synthesis of **145**, an important intermediate in the synthesis of prostaglandins. As shown in Scheme 22, epoxide **142**[58] gave **143** with 91% ee in 90% yield by reaction with *t*-BuSH (**128**) in the presence of (*S*)-GaLB (10 mol%) and MS 4A (rt, 96 h). Oxidation of **143** gave the α-sulfinyl ketone **144**, and subsequent pyrolysis afforded (*R*)-**145** in 77% overall yield.

IV. CATALYTIC ASYMMETRIC SYNTHESIS WITH AN OLIGOMERIC HOMOMETALLIC COMPLEX

A. Preparation of the Ln–BINOL Complex

Alkali metal-free chiral rare earth complexes have been efficiently prepared from Ln(O-*i*-Pr)$_3$ and chiral BINOL derivatives. For example, to a stirred solution of La(O-*i*-Pr)$_3$10 (1.0 mmol) in anhydrous THF was added 1 mol equiv of (*S*)-**2** at 0 °C, which resulted in the formation of a La-**2** complex (**1**). Although the structure of **1** was not unequivocally determined, we examined the catalytic activity of the complex for a number of asymmetric reactions.

B. Catalytic Asymmetric Michael Reaction Promoted by the La-2 Complex (1)

We found that asymmetric Michael reactions proceeded in a highly enantioselective manner (up to 95% ee) with a catalytic amount of **1**.[42] For example, treatment of dibenzyl methylmalonate (**63**) with cyclopentenone (**69**) (1.1 equiv) in anhydrous THF containing 10 mol% of **1** at –20 °C for 48 h gave the Michael adduct **70** of 70% ee in 86% yield. As a solvent, THF gave the best results. A possible mechanism for this Michael reaction is proposed in Scheme 23. It seemed likely that the reaction of **63** with **1** would give the BINOL-lanthanum ester enolate **I**. This lanthanum ester enolate **I** would react with **69**, giving the lanthanum enolate **II** in an enantioselective manner. Further reaction of this enolate **II** with **63** would lead to the Michael adduct **70** together with the lanthanum ester enolate **I** due to the difference in their pK_a values, thus making possible the catalytic cycle. We found that the intermediary lanthanum ester enolate **I**, the real asymmetric catalyst, could be more directly prepared starting with La(O-*i*-Pr)$_3$, dibenzyl methylmalonate (**63**), and BINOL (**2**). Representative results of this Michael reaction are shown in Table 16.

C. Catalytic Asymmetric Epoxidation of Enones Promoted by Ln-BINOL Derivative Complexes

Catalytic asymmetric epoxidation is one of the most important asymmetric processes. An efficient catalytic asymmetric epoxidation of enones with broad

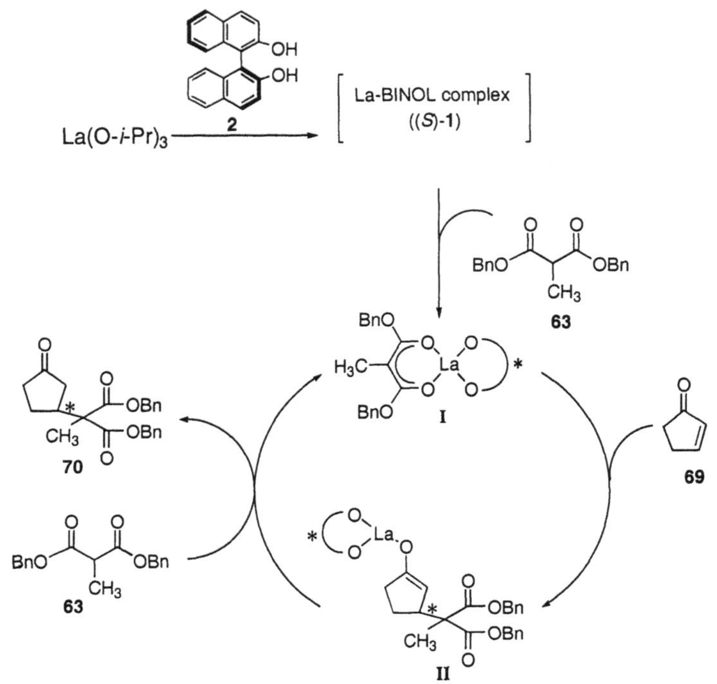

Scheme 23. Proposed mechanism for the La-BINOL complex (**1**) catalyzed asymmetric Michael reaction.

generality, however, has not yet been developed.[58] As described in the preceding chapter we have succeeded in developing efficient catalytic asymmetric Michael reactions using **1** (Scheme 23). We envisioned that **1** would be useful for the asymmetric epoxidation of enones, using hydroperoxides such as *tert*-butyl hydroperoxide (TBHP), since this type of epoxidation proceeds through Michael-type addition of hydroperoxides to enones. As expected, the reaction of **71** with TBHP (1.5 equiv) at rt for 0.5 h catalyzed by **1** (10 mol%), which was prepared in the presence of activated MS 4A powder in THF, afforded **150** with 62% ee in 90% yield. Moreover, the use of cumene hydroperoxide (CMHP) instead of TBHP improved the asymmetric epoxidation, giving **150** with 83% ee in 93% yield (5 mol% of the catalyst **1**).[59]

Although the role of MS 4A is not clear at present, the reactions do appear to be accelerated by their addition. This type of chiral lanthanum catalyst was found to be applicable to a range of enone substrates. Thus, **151** was converted to **152** with 86% ee in 93% yield, and **153** was transformed to **154** with 85% ee in 85% yield (Table 17, entries 1, 4, and 6). After several attempts, we were pleased to find that the use of **147**, prepared from (*R*)-3-hydroxymethyl-BINOL (**146**) instead of **2**, substantially improved the catalytic asymmetric epoxidations (Table 17, entries 2,

Table 16. Catalytic Asymmetric Michael Reactions Promoted by the La-(*S*)-**2** Enolate Complex I (10 mol%)

Entry	*Enone*	*Michael Donor*	*Product*	*Temp. (°C)*	*Time (h)*	*Yield (%)*	*ee (%)*
1	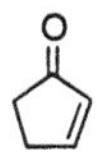 69	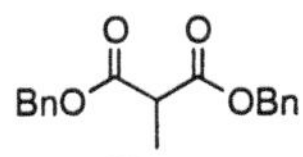 63	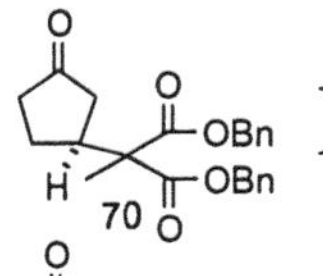 70	−10 −20	84 60	91 97	91 95
2	69	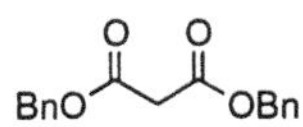 61	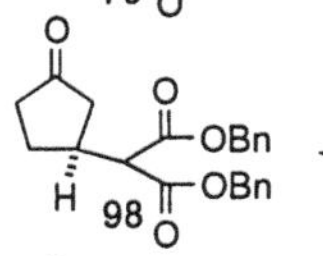 98	−20	72	96	92
3	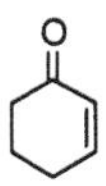 60	63	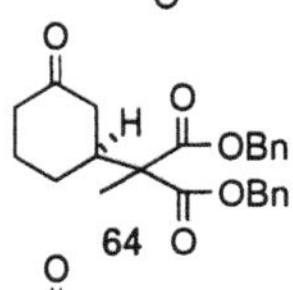 64	0	84	83	87
4	60	61	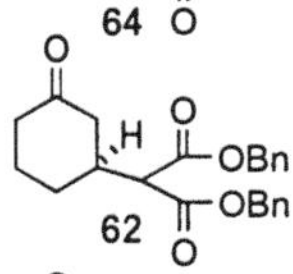 62	−10	84	94	92
5	60	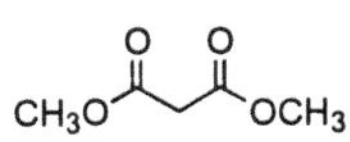 65	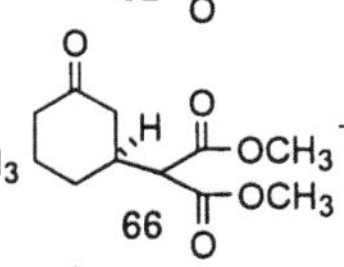 66	−10	84	100	75
6	60	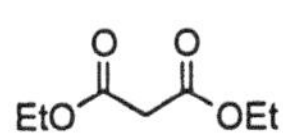 67	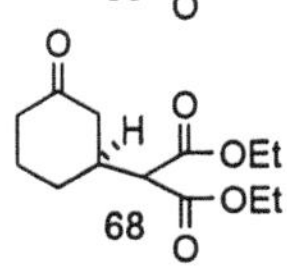 68	−10	84	97	78

3, 5, and 7). Namely, **150, 152**, and **154** were obtained in excellent yields with 91, 94, and 83% ee, respectively.

In contrast to the results presented above, the enones shown in Table 17 (entries 8, 9, 10, and 11) were best converted to the corresponding epoxides using the ytterbium complex **149** generated from Yb(O-*i*-Pr)$_3$, (*R*)-**146**, and MS 4A in THF at 40 °C for 1 h (Scheme 24). That is, treatment of **155** with TBHP (1.5 equiv) in the presence of 5 mol% of **149** in THF at rt for 96 h was found to give **156** with 94% ee in 83% yield. Gratifyingly, **158, 160**, and **162** were obtained in excellent yields with 88, 88, and 91% ee, respectively. In contrast, use of either the **148** catalyst or the La–CMHP system afforded **156** with less satisfactory results. It seems likely that the difference in ionic radius between lanthanum and ytterbium, as well as the difference in Lewis acidities, accounts for the observed center-metal effects.

Table 17. Catalytic Asymmetric Epoxidations Using Alkali Metal-Free Lanthanoid Complexes

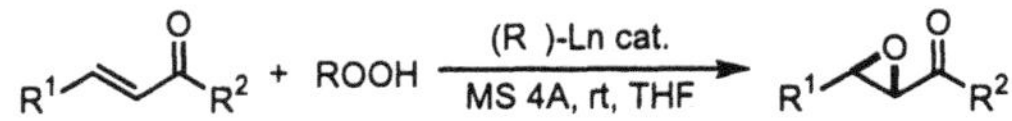

71, 150: R^1 = Ph, R^2 = Ph
151, 152 : R^1 = i -Pr, R^2 = Ph
153, 154 : R^1 = Ph, R^2 = o -MOMOC_6H_4
155, 156: R^1 = Ph, R^2 = CH_3
157, 158: R^1 = Ph, R^2 = i -Pr
159, 160: R^1 = $Ph(CH_2)_2$, R^2 = CH_3
161, 162: R^1 = $CH_3(CH_2)_4$, R^2 = CH_3

Entry	*Enone*	*Epoxide*	*Ln cat. (mol %)*	*ROOH (1.5 eq)*	*Time (h)*	*Yield (%)*	*ee (%)*[a]
1	71	150	1 (5)	CMHP	6	93	83
2	71	150	147 (5)	CMHP	7	93	91
3	71	150	147 (1)	CMHP	44	95	89
4	151	152	1 (5)	CMHP	12	93	86
5	151	152	147 (5)	CMHP	7	95	94
6	153	154	1 (5)	CMHP	20	85	85
7	153	154	147 (5)	CMHP	96	78	83
8	155	156	149 (5)	TBHP	96	83	94
9	157	158	149 (8)	TBHP	159	55	88
10	159	160	149(8)	TBHP	118	91	88
11	161	162	149 (8)	TBHP	67	71	91

Note: [a]Absolute configurations were determined to be (αS, βR).

Although we have not succeeded in determining the structure, it was found that a 1:1–1:1.4 ratio of Ln(O-*i*-Pr)$_3$ (Ln = La or Yb) and **2** gave the maximum enantiomeric excesses (Figure 11). The ^{13}C NMR spectrum of **1** was quite obscure, suggesting that both the chiral catalyst **148** and catalyst **1** exist as oligomers. Moreover, we have succeeded in obtaining an asymmetric amplification of the catalytic asymmetric epoxidation. We believe that the oligomeric structure of these lanthanoid–BINOL catalysts may play a key role in these catalytic asymmetric epoxidations of enones. That is, a Ln–alkoxide moiety in the catalysts appears to act as a Brϕnsted base, activating a hydroperoxide moiety so as to make possible a

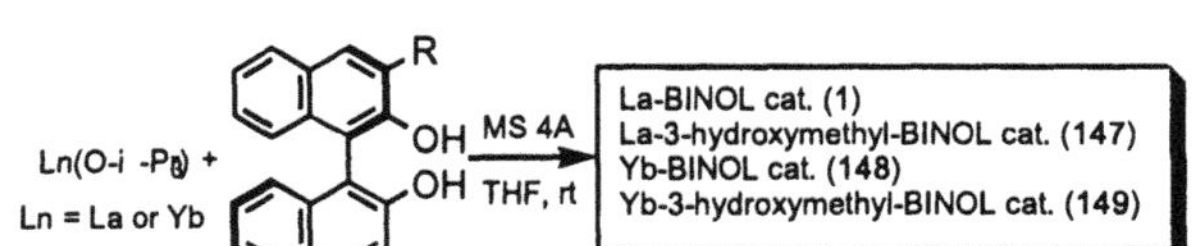

2: R = H (1 equiv)
146: R = CH_2OH (1.25-1.4 equiv)

Scheme 24. Preparation of chiral rare earth-BINOL derivative catalysts.

Michael reaction. At the same time, another Ln metal ion seems to act as a Lewis acid, both activating and controlling the orientation of the enone. Such a mechanism, analogous to those for enzymatic methods, may explain why various epoxides can be synthesized with good enantiomeric excesses even at room temperature.

V. SUMMARY

As has been discussed in the previous sections, we have succeeded in developing conceptually new heterobimetallic asymmetric catalysts such as LnMB, AMB, and GaMB. In addition, we have also developed several homometallic rare earth complexes. Because these complexes are considered to be two- (or more) center catalysts, in order to achieve a high degree of control for the mode of enantioselection, the selection of appropriate metals and chiral ligands is very important.

A visible difference in the catalytic activity of these heterobimetallic complexes can be seen in the TLC analysis of a catalytic asymmetric direct aldol reaction, starting from aldehydes and *unmodified* ketones. Such reactions are known in enzyme chemistry, with the fructose-1,6-bisphosphate and/or DHAP aldolases being characteristic examples. A cooperative mode of action should be important in the reactions mediated by any of several heterobimetallic asymmetric catalysts having both Lewis acidity and Brønsted basicity. Although, as shown in Figure 12, only LLB can efficiently promote the catalytic asymmetric direct aldol reaction of **163** with **164** to give **165** in up to 94% ee, other heterobimetallic complexes gave the desired **165** in trace to low yield.[61]

We believe that the successful development of the heterobimetallic concept has opened up a new field in asymmetric catalysis. We hope that the findings discussed in this chapter will prove to be a significant landmark in the development of the field of catalytic asymmetric synthesis.

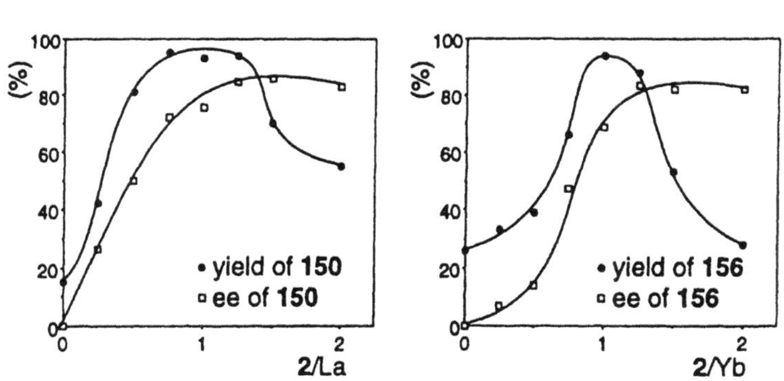

Figure 11. Influence of the ratio of Ln(O-*i*-Pr)$_3$ and **2** on yield and ee. Left: epoxidation of **71** catalyzed by the La-**2** catalyst (**1**). Right: epoxidation of **155** catalyzed by the Yb-**2** catalyst (**148**).

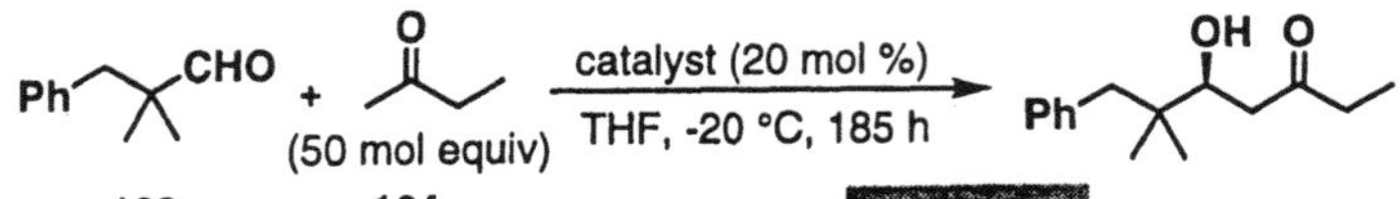

163 164

catalyst	y (%)	ee (%)
LLB	71	94
LSB	0	-
LPB	trace	N. D.
ALB	0	-
GaLB	0	-

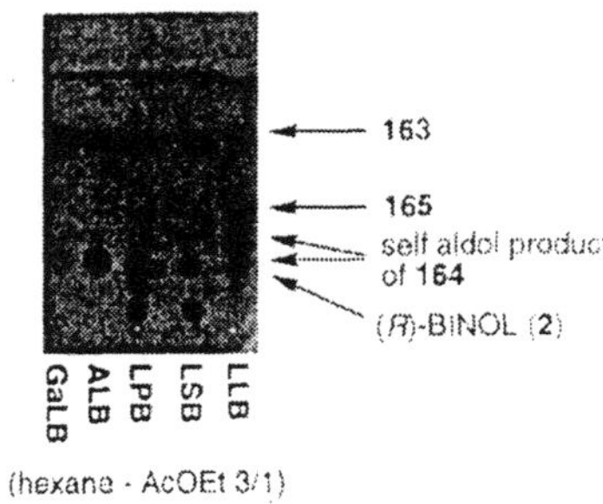

Figure 12. Effects of various heterobimetallic catalysts on direct asymmetric aldol reaction of aldehyde **163** with 2-butanone (**164**).

REFERENCES AND NOTES

1. (a) *Applied Homogeneous Catalysis with Organometallic Compounds*; Herrmann, W. A.; Cornils, B., Eds. VCH: Weinheim, 1996; (b) Noyori, R. In *Asymmetric Catalyst in Organic Synthesis*; John Wiley & Sons: New York, 1994; (c) *Catalytic Asymmetric Synthesis;* Ojima, I. Ed.; VCH: New York, 1994; (d) *Asymmetric Catalysis*; Bosnich, B. Ed; Martinus Nijhoff Publishers: Dordrecht, 1986; (e) *Asymmetric Synthesis*; Morrison, J. D. Ed.; Academic Press: Orlando, 1985, Vol. 5.
2. Steinhagen, H.; Helmchen, G. *Angew. Chem., Int. Ed. Engl.* **1996**, *35*, 2339–2342.
3. Sasai, H.; Suzuki, T.; Arai, S. Arai, T.; Shibasaki, M.; *J. Am. Chem. Soc.* **1992**, *114*, 4418–4420.
4. Sasai, H.; Suzuki, T.; Itoh, N.; Shibasaki, M. *Tetrahedron Lett.* **1993**, *34*, 851–854.
5. Sasai, H.; Suzuki, T.; Itoh, N.; Arai, S.; Shibasaki, M. *Tetrahedron Lett.* **1993**, *34*, 2657–2660.
6. Sasai, H.; Watanabe, S.; Shibasaki, M. *Enantiomer* **1997**, *2*, 267–271.
7. Sasai, H.; Suzuki, T.; Itoh, N.; Tanaka, K.; Date, T.; Okamura, K.; Shibasaki, M. *J. Am. Chem. Soc.* **1993**, *115*, 10372–10373.
8. Sasai, H.; Arai, T.; Satow, Y.; Houk, K. N.; Shibasaki, M. *J. Am. Chem. Soc.* **1995**, *117*, 6194–6198.
9. Takaoka, E.; Yoshikawa, N.; Yamada, Y. M. A.; Sasai, H.; Shibasaki, M. *Heterocycles* **1997**, *46*. In press.
10. Purchased from Kojundo Chemical Laboratory Co. Saitama, Japan.
11. Henry, L.; Hebd, C. R. *Seances Acad. Sci.* **1895**, *120*, 1265.
12. For representative papers see: (a) Seebach, D.; Colvin, E. W.; Lehr, F.; Weller, T. *Chimia* **1979**, *33*, 1–18; (b) Rosini, G. In *Comprehensive Organic Synthesis*; Trost, B. M.; Heathcock, C. H., Eds.; Pergamon: Oxford, 1991, Vol. 2, pp. 321–340.
13. Sasai, H.; Itoh, N.; Suzuki, T.; Shibasaki, M. *Tetrahedron Lett.* **1993**, *34*, 855–858.
14. Sasai, H.; Yamada, Y. M. A.; Suzuki, T.; Shibasaki, M. *Tetrahedron* **1994**, *50*, 12313–12318.
15. Sasai, H.; Suzuki, T.; Itoh, N.; Shibasaki, M. *Appl. Organomet. Chem.* **1995**, *9*, 421–426.
16. Sasai, H.; Kim, W.-S.; Suzuki, T.; Shibasaki, M.; Mitsuda, M.; Hasegawa, J.; Ohashi, T. *Tetrahedron Lett.* **1994**, *35*, 6123–6126.
17. (a) Mimoto, T.; Imai, J.; Kisanuki, S.; Enomoto, H.; Hattori, N.; Akaji, K.; Kiso, Y. *Chem. Pharm. Bull.* **1992**, *40*, 2251–2253; (b) Kageyama, S.; Mitsuto, T.; Murakawa, T.; Nomizu, M.; Ford Jr., H.; Shirasaka, T.; Gulnik, S.; Erickson, J.; Takada, K.; Hayashi, H.; Broder, S.; Kiso, Y.; Mitsuya, H. *Antimicrob. Agents Chemother.* **1993**, *37*, 810–817.
18. Lingfelter, D. S.; Hegelson, R. C.; Cram, D. J. *J. Org. Chem.* **1981**, *46*, 393–406.

19. Maruoka, K.; Itoh, T.; Araki, Y.; Shirasaka, T.; Yamamoto, H. *Bull. Chem. Soc. Jpn.* **1988**, *61*, 2975–2976.
20. Yamamoto, K.; Noda, K.; Okamoto, Y. *J. Chem. Soc. Chem. Commun.* **1985**, 1065–1066.
21. A 6,6′-substituted BINOL was also utilized independently in asymmetric Diels–Alder reactions. See: Terada, M.; Motoyama, Y.; Mikami, K. *Tetrahedron Lett.* **1994**, *35*, 6693–6696.
22. Seebach, D. *Angew. Chem., Int. Ed. Engl.* **1990**, *29*, 1320-1367.
23. Iseki, K.; Oishi, S.; Sasai, H.; Shibasaki, M. *Tetrahedron Lett.* **1996**, *37*, 9081–9084.
24. Sasai, H.; Tokunaga, T.; Watanabe, S.; Suzuki, T.; Itoh, N.; Shibasaki, M. *J. Org. Chem.* **1995**, *60*, 7388–7389.
25. Sasai, H., S. Watanabe, Suzuki, T., Shibasaki, M. *Organic Syntheses.* Submitted.
26. Schwartz, G. K.; Jiang, J.; Kelsen, D.; Albino, P. *J. Nat. Cancer Inst.* **1993**, *85*, 402–407.
27. Arai, T.; Yamada, Y. M. A.; Yamamoto, N.; Sasai, H.; Shibasaki, M. *Chem. Eur. J.* **1996**, *2*, 1368–1372.
28. (a) Young, M.; Pan, W.; Wiesner, J.; Bullough, D.; Browne, G.; Balow, G.; Potter, S.; Metzner, K.; Mullane, K. *Drug Dev. Res.* **1994**, *32*, 19–28; (b) Hammond, H. K.; McKirnan, M. D. *J. Am. Coll. Cardiol.* **1994**, *23*, 475–482.
29. Sasai, H.; Hiroi, M.; Yamada, Y. M. A.; Shibasaki, M.; *Tetrahedron Lett.* **1997**, *38*. In press.
30. (a) Helder, R.; Wynberg, H. *Tetrahedron Lett.* **1975**, 4057–4060; (b) Hermann, K.; Wynberg, H. *J. Org, Chem.* **1979**, *44*, 2238–2244; (c) Cram, C. J.; Sogah, G. D. Y. *J. Chem. Soc., Chem. Commun.* **1981**, 625–627; (d) Brunner, H.; Hammer, B. *Angew. Chem., Int. Ed. Engl.* **1984**, *23*, 312–313; (e) Sera, A.; Takagi, K.; Katayama, H.; Yamada, H.; Matsumoto, K. *J. Org. Chem.* **1988**, *53*, 1157–1161; (f) Tamai, Y.; Kamifuku, A.; Koshiishi, E.; Miyano, S. *Chem. Lett.* **1995**, 957–958; (g) Desimoni, G.; Dusi, G.; Faita, G.; Quadrelli, P.; Righetti, P. *Tetrahedron* **1995**, *51*, 4131–4144; (h) Aoki, S.; Sasaki, S.; Koga, K. *Tetrahedron Lett.* **1989**, *30*, 7229–7230; (i) Sawamura, M.; Hamashima, H.; Shinoto, H.; Ito, Y. *Tetrahedron Lett.* **1995**, *36*, 6479–6482 and references cited therein; (j) Takasu, M.; Wakabayashi, H.; Furuta, K.; Yamamoto, H. *Tetrahedron Lett.* **1988**, *29*, 6943–6946; (k) Aoki, S.; Sasaki, S.; Koga, K. *Heterocycles* **1992**, *33*, 493–495; (l) Yamaguchi, M.; Shiraishi, T.; Hirama, M.; *Angew. Chem. Int. Ed. Engl.* **1993**, *32*, 1176–1178; (m) Yamaguchi, M.; Shiraishi, T.; Igarashi, Y.; Hirama, M. *Tetrahedron Lett.* **1994**, *35*, 8233–8236; (n) Kawara, A.; Taguchi, T. *Tetrahedron Lett.* **1994**, *35*, 8805–8808.
31. Cockerill, A. F.; Davies, L. O.; Harden, R. C.; Rackham, D. M. *Chem. Rev.* **1973**, *73*, 553–588 and references cited therein.
32. (a) Rappé, A. K.; Casewit, C. J.; Colwell, K. S. Goddard III, W. A.; Skiff, W. M.; *J. Am. Chem. Soc.* **1992**, *114*, 10024–10035; (b) Casewit, C. J. Colwell, K. A.; Rappé, A. K.; *ibid.* **1992**, *114*, 10035–10046; (c) Casewit, C. J.; Colwell, J.; K. S. Rappé, A. K.; *ibid.* **1992**, *114*, 10046–10053; (d) Rappé, A. K.; Colwell, K. S.; Casewit, C. J. *Inorg. Chem.* **1993**, *32*, 3438–3450.
33. Sasai, H.; Emori, E.; Arai, T.; Shibasaki, M. *Tetrahedron Lett.* **1996**, *37*, 5561–5564.
34. (a) Burke Jr., T. R.; Barchi Jr., J. J.; George, C.; Wolf, G.; Shoelson, S. E.; Yan, X. *J. Med. Chem.* **1995**, *38*, 1386–1396; (b) Kafarski, P.; Lejczak, B. *Phosphorus, Sulfur Silicon Relat. Elem.* **1991**, *63*, 193–215.
35. Sasai, H., Arai, S., Tahara, Y.; Shibasaki, M. *J. Org. Chem.* **1995**, *60*, 6656–6657.
36. (a) Girard, P.; Namy, J. L.; Kagan, H. B. *J. Am. Chem. Soc.* **1980**, *102*, 2693–2698; (b) Imamoto, T.; Kusumoto, T.; Tawarayama, Y.; Sugiura, Y.; Hatanaka, Y; Yokoyama, M. *J. Org. Chem.* **1984**, *49*, 3904–3912.
37. Gröger, H.; Saida, Y.; Arai, S.; Martens, J.; Sasai, H.; Shibasaki, M. *Tetrahedron Lett.* **1996**, *37*, 9291–9292.
38. For previously reported asymmetric Al-BINOL reagents and related compounds, see: (a) Noyori, R. *Tetrahedron* **1994**, *50*, 4259–4292 and references therein; (b) Maruoka, K.; Yamamoto, H. In *Catalytic Asymmetric Synthesis*; Ojima, I., Ed.; VCH: New York, 1993, pp. 413–440, and references therein.

39. Arai, T.; Sasai, H.; Aoe, K.; Okamura, K.; Date, T.; Shibasaki, M. *Angew. Chem. Int. Ed. Engl.* **1996**, *35*, 104–106.
40. Prepared from diisobutylaluminum hydride, two equivalents of (*R*)-BINOL, and one equivalent of NaO-*t*-Bu, $KN(TMS)_2$, and/or $Ba(O\text{-}t\text{-}Bu)_2$.
41. (a) Canet, D.; Delpuech, J. J. Khaddar, M. R.; Rubini, P. R.; *J. Mag. Res.* **1973**, *9*, 329–330; (b) Delpuech, J. J.; Khaddar, M. R. Peguy, A. A.; Rubini, P. R.; *J. Chem. Soc. Chem. Commun.* **1974**, 154–155; (c) *Idem, J. Am. Chem. Soc.* **1975**, *97*, 3373–3379.
42. Sasai, H.; Arai, T.; Shibasaki, M. *J. Am. Chem. Soc.* **1994**, *116*, 1571–1572.
43. Yokomatsu, T.; Yamagishi, T.; Shibuya, S. *Tetrahedron: Asymmetry* **1993**, *4*, 1783–1784.
44. Rath, N. P.; Spilling, C. D. *Tetrahedron Lett.* **1994**, *35*, 227–230.
45. Sasai, H.; Bougauchi, M.; Arai, T.; Shibasaki, M. *Tetrahedron Lett.* **1997**, *38*, 2717–2720.
46. Arai, T.; Bougauchi, M.; Sasai, H.; Shibasaki, M. *J. Org. Chem.* **1996**, *61*, 2926–2927.
47. Coordination of cyclohexenone appears to be also supported by the fact that BSB does not promote Michael reactions.
48. Shimizu, S.; Yamada, K.; Arai, T.; Sasai, H.; Shibasaki, M.; Abstract of *the 39th Symposium on the Chemistry of Natural Products*, Sapporo, 1997, pp. 25.
49. For a recent review on the enantioselective ring opening of symmetrical epoxides, see: Hodgson, D. M.; Gibbs, A. R.; Lee, G. P. *Tetrahedron* **1996**, *52*, 14361–14384.
50. Whitesell, J. K.; Felman, S. W. *J. Org. Chem.* **1980**, *45*, 755–756.
51. (a) Pluim, H.; Wynberg, H. *Rec. J. R. Neth. Chem. Soc.* **1984**, *103*, 36–37; (b) Yamashita, H.; Mukaiyama, T. *Chem. Lett.* **1985**, 1643–1646; (c) Yamashita, H. *Chem. Lett.* **1987**, 525–528; (d) Su, H.; Walder, L.; Zhang, Z.; Scheffold, R. *Helv. Chim. Acta* **1988**, *71*, 1073–1078; (e) Chan, A. S. C.; Coleman, J. P. *J. Chem. Soc., Chem. Commun.* **1991**, 535–536; (f) Hayashi, M.; Kohmura, K.; Oguni, N. *Synlett* **1991**, 774–776; (g) Nugent, W. A. *J. Am. Chem. Soc.* **1992**, *114*, 2768–2769; (h) Asami, M.; Ishizaki, T.; Inoue, S. *Tetrahedron: Asymmetry* **1994**, *5*, 793–796; (i) Martínez, L. E.; Leighton, J. L.; Carsten, D. H.; Jacobsen, E. N. *J. Am. Chem. Soc.* **1995**, *117*, 5897–5898; (j) Cole, B. M.; Shimizu, K. D.; Krueger, C. A.; Harrity, J. P. A.; Snapper, M. L.; Hoveyda, A. H. *Angew. Chem., Int. Ed. Engl.* **1996**, *35*, 1668–1671.
52. Recently Jacobsen has reported that the (salen)Cr ring-opening functions with HN_3 through a (salen)Cr bimetallic transition state. See: Hansen, K. B.; Leighton, J. L.; Jacobsen, E. N. *J. Am. Chem. Soc.* **1996**, *118*, 10924–10925.
53. Iida, T.; Yamamoto, N.; Sasai, H.; Shibasaki, M. *J. Am. Chem. Soc.* **1997**, *119*, 4783–4784.
54. The absolute configuration of **127** was determined by comparison with the specific rotation reported in ref. 51b.
55. GaLB would be regenerated as the outcome of a decrease in the concentration of **126**, since the ligand exchange seems to be a reversible reaction.
56. Molecular sieves (MS) 4A were dried at 180 °C for 6 h under reduced pressure prior to use.
57. A flexible complex, in which Ga-O and Li-O bonds lengthen due to partial dissociation, might enable an intramolecular S_N2-like process to occur, although this is merely a hypothesis at the present time.
58. Asami, M.; *Tetrahedron Lett.* **1985**, *26*, 5803–5806.
59. (a) Juliá, S.; Masana, J.; Vega, J. C. *Angew. Chem., Int. Ed. Engl.* **1980**, *19*, 929–931; (b) Kroutil, W.; Mayon, P.; Lasterra-Sánchez, M. E.; Maddrell, S. J.; Roberts, S. M.; Thornton, S. R.; Todd, C. J.; Tüter, M. *Chem. Commun.* **1996**, 845–846 and references cited therein; (c) Helder, R.; Hummelen, J. C.; Laane, R. W. P. M.; Wiering, J. S.; Wynberg, H. *Tetrahedron Lett.* **1976**, 1831–1834; (d) Colonna, S.; Gaggero, N.; Manfredi, A.; Spadoni, M.; Casella, L.; Carrea, G.; Pasta, P. *Tetrahedron* **1988**, *44*, 5169–5178; (e) Colonna, S.; Manfredi, A.; Annunziata, R.; Gaggero, N. *J. Org. Chem.* **1990**, *55*, 5862–5866; (f) Baccin, C.; Gusso, A.; Pinna, F.; Strukul, G.; *Organometallics* **1995**, *14*, 1161–1167; (g) Kumar, A.; Bhakuni, V. *Tetrahedron Lett.* **1996**, *37*, 4751–4754; (h) Enders, D.; Zhu, J.; Raabe, G. *Angew. Chem., Int. Ed. Engl.* **1996**, *35*, 1725–1728;

(i) For an impressive catalytic asymmetric epoxidation of cinnamate esters, see: Jacobsen, E. N.; Deng, L.; Furukawa, Y.; Martinez, L. E. *Tetrahedron* **1994**, *50*, 4323–4334.
60. Bougauchi, M.; Watanabe, S.; Arai, T.; Sasai, H.; Shibasaki, M. *J. Am. Chem. Soc.* **1997**, *119*, 2329–2330.
61. Yamada, Y. M. A.; Yoshikawa, N.; Sasai, H.; Shibasaki, M. *Angew. Chem., Int. Ed. Engl.* **1997**, *36*. In press.

PALLADIUM-CATALYZED ENANTIOSELECTIVE ALLYLIC SUBSTITUTION REACTIONS

Simon J. Sesay and Jonathan M. J. Williams

Advances in Asymmetric Synthesis
Volume 3, pages 235–271.

ISBN: 0-7623-0106-6

I. INTRODUCTION

This chapter discusses some of the recent progress in enantioselective variants of palladium-catalyzed allylic substitution reactions. While the results from this research group are particularly highlighted, the results from other groups are also described in some detail. Furthermore, the chapter would not be complete without a reasonable discussion of the basic aspects of palladium-catalyzed allylic substitution reactions. There are several reviews published of palladium-catalyzed allylic substitution,[1] including those concentrating on the stereochemical aspects of the reaction.[2]

II. THE BASIC PROCESS OF PALLADIUM-CATALYZED ALLYLIC SUBSTITUTION

The nucleophilic substitution of allylic substrates may be catalyzed by palladium complexes. The reaction has been applied to many systems, but in essence, the transformation is summarized by the conversion of **1** into **2**, using a nucleophile and a palladium catalyst.

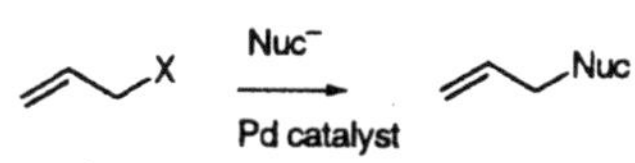

1 2

The mechanism of the reaction when soft nucleophiles are being employed is generally accepted to be as depicted in Scheme 1. The palladium(0) catalyst associates with the alkene substrate **1** to provide a palladium–alkene complex **3**. The palladium is now able to undergo an oxidative addition reaction to form an η^3-allyl complex **4**. There is an equilibrium between the neutral complex **4** and the cationic complex **5**. Bidentate ligands steer the equilibrium towards the formation of the cationic complex, which is presumably more susceptible to attack by a nucleophile, and thereby affords the alkene complex **6**. Dissociation of the palladium liberates the substitution product **2** and regenerates the active catalyst.

The reaction has been conducted with a wide range of substrates and nucleophiles. The leaving group (X) is most commonly acetate (OAc), but alternatively, carbonates (OCO_2R)[3] can be used, as well as halides.[4] Alcohols require activation to act as good leaving groups,[5] while ethers[6] can function effectively as leaving groups especially when the substrate is an epoxide[7] or a phenyl ether.[8] Less commonly, other leaving groups have been employed including phosphates,[9] sulfones,[10] amines,[11] and ammonium salts.[12] Even carbon has behaved as a leaving group when it is well-stabilized[13] or assisted by the release of ring strain.[14] Nucleophiles which have been used in palladium-catalyzed allylic substitution reactions are often soft nucleophiles such as the enolates derived from β-dicarbonyl

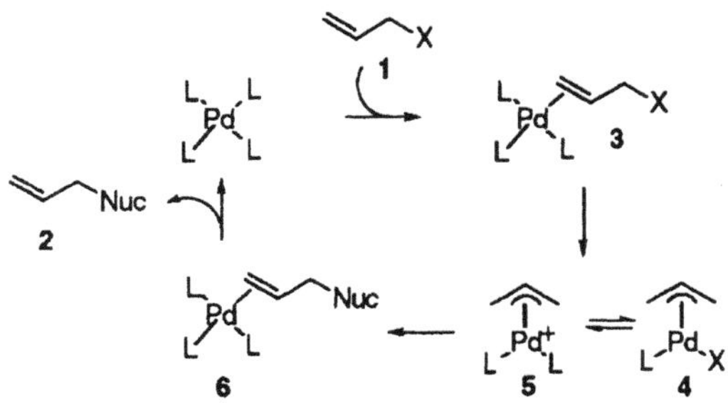

Scheme 1. Mechanism of palladium-catalyzed allylic substitution.

compounds. However, other carbon-based nucleophiles may also be used, including cyanide,[15] simple enolates,[16] bulky stabilized nucleophiles derived from bisphosphonates,[17] as well as harder nucleophiles.[18] Sulfur,[19] nitrogen,[20] oxygen,[21] silicon,[22] tin,[23] and hydride[24] nucleophiles have also been used in palladium-catalyzed substitution reactions.

The palladium catalyst employed is often $Pd(PPh_3)_4$ with additional phosphine. The catalyst is commercially available, but it is air sensitive, and so freshly prepared material can be more effective.[25] Dibenzylidene acetone complexes of palladium(0) ($Pd(dba)_2$ and $Pd_2(dba)_3$) are reasonably air stable, and these may be employed in conjunction with a phosphine ligand.[26] Allylpalladium chloride dimer is in oxidation state +2, and is reasonably air stable.[27] In the presence of a phosphine ligand and a nucleophile, this complex is converted into a palladium(0) complex which enters the catalytic cycle. Alkylphosphines can provide a more reactive catalyst,[28] although these ligands are more air sensitive. The use of bidentate ligands generally provides a more reactive catalyst,[29] and such ligands can be achiral or enantiomerically pure (vide infra). Aqueous[30] and heterogeneous catalysts have also been reported.[31]

Several groups have conducted elegant experiments to determine the basic stereochemical consequences of palladium-catalyzed allylic substitution reactions.[32] The main findings of this work demonstrate that for soft nucleophiles a retention of stereochemistry mechanism is operative. Thus, with the substrate **7**, the allylpalladium complex **8** formed in the catalytic cycle is generated with inversion of configuration.[33] Nucleophilic attack to this complex occurs from the *exo*-face, again with inversion of configuration to give the substitution product **9**. The two inversion steps provide the basis for the overall retention of configuration.

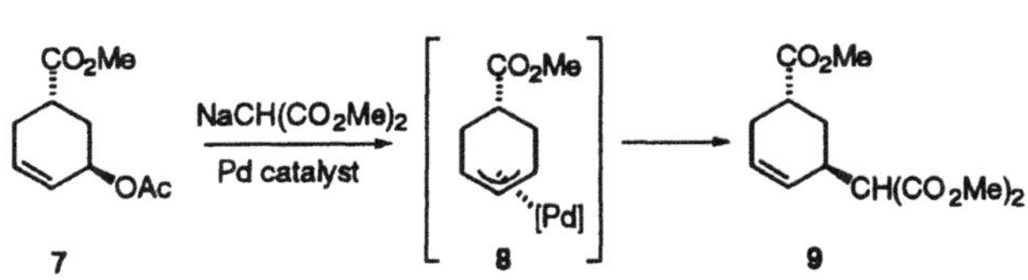

7 8 9

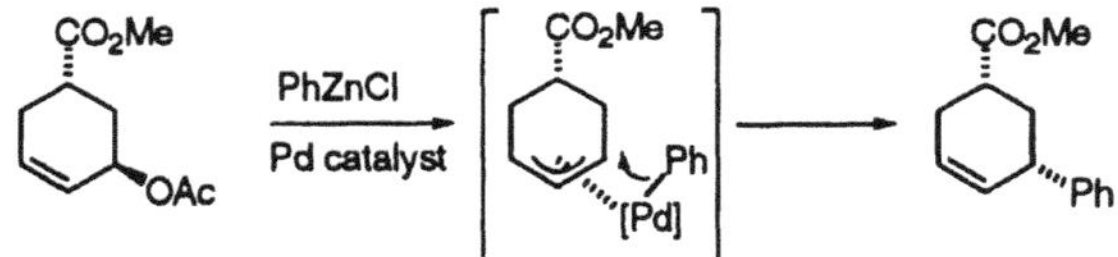

7 10 11

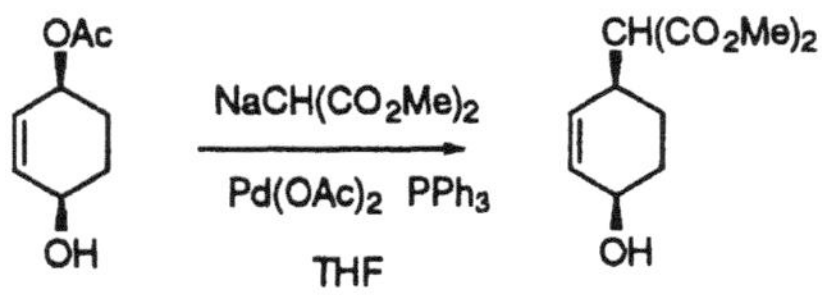

12 13

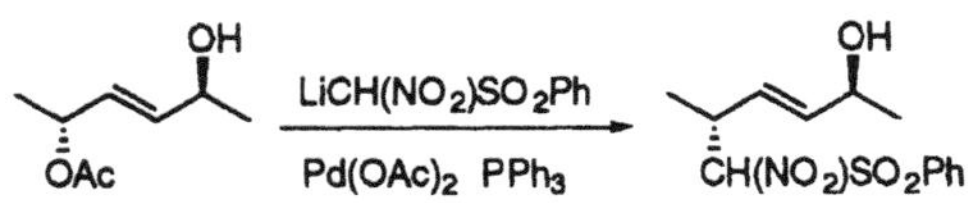

14 15

However, for hard nucleophiles, the situation is different from that shown above. The allylpalladium complex formation takes place with inversion of configuration as before, but the nucleophile then adds to the palladium to give complex **10**.[34] Delivery of the nucleophile from the palladium to the associated allyl moiety then takes place with retention of configuration to give the substitution product **11**. The overall stereochemical consequence in this instance is net inversion of stereochemistry.

Typically, it is expected that an enantiomerically pure substrate should then be converted into an enantiomerically pure product, and this is sometimes the case.[35] Thus, the substrate **12** is converted into the product **13** with complete transmission of stereochemistry,[36] and this is also seen for the conversion of compound **14** into the product **15**.[37]

However, there are also many examples where there is a loss of stereochemistry during the allylic substitution reaction. This is of importance in designing an enantioselective palladium-catalyzed allylic reaction, since if stereochemistry were always preserved, then it would be expected that a racemic substrate would always afford a racemic product. Fortunately this is not the case! Tsuji and coworkers have

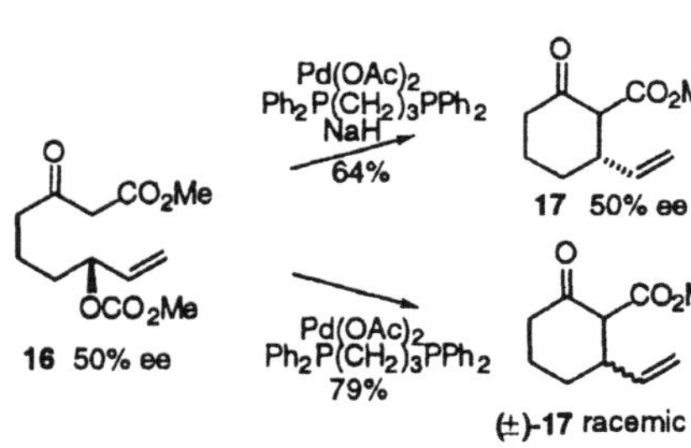

shown that the cyclization of **16** into **17** proceeds with retention of configuration when the nucleophile is preformed.[38] However, if the pro-nucleophile was deprotonated during the course of the reaction, the product was formed as the racemate (±)-**17**. The slower attack by the nucleophile allows the intermediate allylpalladium complexes time to equilibrate via the π-σ-π mechanism (vide infra).

III. REGIOCONTROL IN PALLADIUM-CATALYZED ALLYLIC SUBSTITUTION

Generally, there is an issue of regioselectivity to consider in palladium-catalyzed allylic substitution reactions. This point is nicely illustrated by the work of Keinan and Sahai.[39] These workers have reported that the reaction of **18** affords the regioisomer **20** with high selectivity over the alternative **21**. The incoming morpholine has attacked the allylpalladium intermediate **22** at the least sterically demanding terminus. It is generally considered that the regiochemistry of the starting material is lost upon formation of the allylpalladium intermediate, and so there is no memory of which regioisomeric starting material was employed. In fact this may not be an inviolate rule.[40] However, when a hard nucleophile is employed, the regiochemistry is reversed, and the opposite regioisomer **24** is formed preferentially over the regioisomer **23**. This dichotomy can be rationalized by the fact that hard nucleophiles react by a different mechanism, and by the involvement of intermediate **25**. The regiochemistry of the palladium-catalyzed substitution reactions can also be controlled by the electronic nature of the substituents,[41] as well as the nature of the ligand.[42]

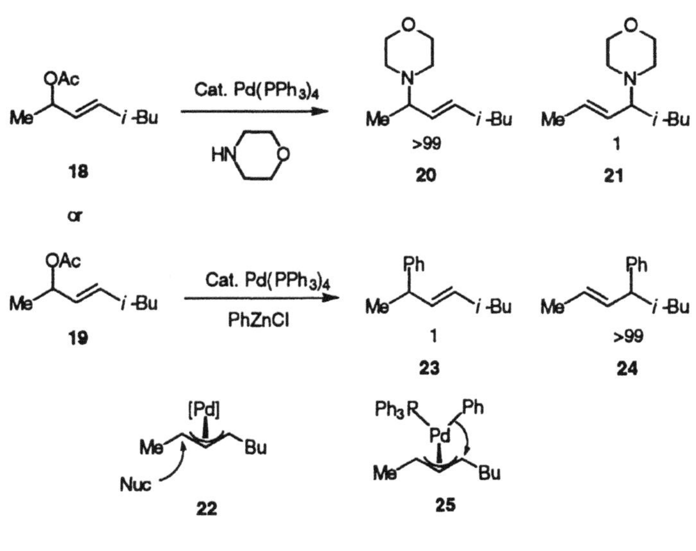

IV. DIASTEREOCONTROL IN PALLADIUM-CATALYZED ALLYLIC SUBSTITUTION

Several diastereoselective reactions have been reported. For example, Trost and Lee have reported that the cyclization of **26** can occur with up to 6.9:1 in favor of the *cis*-product **27**, although by changing the solvent, the selectivity could be altered such that the other diastereomer **28** was somewhat favored.[43]

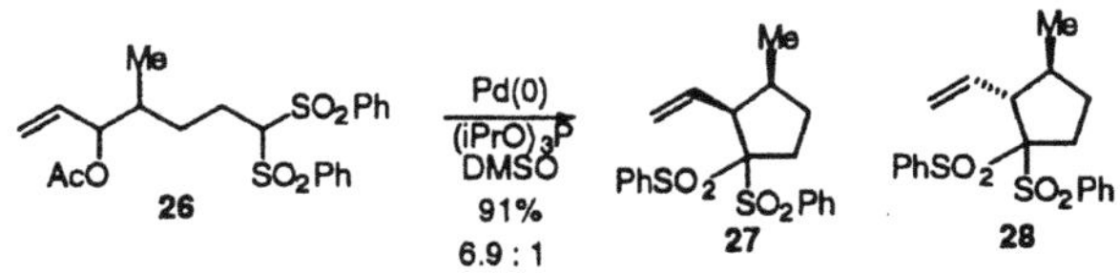

In another recent example, Hiroi and coworkers have employed the 1:1 diastereomeric mixture **29** as a substrate for palladium-catalyzed allylic substitution.[44] The stereochemistry associated with the sulfoxide exerts control over the reaction of the incoming nucleophile, and the product **30** is obtained with up to 79% de when dpph (diphenylphosphinohexane) is employed as the ligand.

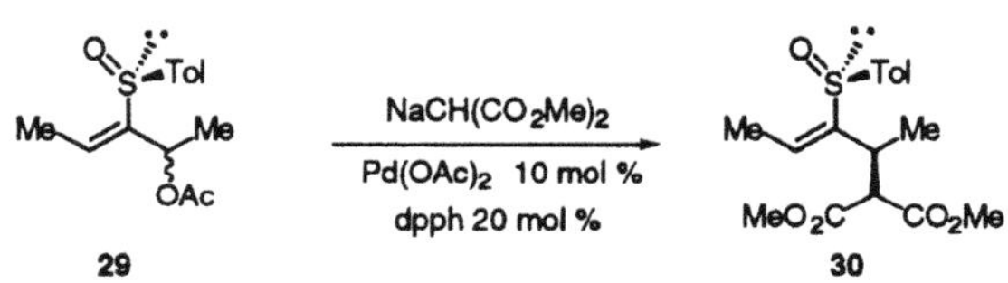

An interesting relay of stereochemical information has been described.[45] The proline-derived substrate **31** acts as both the allylating agent and as an enantiopure nucleophile. For this particular reaction, very high enantioselectivity is observed in the product **32**, although typically the enantioselectivities for related reactions are lower.

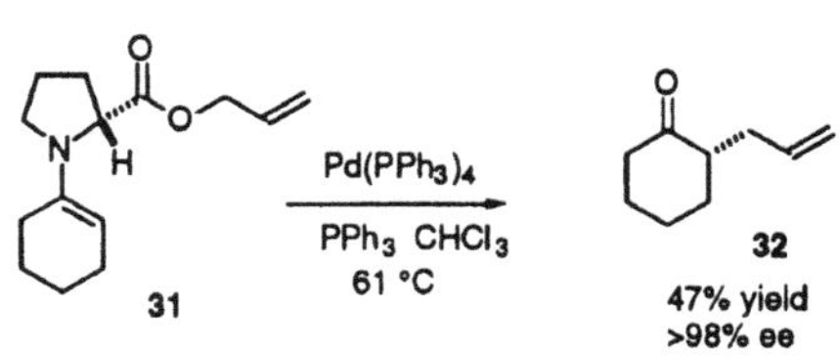

V. ENANTIOCONTROL IN PALLADIUM-CATALYZED ALLYLIC SUBSTITUTION REACTIONS

The enantiocontrol of palladium-catalyzed allylic substitution reactions has been achieved in a number of ways. The different approaches are summarized in Scheme 2, and these are then considered in turn.

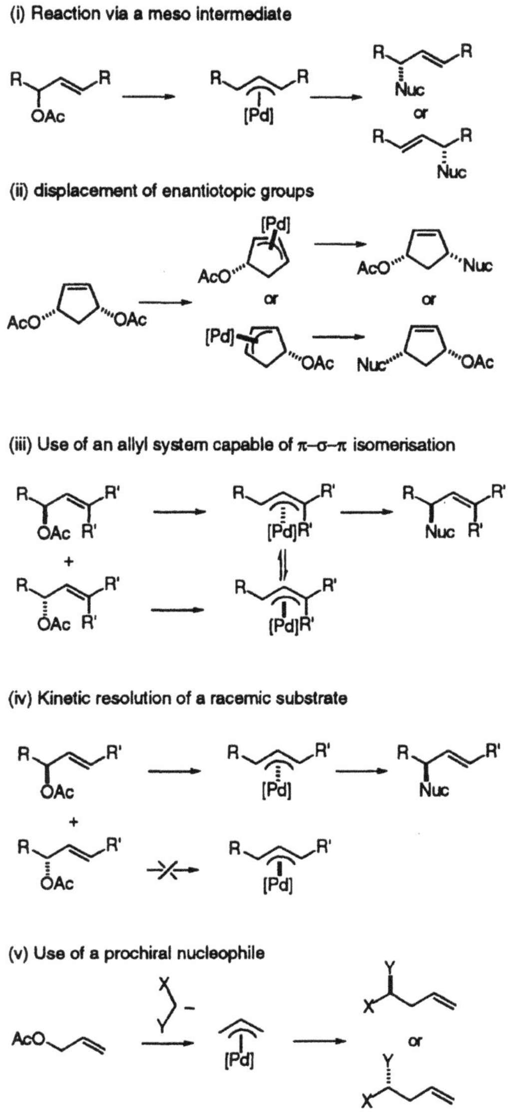

Scheme 2. Various approaches to enantioselective reactions.

The majority of research into enantioselective palladium-catalyzed allylic substitution reactions has investigated examples where the reaction proceeds via a *meso*-intermediate.[46] The principle is to use a substrate which leads to a *meso*-allylpalladium complex. The enantioselectivity of the reaction is then determined

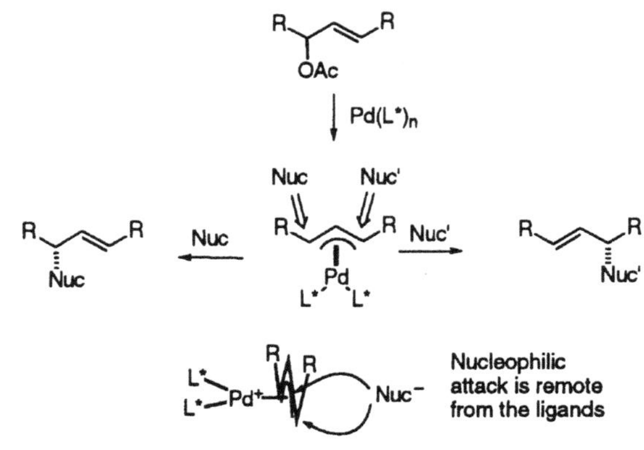

Figure 1. The problem of inducing asymmetry in allylpalladium chemistry.

by which end of the allyl moiety is attacked by the incoming nucleophile, as identified in Figure 1. There is a difficulty in making the nucleophilic addition step highly enantioselective, since the approach of the nucleophile is remote from the metal center and associated ligand. The first example of an asymmetric reaction of this type was reported by Trost in 1973[47] where a stoichiometric amount of palladium was employed and the substitution product was afforded with up to 24% ee.

Several ideas have been proposed to try to overcome this problem. Hayashi and coworkers developed a strategy employing ligands such as **33** and **34** which are designed to reach around to the nucleophile and thereby direct its course, as identified in Figure 2.[48] Using these ligands in the test reaction of the conversion of **35** into **36** afforded good levels of enantioselectivity in many cases. The ligand **37** uses a similar design principle.[49] The ligands **38** and **39** may also be considered in this category, although there are several possible explanations for the sense of asymmetric induction obtained in these reactions.[50] For example, electronic repul-

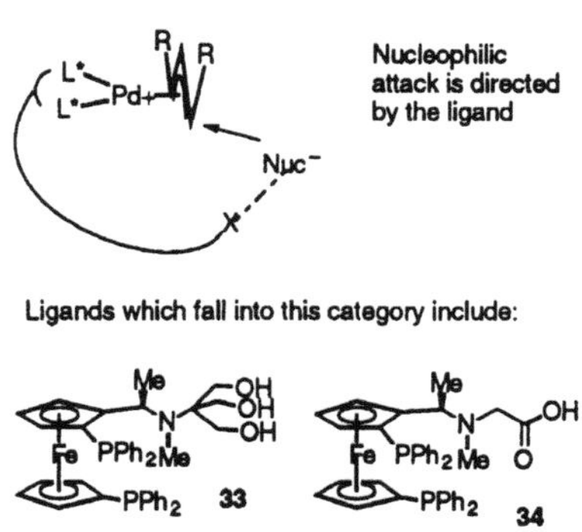

Figure 2. Ligands designed to direct the approach of the nucleophile.

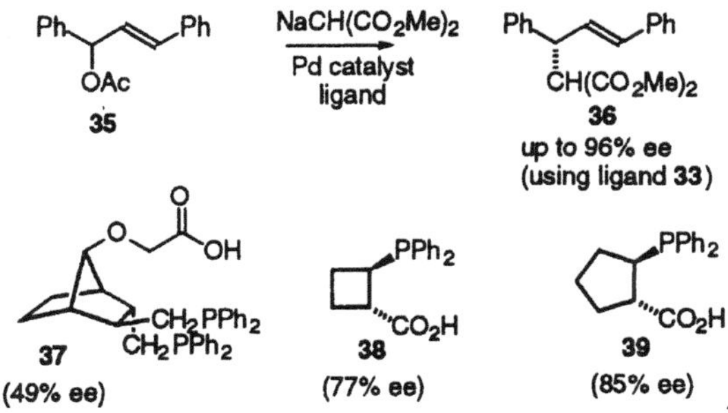

sion between the incoming nucleophile and the carboxylate function or alternatively that the carboxylate adds to the allyl moiety and is subsequently displaced by the incoming nucleophile in an S_N2 fashion. Another possibility is that the carboxylate can act as a ligand for palladium, thereby functioning as a bidentate ligand (vide infra - ligand **96**).

In an alternative approach, a ligand may be designed to be able to sterically affect the enantioselectivity of the reaction, either by blocking the approach of the incoming nucleophile to one of the allylic termini or by perturbing the symmetry of the allyl moiety. Thus, by forcing one end of the allyl moiety away from the palladium, this should make the allyl unit electronically unsymmetrical, with more positive charge character at the terminus which is being forced away from the palladium, as illustrated in Figure 3. It is certainly reasonable to suppose that C_2-symmetric ligands induce asymmetry in this way.

Examples of such ligands include bidentate phosphine ligands **40**,[51] **41**,[52] **42**,[53] **43**,[54] **44**,[55] **45**,[56] and bidentate nitrogen ligands, **46**,[57] **47**,[58] **48**,[59] **49**,[60] **50**,[61] **51**,[62] **52**,[63] and **53**.[64] Many of these ligands have provided high levels of enantioselectivity in palladium-catalyzed asymmetric allylic substitution reactions. However, since different reaction conditions and substrates have been used, a direct comparison is not particularly informative. Additionally, ligands which are not C_2 symmetric may also be able to induce enantioselectivity in a similar manner (in fact ligands **46** and **53** above are not strictly C_2 symmetric). Examples of these ligands include **54**,[65] **55**,[66] **56**, **57**,[67] and **58**.[68] Ligand **56** is of particular note since it is monodentate, and yet still provides a high enantiomeric excess.

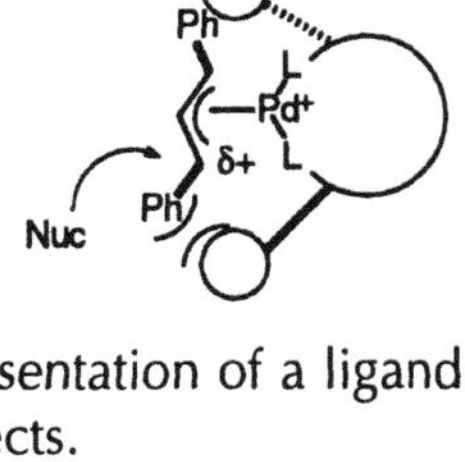

Figure 3. Schematic representation of a ligand perturbing the symmetry of an allyl ligand by steric effects.

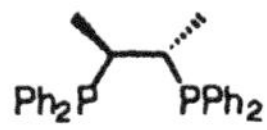

40

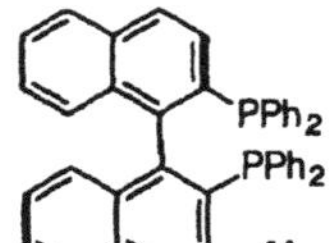

41

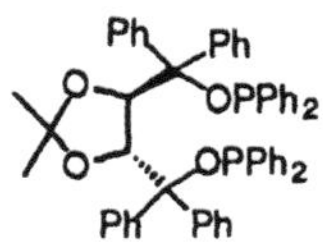

42

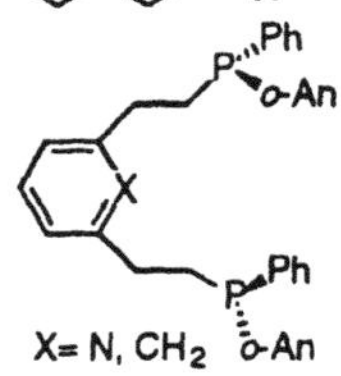

43

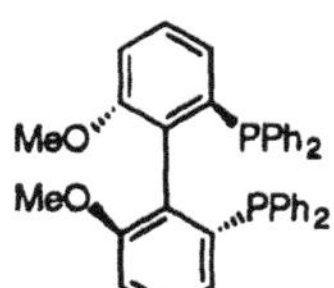

44

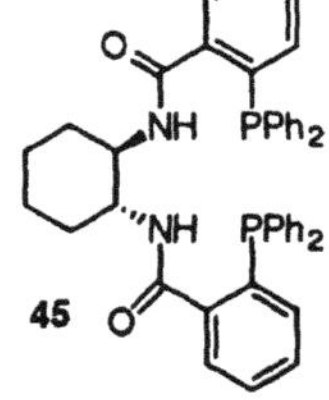

45

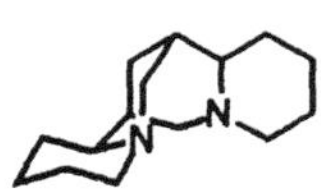

46

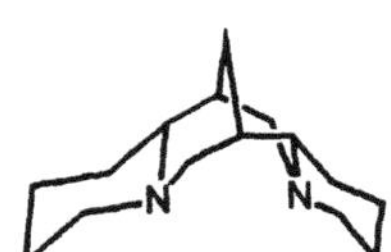

47

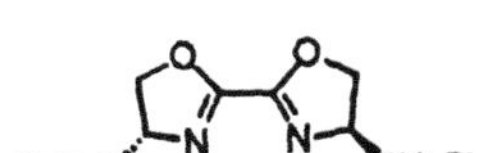

48

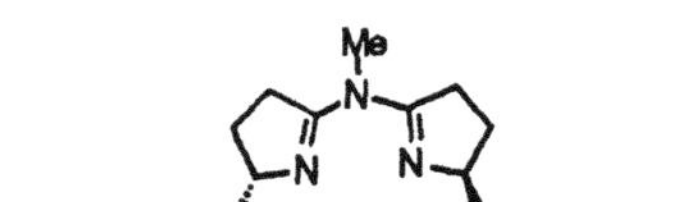

49

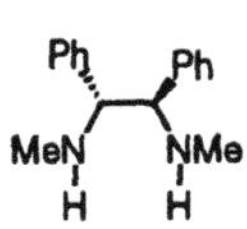

50

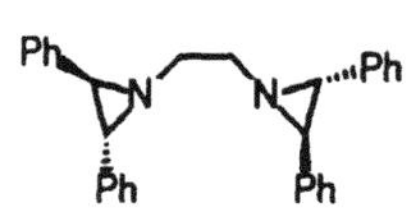

51

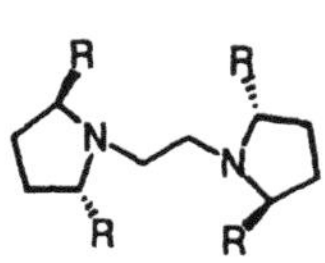

52

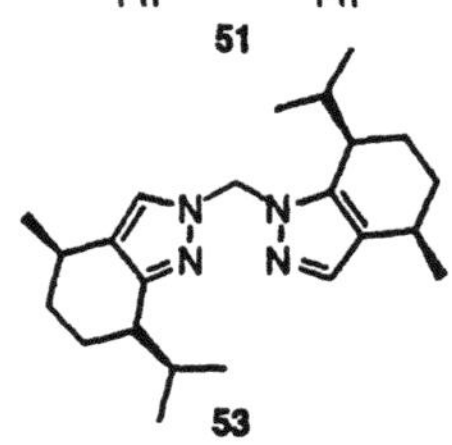

53

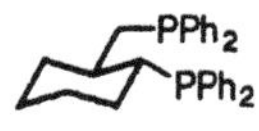

54

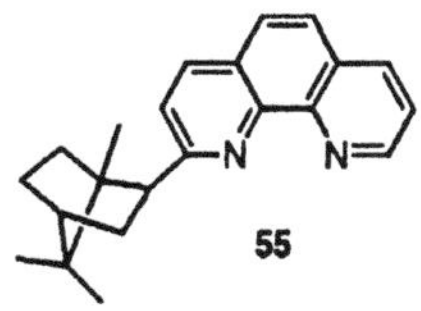

55

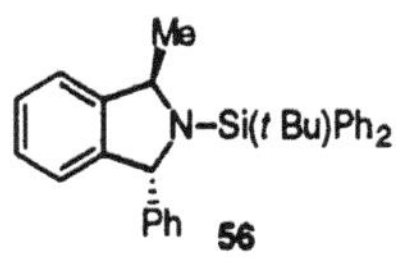

56

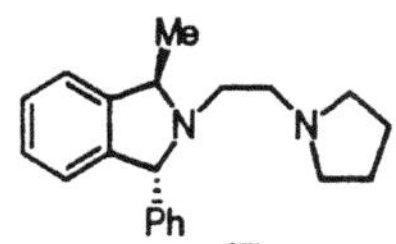

57

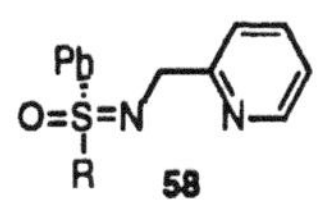

58

We considered that it may be possible to electronically perturb the symmetry of the allyl moiety in a different manner By employing a ligand which contains two different donor atoms, it was anticipated that the donor atom which functioned as the better π-acceptor would create a *trans* influence thereby weakening the Pd–C bond in the *trans* position. This carbon should then be more susceptible to nucleophilic addition. The concept that P,N donor ligands can perturb the electronics of allylpalladium complexes had already been demonstrated very elegantly by Åkermark, Vitagliano, and coworkers, although this group had not extended their work to asymmetric catalytic reactions.[69]

As shown in Figure 4, the idea is that the nucleophile will attack *trans* to the better π-acceptor (in this case the phosphorus). Depending on which configuration of allyl complex is attacked preferentially will then determine the sense of asymmetric induction. Thus, a relatively straightforward ligand design can be considered whereby two different donor atoms are employed, and that the stereochemical environment created by the ligand will dictate whether the M-shaped or W-shaped intermediate predominates. Hopefully the predominant intermediate will also be

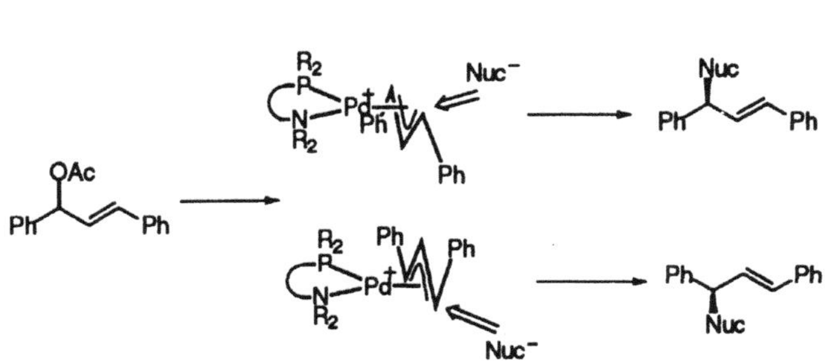

Figure 4. Nucleophilic attack *trans* to the better π-acceptor.

the more reactive to nucleophilic addition. To this effect, our group prepared the oxazoline ligands **59–69**,[70] and the groups of Helmchen[71] and Pfaltz[72] have also reported some of these ligands for palladium–catalyzed allylic substitution reactions. Additionally, these groups have conducted interesting NMR and X-ray crystallographic studies of the allylpalladium intermediates involved with the phosphine-containing oxazoline ligands **68**.[73] The ferrocenyl phosphine oxazoline ligands **70** have also been reported for palladium-catalyzed allylic substitution reactions.[74]

The stereochemistry of the substitution product was found to be opposite to that which we had originally predicted. As shown in Figure 5, we had expected that the major diastereomer would be the one where the oxazoline substituent and the allyl group are out of each other's way. We knew that there was a problem with this explanation, since it predicts the wrong enantiomer of product! However, the work from Helmchen's group has shown from crystallographic and solution NMR studies that it is the other diastereomer which is the major intermediate. In fact the two diastereomers are in very rapid equilibrium, and therefore it must be the relatively faster rate of nucleophilic addition to the major diastereomer which forms the basis for the observed enantioselectivity. The reason for the preferential formation of this diastereomer can be rationalized by the fact that the oxazoline substituent is

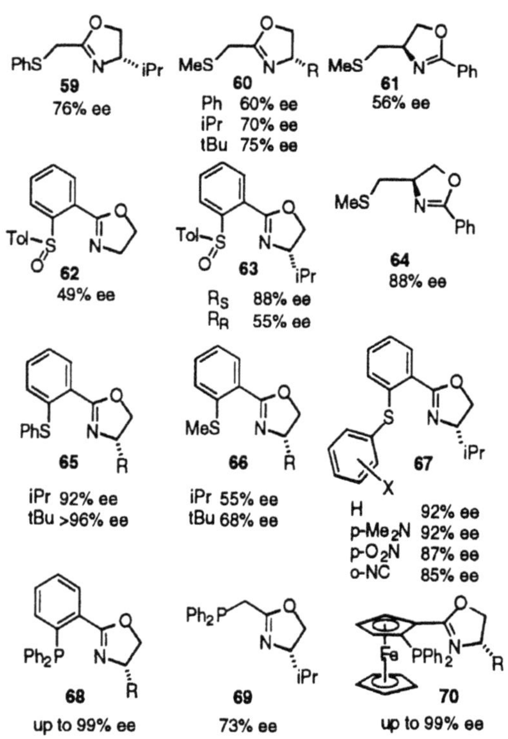

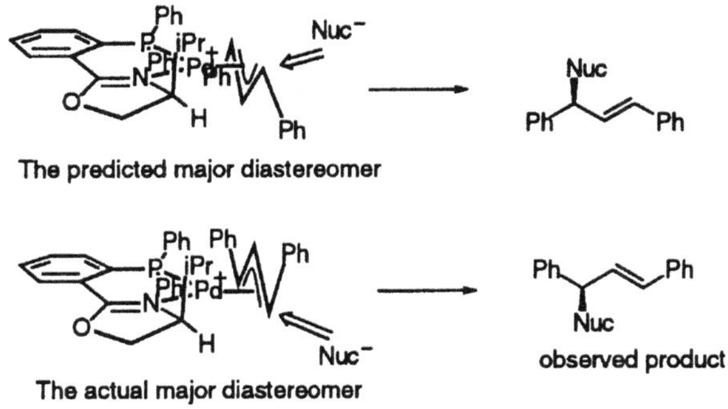

Figure 5.

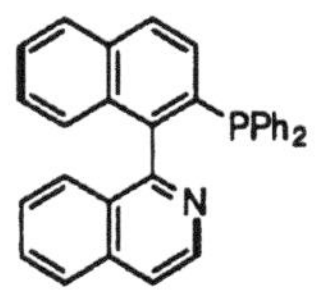

71
up to 98% ee

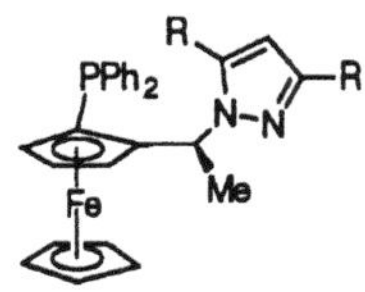

72

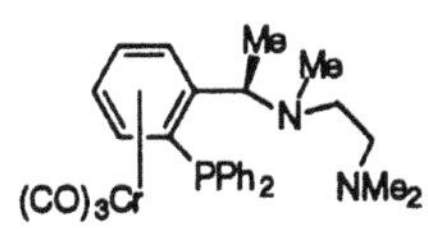

73

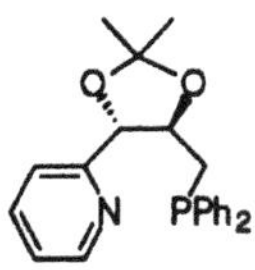

74

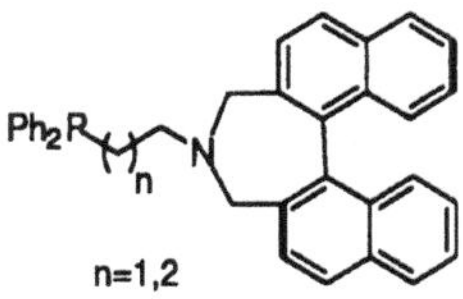

75

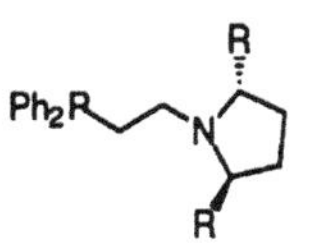

76

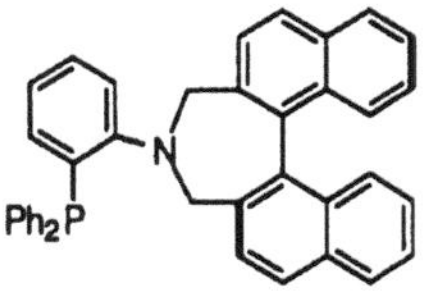

77

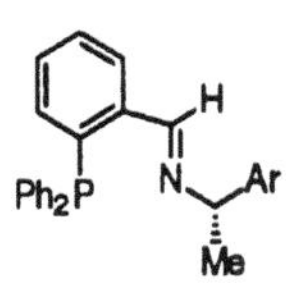

78

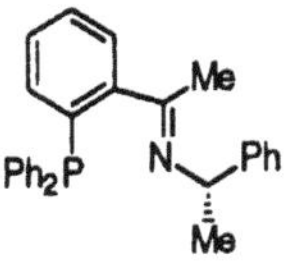

79

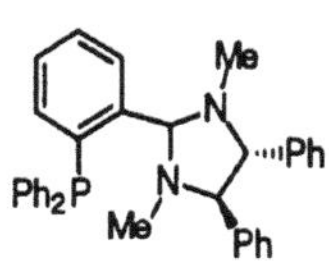

80

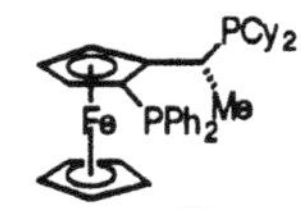

81

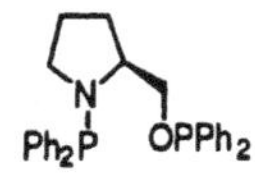

82

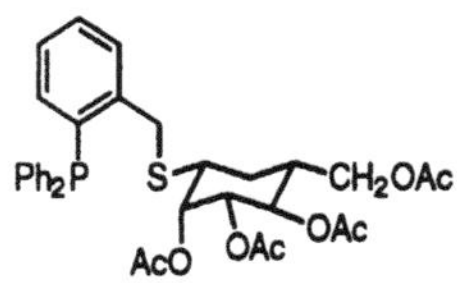

83

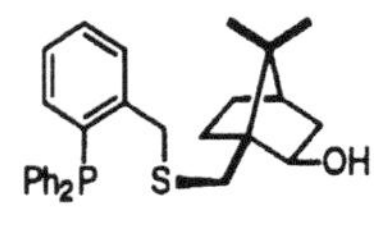

84

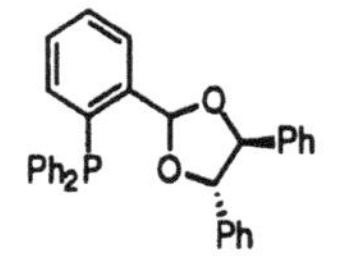

85

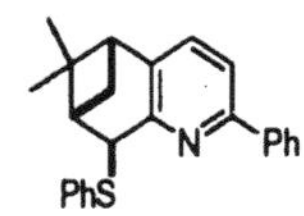

86

pseudoaxial, and therefore does not interact badly with the allyl group. The pseudoequatorial hydrogen therefore has a stronger influence over orienting the allyl group, as well as the twist in the ligand backbone which means that the phenyl groups attached to the phosphorus atom also present an asymmetric environment which can control the orientation of the allyl group.

The idea of using ligands which contain two different donor atoms is by no means unique to the oxazoline family, and many other groups had thought of alternative ligand designs at about the same time. These could either be P,N ligands including **71**,[75] **72**,[76] **73**,[77] **74**,[78] **75**, **76**,[79] **77**,[80] **78**,[81] **79** and **80**,[82] or other combinations of ligating atoms, including **81**,[83] **82**,[84] **83**, **84**,[85] **85**,[86] **86**.[87] Again, many of these ligands have provided formidable levels of enantioselectivity in various palladium-catalyzed allylic substitution reactions.

VI. CYCLIC SUBSTRATES

Suitable cyclic substrates can also proceed via a *meso*-intermediate. Thus, cycloalkenyl acetates **87** will proceed via the *meso*-complexes **88**, and afford the substitution products **89** and *ent*-**89**.

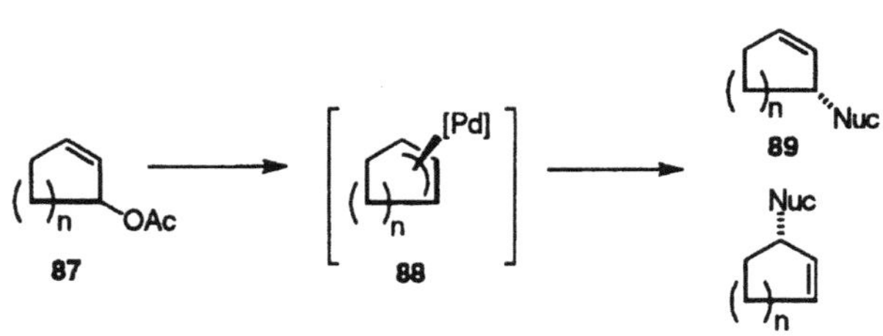

ent-89

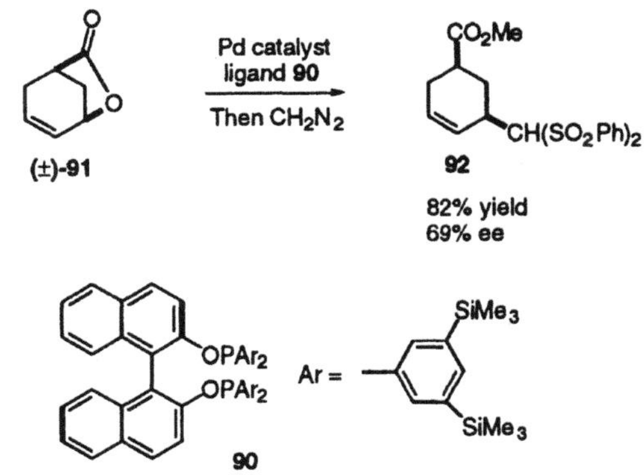

Many of the ligands reported for other reactions have been examined for their selectivity in palladium-catalyzed allylic substitution reactions. However, in comparison with the diphenylpropenyl substrates (such as compound **35**), relatively few ligands afford highly enantioselective reactions. In 1985 Trost and Murphy reported the first example of a ligand which gave good enantioselectivity.[88] The bulky ligand **90** was designed to create a "chiral pocket" and allowed the palladium-catalyzed conversion of the racemic bicyclic substrate **91** into the substitution product **92** with 69% ee.

Helmchen has reported the use of the oxazoline ligand **93**, which under optimized conditions affords 50% ee in the conversion of **94** into **95** (although higher

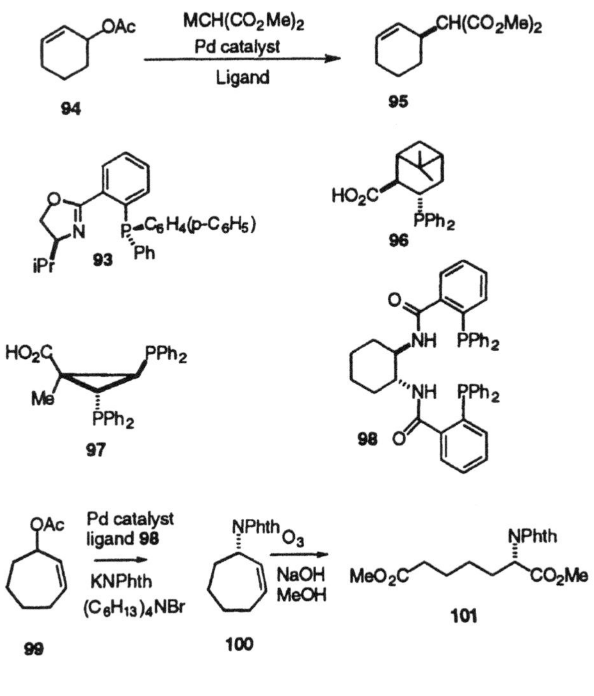

enantioselectivities were obtained for the corresponding seven-membered ring system).[89] The same research group has also reported that the ligand **96** is highly effective for the reaction, affording 98% ee in the same conversion.[90] Minami and coworkers have used the interesting ligand **97** to effect a substitution reaction on substrate **94** with up to 61% ee using an enantiomerically pure nucleophile.[91] However, the Trost ligand **98** has been widely applied to a range of cyclic substrates, and has given 96% ee in the conversion of **94** into **95**.[92] The same ligand has been used to effect the transformation of **99** into **100**, which was subsequently converted into the (*S*)-2-aminopimelic acid derivative **101**. Apart from the high asymmetric induction which ligand **98** imparts, this class of ligand is appealing since there are many simple modifications which can be made for tuning selectivity (vide infra).

VII. DISPLACEMENT OF ONE ENANTIOTOPIC GROUP

There have been several examples of palladium-catalyzed enantioselective allylic substitution of *meso*-substrates. For example, the *meso*-cyclic di-acetate **102** has been exploited in this reaction. Enantioselectivity results as a selection of one or other of the enantiotopic acetates leaving selectively, although since the intermediates can equilibrate, the selectivity of nucleophilic addition is also important.

Trost and coworkers have used the ligands **98**, **104–107** and related systems for enantioselective reactions involving the displacement of one enantiotopic group. Thus, the dibenzoate **108** on treatment with dione **109** and DBU in the presence of a palladium catalyst and ligand **98** affords the substitution product **110** with 98% ee.[93] The diol **111** can be converted into the activated substrate **112**, which on treatment with a palladium catalyst and ligand **104** affords the substitution product **113** with 88% ee.

Mori and coworkers have used ligand **114** with cyclic substrates,[94] and used this methodology in the synthesis of (+)-γ-lycorane.[95] Trost and coworkers have also shown that the diacetate **115** can be used as a suitable substrate for asymmetric induction. Thus, compound **115** is converted into the substitution product **116** with excellent enantioselectivity in the presence of a palladium catalyst and ligand **98**.[96]

Simple achiral substrates of the general type **117** have also been used in enantioselective palladium-catalyzed allylic substitution reactions. By using suitable conditions, two allylic substitution events can take place to give a cyclic product. Thus, compound **117** has been converted into the dihydrofuran **118** with good enantioselectivity using the ruthenocene-based ligand **119**.[97] Related reactions have

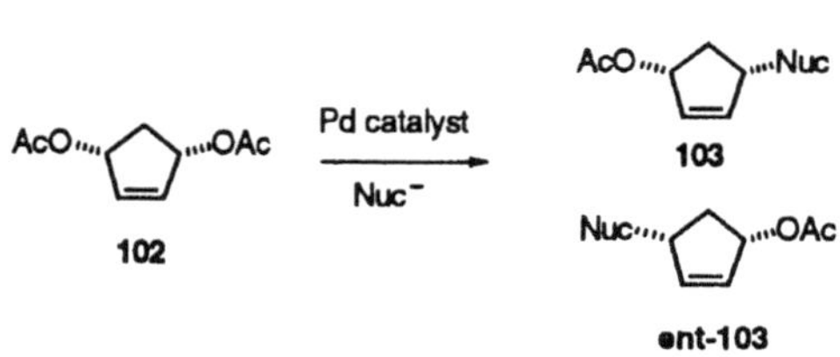

102 103 ent-103

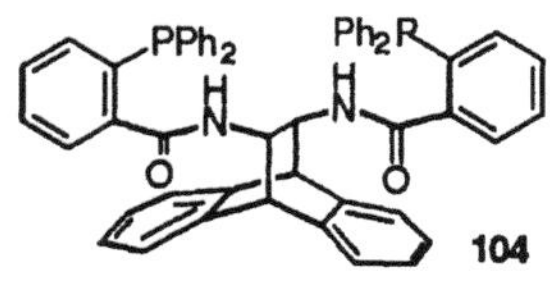

104

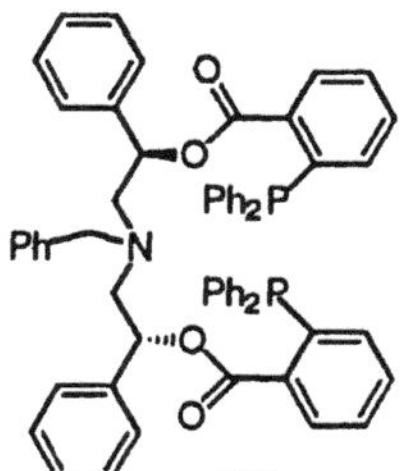

105

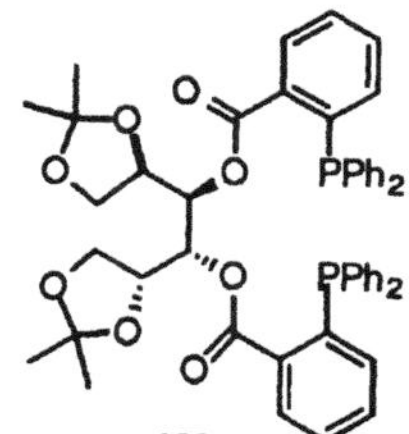

106

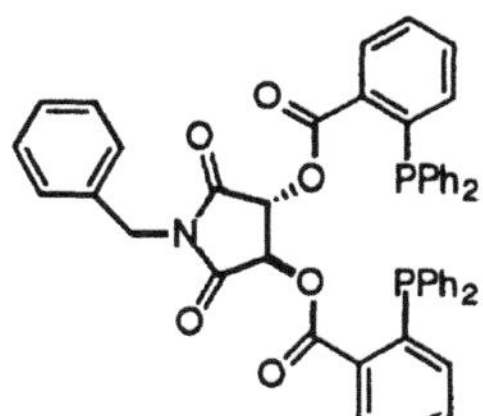

107

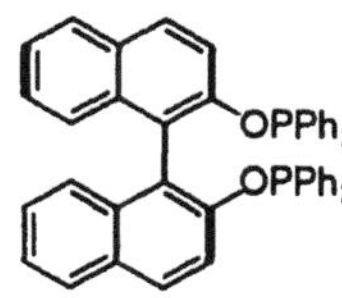

114

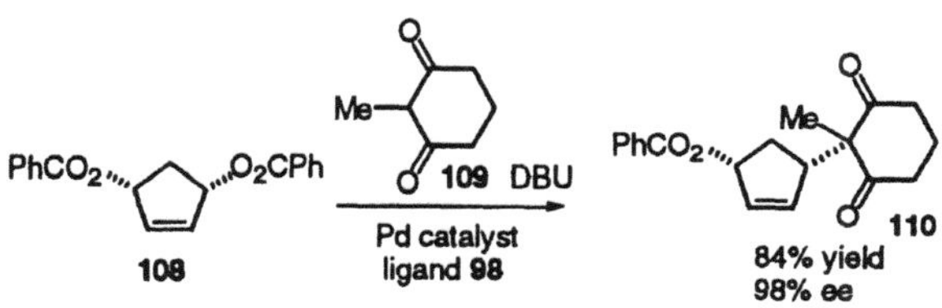

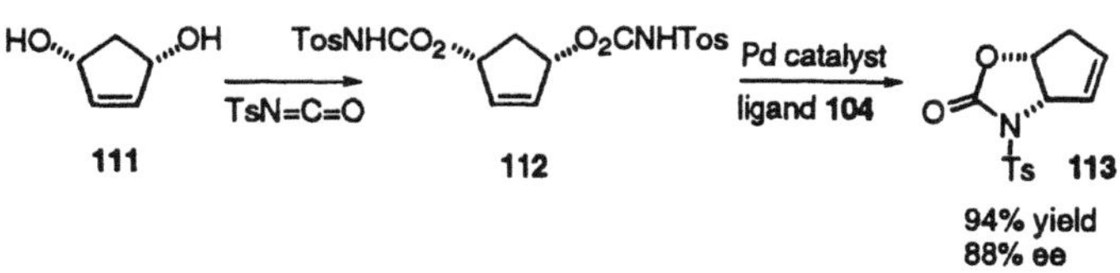

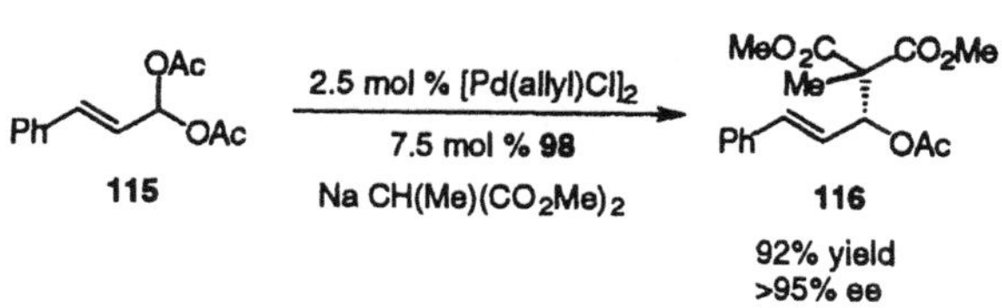

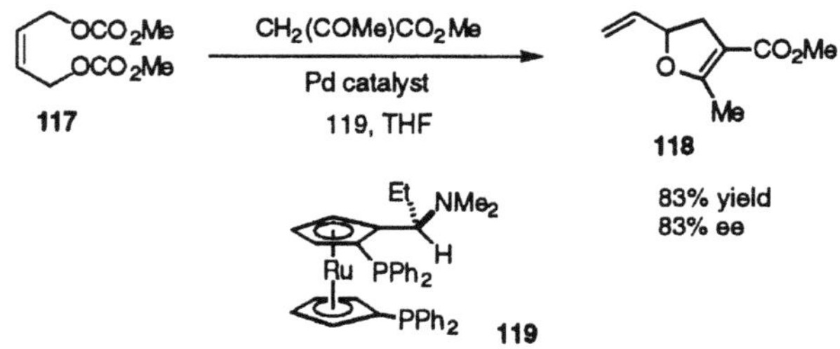

also been performed to generate enantiomerically enriched morpholines and piperazines (vide infra).

VIII. INDUCTION OF AXIAL CHIRALITY

Fiaud and coworkers have designed substrates which can lead to axial chirality in the formation of the product. Thus the achiral acetate **120** is converted into the substitution product **121**/*ent*-**121** with induction of axial asymmetry.[98] The nature of the leaving group has a strong influence over the level of asymmetric induction afforded, but the reaction has been preformed with up to 90% ee using *p*-methoxybenzoate as the leaving group and BINAP as the ligand.[99] In the absence of a nucleophile, this type of system will undergo elimination to the diene. This process has also been achieved in an asymmetric fashion. Thus acetate **122** is converted into the diene **123** with up to 44% ee.[100]

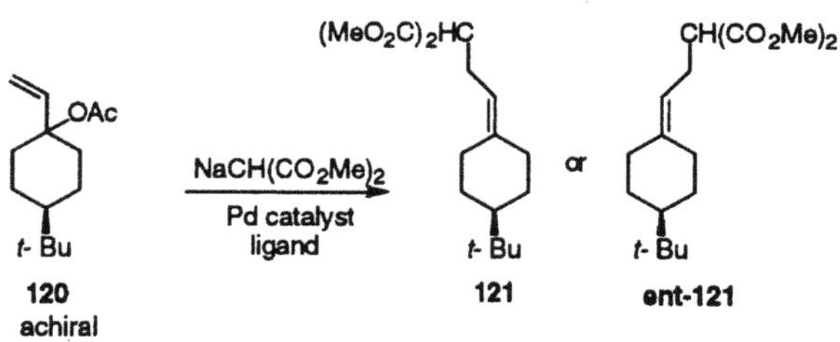

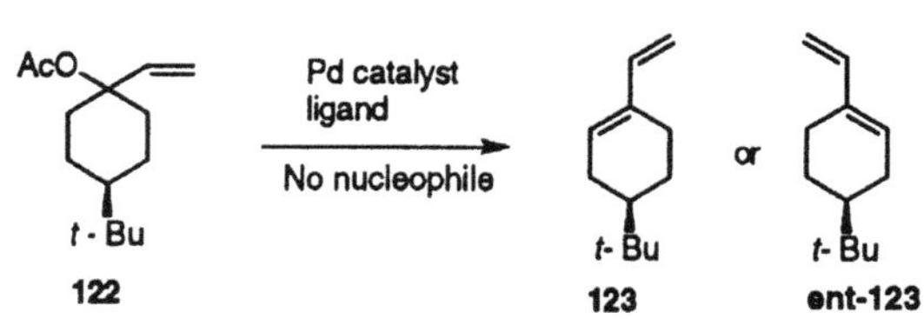

IX. THE USE OF ALLYL SYSTEMS CAPABLE OF ISOMERIZATION

The use of a system where a racemic compound is converted into an enantiomerically pure compound is appealing. Provided that the intermediate allylpalladium

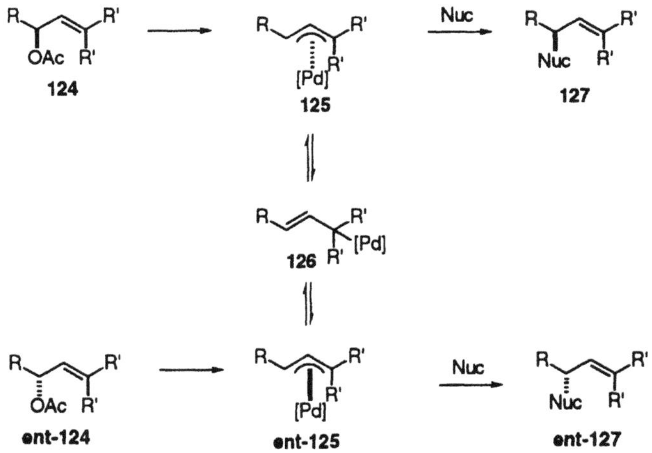

Scheme 3. Interconversion of allylpalladium intermediates.

complexes are able to interconvert rapidly, both enantiomers of a substrate can react to provide the product with 100% conversion and with one enantiomer predominating. Suitable substrates include those which are able to interconvert the allylpalladium complexes by a π-σ-π mechanism. Thus, the allyl acetate **124** will form the palladium complex **125**, whereas its enantiomer *ent*-**124** will form the enantiomeric complex *ent*-**125** (These species are enantiomers when achiral ligands are employed, and diastereomers when chiral ligands are employed). However, as shown in Scheme 3, π-σ-π rearrangement through the σ-allyl complex **126** allows the enantiomeric complexes to interconvert. In the presence of a suitable method for asymmetric induction, one of the two complexes may react preferentially to give the product enriched in one enantiomer **127** or *ent*-**127**.

The ideal system as far as enantioselective palladium-catalyzed allylic substitution reactions are concerned would be the transformation of racemic **128** (where R ≠ H) into enantiomerically pure **129**. It would be expected that the achiral substrate **130** would proceed through a common intermediate **131** to provide the same product **129**. The problem is that the regioselectivity of the reaction tends to favor the formation of the product derived from nucleophilic attack from the less substituted end of the allyl unit to provide the "wrong" achiral substitution product

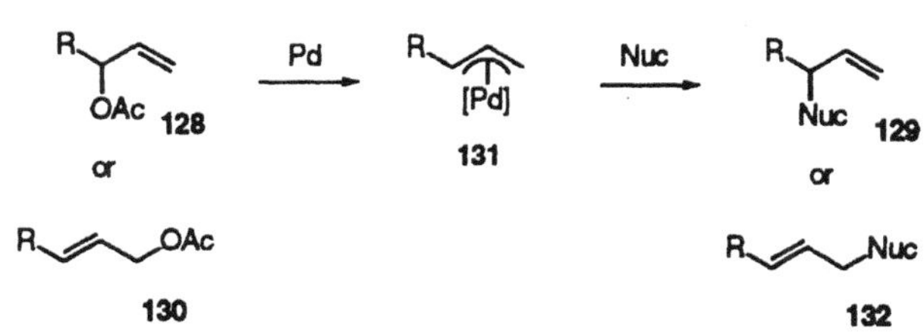

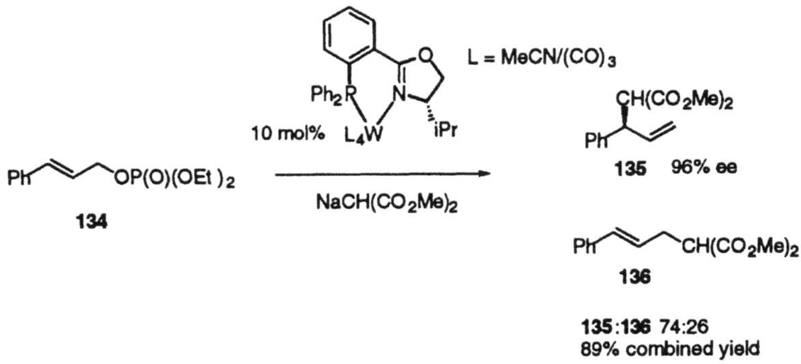

132. Pfaltz and coworkers have addressed this issue by employing tungsten catalysts which generally provide the opposite regiochemistry to palladium catalysts.[101] In their work, they employ the enantiomerically pure tungsten catalyst **133** to effect the substitution reaction of phosphate **134** which affords the substitution products **135** and **136**. The chiral product **135** is generated with 96% ee, although a different mechanism is invoked for the tungsten-catalyzed reaction.

Bosnich and coworkers first reported the use of substrates of type **137** and **138**.[102] Using CHIRAPHOS **40** as the ligand affords the substitution product **139** with up to 86% ee. Subsequently, Togni[57] and Brown[75] have also investigated these systems using ligands **46** and **71**. We also chose to examine this type of system.[103]

Using the oxazoline ligand **68**, we were pleased to find that the substitution product **141** could be formed with very high enantioselectivity and yield for a number of substrates **140**. We did not observe any regioisomeric product where the nucleophile has added adjacent to the two phenyl groups. We assume that equilibration of the intermediates occurs as described in Scheme 3.

There are two advantages to using this type of system. First, a wider range of substrates can be prepared in a straightforward fashion. This is achieved by addition of a suitable Grignard or alkyllithium reagent to commercially available β-phenylcinnamaldehyde. Second, the enantioselectivity is at least as good, if not better than for the corresponding substrate with equivalent termini. Thus, comparing the methyl-containing substrate **142** with the corresponding methyl-containing substrate **144**, it is clear that there are advantages in terms of yield and enantioselectivity (95% ee versus 62% ee). For some synthetic operations, such as formation of a

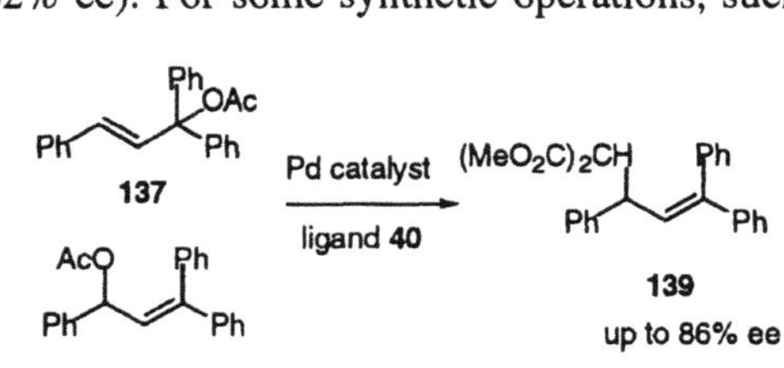

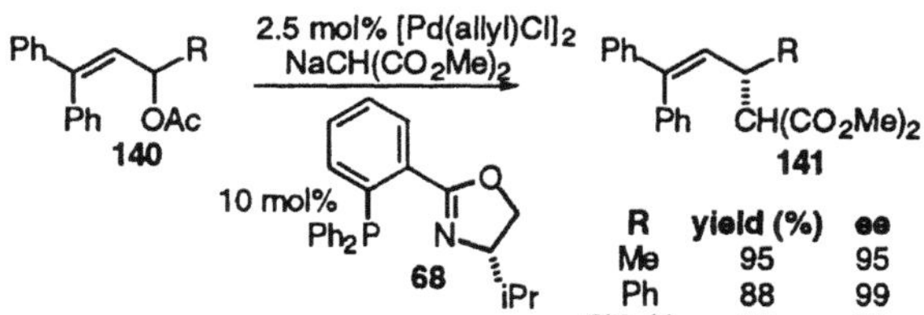

R	yield (%)	ee
Me	95	95
Ph	88	99
ClC_6H_4-	91	>95
2-Py	89	92
1-Naphth	96	>95
Mesityl	91	98

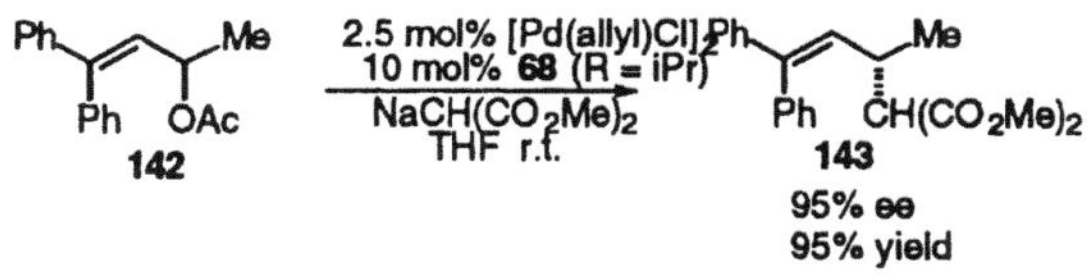

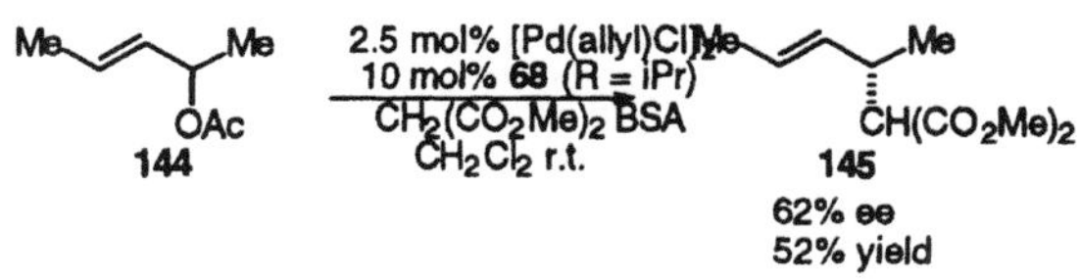

carboxylic acid from the alkene function, it is of little consequence which substrate is employed.

This methodology leads to the synthesis of a number of enantiomerically enriched substitution products. The synthetic utility of the reaction can be further demonstrated by the conversion of the products into more useful structures such as succinic acids, γ-lactones, and β-amino acids.[104] Thus, subjection of the substitution product **139** to oxidative cleavage with chromium(VI), and conversion of the malonate group into a carboxylic acid affords the succinic acid **146** without any significant loss of stereochemical integrity.

Using the *p*-chlorophenyl substitution product **147**, decarboxylation using the Krapcho conditions affords the monoester **148**. Ozonolysis with a sodium borohydride work-up affords the γ-lactone **149**, again with retention of stereochemistry. After Krapcho decarboxylation, as above, the monoester **150** containing the bulky mesityl substituent underwent oxidative cleavage to the corresponding carboxylic

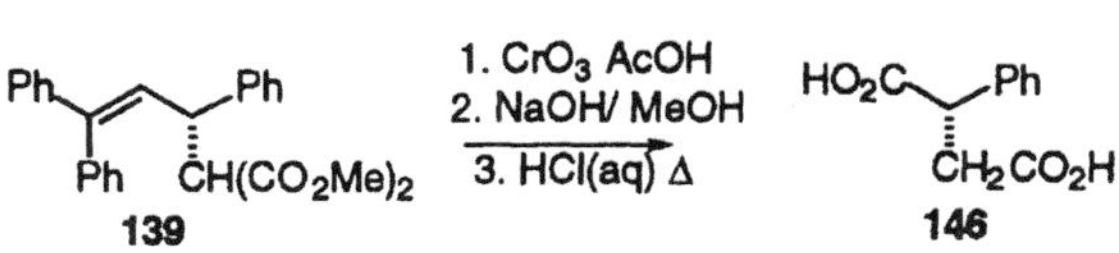

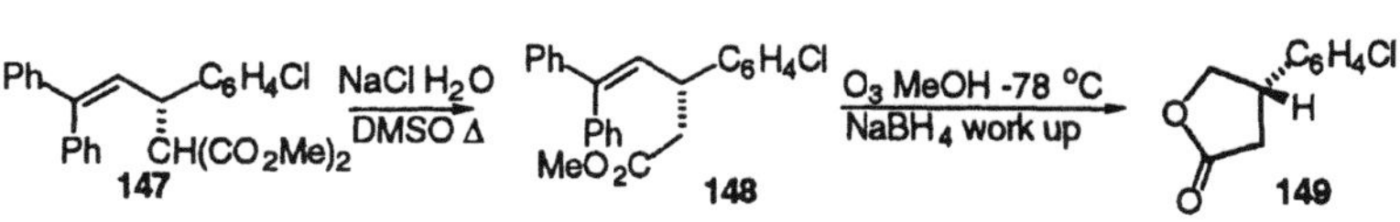

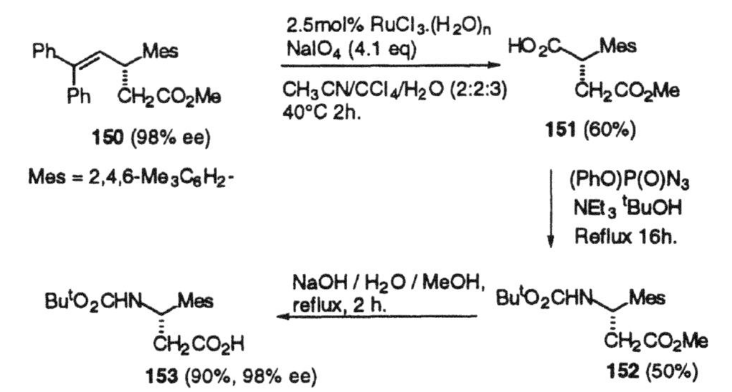

acid **151**. Treatment with $(PhO)_2P(O)N_3$ effected a Curtius-type rearrangement to give the Boc-protected amino ester **152**, and simple basic hydrolysis afforded the Boc-protected amino acid **153**, which had preserved stereochemistry.

By changing the order of the transformation it was also possible to prepare α-substituted β-amino acids. These results are reported for the preparation of four different Boc-protected amino acids, two of which were further transformed into the free β-amino acids. Thus the enantiomerically enriched substitution products **143**, **139**, **154**, and **155** underwent Krapcho decarboxylation to give the monoesters

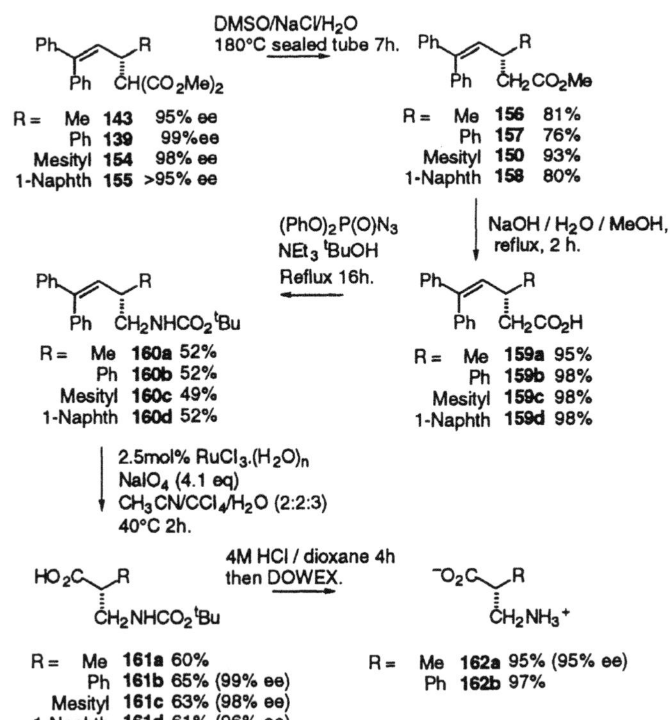

150, 156–158 in good yield, and further converted into the monoacids **159a–d** in excellent yields (although this procedure could be conducted in a one-pot process, higher yields were obtained by this route). The carboxylic acids underwent the modified Curtius reaction in modest yield to give the Boc-protected allylamines **160a–d**. Ruthenium-catalyzed oxidative cleavage of the alkene afforded the Boc-protected β-amino acids, which were analyzed by chiral hplc and found to have retained their stereochemistry. Finally, acid hydrolysis followed by isolation on a Dowex column provided the free β-amino acids **162a** and **162b**.

Mechanistically, understanding the origin of enantioselectivity with substrates which do not possess identical termini is harder than for those which do. Essentially, there are four possible diastereomeric allylpalladium intermediates. Enantioselectivity comes about as a consequence of nucleophilic addition taking place selectively to (presumably) one of these intermediates. Assuming that there is an early transition state which resembles one of these intermediates, then it must be either complex **163** or **164** which is the reactive intermediate, since the stereochemistry of the product is known. Nucleophilic addition to complex **166** or **167** would afford the wrong enantiomer of substitution product. Either the nucleophile approaches *trans* to the phosphorus, as shown in complex **163**, or it approaches *trans* to the nitrogen, as shown in complex **164**. Based on the earlier arguments, it seems likely that the nucleophile will approach *trans* to the better π-acceptor (the phosphorus). However, in complex **163** there is considerable crowding, with four phenyl groups in close proximity, and this may disfavor this process. Brown and coworkers have examined the diastereomeric allylpalladium complexes based on the related ligand **71**, and from NMR studies conclude that although such an arrangement is disfavored, this is still the most reactive complex.[75]

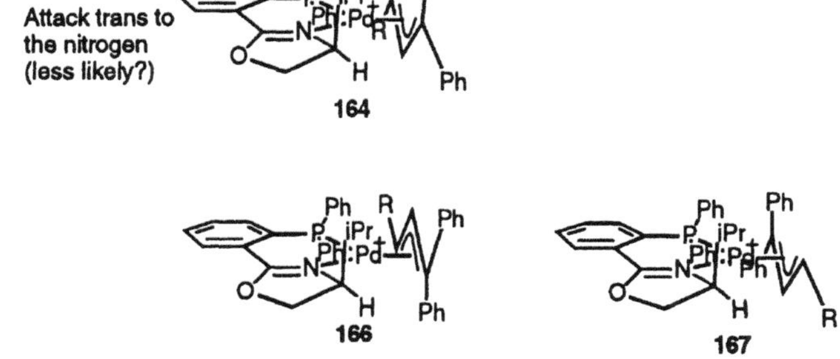

X. ALLYLPALLADIUM INTERMEDIATES WHICH DO NOT READILY INTERCONVERT

There have been a few reports of substrates which give allylpalladium intermediates which are not able to interconvert readily. In fact, one of the very early examples of catalytic asymmetric induction in palladium allyl chemistry involved such a system. In 1977, Trost and Sterge reported that the racemic substrate **168** could be converted into the product **169** with up to 46% ee.[105] The use of the bulky nucleophile **170** was important, and this was attributed to the fact that this may allow time for the allylpalladium intermediates to equilibrate prior to nucleophilic addition. In general, intermediates of type **171** are slow to racemize/epimerize. The slower mechanism of interconversion which must still exist for such intermediates has been attributed to either *syn* attack by acetate[106] or to incoming palladium displacing the original palladium from the allyl complex.[107]

An interesting asymmetric palladium-catalyzed allylic substitution of racemic 1-naphthylethyl esters **173** has been reported by a French group.[108] In the presence of (*R*)-BINAP, up to 61.5% ee could be obtained in the substitution product **174** albeit with a moderate yield.

A Japanese group has reported a kinetic resolution of a racemic allyl acetate using an enantiomerically pure palladium catalyst.[109] One of the enantiomers of the racemic substrate **175** reacts more quickly than the other one. Thus, the unreacted substrate **175** is recovered in high enantiomeric excess, and the substitution products **176** and **177** are also isolated with good enantioselectivity when ligand **178** is employed. The fact that the regioisomeric substitution products **176** and **177** are

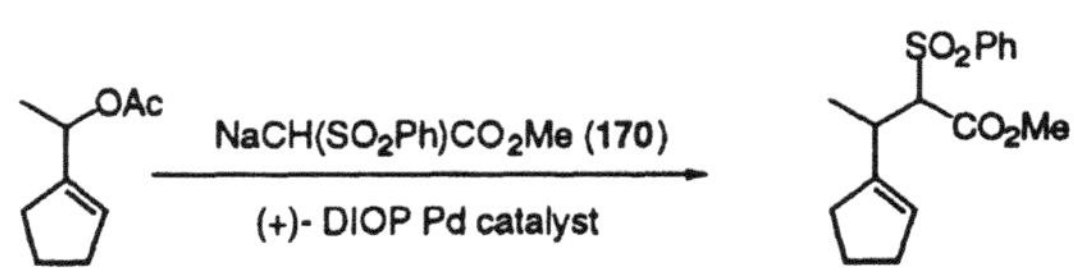

168 169

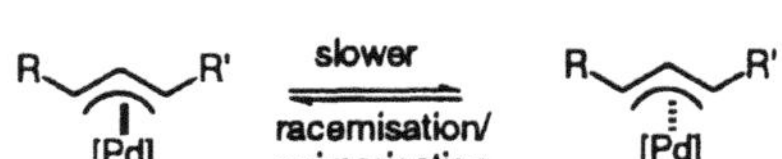

171 172

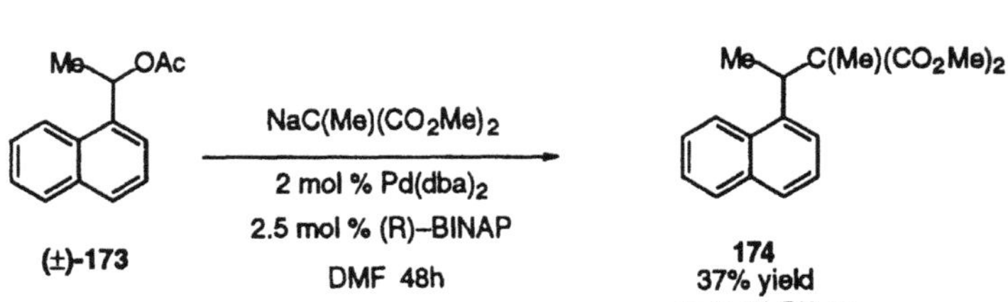

(±)-173

174
37% yield
up to 61.5% ee

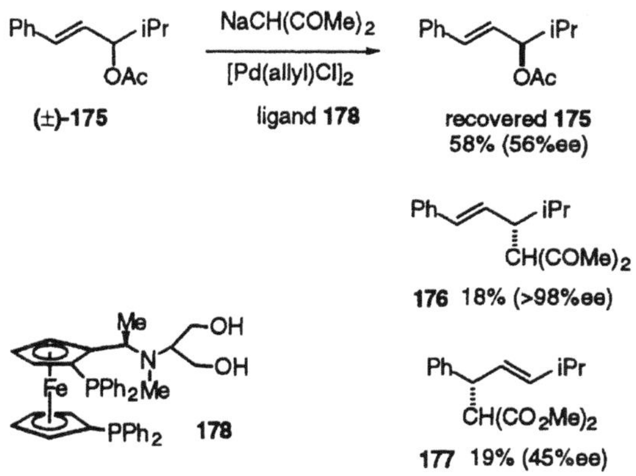

formed with different enantioselectivities indicates that the diastereomeric allylpalladium intermediates exhibit differing regioselectivities from each other.

XI. ASYMMETRIC INDUCTION USING A PROCHIRAL NUCLEOPHILE

Very high levels of asymmetric induction have yet to be achieved with any consistency for palladium-catalyzed reactions such as the conversion of allyl acetate **179** and the achiral nucleophile **180** into the substitution product **181**, where the asymmetric center resides in the nucleophilic component of the product.[110] The first ligand to provide reasonable enantioselectivity in this type of process was ligand **182**, which was designed to provide an asymmetric environment in the vicinity of the incoming nucleophile by association. Thus, 52% ee was achieved in the conversion of diketone **183** into the product **184**.[111] Using the ligand **185**, the same reaction has been achieved with up to 75% ee,[112] and with the ligand **186**, 81% ee has been achieved under optimized conditions.[113] Again, the design principle with ligands **185** and **186** is that the ligand is able to associate with the incoming nucleophile and thereby control its approach to the allyl moiety.

Genet and coworkers have employed the Schiff bases of amino esters, such as compound **187** as the incoming nucleophile.[114] This group has achieved the conversion of allyl acetate **179** into the substitution product **188** with up to 68% ee.[115] By using enantiomerically pure Schiff bases, very high diastereoselectivities can be obtained.[116]

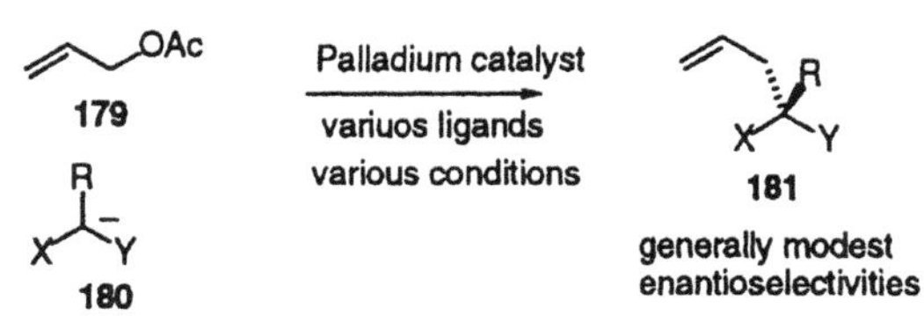

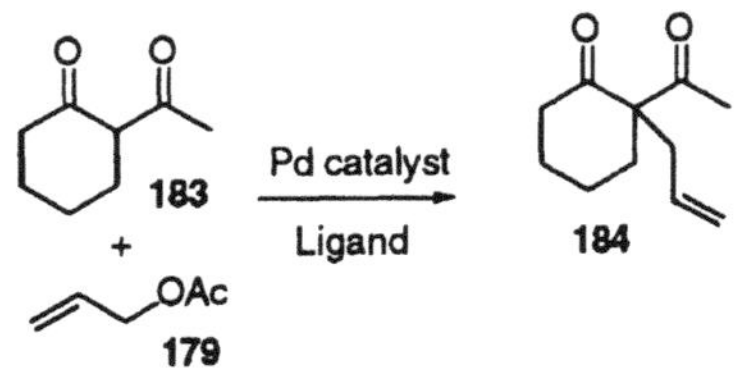

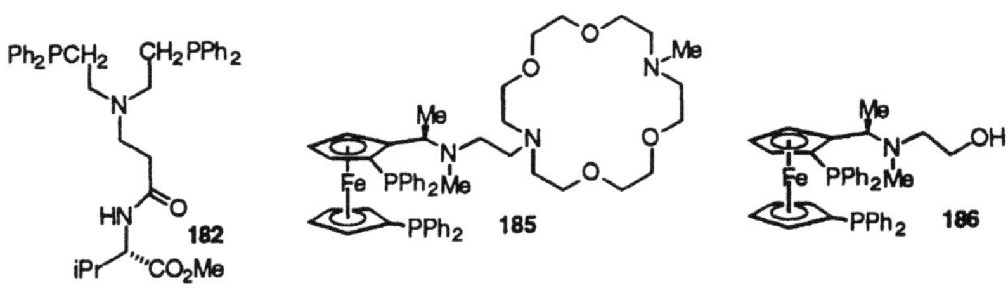

Our own efforts in this area led to disappointing levels of asymmetric induction. Thus, the conversion of allyl acetate **179** into the substitution product **189** was achieved with only 12% ee in the presence of a palladium catalyst and ligand **68**.[117] However, by using the same nucleophiles with the diphenylpropenyl acetate substrate **35**, good levels of selectivity could be obtained.[118] Thus acetate **35** is converted into the imino ester **191** and into the imino phosphonate **192**. In these cases, two new chiral centers are formed in the reaction. In each case both diastereomers are formed with high enantioselectivity. We believe that the high enantioselectivity is provided in a similar fashion to the approach of other nucleophiles. The reasonable diastereoselectivity does not derive from the use of an enantiomerically pure ligand, since similar diastereoselectivity is observed when an achiral ligand is employed and the product formed is racemic.

There is therefore still scope for the generation of a simple ligand which will provide excellent enantioselectivity when a prochiral nucleophile is used in palla-

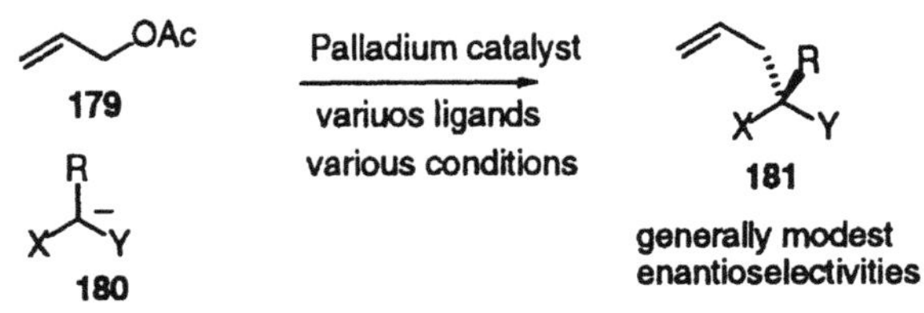

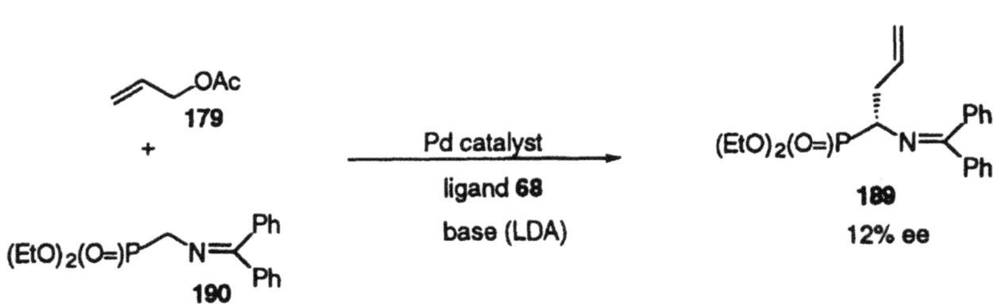

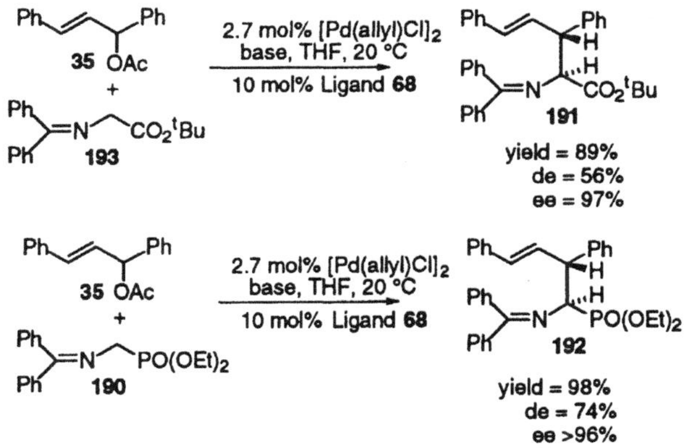

dium-catalyzed allylic substitution reactions. It may well be that the use of an enantiomerically pure ligand on the palladium is not the most convenient way to achieve this goal.

XII. HETEROATOM NUCLEOPHILES

The previous sections have discussed different methods and substrates for inducing asymmetry in palladium-catalyzed allylic substitution reactions. However, these have been predominantly carbon nucleophiles, especially stabilized enolates. This section describes some of the research work which has investigated heteroatom nucleophiles.

Perhaps the most unusual use of heteroatom nucleophiles is that reported by Trost and Organ.[119] The racemic **193** was treated with sodium pivalate and a catalytic system comprising of ligand **45** and allylpalladium chloride dimer. With careful control of temperature, the product ester **194** was isolated with 97% ee. This process effectively functions as a de-racemization procedure.

Trost has also employed the same ligand for enantioselective amination (vide supra). The reaction has been conducted on the bis-carbonate **195** using trimethylsilylazide as the nitrogen source. In this case, the product was formed highly enantioselectively (the other enantiomer was not observed), and the initial product **196** was used as the basis for the synthesis of (+)-pancratistatin **197**.[120]

The ferrocenylphosphine ligand **178** has been employed in an asymmetric amination procedure for the conversion of the carbonate **198** into the allylamine **199**,[121] as well as conversion of the acetate **200** into the substitution product **201**.

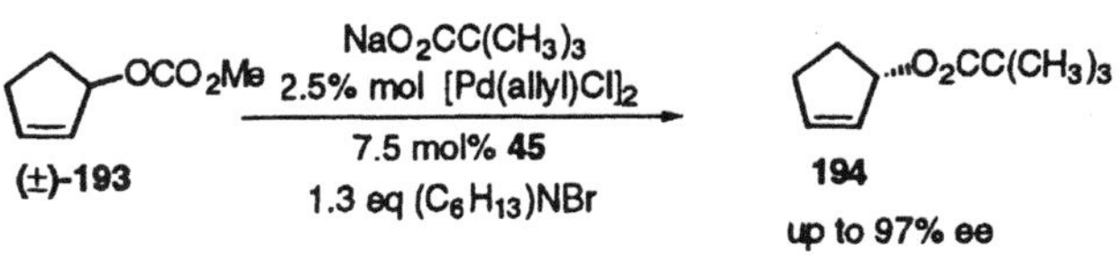

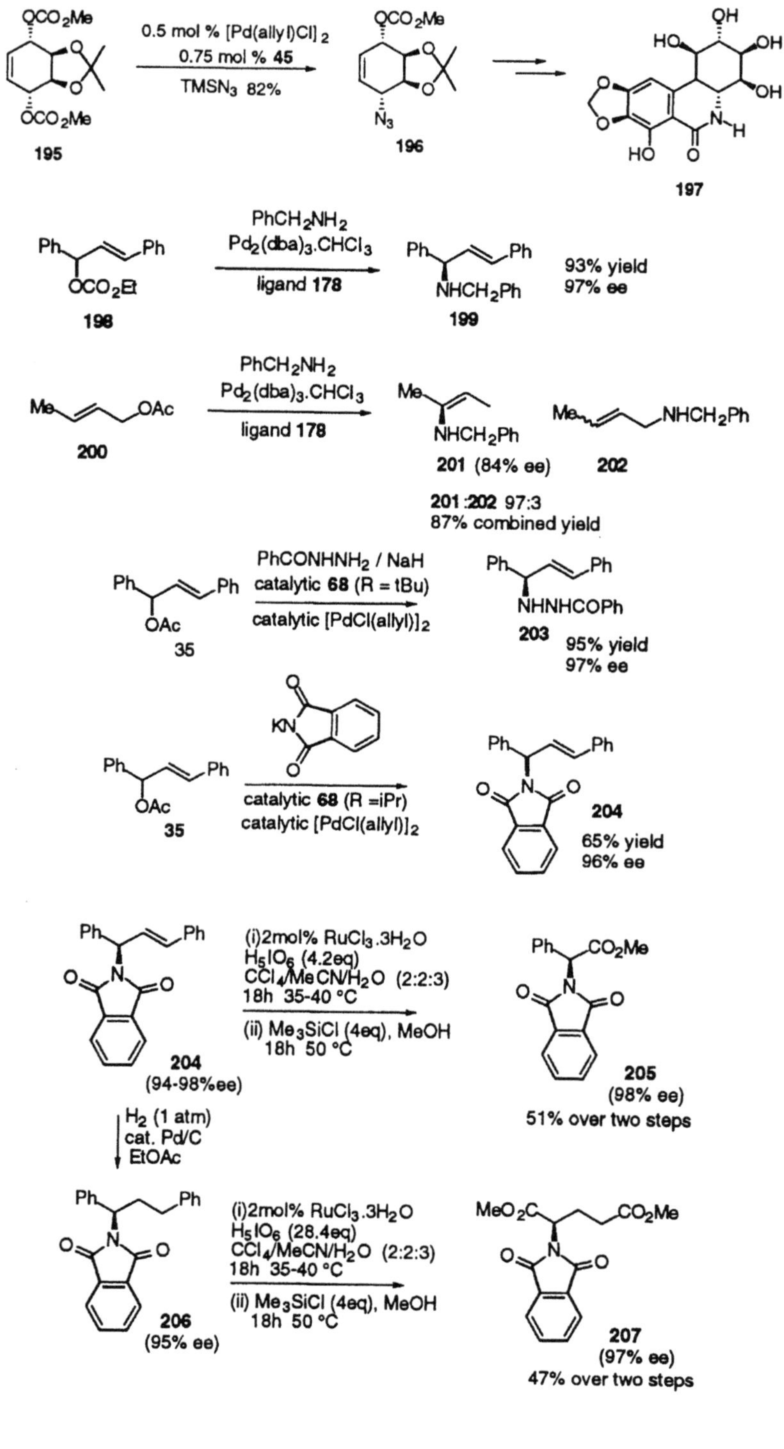

0.5 mol % [Pd(allyl)Cl]2
0.75 mol % 45
TMSN3 82%
195
196
197
PhCH2NH2
Pd2(dba)3.CHCl3
ligand 178
198
199
93% yield
97% ee
200
201 (84% ee)
202
201:202 97:3
87% combined yield
PhCONHNH2 / NaH
catalytic 68 (R = tBu)
catalytic [PdCl(allyl)]2
35
203
95% yield
97% ee
catalytic 68 (R =iPr)
catalytic [PdCl(allyl)]2
204
65% yield
96% ee
(i)2mol% RuCl3.3H2O
H5IO6 (4.2eq)
CCl4/MeCN/H2O (2:2:3)
18h 35-40 °C
(ii) Me3SiCl (4eq), MeOH
18h 50 °C
204
(94-98%ee)
205
(98% ee)
51% over two steps
H2 (1 atm)
cat. Pd/C
EtOAc
(i)2mol% RuCl3.3H2O
H5IO6 (28.4eq)
CCl4/MeCN/H2O (2:2:3)
18h 35-40 °C
(ii) Me3SiCl (4eq), MeOH
18h 50 °C
206
(95% ee)
207
(97% ee)
47% over two steps

In this reaction the achiral regioisomeric product **202** was only produced as a minor by-product.[122]

There are two reports on the use of oxazoline ligands **68** in amination reactions. Pfaltz and Helmchen have examined several nitrogen nucleophiles with these ligands, and in the case of benzoylhydrazine, 97% ee was achieved in the synthesis of compound **203** from the class substrate **35**.[123] Typically, other substrates afforded lower levels of enantioselectivity. Our group reported similar results, including the use of potassium phthalimide as the nitrogen nucleophile.[124] The substitution product **204** was formed with 96% ee, and was transformed into the L-phenylglycine derivative **205** by oxidative cleavage of the alkene and esterification. Alternatively, hydrogenation of the alkene afforded compound **206**, which underwent ruthenium-catalyzed oxidative degradation of the phenyl groups followed by esterification to give the D-glutamic acid derivative **207**.

The formation of new C–O and C–N bonds has been reported by Hayashi and coworkers, using the diacetate **208** and an amino alcohol **209**. The cyclization product **210** was obtained in up to 65% ee, using (*R*)-BINAP as the ligand.[125] Achiwa and Yamazaki reported a similar reaction using ligand **211**, and obtained up to 83% ee.[126]

The formation of a C–S bond using asymmetric palladium catalyzed allylic substitution was first reported by Hiroi and Makino. The allyl acetate **200** was converted into the allyl sulfone **212** with 88% ee using (–)-DIOP as the ligand in the presence of sodium *p*-toluenesulfonate.[127] Using the oxazoline ligand **68** (R =

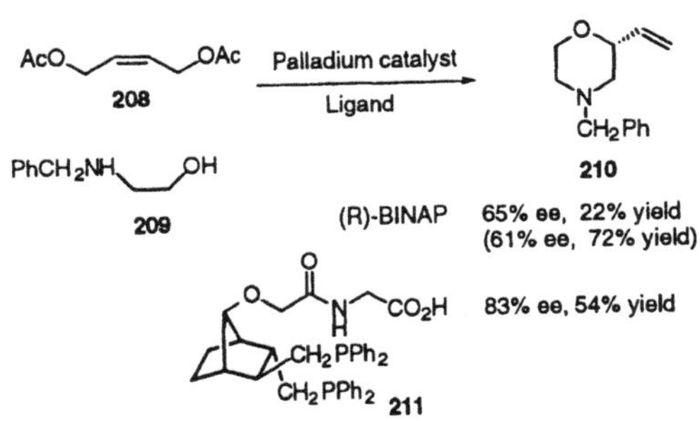

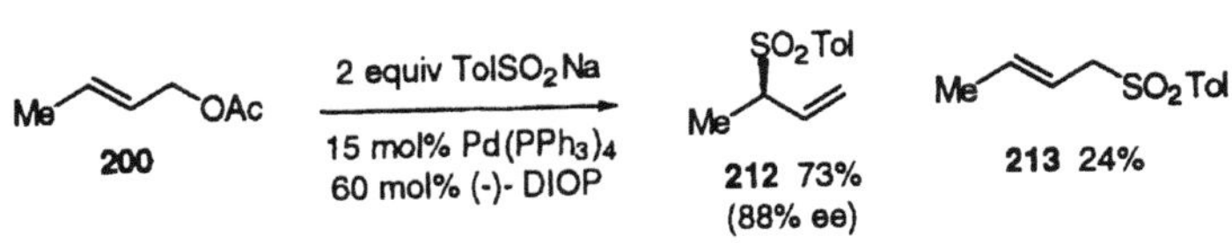

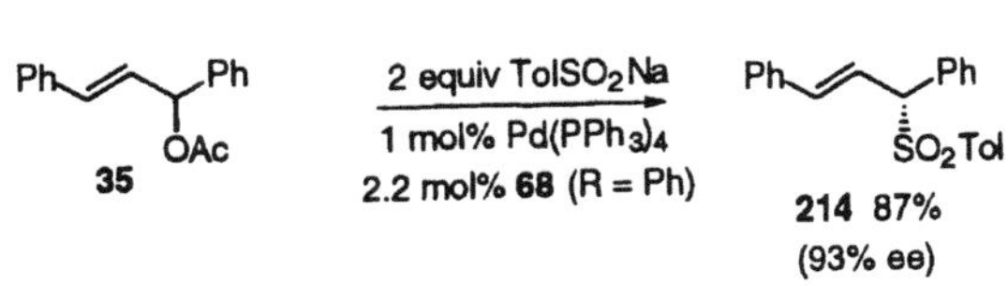

Ph), Eichelmann and Gais converted the allyl acetate **35** into the substitution product **214** with 93% ee.[128] Trost and coworkers have looked at the formation of cyclic allylic sulfones, and in some cases obtained high levels of asymmetric induction.[129]

XIII. ASYMMETRIC ALLYLIC SUBSTITUTION WITH HARD NUCLEOPHILES

The mechanism of palladium-catalyzed allylic substitution with hard nucleophiles involves transfer of the nucleophile from the metal to the allyl moiety rather than direct attack of the free nucleophile. Therefore, it might be expected that the mechanism for asymmetric induction differs. This is indeed true, and ligands that have been examined tend to give low levels of asymmetric induction. Thus, Buono and coworkers have shown the asymmetric coupling between cyclohexenyl acetate **94** and the vinyl Grignard reagent **215** afforded the substitution product **216** with 33% ee under the best conditions reported.[130] Using the same substrate, Fiaud has obtained low enantiomeric excess in the formation of the substitution product **217** with an organozinc reagent in the presence of a palladium catalyst and the monodentate ligand NMDPP (neomethyldiphenylphosphine).[131] However, the use of a monodentate phosphine may be more appropriate because a bidentate ligand may not be able to remain fully coordinated to the allylpalladium moiety once the nucleophile has added to the palladium. With this consideration, Fiaud and Legros designed the ligand **218**, but obtained a similar low enantioselectivity.[132]

However, the use of hydride (in the form of formic acid) as the incoming nucleophile has been employed with spectacular success by Hayashi and coworkers. A family of ligands related to the MOP ligand **219** has been used with great

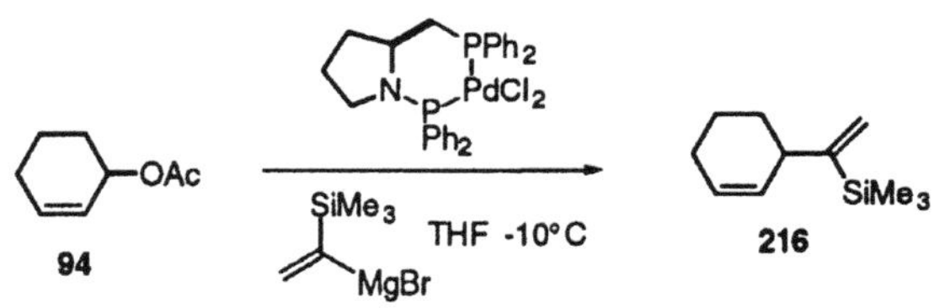

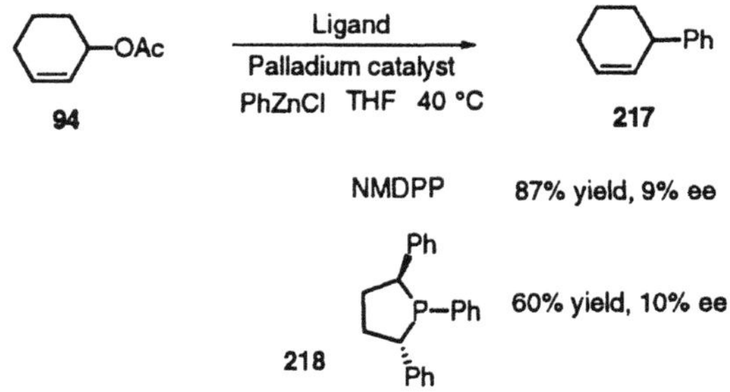

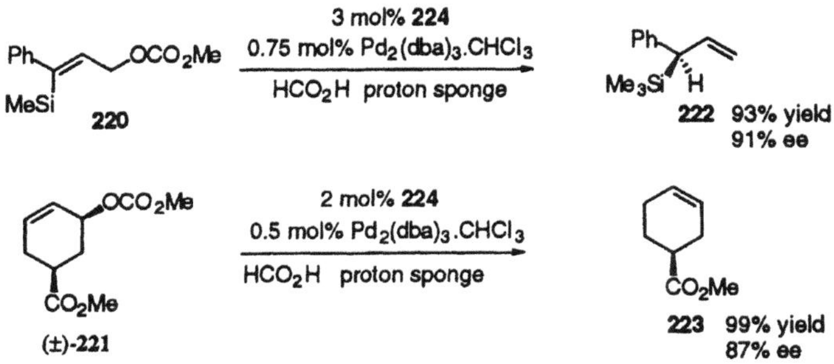

proton sponge = 1,8-bis(dimethylamino)naphthalene

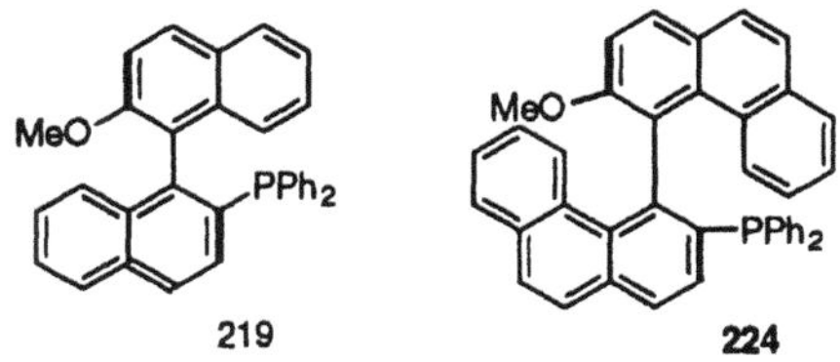

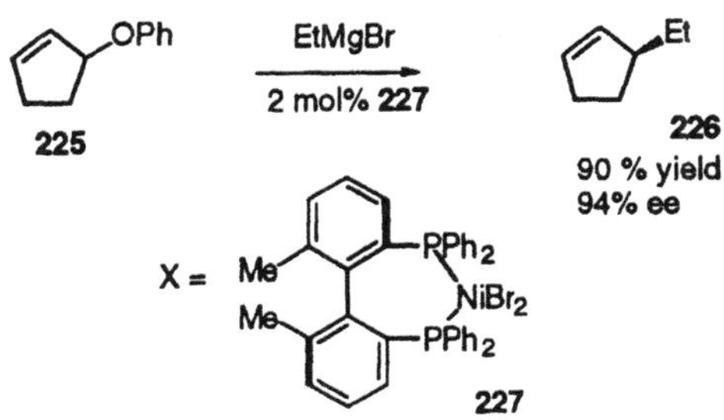

success in the asymmetric reduction of allylic esters. These ligands are believed to function as monodentate ligands. Thus, compounds **220** and **221** are converted into the products **222** and **223** using ligand **224** in the presence of a palladium catalyst and formic acid.[133]

Although outside the main scope of this review, nickel-catalyzed asymmetric allylic substitution has proved to be successful with hard nucleophiles.[134] Thus cyclopentenyl phenyl ether **225** is converted into 3-ethylcyclopentene **226** in 94% ee using the enantiomerically pure nickel catalyst **227**.[135] Recently, Mortreux and co-workers have shown that nickel catalyzed allylic substitution can also be achieved asymmetrically with soft nucleophiles (malonate).[136]

XIV. ASYMMETRIC CYCLIZATION REACTIONS

Despite the enormous synthetic potential, there have been relatively few examples of palladium-catalyzed asymmetric allylic substitution reactions which involve a

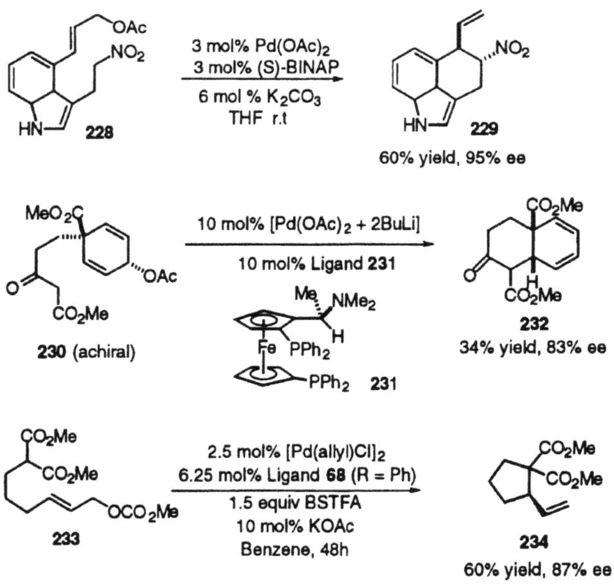

BSTFA = *N,O*-bis(trimethylsilyl)trifluoroacetamide

cyclization process. Kardos and Genet have reported the cyclization of **228** into **229** with up to 95% diastereoselectivity and enantioselectivity.[137] The cyclization product **229** was further elaborated into (−)-chanoclavine I. Shibasaki and coworkers have used the substrate **230**, where the enantioselectivity of the cyclization comes about by virtue of selection of one of the enantiotopic alkene functionalities. Thus, using ligand **231**, the cyclization product **232** is obtained with up to 83% ee.[138] Koch and Pfaltz have recently reported the use of the oxazoline ligand **68** (R = Ph) in the palladium-catalyzed cyclization of **233** into **234** with 87% ee.

XV. CONCLUSION

Palladium-catalyzed allylic substitution reactions have been performed by many research groups around the world. In the last few years there has been particular activity in the enantioselective variants of this reaction. This activity has produced useful results. There are now many ligands available for enantioselective palladium-catalyzed reactions, and often enantioselectivities in excess of 90% ee have been reported. There are still areas where these very high enantioselectivities have yet to be achieved. The future holds some exciting challenges!

REFERENCES AND NOTES

1. (a) Godleski, S. A. In *Comprehensive Organic Synthesis*; Trost, B. M.; Fleming, I.; Semmelhack, M. F., Eds.; Pergamon Press: Oxford, 1991; Vol. 4, pp. 585–661; Consiglio, G.; Waymouth, R. M.

Chem. Rev. **1989**, *89*, 257–276. See also: (c) Trost, B. M. *Angew. Chem., Int. Ed. Engl.* **1989**, *28*, 1173–1192; (d) Tsuji, J. *Tetrahedron* **1986**, *42*, 4361–4410; (e) Trost, B. M. *J. Organomet. Chem.* **1986**, *300*, 263–280; (f) Tsuji, J. *Pure Appl. Chem.* **1982**, *54*, 197–206; (g) Trost, B. M. *Pure Appl. Chem.* **1981**, *53*, 2357–2370; (h) Trost B. M. *Acc. Chem. Res.* **1980**, *13*, 385–393; (i) Trost B. M. *Tetrahedron* **1977**, *33*, 2615–2649.

2. (a) Trost, B. M.; Van Vranken, D. L. *Chem. Rev.* **1996**, *96*, 395–422; (b) Frost, C. G.; Howarth. J.; Williams. J. M. J. *Tetrahedron: Asymmetry* **1992**, *3*, 1089–1122; (c) Williams, J. M. J. In *Advanced Asymmetric Synthesis*; Stephenson, G. R., Ed.; Chapman and Hall: London, 1996; (d) Hayashi, T. In *Catalytic Asymmetric Synthesis*; Ojima, I., Ed.; VCH: New York, 1993.
3. (a) Tsuji, J.; Shimizu, I.; Minami, I.; Ohashi, Y.; Sugiura, T.; Takahashi, K. *J. Org. Chem.* **1985**, *50*, 1523–1529; (b) Tsuji, J.; Minami, I. *Acc. Chem. Res.* **1987**, *20*, 140–145.
4. Sheffy, F. K.; Stille, J. K. *J. Am. Chem. Soc.* **1983**, *105*, 7173.
5. Stary, I.; Stará, I. G.; Kocovsky, P. *Tetrahedron* **1994**, *50*, 529–537.
6. Takahashi, K.; Miyake, A.; Hata, G. *Bull. Chem. Soc. Jpn.* **1972**, *45*, 230–236.
7. (a) Trost, B. M.; Molander, G. A. *J. Am. Chem. Soc.* **1981**, *103*, 5969-5972; (b) Trost B. M.; Hurnaus, R. *Tetrahedron Lett.* **1989**, *30*, 3893–3896.
8. Hutchins, R. O.; Learn, K. *J. Org. Chem.* **1982**, *47*, 4380–4382.
9. Tanigawa, Y.; Nishimura, K.; Kawasaki, A.; Murahashi, S.-I. *Tetrahedron Lett.* **1982**, *23*, 5549–5552.
10. Trost, B. M.; Schmuff, N. R. *J. Am. Chem. Soc.* **1985**, *107*, 396–405.
11. Lemaire-Audoire, S.; Savignac, M.; Genêt, J. P. *Tetrahedron Lett.* **1995**, *36*, 1267–1270.
12. Hirao, T.; Yamada, N.; Ohshiro, Y.; Agawa, T. *J. Organomet. Chem.* **1982**, *236*, 409–414.
13. Nilsson, Y. I. M.; Andersson, P. G.; Bäckvall, J.-E. *J. Am. Chem. Soc.* **1993**, *115*, 6609–6613.
14. Shimizu, I.; Ishikawa, T. *Tetrahedron Lett.* **1994**, *35*, 1905–1908; (b) Satake, A.; Shimizu, I.; Yamamoto, A. *Synlett* **1995**, 64–68.
15. Tsuji, Y.; Yamada, N.; Tanaka, S. *J. Org. Chem.* **1993**, *58*, 16–17.
16. Fiaud, J.-C.; Malleron J.-L. *J. Chem. Soc., Chem. Commun.* **1981**, 1159–1160; (b) Negishi, E.-I.; John, R. A. *J. Org. Chem.* **1983**, *48*, 4098–4102.
17. Sulsky, R.; Magnin, D. R. *Synlett* **1993**, 933–934.
18. For an account of different hard nucleophiles, including Grignard reagents, organotin and organozinc reagents, see: ref. 1(a), pp. 594–596.
19. Goux, C.; Lhoste, P.; Sinou, D. *Tetrahedron* **1994**, *50*, 10321–10330; (b) Goux, C.; Lhoste, P.; Sinou, D. *Tetrahedron Lett.* **1992**, *33*, 8099–8102.
20. For a representative selection, see: (a) Atkins, K. E.; Walker, W. E.; Manyik, R. M. *Tetrahedron Lett.* **1970**, *43*, 3821–3824; (b) Byström, S. E.; Aslanian, R.; Backvall, J.-E. *Tetrahedron Lett.* **1985**, *26*, 1749–1752; (c) Connell, R. D.; Rein, T.; Akermark, B.; Helquist, P. *J. Org. Chem.* **1988**, *53*, 3845–3849; (d) Takagi, M.; Yamamoto, K. *Chem. Lett.* **1989**, 2123–2124; (e) Murahashi, S.-I.; Taniguchi, Y.; Imada, Y.; Tanigawa, Y. *J. Org. Chem.* **1989**, *54*, 3292–3303; (f) Genet, J.-P.; Thorimbert, S.; Touzin, A.-M. *Tetrahedron Lett.* **1993**, *34*, 1159–1162; (g) Jumnah, R.; Williams, J. M. J.; Williams, A. C. *Tetrahedron Lett.* **1993**, *34*, 6619–6622.
21. (a) Keinan, E.; Sahai, M.; Roth, Z.; Nudelman, A.; Herzig, J. *J. Org. Chem.* **1985**, *50*, 3558–3566; (b) Goux, C.; Lhoste, P.; Sinou, D. *Synlett* **1992**, 725–727; (c) Trost, B. M.; Ito, N.; Greenspan, P. D. *Tetrahedron Lett.* **1993**, *34*, 1421–1424.
22. Tsuji, Y.; Kajita, S.; Isobe, S.; Funato, M. *J. Org. Chem.* **1993**, *58*, 3607–3608.
23. Trost, B. M.; Herndon, J. W. *J. Am. Chem. Soc.* **1984**, *106*, 6835–6837.
24. For a recent review, see: Tsuji, J.; Mandai, T. *Synthesis* **1996**, 1.
25. Coulson, D. R. *Inorg. Synth.* **1990**, *28*, 107.
26. Bäckvall, J.-E.; Nystrom, J.-E.; Nordberg, R. E. *J. Am. Chem. Soc.* **1985**, *107*, 3676.
27. Ukai, T.; Kawazura, H.; Ishii, Y.; Bonnett, J. J.; Ibers, J. A. *J. Organomet. Chem.* **1974**, *65*, 253.
28. Mandai, T.; Matsumoto, T.; Tsuji, J.; Saito, S. *Tetrahedron Lett.* **1993**, *34*, 2513–2516.
29. Takemoto, T.; Nishikimi, Y.; Sodeoka, M.; Shibasaki, M. *Tetrahedron Lett.* **1992**, *33*, 3527.

30. Blart, E.; Genêt, J. P.; Safi, M.; Savignac, M.; Sinou, D.; *Tetrahedron* **1994**, *50*, 505–514.
31. (a) Bergbreiter, D. E.; Chen, B. *J. Chem. Soc., Chem. Commun.* **1983**, 1238–1239; (b) Bergbreiter, D. E.; Weatherford, D. A. *J. Org. Chem.* **1989**, *54*, 2726–2730; (c) Schneider, P.; Quignard, F.; Choplin, A.; Sinou, D. *New J. Chem.* **1996**, *20*, 545–547.
32. (a) Trost, B. M.; Weber, L. *J. Am. Chem. Soc.* **1975**, 1611–1612; (b) Trost, B. M.; Verhoeven, T. R. *J. Am. Chem. Soc.* **1980**, *102*, 4730–4743; (c) Bäckvall, J.-E.; Nordberg, R. E.; Vågberg, J. *Tetrahedron Lett.* **1983**, *24*, 411–412: (d) Keinan, E.; Roth, Z. *J. Am. Chem. Soc.* **1983**, *48*, 1771–1772; (e) Hayashi, T.; Konishi, M.; Kumada, M. *J. Chem. Soc., Chem. Commun.* **1984**, 107–108; (f) Fiaud, J.-C.; Legros, J.-Y. *J. Org. Chem.* **1987**, *52*, 1907–1911; (g) Greenspoon, N.; Keinan, E. *J. Org. Chem.* **1988**, *53*, 3723–3731; (h) Goux, C.; Massacret, M.; Lhoste, P.; Sinou, D. *Organometallics* **1995**, *14*, 4585–4593.
33. Trost, B. M.; Verhoeven, T. R. *J. Org. Chem.* **1976**, *41*, 3215.
34. (a) Matsushita, H.; Negishi, E.-I. *J. Chem. Soc., Chem. Commun.* **1982**, 160; (b) Trost, B. M.; Herndon, J. W. *J. Am. Chem. Soc.* **1984**, *106*, 6835.
35. Hayashi, T.; Yamamoto, A.; Hagihara, T. *J. Org. Chem.* **1986**, *51*, 723–727.
36. Bäckvall, J.-E.; Gatti, R.; Schink, H. E. *Synthesis* **1993**, 343–348.
37. Schink, H. E.; Bäckvall, J.-E. *J. Org. Chem.* **1992**, *57*, 1588–1591.
38. Yamamoto, K.; Deguchi, R.; Ogimura, Y.; Tsuji, J. *Chem. Lett.* **1984**, 1657–1660.
39. Keinan, E.; Sahai, M. *J. Chem. Soc., Chem. Commun.* **1984**, 648–650.
40. Trost, B. M.; Bunt, R. C. *J. Am. Chem. Soc.* **1996**, *118*, 235–236.
41. (a) Moreno-Mañas, M.; Ribas, J. *Tetrahedron Lett.* **1989**, *30*, 3109–3112; (b) Tanikaga, R.; Jun T. X.; Kaji, A. *J. Chem. Soc., Perkin Trans. 1* **1990**, 1185–1191.
42. (a) Åkermark, B.; Hansson, S.; Krakenberger, B.; Vitagliano, A.; Zetterberg, K. *Organometallics* **1984**, *3*, 679; (b) Äkermark, B.; Zetterberg, K.; Hansson, S.; Krakenberger, B.; Vitagliano, A. *J. Organomet. Chem.* **1987**, *335*, 133–142.
43. Trost, B. M.; Lee, P. H. *J. Am. Chem. Soc.* **1991**, *113*, 5076.
44. Hiroi, K.; Onuma, H.; Arinaga, Y. *Chem. Lett.* **1995**, 1099–1100.
45. Hiroi, K.; Abe, J.; Suya, K.; Sato, S.; Koyama, T. *J. Org. Chem.* **1994**, *59*, 203–213.
46. The allylpalladium intermediates are only *meso* in the absence of chiral ligands, but nevertheless, this provides a convenient classification.
47. Trost, B. M.; Dietsche, T. J. *J. Am. Chem. Soc.* **1973**, 8200–8201.
48. (a) Hayashi, T.; Yamamoto, A.; Hagihara, T.; Ito, Y. *Tetrahedron Lett.* **1986**, *27*, 191–19; (b) Hayashi, T.; Yamamoto, A.; Ito, Y. *Chem. Lett.* **1987**, 177–180; (c) Hayashi, T. *Pure Appl. Chem.* **1988**, *60*, 7–12.
49. (a) Yamazaki, A.; Morimoto, T.; Achiwa, K. *Tetrahedron: Asymmetry* **1993**, *4*, 2287–2290; (b) Yamazaki, A.; Achiwa, K. *Tetrahedron: Asymmetry* **1995**, *6*, 51–54.
50. Okado, Y.; Minami, T.; Sasaki, Y.; Umezu, Y.; Yamaguchi, M. *Tetrahedron Lett.* **1990**, *31*, 3905–3908; (b) Okado, Y.; Minami, T.; Umezu, Y.; Nishikawa, S.; Mori, R.; Nakayama, Y. *Tetrahedron: Asymmetry* **1991**, *2*, 667–682.
51. Auburn, P. R.; Mackenzie, P. B.; Bosnich, B. *J. Am. Chem. Soc.* **1985**, *107*, 2033.
52. Yamaguchi, M.; Shima, T.; Yamagishi, T.; Hida, M. *Tetrahedron Lett.* **1990**, *31*, 5049–5052.
53. Seebach, D.; Devaquet, E.; Ernst, A.; Hayakawa, M.; Kuhnle, F. N. M.; Schweizer, W. B.; Weber, B. *Helv. Chim. Acta* **1995**, *78*, 1636–1650.
54. Zhu, G.; Terry, M.; Zhang, X. *Tetrahedron Lett.* **1996**, *37*, 4475–4478.
55. Bolm, C.; Kaufmann, D.; Gessler, S.; Harms, K. *J. Organomet. Chem.* **1995**, *502*, 47–52.
56. (a) Trost, B. M.; Breit, B.; Peukert, S.; Zambrano, J.; Ziller, J. W. *Angew. Chem., Int. Ed. Engl.* **1995**, *34*, 2386–2388; (b) Trost, B. M.; Bunt, R. C. *Angew. Chem., Int. Ed. Engl.* **1996**, *35*, 99–102.
57. (a) Togni, A.; Rihs, G.; Pregosin, P. S.; Ammann, C. *Helv. Chim. Acta* **1990**, *73*, 723–732; (b) Togni, A. *Tetrahedron: Asymmetry* **1991**, *2*, 683–690.
58. Kang, J.; Cho, W. O.; Cho, H. G. *Tetrahedron: Asymmetry* **1994**, *5*, 1347–1352.

59. (a) Müller, D.; Umbricht, G.; Weber, B.; Pfaltz, A. *Helv. Chim. Acta* **1991**, *74*, 232–240; (b) von Matt, P.; Lloyd-Jones, G. C.; Minidis, A. B. E.; Pfaltz, A.; Macko, L.; Neuburger, M.; Zehnder, M.; Rüegger, H.; Pregosin, P. S. *Helv. Chim. Acta* **1995**, *78*, 265–284. See also; (c) Ohkita, K.; Kurosawa, H.; Hasegawa, T.; Hirao, T.; Ikeda, I. *Organometallics* **1993**, *12*, 3211–3215.
60. (a) Leutenegger, U.; Umbricht, G.; Fahrni, C.; von Matt, P.; Pfaltz, A. *Tetrahedron* **1992**, *48*, 2143–2156; (b) Pfaltz, A. *Acc. Chem. Res.* **1993**, *26*, 339–345.
61. Gamez, P.; Dunjic, B.; Fache, F.; Lemaire, M. *J. Chem. Soc., Chem. Commun.*, **1994**, 1417–1418.
62. Tanner, D.; Andersson, P. G.; Harden, A.; Somfai, P. *Tetrahedron Lett.* **1994**, *35*, 4631–4634.
63. Kubota, H.; Nakajima, M.; Koga, K. *Tetrahedron Lett.* **1993**, *34*, 8135–8138.
64. Bovens, M.; Togni, A.; Venanzi, L. M. *J. Organomet. Chem.* **1993**, *451*, C28–C31.
65. Halterman, R. L.; Nimmons, H. L. *Organometallics* **1990**, *9*, 273–275.
66. Peña-Cabrera, E.; Norrby, P.-O.; Sjögren, M.; Vitagliano, A.; De Felice, V.; Oslob, J.; Ishii, S.; O'Neill, D.; Åkermark, B.; Helquist, P. *J. Am. Chem. Soc.* **1996**, *118*, 4299–4315.
67. Brenchley, G.; Fedouloff, M.; Mahon, M. F.; Molloy, K. C.; Wills, M. *Tetrahedron* **1995**, *51*, 10581–10592; (b) Brenchley, G.; Merifield, E.; Wills, M.; Fedouloff, M. *Tetrahedron Lett.* **1994**, *35*, 2791–2794.
68. Bolm, C.; Kaufmann, D.; Zehnder, M.; Neuburger, M. *Tetrahedron Lett.* **1996**, *37*, 3985–3988.
69. Åkermark, B.; Krakenberger, B.; Hansson, S. *Organometallics* **1987**, *6*, 620. See also: Malet, R.; Moreno-Mañas, M.; Parella, T.; Pleixats, R. *J. Org. Chem.* **1996**, *61*, 758–763.
70. (a) Frost, C. G.; Williams, J. M. J. *Tetrahedron Lett.* **1993**, *34*, 2015–2018; (b) Dawson, G. J.; Frost, C. G.; Williams, J. M. J.; Coote, S. J. *Tetrahedron Lett.* **1993**, *34*, 3149–3150; (c) Frost, C. G.; Williams, J. M. J. *Tetrahedron: Asymmetry* **1993**, *4*, 1785–1788; (d) Dawson, G. J.; Frost, C. G.; Martin, C. J.; Williams, J. M. J.; Coote, S. J. *Tetrahedron Lett.* **1993**, *34*, 7793–7796; (e) Allen, J. V.; Dawson, G. J.; Frost, C. G.; Williams, J. M. J.; Coote, S. J. *Tetrahedron* **1994**, *50*, 799–808; (f) Allen, J. V.; Coote, S. J.; Dawson, G. J.; Frost, C. G.; Martin, C. J.; Williams, J. M. J. *J. Chem. Soc., Perkin Trans. 1* **1994**, 2065–2072; (g) Allen, J. V.; Bower, J. F.; Williams, J. M. J. *Tetrahedron: Asymmetry* **1995**, *5*, 1895–1898.
71. (a) Sprinz, J.; Helmchen, G. *Tetrahedron Lett.* **1993**, *34*, 1769–1772; (b) Rieck, H.; Helmchen, G. *Angew. Chem., Int. Ed. Engl.* **1995**, *34*, 2687–2689.
72. von Matt, P.; Pfaltz, A. *Angew. Chem., Int. Ed. Engl.* **1993**, *32*, 566–568.
73. (a) Sprinz, J.; Kiefer, M.; Helmchen, G.; Reggelin, M.; Huttner, G.; Walter, O.; Zsolnai, L. *Tetrahedron Lett.* **1994**, *35*, 1523–1526; (b) Pfaltz, A. *Acta Chem. Scand.* **1996**, *50*, 189–194 and references therein.
74. Zhang, W.; Hirao, T.; Ikeda, I. *Tetrahedron Lett.* **1996**, *37*, 4545–4548.
75. Brown, J. M.; Hulmes, D. I.; Guiry, P. J. *Tetrahedron* **1994**, *50*, 4493–4506.
76. Togni, A.; Burckhardt, U.; Gramlich, V.; Pregosin, P. S.; Salzmann, R. *J. Am. Chem. Soc.* **1996**, *118*, 1031–1037.
77. Hayashi, Y.; Sakai, H.; Kaneta, N.; Uemura, M. *J. Organomet. Chem.* **1995**, *503*, 143–148.
78. Chelucci, G.; Cabras, M. A.; Botteghi, C.; Basoli, C.; Marchetti, M. *Tetrahedron: Asymmetry* **1996**, *7*, 885–895.
79. Kubota, H.; Koga, K. *Tetrahedron Lett.* **1994**, *35*, 6689–6692.
80. Wimmer, P; Widhalm, M. *Tetrahedron: Asymmetry* **1995**, *6*, 657–660.
81. (a) Barbaro, P.; Pregosin, P. S.; Salzmann, R.; Albinati, A.; Kunz, R. W. *Organometallics* **1995**, *14*, 5160–5170; (b) Sesay, S. J.; Williams, J. M. J. Manuscript in preparation.
82. Sesay, S. J.; Williams, J. M. J. Manuscript in preparation.
83. For ligand **81** and related structures, see: (a) Togni, A.; Breutel, C.; Schnyder, A., Spindler, F.; Landert, H.; Tijani, A. *J. Am. Chem. Soc.* **1994**, *116*, 4062–4066: (b) Pregosin, P. S.; Salzmann, R.; Togni, A. *Organometallics* **1995**, *14*, 842–847; (c) Abbenhuis, H. C. L.; Burckhardt, U.; Gramlich, V.; Martelletti, A.; Spencer, J.; Steiner, I.; Togni, A. *Organometallics* **1996**, *15*, 1614–1621; (d) Abbenhuis, H. C. L.; Burckhardt, U.; Gramlich, V.; Kollner, C.; Pregosin, P. S.;

Salzmann, R; Togni, A. *Organometallics* **1995**, *14*, 759–766; (e) Spencer, J.; Gramlich, V.; Häusler, R.; Togni, A. *Tetrahedron: Asymmetry* **1996**, *7*, 41–44.
84. Cesarotti, E.; Grassi, M.; Prati, L.; Demartin, F. J. *Chem. Soc., Dalton Trans. 1* **1991**, 2073–2082.
85. Barbaro, P.; Currao, A.; Herrmann, J.; Nesper, R.; Pregosin, P. S.; Salzmann, R. *Organometallics* **1996**, *15*, 1879–1888.
86. Frost, C. G.; Williams, J. M. J. *Synlett* **1994**, 551–552.
87. Chelucci, G.; Cabras, M. A. *Tetrahedron: Asymmetry* **1996**, *7*, 965–966.
88. Trost, B. M.; Murphy, D. J. *Organometallics* **1985**, *4*, 1143–1145.
89. Sennhenn, P.; Gabler, B.; Helmchen, G. *Tetrahedron Lett.* **1994**, *35*, 8595–8598.
90. Knühl, G.; Sennhenn, P.; Helmchen, G. *J. Chem. Soc., Chem. Commun.* **1995**, 1845–1846.
91. Okada, Y.; Minami, T.; Yamamoto, T.; Ichikawa, J. *Chem. Lett.* **1992**, 547–550.
92. (a) Trost, B. M.; Bunt, R. C. *J. Am. Chem. Soc.* **1994**, *116*, 4089–4090; (b) Trost, B. M.; Breit, B.; Organ, M. G. *Tetrahedron Lett.* **1994**, *35*, 5817–5820.
93. (a) Trost, B. M.; Van Vranken, D. L. *Angew. Chem., Int. Ed. Engl.* **1992**, *31*, 228–230; (b) Trost, B. M.; Van Vranken, D. L.; Bingel, C. *J. Am. Chem. Soc.* **1992**, *114*, 9327–9343.
94. Mori, M.; Nukui, S.; Shibasaki, M. *Chem. Lett.* **1991**, 1797–1800.
95. Yoshizaki, H.; Satoh, H.; Sato, S.; Nukui, S.; Shibasaki, M.; Mori, M. *J. Org. Chem.* **1995**, *60*, 2016–2021.
96. Trost, B. M.; Lee, C. B.; Weiss, J. M. *J. Am. Chem. Soc.* **1995**, *117*, 7247–7248.
97. (a) Hayashi, T.; Ohno, A.; Lu, S.; Matsumoto, Y.; Fukuyo, E.; Yanagi, K. *J. Am. Chem. Soc.* **1994**, *116*, 4221–4226; (b) Ohno, A.; Yamane, M.; Hayashi, T.; Oguni, N.; Hayashi, M. *Tetrahedron: Asymmetry* **1995**, *6*, 2495–2502. See also: (c) Hayashi, T.; Yamamoto, A.; Ito, Y. *Tetrahedron Lett.* **1988**, *29*, 99–102.
98. Fiaud, J. C.; Legros, J. Y. *Tetrahedron Lett.* **1988**, *29*, 2959–2962.
99. (a) Fiaud, J.-C.; Legros, J.-Y. *J. Org. Chem.* **1990**, *55*, 4840–4846; (b) Legros, J.-Y.; Fiaud, J.-C. *Tetrahedron* **1994**, *50*, 465–474; (c) Gil, R.; Fiaud, J.-C. *Bull. Chem. Soc. Fr.* **1994**, *131*, 584–589.
100. Hayashi, T.; Kishi, K.; Uozurni, Y. *Tetrahedron: Asymmetry* **1991**, *2*, 195–198.
101. (a) Lloyd-Jones, G. C.; Pfaltz, A. *Angew. Chem., Int. Ed. Engl.* **1995**, *34*, 462–464; (b) Lloyd-Jones, G. C.; Pfaltz, A. *Z. Naturforsch*, **1995**, *50b*, 361–367.
102. (a) Auburn, P. R.; Mackenzie, P. B.; Bosnich, B. *J. Am. Chem. Soc.* **1985**, *107*, 2033–2046; (b) Mackenzie, P. B.; Whelan, J.; Bosnich, B. *J. Am. Chem. Soc.* **1985**, *107*, 2046–2054; (c) Farrar, D. H.; Payne, N. C. *J. Am. Chem. Soc.* **1985**, *107*, 2054–2058.
103. (a) Dawson G. J.; Williams, J. M. J. *Tetrahedron Lett.* **1995**, *36*, 461–462; (b) Dawson G. J.; Williams, J. M. J. *Tetrahedron: Asymmetry* **1995**, *6*, 2535–2546.
104. Bower, J. F.; Williams, J. M. J. *Synlett* **1996**, 685–686.
105. Trost, B. M.; Strege, P. E. *J. Am. Chem. Soc.* **1977**, 1649–1651.
106. Grennberg, H.; Langer, V.; Bäckvall, J.-E. *J. Chem Soc., Chem. Commun.* **1991**, 1190–1192.
107. Granberg, K. L.; Bäckvall, J.-E. *J. Am. Chem. Soc.* **1992**, *114*, 6858–6863.
108. Legros, J.-Y.; Toffano, M.; Fiaud, J.-C. *Tetrahedron: Asymmetry* **1995**, *6*, 1899–1902.
109. Hayashi, T.; Yamamoto, A,; Ito, Y. *J. Chem Soc., Chem. Commun.* **1986**, 1090.
110. Fiaud, J. C.; Hibon de Gournay, A.; Larcheveque, M.; Kagan, H. B. *J. Organomet. Chem.* **1978**, *154*, 175.
111. Hayashi, T.; Kanehira, K.; Tsuchiya, H.; Kumada, M. *J. Chem. Soc., Chem. Commun.* **1982**, 1162–1164.
112. Sawamura, M.; Nagata, H.; Sakamoto, H.; Ito, Y. *J. Am. Chem. Soc.* **1992**, *114*, 2586–2592.
113. Hayashi, T.; Kanehira, K.; Hagihara, T.; Kumada, M. *J. Org. Chem.* **1988**, *53*, 113–120.
114. (a) Genet, J. P.; Ferroud, D.; Juge, S.; Montes, J. R. *Tetrahedron Lett.* **1986**, *27*, 4573–4576; (b) Genet, J.-P.; Juge, S.; Ruiz Montès, J.; Gaudin, J.-M. *J. Chem. Soc., Chem. Commun.* **1988**, 718–719.
115. Genet, J.-P.; Juge, S.; Achi, S.; Mallart, S.; Ruiz Montes, J.; Levif, G. *Tetrahedron* **1988**, *44*, 5263–5275.

116. (a) Genet, J. P.; Kopola, N.; Juge, S.; Ruiz-Montes, J.; Antunes, O. A. C.; Tanier, S. *Tetrahedron Lett.* **1990**, *31*, 3133–3136; (b) Voigt, K.; Stolle, A.; Salaün, J.; de Meijere, A. *Synlett* **1995**, 226–228.
117. Baldwin, I. C.; Williams, J. M. J.; Beckett, R. P. *Tetrahedron: Asymmetry* **1995**, *6*, 679–682.
118. Baldwin, I. C.; Williams, J. M. J.; Beckett, R. P. *Tetrahedron: Asymmetry* **1995**, *6*, 1515–1518.
119. Trost, B. M.; Organ, M. G. *J. Am. Chem. Soc.* **1994**, *116*, 10320–10321.
120. Trost, B. M.; Pulley, S. R. *J. Am. Chem. Soc.* **1995**, *117*, 10143–10144.
121. Hayashi, T.; Yamamoto, A.; Ito, Y.; Nishioka, E.; Miura, H.; Yanagi, K. *J. Am. Chem. Soc.* **1989**, *111*, 6301–6311.
122. Hayashi, T.; Kishi, K.; Yamamoto, A.; Ito, Y. *Tetrahedron Lett.* **1990**, *31*, 1743–1746.
123. von Matt, P.; Loiseleur, O.; Koch, G.; Pfaltz, A.; Lefeber, C.; Feucht, T.; Helmchen, G. *Tetrahedron: Asymmetry* **1994**, *5*, 573–584.
124. Jumnah, R.; Williams, A. C.; Williams, J. M. J. *Synlett* **1995**, 821–822.
125. Uozumi, Y.; Tanahashi, A.; Hayashi, T. *J. Org. Chem.* **1993**, *58*, 6826–6832.
126. Yamazaki, A.; Achiwa, K. *Tetrahedron: Asymmetry* **1995**, *6*, 1021–1024.
127. Hiroi, K.; Makino, K. *Chem. Lett.* **1986**, 617–620.
128. Eichelmann, H.; Gais, H.-J. *Tetrahedron: Asymmetry* **1995**, 643–646.
129. Trost, B. M.; Organ, M. G.; O'Doherty, G. A. *J. Am. Chem. Soc.* **1995**, *117*, 9662–9670.
130. Fotiadu, F.; Cros, P.; Faure, B.; Buono, G. *Tetrahedron Lett.* **1990**, *31*, 77–80.
131. Fiaud, J.-C.; Aribi-Zouioueche, L. *J. Organomet. Chem.* **1985**, *295*, 383–387.
132. Fiaud, J.-C.; Legros, J.-Y. *Tetrahedron Lett.* **1991**, *32*, 5089–5092.
133. (a) Hayashi, T.; Iwamura, H.; Uozumi, Y. *Tetrahedron Lett.* **1994**, *35*, 4813–4816; (b) Hayashi, T.; Iwamura, H.; Naito, M.; Matsumoto, Y.; Uozumi, Y.; Miki, M.; Yanagi, K. *J. Am. Chem. Soc.* **1994**, *116*, 775–776; (c) Hayashi, T.; Iwamura, H.; Uozumi, Y.; Matsumoto, Y.; Ozawa, F. *Synthesis* **1994**, 526–532.
134. (a) Consiglio, G.; Morandi, F.; Piccolo, O. *J. Chem Soc., Chem. Commun.* **1983**, 112–114; (b) Hiyama, T.; Wakasa, N. *Tetrahedron Lett.* **1985**, *26*, 3259–3262; (c) Indolese, A. F.; Consiglio, G. *Organometallics* **1994**, *13*, 2230–2234.
135. Consiglio, G.; Indolese, A. *Organometallics* **1991**, *10*, 3425–3427.
136. Bricout, H.; Carpentier, J-F.; Mortreux, A. *Tetrahedron Lett.* **1996**, *37*, 6105–6108.
137. Kardos, N.; Genêt, J.-P.; *Tetrahedron: Asymmetry* **1994**, *5*, 1525–1533.
138. Takemoto, T.; Nishikimi, Y.; Sodeoka, M.; Shibasaki, M. *Tetrahedron Lett.* **1992**, *33*, 3531–3532.

NEW ACHIEVEMENTS IN ASYMMETRIC SYNTHESIS OF ORGANOPHOSPHORUS COMPOUNDS

Oleg I. Kolodiazhnyi

Advances in Asymmetric Synthesis
Volume 3, pages 273–357.

ISBN: 0-7623-0106-6

I. INTRODUCTION

The first optical resolutions of chiral organophosphorus compounds were reported many years ago. However the asymmetric synthesis of organophosphorus compounds is a relatively new field, which has developed mostly during the past two decades. The major factor stimulating the rapid development of asymmetric synthesis of P-chiral organophosphorus compounds is their great practical value as ligands in catalysts for asymmetric organic synthesis.[162,163,224,235,242] Very recently, chiral organic phosphorus compounds have attracted much attention as efficient reagents in asymmetric synthesis of organic and organophosphorus compounds. Optically active phosphorus compounds are also important objects for studies of mechanism and stereochemical course of reactions at chiral phosphorus.

Many previous reviews and literature compilations have dealt with phosphorus stereochemistry.[91,134,154,162,168,218] The most interesting of them is certainly the review of Valentine Jr.[286] describing chemistry of compounds containing chiral phosphorus centers, including a short description of asymmetric synthesis of organophosphorus compounds.

The lack of a special review dedicated to modern asymmetric synthesis of organophosphorus compounds induced us to write this report with emphasis on the most recent findings. The chapter consists of six major parts and focuses on those compounds having demonstrated or potential use as a reagent in asymmetric synthesis. The chapter covers only the asymmetric synthesis of chiral compounds that are cited in the literature by virtue of their activity. And to keep the review to an acceptable length, only references to recent papers and review articles on special topics are given.

II. BASIC PROBLEMS OF THE STEREOCHEMISTRY OF CHIRAL PHOSPHORUS COMPOUNDS

Numerous types of organophosphorus compounds have been obtained and different resolution methods have been successfully employed. They include resolution through separation of diastereomers as well as kinetic resolution and asymmetric synthesis. Moreover, the stereospecificity of substitution reaction at the phosphorus atom makes possible the synthesis of various optically active compounds starting from resolved organophosphorus compounds. In this chapter we will first review the kinetic resolution and asymmetric induction at the phosphorus atom in organophosphorus compounds.

A. Asymmetric Synthesis

"Asymmetric synthesis" is a term first used in 1894 by Emil Fisher and defined in 1904 by W. Markwald[214] as a reaction which produces optically active substances from symmetrically constituted compounds with the intermediate use of optically active materials, but with the exclusion of all analytic processes. Morrison and Mosher[224] considered the asymmetric synthesis as a reaction in which an achiral unit in an ensemble of substrate molecules is converted by a reactant into a chiral unit in such manner that the stereoisomeric products are formed in unequal amounts.

The most promising candidates for the asymmetric synthesis itself are reactions of known mechanisms which have ordered transition states without accessible symmetry elements and which are already known to proceed generally in a stereospecific manner. We suppose that for an asymmetric synthesis to be preparatively useful it must give at least a 3:1 ratio of enantiomeric products (50% ee). And of course a method for enantiomeric enrichment must exist. Evidently a process which gives the product in high enantiomeric excess but low conversion is not preparatively interesting. Asymmetric synthesis of chiral organophosphorus reagents has received particular attention because of the exceptional versatility of these compounds.

B. Kinetic Resolution

Kinetic resolution is of general potential interest in organophosphorus chemistry. The rates of substitution of the functional group X of the two enantiomeric R^1R^2PX compounds by a chiral reagent Y* will, in general, be different. If the reaction is stopped before completion, unequal amounts of the diastereomeric-substituted organophosphorus compounds should thus be obtained. The remaining unreacted starting material will also exhibit optical activity.

The first kinetic resolution of a phosphorus compounds was performed probably by G. Wittig in partially resolving *P*-biphenyl-1-naphthylphenylphosphine by means of reaction with half-molar amounts of paraformaldehyde (+)-camphorsulfonic acid. The unreacted part of the phosphine was enriched in the (–)-enantiomer.[295]

Kinetic resolution was observed in the partial reduction of racemic phosphine oxides by enantioselective reducing complexes of aluminum hydride, $LiAlH_4$/(*S*)-2-(anilinomethyl) pyrrolidine, and AlH_3/(–)-β-phenylethyl-amine. Incomplete reduction led to optically active phosphine and unreacted phosphine oxides.[57,207]

Kinetic resolution of a preformed aminophosphonium salts was developed by Horner and Jordan by complement 1:1 resolution of the aminophosphonium salt with potassium hydrogen dibenzoyltartrate. They succeeded in isolation of only the less soluble (*S*)-isomer of the phosphonium salt.[295]

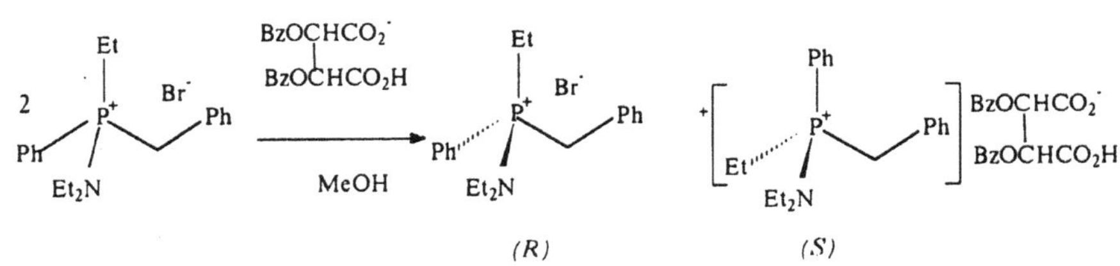

Enzymatic kinetic resolution of P-chiral phosphine derivatives in many cases give excellent results. A series of phosphinylacetates were successfully resolved by PLE-catalyzed hydrolysis providing unreacted esters and acids of high enantiomeric purity.[242] Enantioselective hydrolysis of α-(acyloxy)phosphonates by esterolytic enzymes was described by Hammerschmidt.[203]

C. Configurational Stability

There are several types of organophosphorus compounds. Some of them are conformationally stable, some racemize easily, while others which possess planar or axial symmetry are prochiral compounds, that can be used as starting compounds for asymmetric synthesis (Scheme 1). Tetracoordinate organophosphorus compounds are configurationally stable. Chiral tertiary phosphine oxides, derivatives of phosphorus acids and phosphonium salts, are of high optical stability and racemize usually under action of chemical agents. Functional tetracoordinate organophosphorus compounds are also optically stable. Their racemization is observed only in the presence of nucleophilic agents.

Phosphorus atoms in low-coordinate valent states (mono- and dicoordinate tervalent phosphorus, tricoordinate pentavalent phosphorus) possess axial or planar symmetry and cannot be optically active. Therefore low-coordinate organophosphorus compounds are potential prochiral reagents for the asymmetric synthesis.

Tervalent phosphorus compounds exist in optical active form; however they more or less easily racemize under the action of chemical reagents or upon heating. Nevertheless, many of them are of great practical importance as chiral ligands of transition metal complexes.

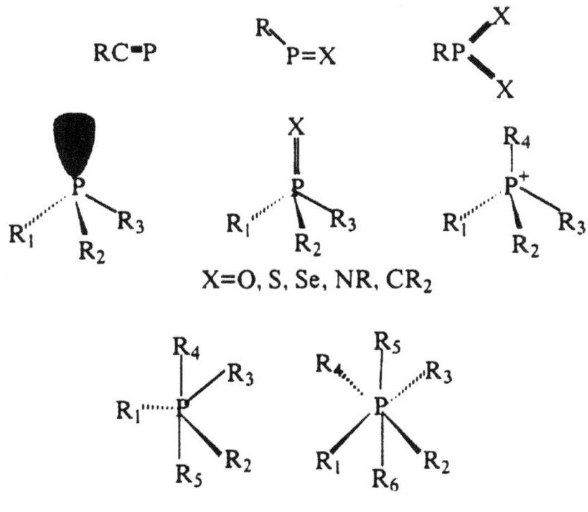

Scheme 1.

Pentacoordinate and hexacoordinate phosphorus compounds are conformationally unstable, though some of them were obtained in optical-active form. Chiral pentacoordinate and hexacoordinate phosphorus compounds are important intermediates in asymmetric syntheses, therefore their stereochemistry has been studied in detail.

In summary, organophosphorus compounds from the point of asymmetrical synthesis can be divided into three groups:

1. Low-coordinate organophosphorus compounds. They can be used as a prochiral starting compound for asymmetric synthesis.
2. Tricoordinate tervalent and tetracoordinate pentavalent organophosphorus compounds. These configurationally stable compounds are the most important in the chemistry of phosphorus and as well in organophosphorus stereochemistry.
3. Pentacoordinate and hexacoordinate organophosphorus compounds. They are usually conformationally unstable; however are very important as intermediates of many chemical and biochemical reactions, including reactions of asymmetric synthesis.

III. LOW-COORDINATE ORGANOPHOSPHORUS COMPOUNDS

Many low-coordinate organophosphorus structures are interesting as starting compounds for asymmetric synthesis.

A. Dicoordinate Phosphorus Compounds

Many low-coordinate organophosphorus structures are interesting as starting compounds for asymmetric synthesis. Some of them are shown in Scheme 2. It is important to note that examples of asymmetric synthesis at low-coordinate phosphorus atoms is limited and many prochiral structures have not been studied up to now.

The first attempt of the diastereoselective addition of optically active alcohols and amines [(–)-menthol, (–)-menthylamine, (–)-α-phenylethylamine)] to λ^3-iminophosphines possessing trigonal configuration was performed by Romanenko et al.[213] Stereoselectivity of this reaction was not very high. The maximal degree of stereoselectivity (34% de) was observed in the reaction of λ^3-iminophosphine with (–)-menthol. However in the presence of chiral tertiary amines [(–)-*N*-dimethylmenthylamine or (–)-*N*-dimethyl-1-methylbenzylamine] the reaction of λ^3-iminophosphines with methanol afforded methoxyaminophosphime with 55% optical purity.[219b] Good stereoselectivity (de 80%) was obtained with the addition of benzyl alcoholate to *P*-menthoxy-λ^3-iminophosphine.[257b]

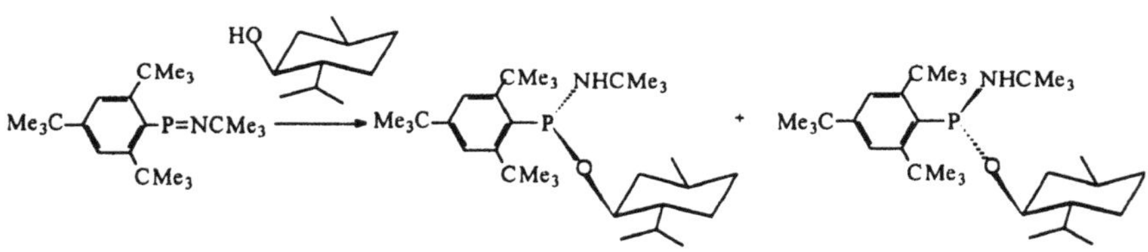

Phosphaalkenes in the form of their tungsten and molybdenum complexes were used by Mathey and coworkers for asymmetric syntheses.[8] Prochiral L-menthyl

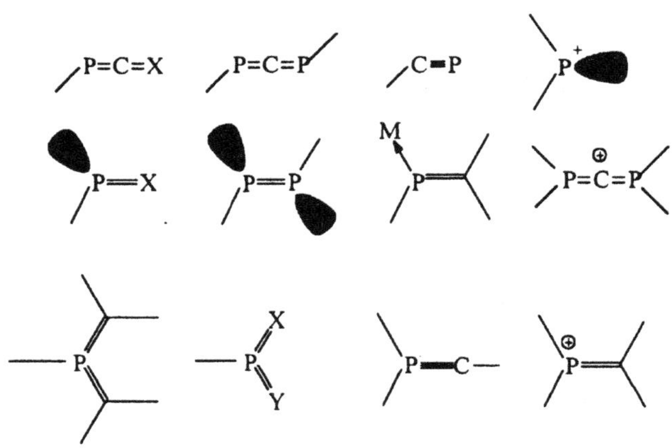

X, Y=O, S, NR, CR_2

Scheme 2.

phosphaalkene complexes **1** were prepared from phospha-Wittig reagents and several aldehydes.

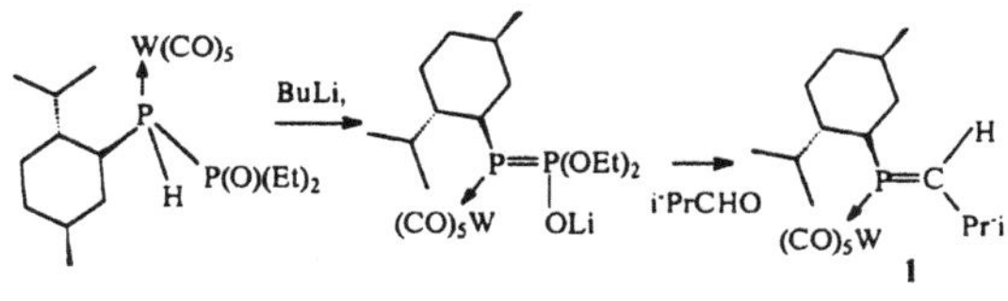

Placement of a menthyl auxiliary on the phosphorus atom promoted a highly diastereoselective (9:1) hydrogenation reaction using achiral catalysts and, ultimately, a completely diastereoselective process under the influence of chiral catalysts.[13] Catalytic hydrogenation of **1**, using RhL_2^+ catalysts and its [2+4] cycloaddition with cyclopentadiene proceeded with full diastereoselectivity. With L_2-DIPHOS, the diastereomeric excess was better than 90% and with L_2 = (–)-CHIRAPHOS only one diastereomer **2** was obtained. At –80 °C alkylation of **2** proceeded with complete stereoselectivity. Complexes **3** were converted into the corresponding oxides **4**, a reaction that has been shown to proceed with full retention of configuration at the phosphorus atom.

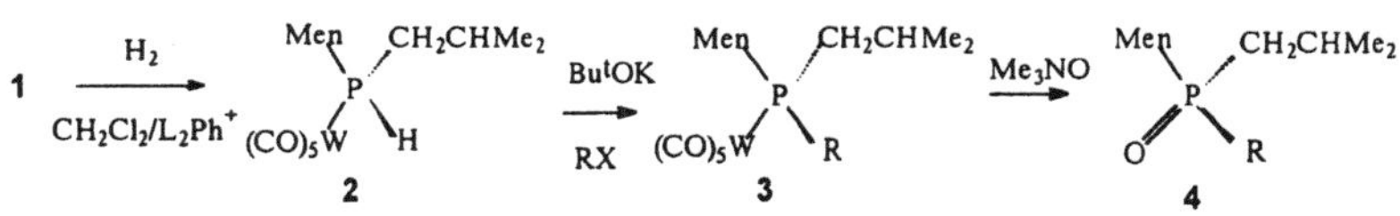

The reaction of isobutyl iodide with menthylphosphine complexed to tungsten or to molybdenum is a diastereoselective process (40% de) as are analogous acylations by phosphorochloridates (50–60% de). It appears that in such reactions involving pyrrolidines of the *R,R* configuration, creation of the s_P center in the first of the two consecutive asymmetric processes is strongly favored.

Mathey introduced the phospha-Wittig reagent also into reaction with arylaldehydes and then trapped, the generated phosphaalkene by means of excess cyclopentadiene to obtain phosphanorbornene derivatives. From the molybdenum derivative the cyclic complex **5** was obtained as a single stereomer. This complex was then converted into optically pure tertiary phosphine **6**.[287]

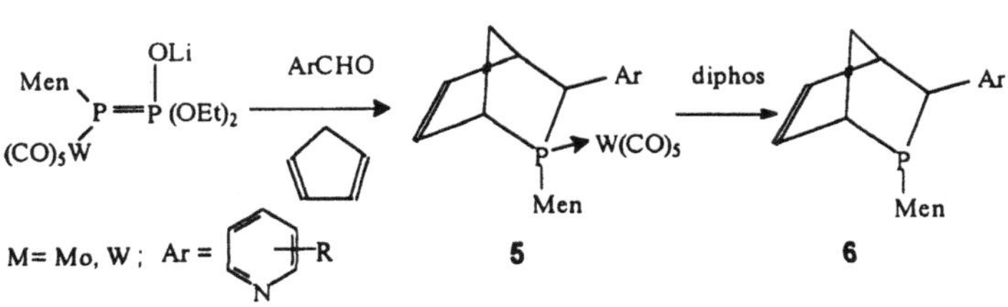

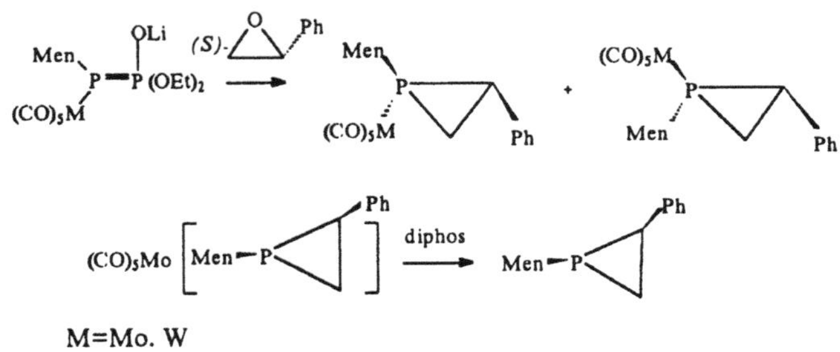

Scheme 3.

Reaction of the anion of pentacarbonyl [(diethoxyphosphoryl) phosphine]tungsten complexes with optically pure *(S)*-styrene oxide (Scheme 3) led to the formation of the first optically active (phosphirane)W$(CO)_5$ complexes (de 24:76). Inversion of the carbon configuration was observed. Total inversion of the carbon configuration supposes an intramolecular S_N2 process.[211,212] Decomplexation of phosphirane diastereomers has been performed by heating of the molybdenum complexes with bis(diphenylphosphino) ethane (DIPHOS). The tervalent phosphiranes have been used as ligands for cationic rhodium complexes. By using the same methodology different optically active phosphirane molybdenum complexes were synthesized.

B. Tricoordinate Pentavalent Phosphorus Compounds

Only two attempts to study asymmetric induction at a prochiral pentavalent tricoordinate phosphorus atom have been reported. One was the asymmetric induction at the phosphorus atom of unstable metathioiminophosphate. The prochiral compound **7** of planar structure at the phosphorus atom was generated by dehydrochlorination of N-substituted amide chlorophosphonate **6**. The addition of *tert*-butylamine to the metathioiminophosphate proceeded stereoselectively with the formation of a diastereomeric mixture. The ratio of diastereomers **8** increases with the decrease in polarity of the solvent from 57: 43 (MeCN) up to 80:20 (cyclohexane).[99,125]

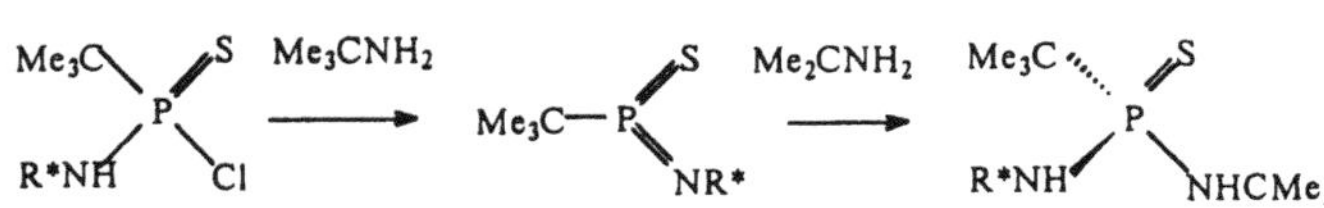

6 **7** **8**

R*= (*S*)-PhMeCH-

Addition of alcohol to prochiral (*S*)-*sec*-butoxy metathiophosphonate **9** possessing a planar-trigonal configuration provides an equimolecular mixture of dias-

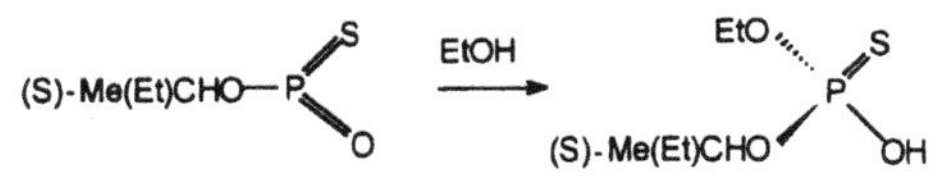

9 10

tereomers **10.** [250] Hence, low-coordinate organophosphorus compounds are interesting starting compounds for asymmetric synthesis. However their stereochemistry has been studied insufficiently. Looking into the future one can expect a wide application of low-coordinate organophosphorus compounds as starting reagents for asymmetric synthesis. Further development in the field of their stereochemistry may be also expected.

IV. ASYMMETRIC SYNTHESIS OF TERVALENT PHOSPHORUS COMPOUNDS

In this section we present a survey of the different types of optically active tervalent organophosphorus compounds and their asymmetric synthesis. Chiral tervalent organophosphorus compounds play a key role in the stereochemistry of phosphorus. A principal application of the tertiary phosphines is their use as chiral ligands in catalysts for asymmetric synthesis. Therefore much effort has been devoted to elaboration of convenient methods for their synthesis.[162,235,242,286] Since the stereochemical behavior of chiral tervalent phosphorus species differs from that of corresponding nitrogen and carbon species, it seems of interest to summarize briefly the current status of the field, making comparison also with related nitrogen and carbon compounds.

A. Configurational Stability and Epimerization

Tervalent phosphorus compounds exhibit fair configurational stability. By comparison nitrogen compounds are less optically stable. Usually the barrier of inversion in acyclic phosphines is about 150 kJ/mol, whereas the barrier of inversion for acyclic amines is about 30 kJ/mol.[252] A trivalent phosphorus atom bonded to three substituents in a pyramidal geometry and possessing one unshared electron pair may spontaneously undergo an inversion of configuration. Such a process of pyramidal atom inversion must involve passage through a transition state in which the lone pair possessing pure *p* character and the bonds from the central atom to the substituents are sp^2.

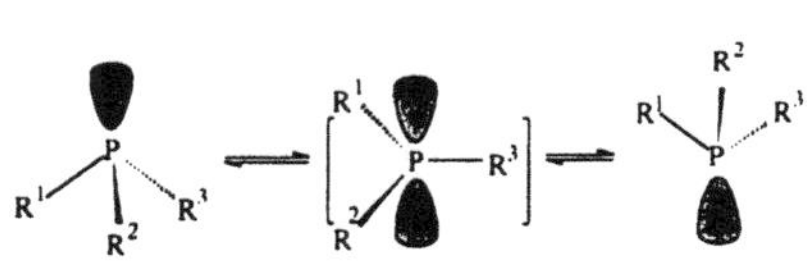

Racemization of trivalent phosphorus compounds depends strongly on their structure, first on electron-accepting substituents at the phosphorus atom, which decrease the configurational stability of compounds. Mislow showed that the barrier of inversion in phosphines depends on the electronegativities of substituents bonded to the phosphorus atom.[13] Phosphorus inversion is accelerated by $(p\text{–}p)_\pi$ conjugation, as shown in the substituted phenyl series. In many cases, compounds bearing electron-accepting groups at the phosphorus atom are racemized. For instance, electron-accepting groups in the *para*-position of the benzene ring in arylphosphines reduce the barrier of inversion.[12]

$$(CH_3)(p\text{-}XC_6H_4)(C_6H_5)P \quad \Delta G^{\neq}\ (\text{kkcal/mol}) = 30.8$$
$$(X = CH_3O),\ 30.3\ (X = CH_3),\ 29.1\ (X = CF_3)$$

Moreover, chiral chlorophosphines are conformationaly labile compounds existing as an equilibrium racemic mixture of (*R*)- and (*S*)-enantiomers. Although calculations indicate substantial pyramidal stability at phosphorus in halophosphines of the type R_1R_2PX, attempts to isolate enantiomerically pure chlorophosphines were unsuccessful.[138,237,239] Thus, *tert*-butylphenylchlorophosphine of 49.4% optical purity, prepared by Omelanczuk, lost its optical activity over 20 h in the polarimeter cell.[237] Recently Wild and coworkers showed that isopropylphenylchlorophosphine in palladium (II) complex **11** can be resolved; however liberation of epimeric chlorophosphines from complexes led to complete racemization of the free phosphine within 5 min.[239]

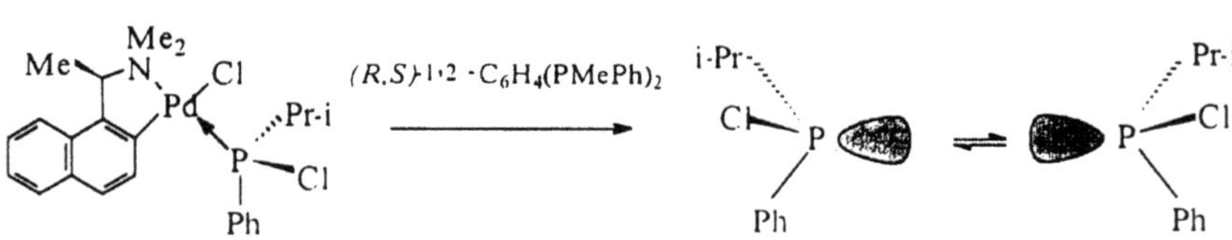

11

The esters of chiral tervalent phosphorus acid R_2POR are more stable than chlorophosphines and can be isolated in enantiomerically pure state.[239] However racemization of P-chiral phosphorus esters R_2POR occurs with measurable rates at room temperature. Acids catalyze racemization of chiral phosphonic esters and may involve not only the pyramidal inversion but also exchange of the ester group.[2,136,286]

Tertiary alkylaryl- and diarylphosphines are more or less configurationally stable. They have a barrier of pyramidal inversion of 30–35 kcal/mol and may be obtained as individual enantiomers at or near ambient temperature. However at high temperature they can racemizate. They can be isolated and in some cases purified by distillation under vacuum. Syntheses of P-chiral teriary phosphines must take into account the moderate epimeric stability of these substances.

B. Synthesis

Until now chiral tervalent organophosphorus compounds have been obtained in the following ways:

1. Asymmetric synthesis
2. Kinetic resolution
3. Stereospecific synthesis

Asymmetric synthesis is an attractive route to chiral tervalent phosphorus compounds. In recent years the asymmetric synthesis of tervalent phosphorus compounds was developed with great success. The successful development of the asymmetric version of many transition metal-catalyzed reactions is dependent upon the design and synthesis of new chiral ligands and inter alia of new chiral phosphines. An attractive way to chiral tertiary phosphines is nucleophilic substitution at the tervalent phosphorus atom in the presence of tertiary bases.

Nucleophilic Substitution at the Tervalent Phosphorus Atom

The most frequently encountered reactions in organic phosphorus chemistry are nucleophilic displacement reactions. The mechanism and steric course of S_NP reactions have been intensively studied. The results of these investigations have been discussed exhaustively in reviews on phosphorus stereochemistry. In the overwhelming majority of cases S_N2 nucleophilic substitution **12** ⇌ **13** at chiral tricoordinate tervalent phosphorus results in inversion of configuration that assumes the formation of pentacoordinate anionic transition state **A**, stabilized by an extension of the valency shell, in which the free pair of electrons resides in an equatorial position.[42,43,46,150]

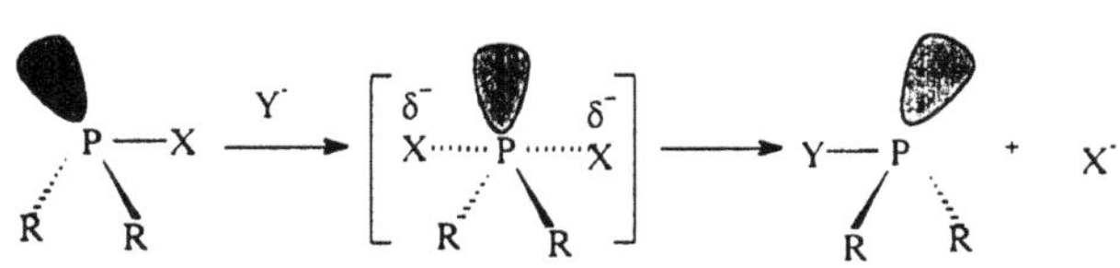

12 A 13

The simplest potential asymmetric synthesis is the reaction of mixed phosphinic acid chlorides with chiral substrates. In the literature one can find different examples of such reactions, though stereoselectivity during these syntheses is, as a rule, not very high. In many cases tervalent phosphorus products were not isolated from a reaction mixture, and often diastereomer ratios were not determined. The pioneering work in this area is due to Mislow, who obtained menthyl esters **14** of phosphinic acids in low stereochemical yield.[194–196,199,200]

Mikolajczyk and coworkers have prepared several partially P-resolved alkyl phosphinites and thiophosphinites via an asymmetric route.[221] Enantiomeric yields

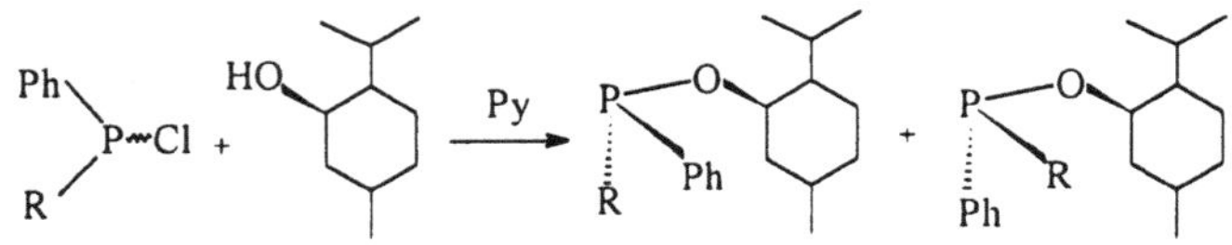

14

were not very high and up to now the development of this theoretically interesting reaction has been limited. The reaction of asymmetric chlorophosphines with simple alcohols in the presence of chiral (–)-*N,N*-dimethyl-(1-phenylethyl)amine gave slightly enantiomerically enriched mixtures of both epimers. The optical purity of the obtained compounds was only 10%.[221] However, possibly the choice of reaction conditions and of chiral tertiary base could give higher stereoselectivity.

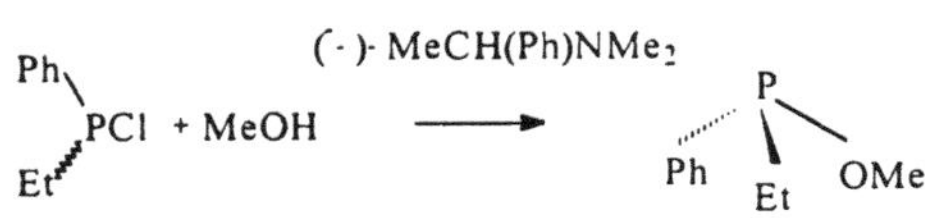

(+)-R **15**

Chodkiewicz has completed a synthesis of a homochiral phosphine using (+)-cinchonine as a chiral auxiliary (Scheme 4). Consecutive substitutions of chlorine in dichlorophenylphosphine by lithium cinchoninate and arylcyanocuprates led stereoselectively to the corresponding P(III)esters. The reaction of phenylalkylchlorophosphines with (+)-cinchonine proceeds analogously. The P-resolved phosphinites were then converted by reaction with alkyllithium into chiral tertiary phosphines. However, details of this methodology have not been published and since it relies on kinetic discrimination between the two diastereomeric forms in the subsequent nucleophilic displacement step, optical purity is not certain.[58–60]

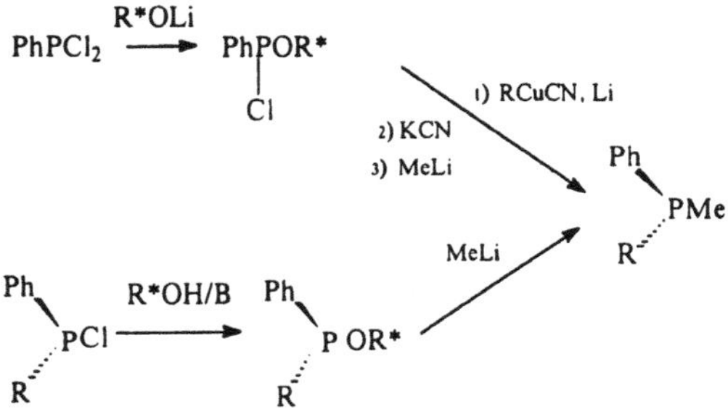

R*OH = (+)-cinchonine

Scheme 4.

The best results were achieved in our laboratory with 1,2:5,6-disubstituted derivatives of glucofuranose as a very effective inductor of chirality at the phosphorus atom. The reaction between 1,2:5,6-substituted derivatives of α-D-glucofuranose and nonsymmetrical racemic chlorides of tervalent phosphorus acids in the presence of tertiary bases proceeds, depending on reaction conditions, with high stereoselectivity to give enantiomerically pure phosphinic acid esters. Careful observance of reaction conditions between alkylarylchlorophosphines and (–)-1,2:5,6-diisopropylidene- or (–)-1,2:5,6-dicyclohexylidene-α-D-glucofuranoses **16** proceeds with very high stereoselectivity and can be an excellent method for preparation of enantiomerically pure phosphinates.[175–177,179,182] Subsequent substitution of alkoxy groups in phosphinites by organolithium compounds occurs with inversion of configuration at the tervalent phosphorus atom to give tertiary phosphines of the *S* configuration.

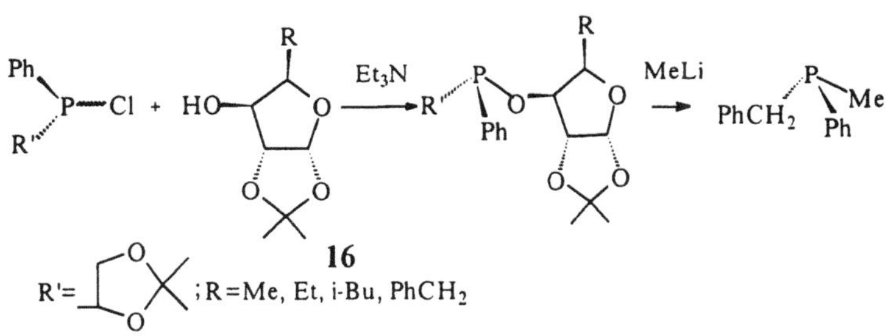

The stereoselectivity of the reaction depends on the nature of the base and the solvent. Thus, the addition of racemic chlorophosphines to a solution of (–)-1,2:5,6-diisopropylidene-D-glucofuranose in toluene at 20 °C in the presence of such strong tertiary bases as 1,4-diazabicyclo[2,2,2]octane (DABCO) or triethylamine furnishes in good yields the enantiomerically pure phosphinite. Sterically hindered tertiary amines, such as DABCO or triisopropylamine, in most cases increase the stereoselectivity of the reaction. However the more accessible triethylamine provides also a good diastereomeric yield of the reaction. The highest de values were obtained in toluene as a solvent. The stereoselectivity of the reaction was raised when an excess of chlorophosphine was used.[179] In the presence of triisopropylamine or DABCO the reaction of glucofuranose with chlorophosphines leads in 100% stereochemical purity to the levorotatoring phosphinate having the *S* configuration at the phosphorus atom. In the presence of weak bases, such as dimethylaniline in ether, a mixture of two diastereoisomeres in the ratio 50:50 was obtained. Surprisingly, in the presence of pyridine instead of DABCO as the base the minor (*R*)-diastereomer becomes major.

The reaction proceeds probably under kinetic control via the formation of a five-membered ring intermediate complex having the structure of a *threo*- or *erythro*-diastereomer, respectively.[181] The stereoselectivity of the reaction of chlo-

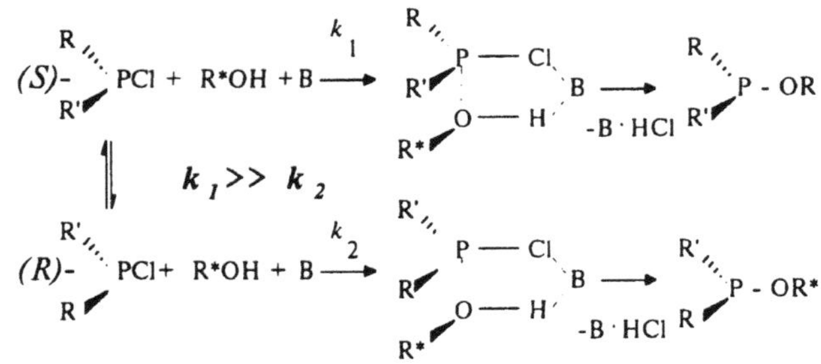

rophosphines with other chiral secondary alcohols is not as high as with (–)-1,2:5,6-diisopropylidene-D-glucofuranose. Thus, the reaction of the twofold excess of *tert*-butyl-isobutyl-chlorophosphine with (–)-1-diethylamino-2-propanol gives a 6:1 mixture of two diastereomers.[179]

The reaction of chlorophosphines with the chiral secondary alcohols [(*R*)-2-diethylaminopropanol, L-menthol, lactates], chiral amines (2-phenylethylamine, natural aminoacid esters) depends strongly on experimental conditions. The ratio of diastereomers of phosphorylated α-phenylethylamine can be raised considerably depending on the base and the excess of chlorophosphine (Table 1).[148,194–196,199,200,264]

The reaction of aminochlorophosphines with chiral aryllithium reagents derived from enantiomeric α-phenylethylamine was used for the synthesis of C,P-chiral aryldialkyl and alkyldiarylphosphines.[137,297] The reaction of prochiral dimenthyl

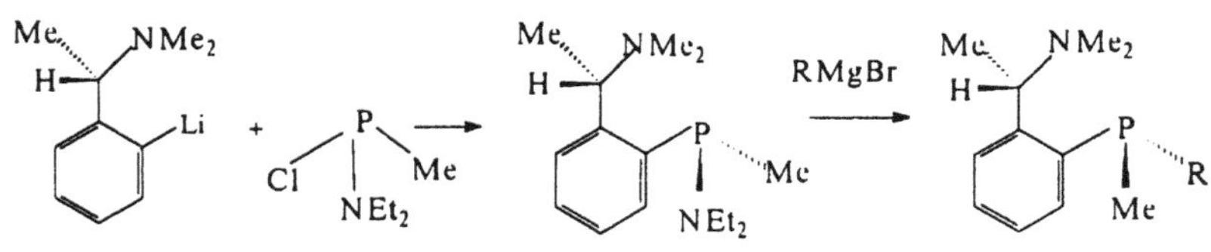

phosphonite with alkyllithium at low temperature proceeds with formation of P-resolved menthyl alkylphenylphosphinites of high diastereomeric purity (90–96% de), The second displacement gives enantiomerically enriched tertiary phosphines (73–79% ee).[194,233]

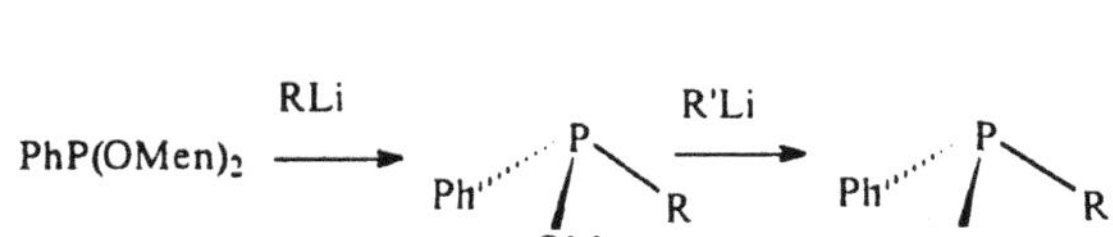

The reaction of racemic isopropylphenylchlorophosphines with *ortho*-metallated (*R*)-[1-(dimethylamino)ethyl]naphthalene–palladium(II) complex **17** in dichlo-

Table 1. Reaction of Chlorophosphines with Chiral Nucleophiles in the Presence of Tertiary Amine Bases

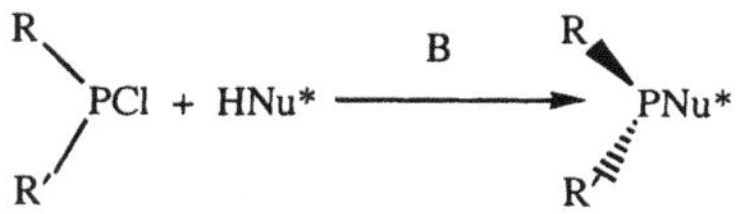

Entry	*R(R')*	*HNu*[a]	*Base*	*Solvent/ Temp., °C*	*de*	*Ref.*
5	Bz(Ph)	GF	Et_3N	Toluene/ −20–+20 °	~96	177
6	*i*-Pr(Ph)	GF	Et_3N	Toluene/ −20–+20°	>98	177
7	*i*-Bu(Ph)	GF	Et_3N	Toluene/ −20– +20°	~98	177
10	*i*-Bu(Ph)	GF	$PhNMe_2$	Ether/+20°	65	177
11	*i*-Bu(Ph)	GF	Py	Ether/+20°	−20	177
12	*i*-Bu(Ph)	GF	NaH	DME/−20°		177
13	*i*-Bu(*t*-Bu)	(−)-HO(Me)CH_2 CH_2NEt_2	Et_3N	Ether/20°	83	177
14	*i*-Bu(Ph)	(−)-Menthole	Et_3N	Toluene/+20°	20	181
15	*i*-Bu(Ph)	HOCH(Me)CH_2CO_2Et	Et_3N	Toluene/+20°	20	181
16	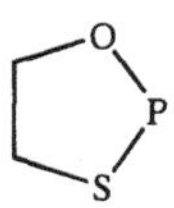	NEA	NEA		35	283
17	*t*-Bu(Ph)	PEA	PEA	Benzene	60	177
18	*t*-Bu(Ph)	PEA	Et_3N	Benzene	80	177
19	*t*-Bu(Ph)	Methyl leucinate	Et_3N	Benzene	15	177,188

Note: [a]GF = (−)-1,2:5,6-diisopropylidene-D-glucofuranose; NEA= (−)-2-naphthylethylamine; PEA = (−)-2-phenylethylamine.

romethane produces a pair of (*R,Rp*)- and (*R,Sp*)-diastereomeric complexes in a 78:22 ratio.[239] Crystallization of the single diastereomer from dichloromethane-diethyl ether proceeds by a second-order asymmetric *trans*-formation because of an equilibrium process and interconversion between (*R,Rp*)- and (*R,Sp*)-isomers. Thus, the configurationally homogenous (*R,Rp*)-diastereomer **17** was separated in 82% overall yield.

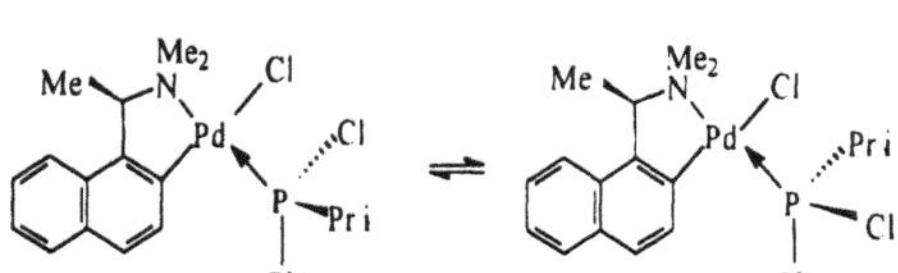

(*R.Rp*)-**17** (*RSp*)-**17**

Addition of the (*R,Rp*)-diastereomer in dichloromethane to an excess of triethylamine and methanol led to the quantitative and completely stereoselective formation of the (*R,Sp*)-phosphonite complex **18**. Substitution of the P-chloride in the (*R,Rp*)-diastereomer **17** by methoxide proceeds with complete inversion at phosphorus. Free configurationally stable (*R*)-**19** can be isolated from this complex in 93% ee by treatment of the complex with DPPE.[239]

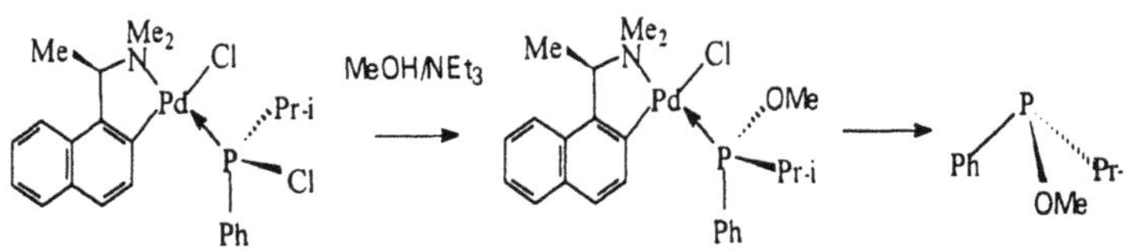

17-(*R,Rp*) 18-(*R,Sp*) 19-(*R*)-(+)

The stereochemical result of nucleophilic substitution at the tervalent phosphorus atom sometimes depends on reaction conditions. Thus the reaction of *trans*-**20** with an excess of phenol and triethylamine at room temperature led exclusively to the *trans*-phenylphosphite **22** (84% de), i.e. reaction proceeded with retention of configuration. Addition of phenol to the chlorophosphine to avoid excess of phenoxide ion during the reaction provides exclusively *cis*-phenylphosphite **21**. This result was explained by postulating a S_N2P type of mechanism and a second attack following an antiperiplanar pathway which produces the thermodynamically more stable *trans*-1-diastereomer **22**.[65]

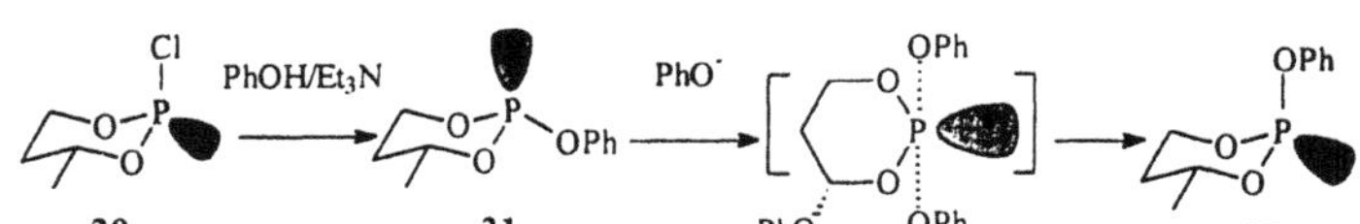

20 **21** **22**

Electrophilic Substitution at the Tervalent Phosphorus Atom

In contrast to nucleophilic substitution, the number of known stereoselective electrophilic reactions at phosphorus proceeding with high asymmetric induction is not very big. These are practically limited to chiral tricoordinate phosphorus compounds that on reactions with electrophilic reagents produce a more stable tricoordinate derivative. It is generally assumed that the electrophilic attack is directed on the lone-electron pair on phosphorus and that the reaction is accompanied by retention of configuration. Thus, alkylation of P-prochiral phosphides was utilized by several groups but in all cases, except for the 25% induction achieved by Naylor and Walker,[232] asymmetric induction was not very high.[52,98,155,165,284]

The reaction of sodium methylphenylphosphide with (+)-(*R*)-1-chloroethylbenzene gives (–)-(*Sp*)-(*Sc*)- phosphine oxide **23** after oxidation. [232] Both (*R*)- and

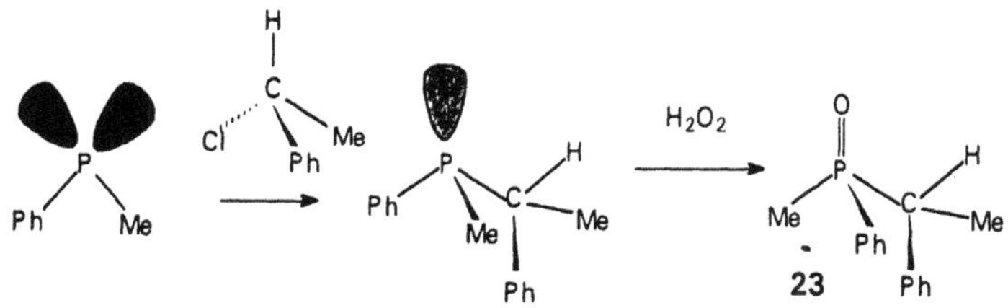

(*S*)-phosphorus epimers of menthylmethylphenylphosphine and its phosphine oxide as well as epimers of neomenthylmethylphenylphosphine were prepared. The menthylmethylphenylphosphines were prepared from neomentylmethylphenylphosphine by a method which is potentially general for the synthesis of phosphines and phosphine oxides having chiral groups at the chiral phosphorus (Scheme 5).[284]

Mosher and Fisher noted that diastereomeres of menthyl methylphenylphosphine epimerized at 120 °C to give a 70:30 equilibrium mixture of epimers.[98] Similar

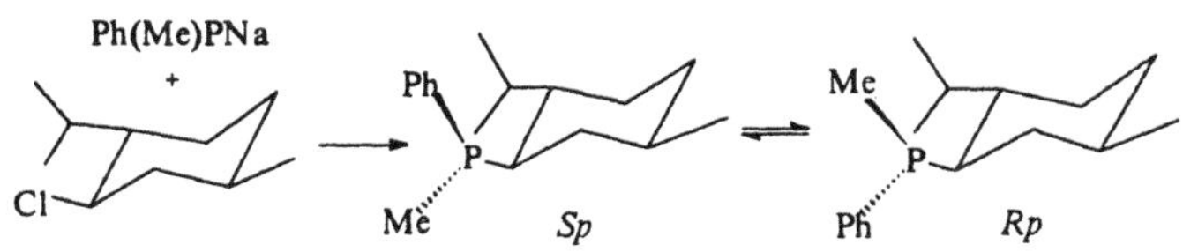

alkylation of phenylphosphine in cationic iron complexes was studied by Wild and coworkers.[61,62] Alkylation of phenylphosphine in cationic iron complexes was also found to be diastereoselective to the extent of ca. 60% de.[61,62,241] At −95 °C deprotonation of the secondary phosphine complex [(*R**,*R**)(*R**)]-**24** (R = H) can be performed with complete stereoselectivity, as demonstrated by the quantitative recovery of the diastereomerically pure starting material following acidification of the reaction mixture at this temperature. Alkylation of the tertiary phosphidoiron complex, generated and maintained at −95 °C also proceeded with retention of

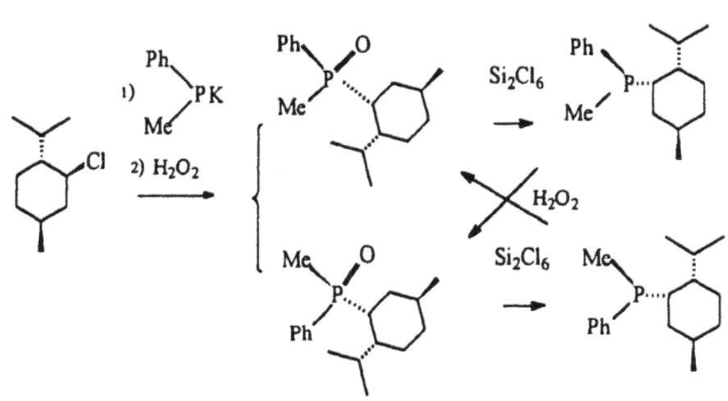

Scheme 5.

configuration and complete stereoselectivity. Reaction above this temperature gave a mixture of thermodynamic products because of the relatively low barrier to inversion of the pyramidal phosphorus stereocenter in the intermediate tertiary phosphido–iron complex.[11,61] Reaction of [(*R**,*R**)-[(η^5-C_5H_5){1,2-$C_6H_4P(MePh)_2$}$FePH_2Ph$]$^+PF_6^-$ with iodo-methane in the presence of triethylamine at 20 °C produced a separable 4:1 mixture of [(*R**,*R**)(*R**)]- **26** and [(*R**,*R**)(*S**)]- **27**.

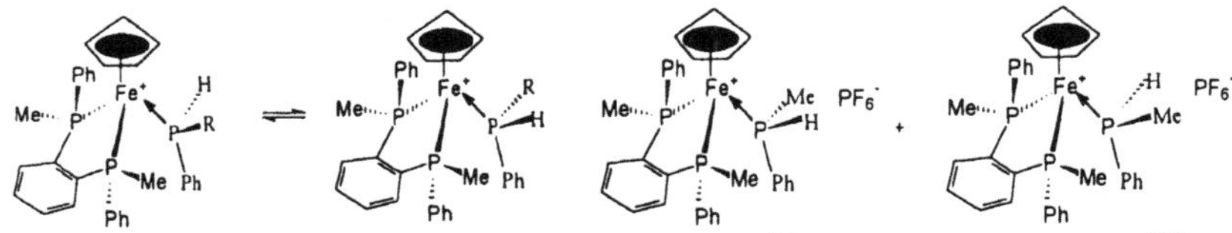

[(*R**,*R**),(*R**)]**24** [(*R**,*R**),(*S**)] **25** [(*R**,*R**),(*R**)]- **26** [(*R**,*R**),(*S**)]-**27**

In case of platinum complexes two diastereomeric methylphenylphosphines have been obtained in a 1:3 ratio and the pure (*S*,*S*,*S*)-diastereomer was isolated by crystallization.[11] Soluble rhodium complexes containing enantiomers of (*R*,*R*)-1,2-

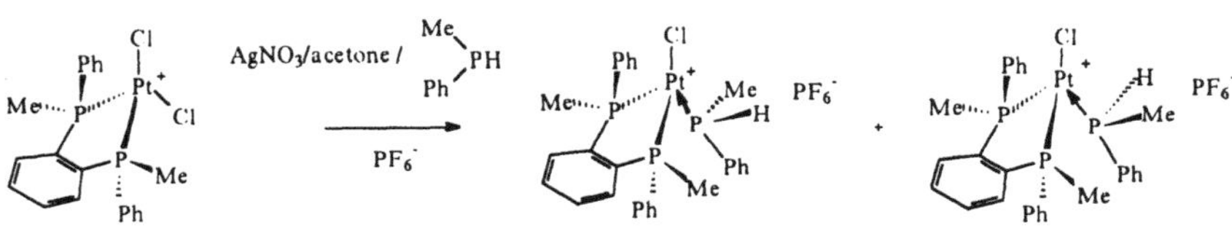

phenylenebis(methylphenyl-phosphine) have been shown to be highly efficient catalysts for the asymmetric hydrogenation of a variety of prochiral *Z*-substituted enamide acids and esters.[62]

The reaction of C_2-symmetric bis(phosphides) with various electrophiles proceeds with the formation of only one (usually *S*,*R*,*R*,*S*) of the three possible products in high predominance (~ 90% ee) of the total isomeric mixture. The isomerically pure [(*PR*,*3R*,*4R*,*P′R*)(*PS*,*3R*,*4R*,*P′S*)]-1-(*t*-butoxycarbonyl)-3,4-bis[(2-cyanoethyl)phenylphosphino] pyrrolidine (**28**) was prepared via the neutral diiodopalladium complex and then split off from the palladium with cyanide ion.[9–12]

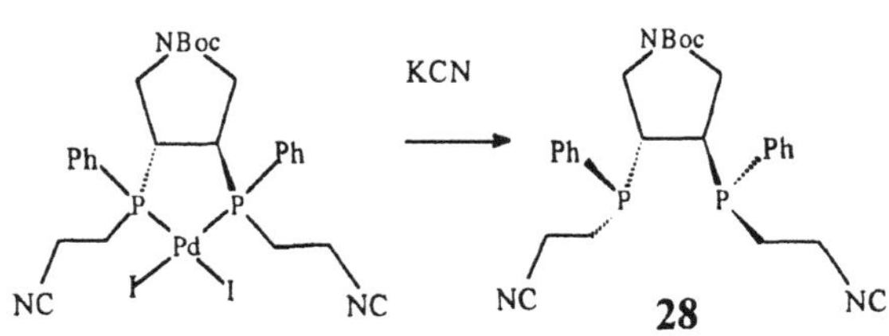

28

Burgess and coworkers performed the synthesis of stereochemically matched biphosphine ligands, representing DIOP-DIPAMP hybrids. Absolute configurations of chiral phosphine ligands **29** were determined via single-crystal X-ray diffraction studies of molybdenum tetracarbonyl derivatives.[52]

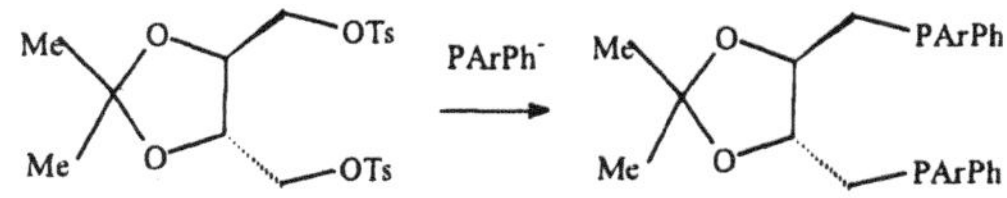

29

The reaction of prochiral phosphides with achiral electrophiles has also found considerable utility in the preparation of chiral tertiary phosphines. King and coworkers obtained diphosphines **30** as a diastereomer mixture by the reaction of neomenthylphenylphosphine with diphenylvinylphosphine in the presence of potassium *tert*-butoxide. The mixture was separated into pure diastereomers by fractional crystallization. These diastereomers were used as ligands of rhodium[I] complexes catalyzing asymmetric homogenous hydrogenation.[165]

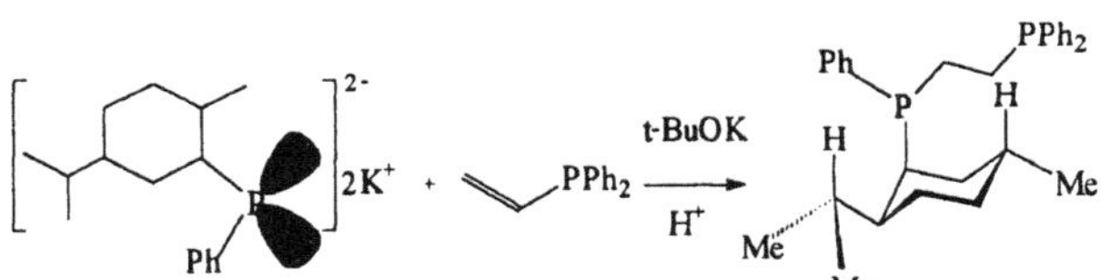

30

Mathey and coworkers reported that deprotonation and subsequent alkylation of (menthylphosphine)pentacarbonyltungsten with *i*-BuI gave two diastereomeric secondary phosphine complexes in the ratio 7:3. Futher alkylation of these complexes led to the preferential formation of tertiary phosphines with a diastereomeric excess of 80%. Diasteromerically pure tertiary phosphines have been isolated after crystallization and oxidative decomplexation.[287]

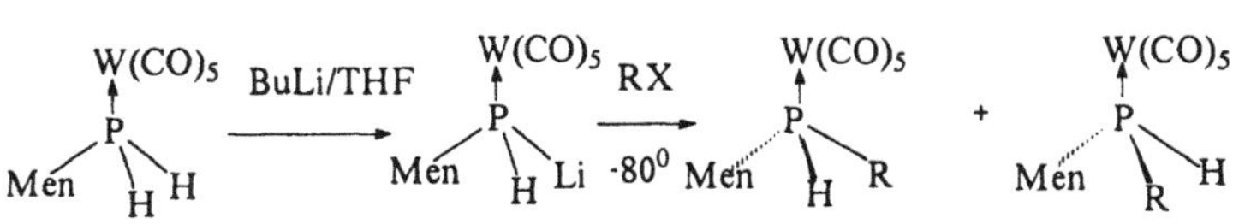

Different examples of alkylations and arylations of prochiral phosphides derived from tartaric acid were described by Nagel.[227–230]

Borate Complexes in the Synthesis of Tertiary Phosphines

Juge and coworkers elaborated a general method for asymmetric synthesis of optically pure tertiary mono- and diphosphine ligands based on the regio- and stereoselectivity of the ring opening of the oxazaphospholidine–borane complex.[155,156] The diastereomerically pure borane complexes **31** were prepared in one step from bis(diethylamino) phenylphosphine, (–)-ephedrine and $BH_3 \cdot SMe_2$. The complex **31** was isolated as a crystalline product of *2R,4S,5R*-configuration. The structure of the complex was confirmed by X-ray analysis. Alkyl and aryl lithium compounds cleanly reacted with the borane complex at low temperature in THF to give the corresponding aminophosphine boranes **32** by P–O bond cleavage with a diastereomeric ratio better than 92:8. The X-ray crystal structure revealed that the ring cleavage of the borane complex proceeds with retention of configuration at the phosphorus atom. The acid methanolysis of the aminophosphine boranes **33** gave quantitatively the corresponding phosphinites. The reaction of the phosphinite **33** with alkyl or aryl lithium compounds in THF leads to the optically active phosphine boranes **34** easily isolated in good yield and stored without any special care. Enantiomeric excesses in this case are 85–100%. The transformaton of the phosphine boranes **34** into the corresponding phosphines was carried out quantitatively without loss of chirality by decomplexation under mild conditions using diethylamine.[157,158]

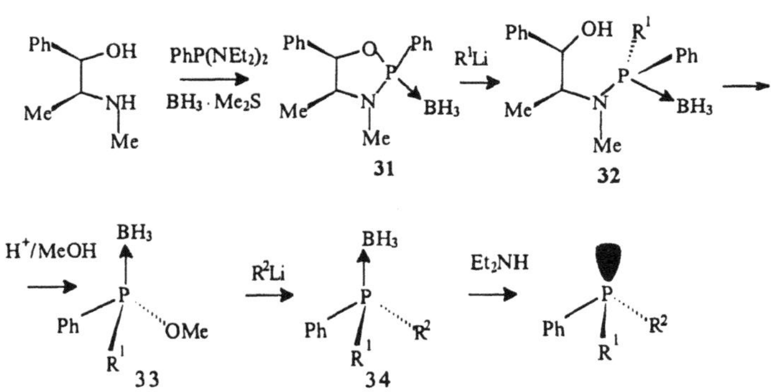

The phosphine boranes could be converted into 1,2-diphosphinoethane complexes by oxidative coupling and afterwards into diphosphines, following the procedure described recently by Imamoto and coworkers[143,145] The reaction of the 3,4-dimethyl-2,5-diphenyl-1,3,2-oxazaphospholidine borane complex with organometallic reagents involves a stereoselective bond cleavage. The P–C bond formation proceeds with a retention of configuration, which is explained by a nucleophilic attack on the less hindered side-face of the pyramidal phosphorus

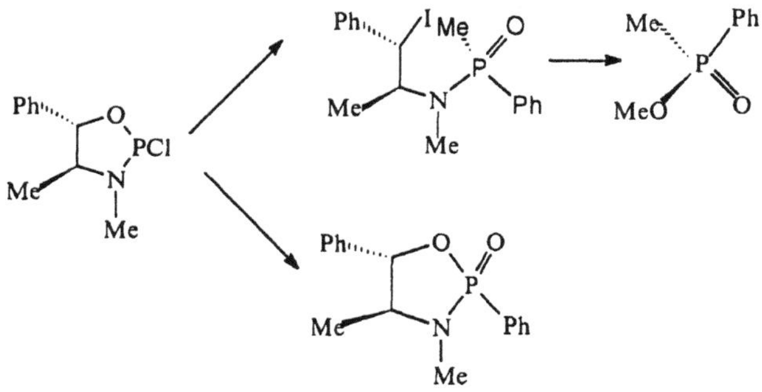

Scheme 6.

atom. Phosphine–boranes could be utilized as electrophiles in a stereospecific sequence through which one aryl and one alkyl group can be introduced sequentially (Scheme 6).[143,146] This route provides the best practical source of the P-chiral diphosphine DIPAMP.[143,146] Antipodal phosphines can be obtained from ephedrine derivatives via the phosphinite boranes and aminophosphine boranes, respectively, prepared from the same starting complex (Scheme 7).

One-pot synthesis of phosphine–boranes from phosphine oxides and a new reagent system, $LiAlH_4$–$NaBH_4$–$CeCl_3$, was described by Imamoto. On the basis of this unique reactivity of phosphine–boranes, a route to optically pure ligands in asymmetric hydrogenation has been developed (Scheme 8).[143]

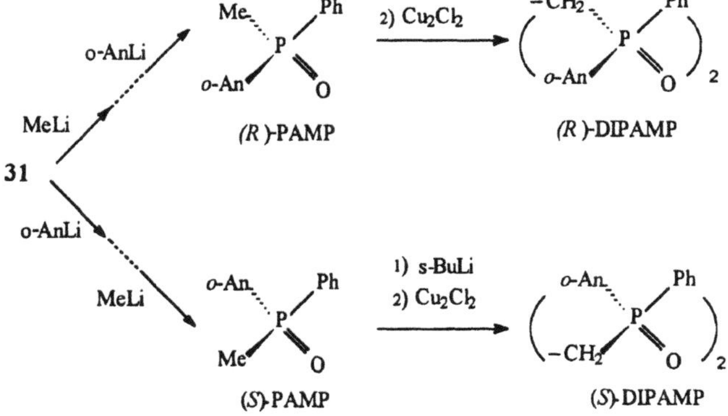

Scheme 7.

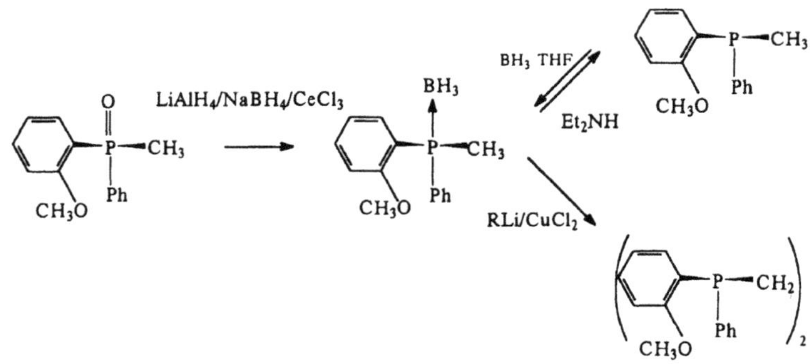

Scheme 8.

C. Chiral Tervalent Phosphorus Compounds

Tertiary Phosphines

Chiral monophosphines are a useful class of reagents for organic synthesis as well as of ligands for enantioselective catalysis. Chiral phosphine ligands are of key importance in catalytic transition metal-mediated processes which attract considerable attention in asymmetric synthesis and the industrial application thereof. The characteristic features of the asymmetric organometallic catalyses are their high stereo- and chemoselectivity applicable to a large-scale production. A number of methods have been developed for their synthesis and a large variety of different structures with one chiral center either at the central phosphorus or at the β-carbon atom in the side chain of an alkyl substituent have been prepared.[235a] A review dedicated to scalemic tertiary phosphines has been recently published.[242]

Cyclic Phosphines and Related Substances

Cyclic phosphines seem to offer some significant advantages over their acyclic analogues in that the restricted number of possible conformations allows a better definition of the relative position of the various substituents and thus a better understanding of the chiral environment at phosphorus. Moreover, the restricted rotational freedom of the ligand is expected to improve enantioselectivity. Several examples of phosphines are known in which a chiral phosphorus atom is incorporated into a ring system.

Marinetti and Ricard developed a simple methodology for the synthesis of *P*-phenylphosphetanes having known stereochemistry at several chiral centers, including phosphorus. When 2,3,3-trimethyl-1-butene was reacted with the MenPCl$_2$-AlCl$_3$ complex the two isomeric phosphetanium salts **35** were formed in equal amount. Hydrolysis afforded the diastereomers of phosphetane oxides **36** which were separated by column chromatography and crystallization. The derived chiral phosphines **37** have been tested as ligands in olefin hydrogenation reactions and in the palladium-catalyzed hydrosilylation of styrene and cyclopentadiene.[209,210]

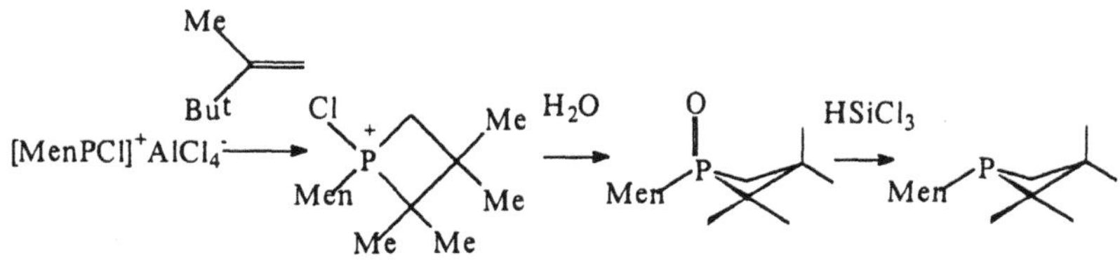

35 **36** **37**

P-Menthylphosphiranes **38,39** have been synthesized by Richter in a reaction involving treatment of dichloromenthylphosphine with achiral butadienylmagnesium at lowered temperature. The phosphiranes were obtained as a mixture of four diasteremers with a *cis/trans* ratio of 9:1. The level of asymmetric induction in the formation of the major isomer, *cis*-2-vinylphosphirane, was 82%.[256]

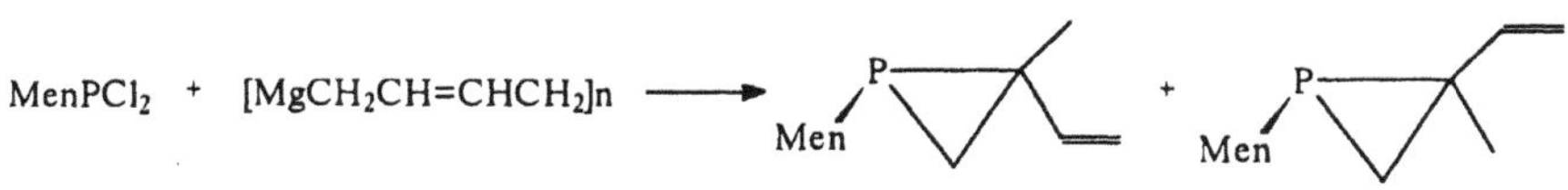

38 **39**

Ephedrine Derivatives. During the last few years the stereochemistry of organophosphorus heterocycles representing derivatives of ephedrine was intensively investigated. A variety of studies have been dedicated to the use of (*1R,2S*)-ephedrine as a chiral auxiliary, following from the early pioneering observations of Inch and coworkers in this field.[112] These workers reported the isolation of compounds in an isomerically pure form as a distillable liquid and they assigned the structure as being monomeric and having the geometry in which the chlorine atom has the *trans* geometry to the CPh and CMe substituents.[113] The chlorophosphoramidate **40** was obtained by reaction of (*1R,2S*)-ephedrine with PCl_3 in the presence of *N*-methyl morpholine at –78 °C (in 90% de). The chloridate **40** was distilled as a single diastereomer (although evidence for the presence of some *cis*-isomer in the crude reaction mixture was obtained), which with methanol in the presence of triethylamine afforded the methyl ester **41**. The reaction between these compounds and sulfur or selenium gave thio(selene) derivatives **42**.[113,63] This simple synthesis demonstrates the feasibility of a route which formally commences with PCl_3.

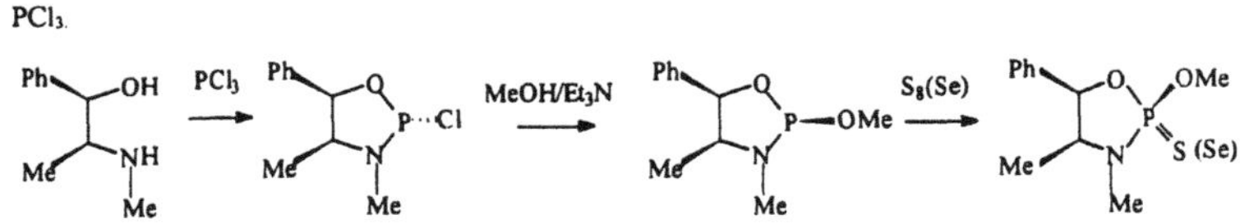

40 41 42

Brown and coworkers prepared oxazaphospholidine **39** by reaction of commercialy available phenyldichlorophosphine with (–)-ephedrine in the presence of two equivalents of *N*-methylmorpholine in toluene at 0 °C. Initialy there was evidence for two diastereomeric compounds in the reaction mixture in ca. 50:50 ratio, but after stirring for a day at ambient temperature only one diastereomer **39** was obtained.[41] This compound was isolated as a low-melting solid after distillation again as a single isomer. The stereoisomerically pure oxazaphospholidine **39** formed by reaction of *(1R,2S)*-ephedrine with $PhPCl_2$ was oxidized by *tert*-butylhydroperoxide to the corresponding P-oxide **40**, shown to have *R* stereochemistry at phosphorus by X-ray analysis. This product reacted stereospecifically with *o*-anysilmagnesium bromide to give the acyclic phosphonite formed by P–O fission with retention of configuration. The ephedrine residue was replaced by O-methyl under acid catalysis with inversion of configuration.[53,157,158,170]

Juge and Genet had prepared pure oxazaphospholidines **39** from bis(diethylamino)phenylphosphine and (–)-ephedrine in toluene at 100 °C, when one pure diastereomer was formed, presumably through thermodynamic control. The reaction provides the phosphines in yields up to 82% de.[156] Reaction of the chlorophos-

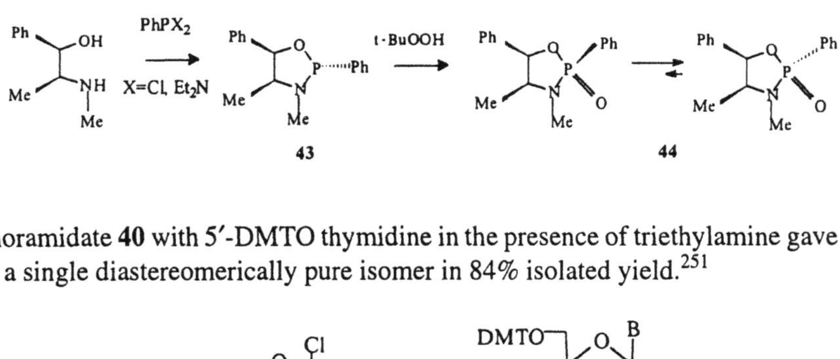

43 44

phoramidate **40** with 5′-DMTO thymidine in the presence of triethylamine gave **45** as a single diastereomerically pure isomer in 84% isolated yield.[251]

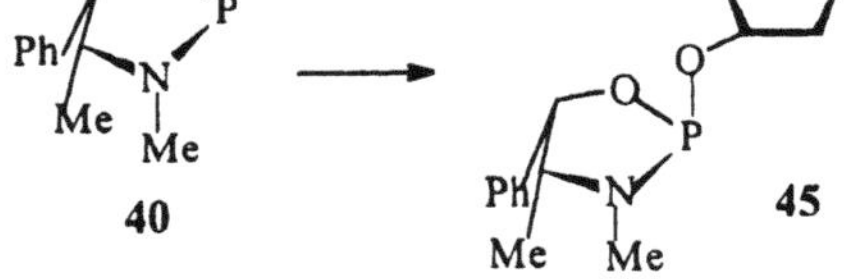

40 45

The conformations of the 2-(*Z*)-phenyl-1,3,2-oxazaphosphorinanes were investigated by ^1H and ^{31}P NMR spectroscopy and X-ray crystallography.[141] X-ray and

NMR studies of the 5,5-dimethyl-2,3-diphenyl-1,3,2-oxazaphosphorinane **46** revealed a chair conformer with the phenyl group attached axially to the phosphorus atom. For the 1,3,2-oxazaphosphorinanes **47** with a Me_2N substituent on phosphorus a chair–chair equilibrium was found in solution that features an 80–90% population of the Me_2N axial conformation. The *cis*-compound displays a conformational equilibrium involving mainly a chair conformer with the *t*-Bu equatorial and Me_2N axial. The axial orientation of the Me_2N group was confirmed also by X-ray analysis. The diastereomeric molecules of 1,3,2-oxazaphosphorinanes with a *cis*- or *trans-tert*-butyl group exist in thermodynamic equilibrium at room temperature.

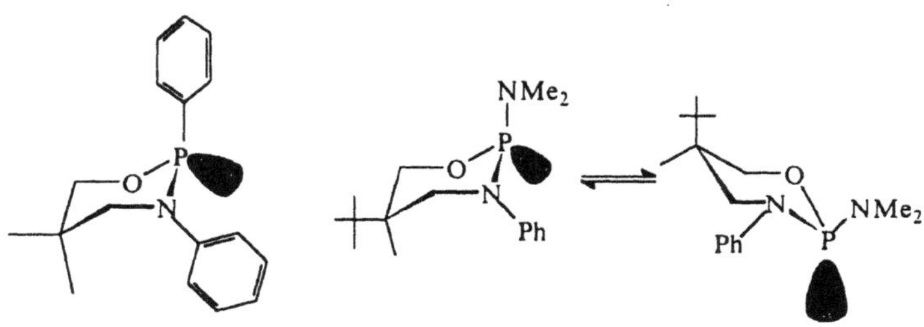

46 **47**

Sum and Kee showed by mass spectroscopy that 1,3,2-oxazaphosphorinanes exist as dimers. The chiral 2-triorganosiloxy-1,3,2-oxazaphospholidines have been prepared via the reaction of (1*R*,2*S*-ephedrine)PCl with R_3SiOH in the presence of Et_3N. These esters exists as an equilibrium mixture of two epimers with a diastereoselectivity that is dependent upon the nature of the silicon substituents.[267,268]

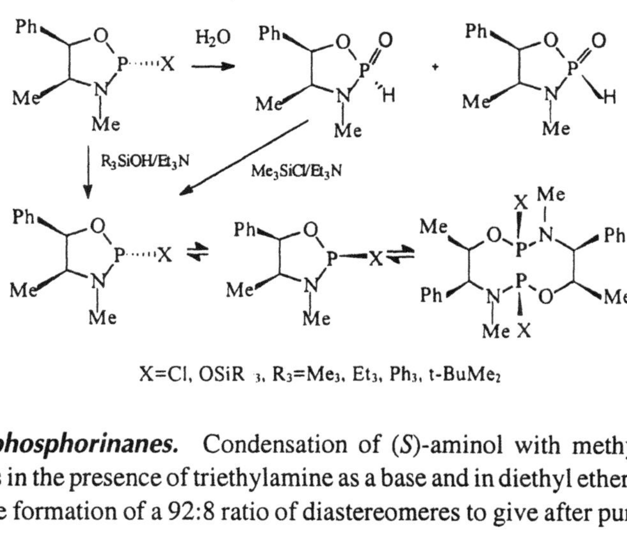

Oxazaphosphorinanes. Condensation of (*S*)-aminol with methyl dichlorophosphites in the presence of triethylamine as a base and in diethyl ether as a solvent leads to the formation of a 92:8 ratio of diastereomeres to give after purification by vacuum distillation *trans*- and *cis*-2-methyl-3-isopropyl-6-methyl-1,3,2-oxazaphosphorinanes **48**. The major product was assigned the *trans* stereochemistry on the basis of its higher field NMR resonance.[107]

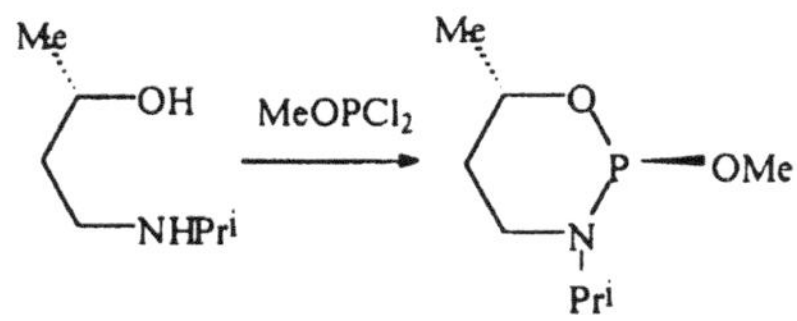

48

Denmark et al.[81] performed the reaction of the aminoalcohol **48** with ethyl dichlorophosphite and triethylamine in refluxing dichloromethane leading to the ratio of diastereomers of 28:1. A chair–chair conformational equilibrium was shown for a number of 1,3,2-oxazaphosphorinanes.[141]

C_2 Symmetric Compounds

In the majority of scenarios for absolute stereochemical control the presence of a C_2 symmetry axis within the chiral auxiliary can serve the very important function of dramatically reducing the number of possible competing, diastereomeric transition states.[296] It is almost universally observed that auxiliaries with C_2 symmetry elements perform in their capacity as stereochemical directors to provide higher levels of absolute stereochemical control as compared to those totally lacking in symmetry. A large collection of molecules with C_2 symmetry were introduced in organophosphorus chemistry as chiral auxiliaries. Some of them will be reviewed in this chapter. At this time we must inform the reader that a part of this area has been quite thoroughly reviewed and therefore will not be surveyed again here (DIOP, CHIRAPHOS, DIPAMP and other ligands, described in the literature).[50,162,163,296]

Some of the most interesting C_2 symmetric auxiliaries are certainly the topologically unique chiral bicyclic phosphonamides **49**,**50**. They are easily available from enantiomeric C_2 symmetrical (*R*,*R*)- or (*S*,*S*)-1,2-diaminoalkanes and diaminocycloalkanes. The synthesis of chiral bicyclic phosphonamides **50** starting from known, optically active (*R*,*R*)-1,2-diaminocyclohexane was described by Hanessian and coworkers.[19]

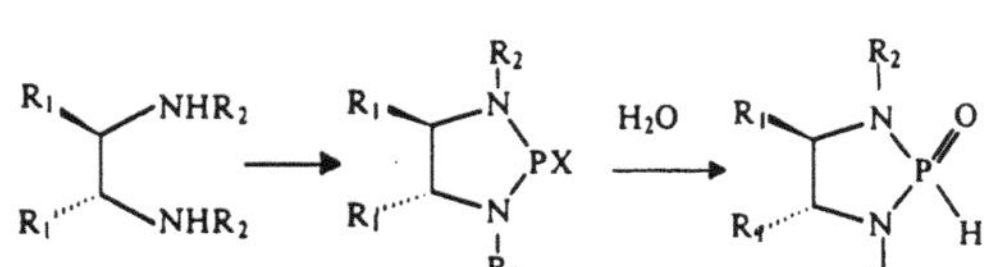

49

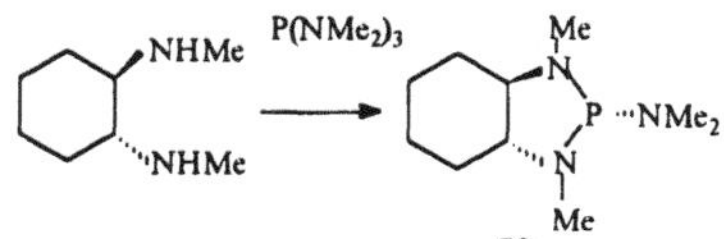

50

(*4R,5R*)-δ-Carboalkoxy-2-chloro 1,3,2-dioxaphospholane (**51**) was prepared by reaction of PCl_3 with a dialkyl tartrate in refluxing THF in very high yield. 1,3,2-Dioxaphospholane (**51**) represents a colorless pentane-soluble liquid. Only a single isomer of this compound was registered by NMR due to the C_2 symmetry of the tartrate ester.[247] The reagents are unique in that either retention or inversion at phosphorus occurs during derivatization.[48]

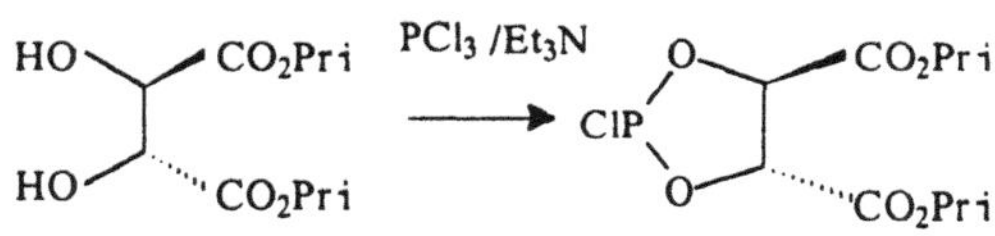

51

Compounds possessing C_2 or higher symmetries and whose chirality originates from the geometry of the whole molecule, rather than from the presence of stereogenic carbon centers are very useful. Starting from taddol Seebach et al.[262] prepared different bicyclic phosphites and phosphonites. These chiral compounds, containing phosphorus atom are useful ligands and auxiliaries for enantioselective synthesis. Bicyclic phosphates and phosphonites have been prepared by reaction from α,α,α,α-tetraaryl-2,2-dimethyl-1,3-dioxolan-4,5-dimethanols with Cl_2PR and Cl_2POR derivatives.[262]

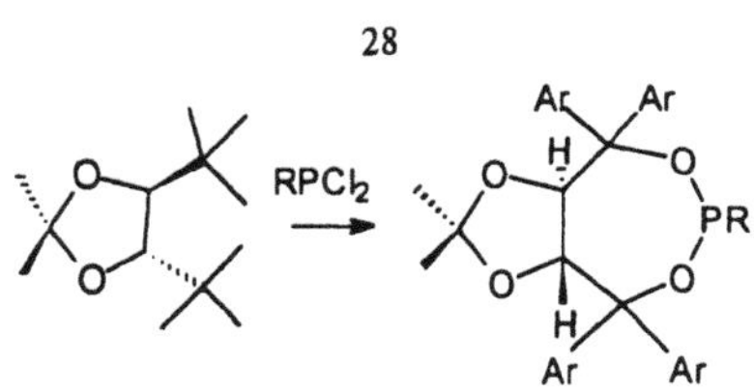

52

Ar=Ph, 2 Napht; R=Me, Ph, MeO, PhO, Ph

Phospholes of type **53** are atropisomerically chiral molecules which may experience two different fluxional processes: pyramidal inversion at phosphorus and flipping of the naphthyl rings.[94] The former is a chirality-invariant high-energy process for which an energy barrier of about 130 kJ/mol can be expected, while the second involves net inversion of configuration. Diastereomeres **53** (R^* = $CH_2CH(Me)Et$, neomenthyl) showed in ^{31}P NMR at low temperature two separate resonances, which coalesced at ca. 10 °C into a single unresolved peak.[90]

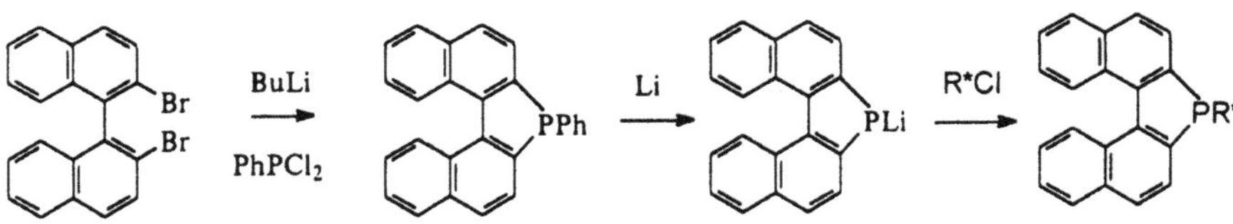

53

Rapid ring inversion (atropisomerization) in bisphenol–phosphorus moieties **54** occurs on the NMR time scale since the expected diastereomers could not be detected by low-temperature phosphorus NMR.[240] The compounds **54** were all stable during purification on silica gel and were isolated as white solids. This contrasts with racemic bis(naphtol)phosphorus chlorides **55**. The intrinsically hindered rotation around the 2,2-dinaphtyl linkage gave rise to mixtures of diastereomers **55**.[51] Sterically hindered bis-phenol-hindered chlorides **56** is configurationaly stable.[14] A novel class of C_2-symmetric ligands has been prepared using an intramolecular cyclization of a diazaphosphole borane complex **57**, which can be applyed to the asymmetric catalysis of hydrogenation, hydrosilylation, and allylic substitution reactions.

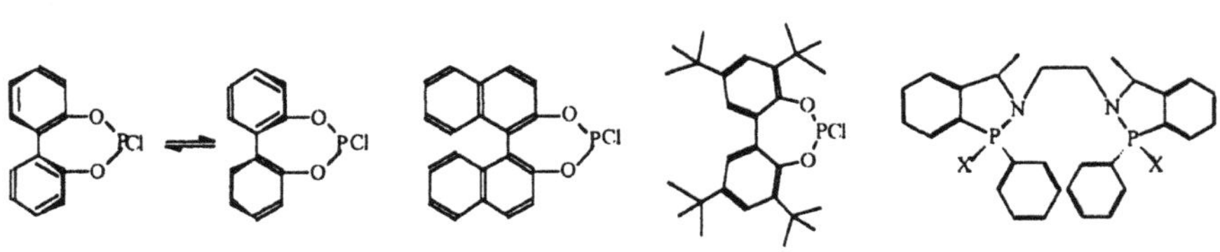

54 **55** **56** **57**

Baker and Pringle[14] described a simple, one-step procedure for the synthesis of an optically active, tetradentate ligand **58** having C_3 symmetry and its platinum(0) coordination chemistry. The phosphino-phenol reacts with chlorophosphite in the presence of Et_3N to give the optically active tridentate ligand which upon addition of [Pt(norbornene)$_3$] forms the bis-chelate platinum(0) complex. Optically active tetraphos ligands of C_3 symmetry have attracted much interest because of their great potential in asymmetyric catalysis.[14]

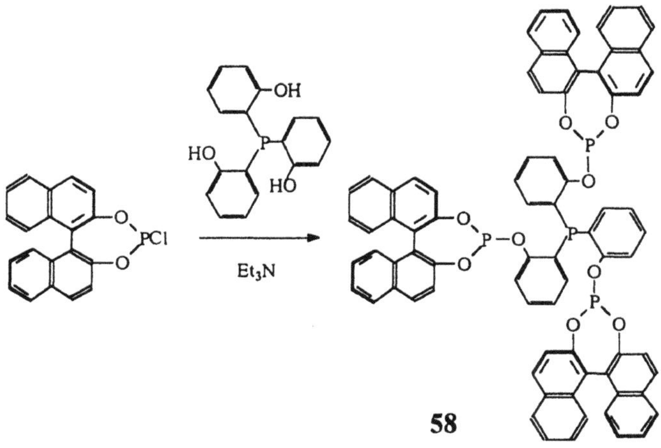

58

Several organophosphorus chiral derivatizing reagents for the analysis of chiral alcohols by ^{31}P NMR have been successfully used. The synthesis and use of enantiomerically pure five-membered rings—phospholanes—have been successfully developed by several groups, who have both specifically targeted species with C_2 symmetry and with achiral phosphorus. Thus, 1,2,3-dioxaphospholane was proposed by Anderson.[8] This reagent is unique in that the phosphorous atom is not chiral due to the C_2 symmetry of the chiral glycol. Therefore either retention or inversion at phosphorous during derivatization of an enantiomerically pure alcohol yields a single diastereomer.

New C_2 symmetric reagents **59–62** useful for simple and efficient determination of enantiomerical purity of alcohols by ^{31}P NMR was proposed by Alexakis and coworkers.[5–7] A large array of primary, secondary, and tertiary alcohols, functionalized or not, were successfully tested. The method is accurate and very general, and no kinetic discrimination is observed. Sulfuration or selenation of the trivalent phosphorus derivatives, carried out in the NMR tube, allows for a second ^{31}P NMR determination, in addition to the ^1H, ^{13}C and ^{19}F NMR spectra which were recorded.

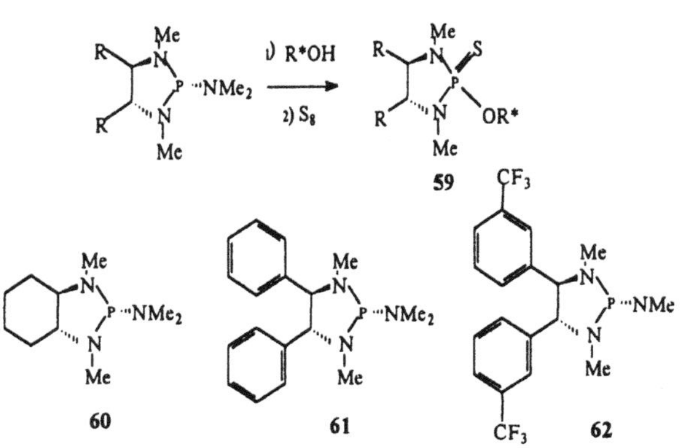

Economic and efficient chiral derivatizing agents for the determination of enantiomeric excesses of chiral alcohols by ^{31}P NMR were developed by Brunel and coworkers on the basis of *(4R,5R)*-δ-carboalkoxy-2-chloro-1,3,2-dioxaphospholane (**51**). The reagents are unique in that either retention or inversion at phosphorus during derivatization of an enantiomerically pure alcohol yields a single diastereomer due to the C_2 symmetry of the dialkyl tartrate ligand on phosphorus.[48]

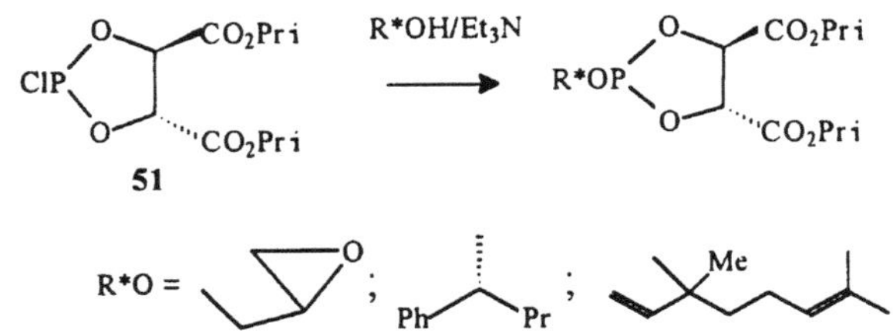

A reagent for the determination of enantiomeric excess of diols and diamines with C_2 symmetry was proposed by Brunel and Faure on the basis of menthyl dichlorophosphite.[49] The derivatizing reagent **63** easily prepared from PCl_3 and (*1R,2S,5R*)-(–)-menthol, allows an accurate analysis of the diastereomeric derivatives by ^{31}P NMR spectroscopy. Sterically hindered bisphenolphosphorus chloride **56** was proposed as a reagent for the determination of enantiomeric composition of alcohols.[51]

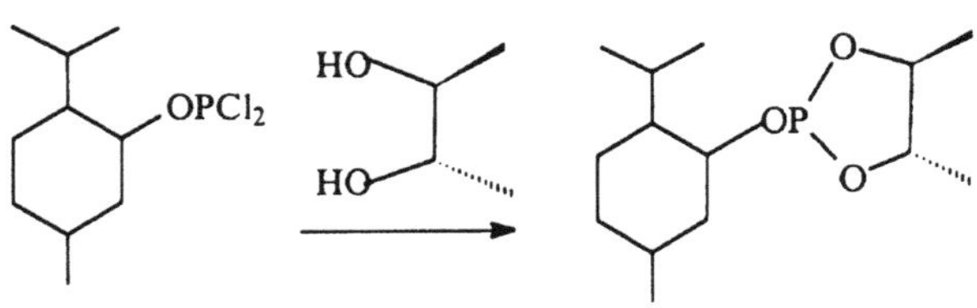

V. ASYMMETRIC SYNTHESIS OF TETRACOORDINATE PHOSPHORUS COMPOUNDS

Compounds of tetracoordinate pentavalent phosphorus are the most important in the chemistry of phosphorus and as well in organophosphorus stereochemistry. These compounds are widely used as biologically active compounds and enantioselective reagents. Tetracoordinate organophosphorus compounds to be effective synthetic tools for asymmetric carbon–carbon bond formation are required in optically active form.

A. Configurational Stability

Tetracoordinate organophosphorus compounds exhibit in general high configurational stability, which depends on the structure of these compounds. Tertiary phosphines are the most stable. Esters of chiral phosphorus acids are configurationally stable, but racemize slowly upon heating. Diastereomers of menthyl benzylphosphonic acid at 120 °C are smoothly epimerized at the phosphorus atom to give a mixture of diastereomers in a ratio of 1:1.[35] Chlorides of phosphorus acids are less stable and can be racemizated at room temperature in the presence of nucleophilic agents. This is related to the formation of bipyramidal pentacoordinated phosphorus species that invert their configuration.

Racemization, which results in the loss of optical activity of chiral compounds, is considered to be one of the fundamental processes in dynamic stereochemistry. Quite generally, racemization can be caused by supplying adequate energy by heating or irradiation, or it may be effected by chemical reactions.

B. Nucleophilic Substitution

The most frequently encountered reaction in organic phosphorus chemistry is the nucleophilic displacement reactions.

Mechanism

The mechanism and steric course of nucleophilic substitution at the tetracoordinate phosphorus atom have been the main points of interest of many research groups: McEwen,[216,217] Mislow,[43,199] DeBruin and Johnson,[43–47] Trippett,[71,282] etc. The results of these studies have been discussed in many reviews on organophosphorus chemistry. Therefore only the most representative examples of nucleophilic substitution at chiral phosphorus concerning stereochemistry are discussed here.[20,32,38,43,45,71,200,282]

Pentacoordinated phosphorus compounds possess the geometry of a trigonal bipyramid, which can be formed via equatorial attack or via apical attack (Scheme 9). It is customary to consider that such reactions occur synchronously by an S_N2P mechanism involving a trigonal-bipyramidal phosphorane intermediate. The latter is formed by addition of the nucleophile (Nu) opposite to the leaving group (L) occupying the apical position and is decomposed before any ligand pseudorotation has taken place.

A stereochemical parthway proposed by McEwen includes synchronous attack with elimination of the anion by second a hydroxide ion or with formation of the unstable hexacoordinated intermediate bearing two hydroxides connected with the phosphorus atom. The reaction scheme includes two intermediates: one is pentacoordinated and the other is hexacoordinated. The stereochemical result is defined by the structure, by the lifetime, and by the reorganization of ligands of these intermediate products (Scheme 10).[218]

Westheimer,[222,274,275,294] Ruedl[106] report that enzymatic reactions at P proceed with inversion and, therefore, occur without pseudorotation. In fact, there is no unambiguous evidence that pseudorotation or adjacent attack at the P atom is a

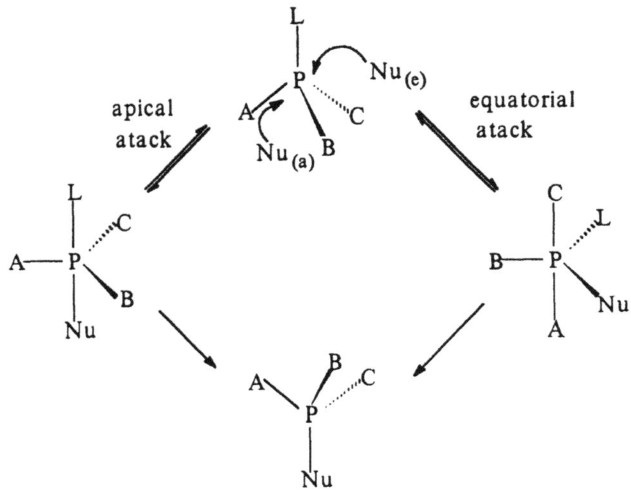

Scheme 9.

process of significance in any biological system, and formal retention is rationalized by a multistereo process with an even number of inversions.

Enantioselective Transformations

Reaction with Organometallic Compounds. The most popular means for preparing of optically active phosphine oxides is the method developed by Mislow[194,199] based on the reaction between the diastereomerically pure (or strongly enriched in one diastereomer) menthylphosphinate ester and Grignard reagents. The nucleophilic displacement of a menthyl phosphinate ester with Grignard reagents proceeds with clean inversion of stereochemistry at phosphorus to afford the homochiral phosphine oxides. The reaction may give racemic products or fail

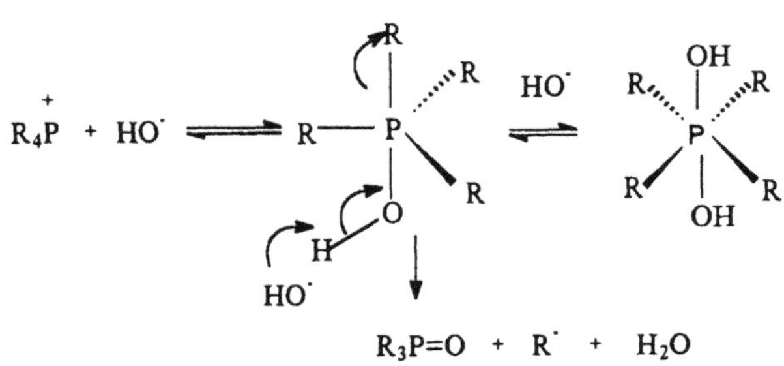

Scheme 10.

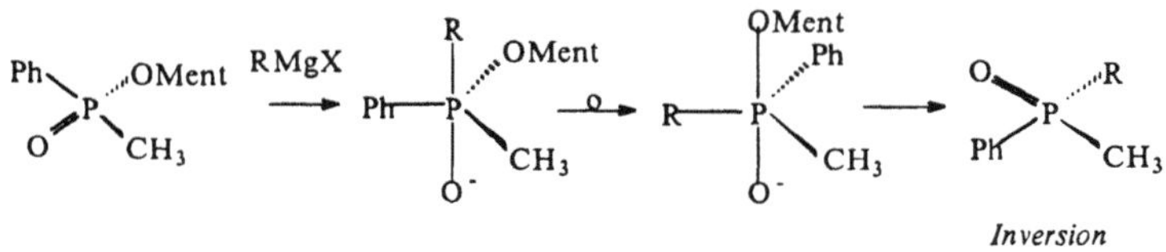

completely under the drastic reaction condition required when there are bulky groups on either phosphorus or the Grignard reagents. Organolithium reagents are more reactive but epimerization at phosphorus may occur.[148,194–196,199,200,298]

The synthesis of optically pure phosphines and phosphine oxides with a stereogenic center at phosphorus has attracted much attention. For example, the reaction of *o*-anisylmagnesium bromide with the optically active phosphinite gives (+)-(R)-O-anisylphenylmethylphosphine oxide (95% ee), which was used for the synthesis of chiral organophosphorus ligands **64** and others.[25,58,63,113,156,296]

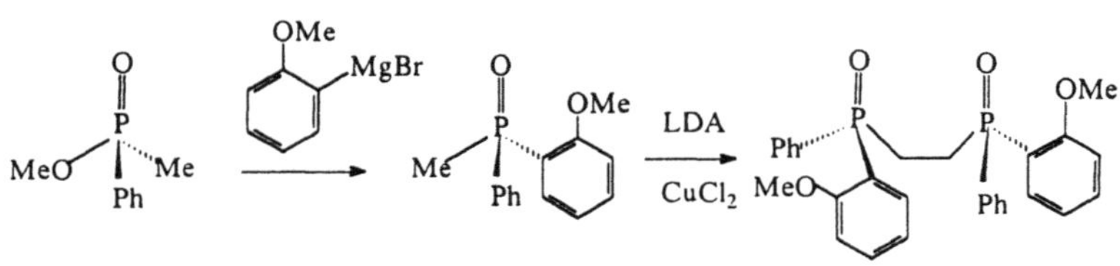

64

During the last 20 years several attempts have been made to improve this route, and to develop alternative approaches to homochiral phosphines and related compounds. The basic chemistry of Mislow was developed by Horner and coworkers who synthesized by this method a variety of monophosphine oxides with different aryl groups in optically active form.[136,138] Inch proposed a synthesis of chiral phosphinates from 1,3,2-dioxaphospholanes **65** prepared from carbohydrates.[112,113]

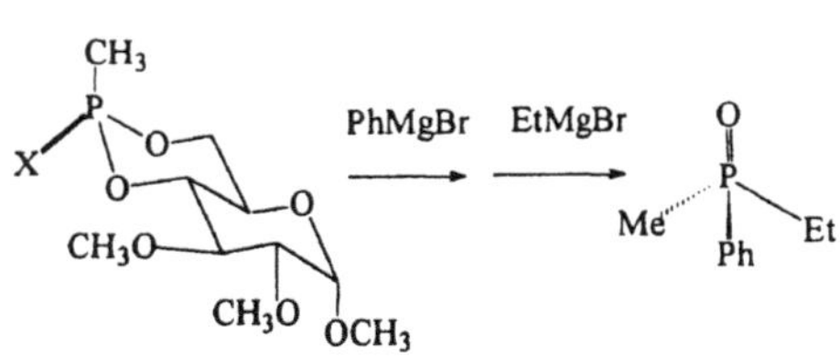

65

The synthesis of enantiomerically pure triaryl- and diarylvinylphosphine oxides from PCl_3 by three sequential nucleophilic displacements at phosphorus has been demonstrated. A single diastereoisomer of the *P*-chlorooxazaphospholidine was treated with an arylmagnesium halide to effect displacement of chloride. The major diastereoisomer is formed with retention of configuration. The stereochemistry of displacement with arylmagnesium halides is consistent with a model in which the organomagnesium reagent is compelled to the ox-group and attacks phosphorus *cis* to it in a direction defined by both electronic and steric factors. Three sequential nucleophilic displacements provide the three new P–C bonds of a chiral triarylphosphine, based on the earlier discovery of the stereospecific P–O cleavage of oxazaphospholidines with arylmagnesium chloride.[41] The cyclic ephedrine derivatives **66** react with organomagnesium reagents in an opposite stereochemical sense to the open-chain examples of Mislow. The reasons for this stereochemical divergence must require five-coordinate intemediates **E, F** in an associative pathway with the opportunity for pseudorotation.[53,156]

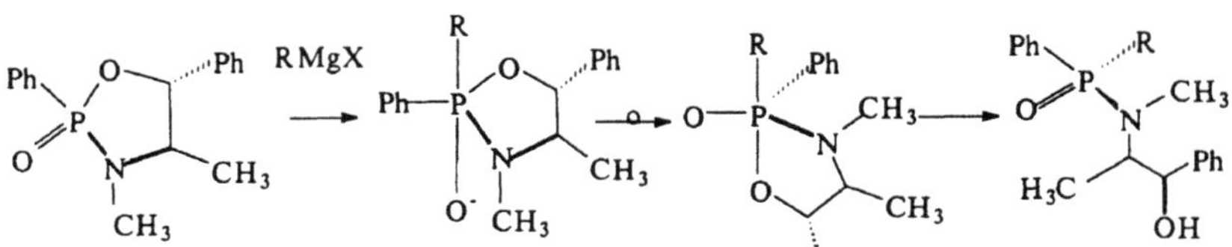

66 **E** **F** **67**

Koizumi and coworkers have studied the diastereomerically pure oxazaphospholidine oxide and sulfide **68** and demonstrated that they react with Grignard reagents. The optical purity of the products **69** was 73–99%.[173]

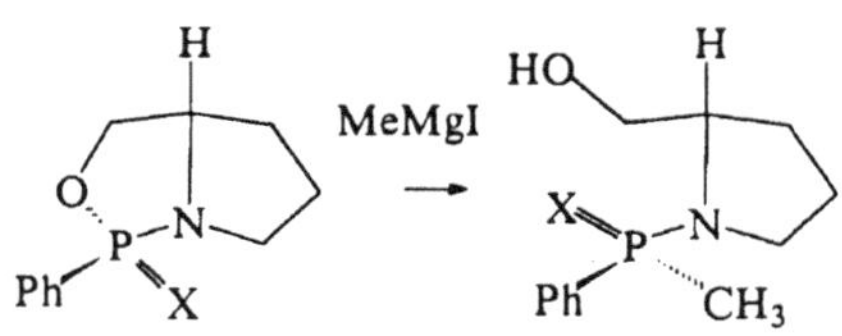

68 **69**

The synthesis of enantiomericaly pure tertiary phosphines using camphor derivatives **70** was developed by Corey.[66]

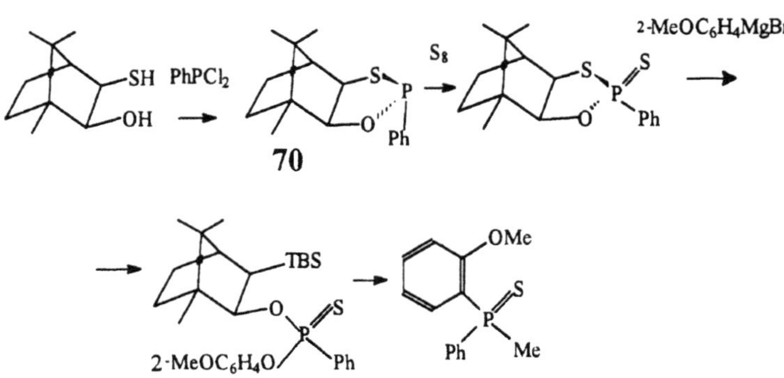

Reaction of Phosphorus(IV) Chlorides with Chiral Substrates. Since the pioneering work of Mislow,[99,194] the synthesis of optically pure phosphines and phosphine oxides with a stereogenic center at phosphorus has attracted much attention. Methods based on the fractional crystallization of the diastereomers of *O*-menthyl phenylphosphinate with stereospecific displacement and reduction have been developed. Nudelmann and Cram[236] for the first time, and then Mislow and coworkers[195,196] demonstrated that unsymmetrically substituted menthylphosphinates could be separated readily into the diastereoisomeres.

Knowles found that the reaction of phenyl(4-methoxyphenyl)phosphinic chloride with (–)-menthol proceeds with the formation of a 4:1 mixture of diastereomers, which can then be separated by crystallization.[193,290] The reaction of

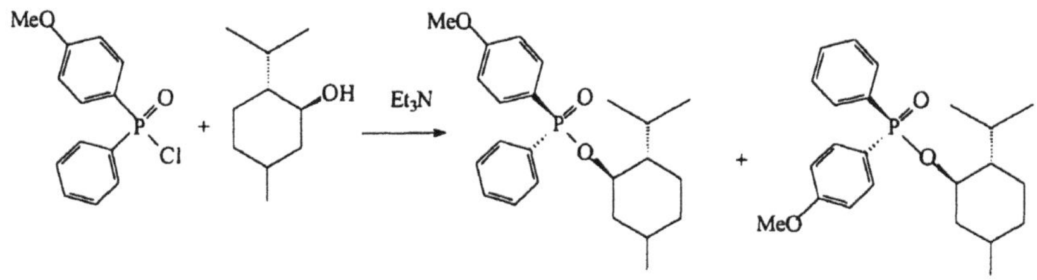

71 72

phenylmethylphosphinothiochloride with (–)-menthol gives a mixture of diastereomers which can be easily separated by crystallization from hexane. The reaction of racemic chlorides with menthol in the presence of base has been reported to give a reaction mixture. The ester had the correct phosphorus configuration for conversion by successive reduction to the desired chiral tertiary phosphine, which is a ligand used to obtain (*S*)-amino acids in a rhodium phosphine complex-catalyzed asymmetric hydrogenation (Scheme 11).[193,284,290]

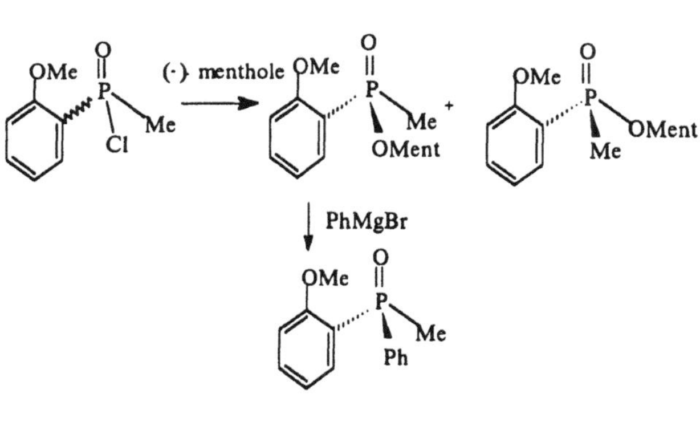

Scheme 11.

While stereochemically reliable, this approach has several limitations. The diastereomeric menthylphosphinate esters are prepared with low kinetic selectivity (2 or 3:1) and require separation before the next step. This becomes very tedious for liquid phosphinates, which includes most alkanephosphinate esters. Therefore, some important classes of phosphinoxides are effectively inaccessible by this method. The low kinetic selectivity observed in the preparation of menthyl methylphenylphosphinate can be circumvented by using an epimerization to equilibrate the diastereomeric esters.

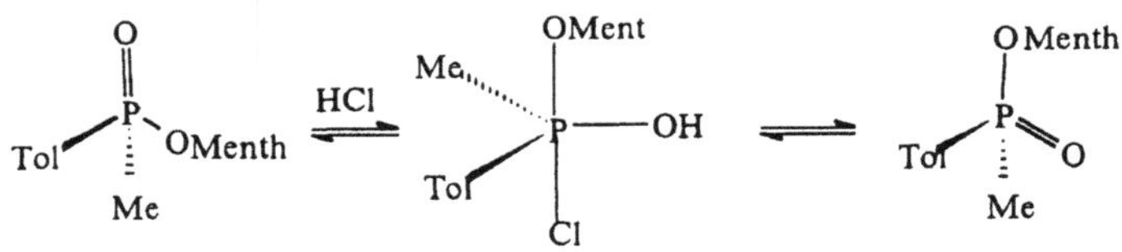

A promising and potentially inexpensive route to homochiral phosphinyl oxides, using diisopropylidene–glucofuranose as the source of chirality was disclosed by the author of the review and his coworkers.[175,176,182,185,187] Either phosphinate diastereomer can be prepared in excellent yield and with high diastereoselectivity by the appropriate choice of base. The reaction of the tetracoordinate phosphorus chlorides with D-glucofuranose in the presence of tertiary amines provides stereochemically pure (*S*)-phosphinates **73**.

37

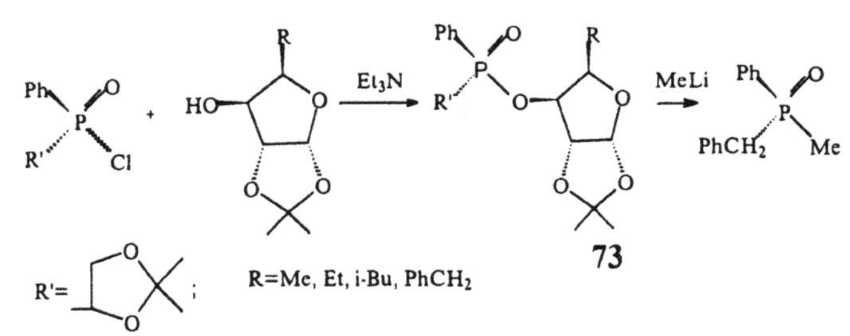

The reaction of phosphinic acid chlorides with D-glucofuranose was performed using an equimolecular ratio of reagents with a one and a half excess of triethylamine in toluene at room temperature for 12–24 h. The yields of phosphinates were 70–75% and the diastereomeric excesses were 80-~100%. The products were purified by crystallization from hexane. The ratio of diastereomers was not dependent on excess of chlorophosphinate, corresponding to thermodynamic reaction control. This was unlike the reaction of chlorides of trivalent phosphorus with glucofuranose(II), which proceeds under kinetic control. The stereoselectivity influenced by the nature of the base and the solvent with the highest stereoselectivity being achieved in toluene with triethylamine as a base.

These studies were later continued by Alcudia and coworkers who performed the reaction with a 10-fold excess of triethylamine and a threefold excess of chlorophosphinate. They prepared phosphinates by reaction with Grignard reagents and converted them into tertiary phosphines oxides.[22]

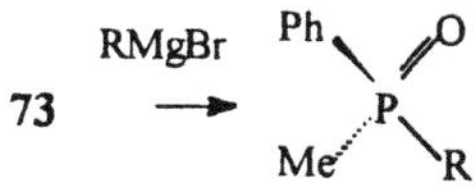

R=Et, i-Pr

Phosphorylated derivatives of 1,2:5,6-disubstituted D-glucofuranose, including derivatives asymmetric at the phosphorus atom, were prepared as well by Nifant'ev.[234] However this reaction proceeded with low stereoselectivity, because optimal experimental conditions were not found.

Reaction of phosphinates with alkyl lithiums proceeds with inversion of configuration at the phosphorus atom to provide optically active homochiral tertiary phosphine oxides; as a chiral auxiliary (1*S*)-borneol was used.[259] Reaction of the racemic methyl(phenyl)phosphinic chloride with (1*S*)-borneol gave a 1:4 diastereomeric mixture of (1*S*)-bornyl (*Sp*)- and (*Rp*)-phosphinates. Apparently a kinetic racemate cleavage accompanied by a second order asymmetric transformation is involved. Accordingly the (*R*)-phosphinic chloride reacted with (1*S*)-borneol considerably more rapidly than the (*S*)-compound **75**, which is continuously rearranged into the (*R*)-phosphinic chloride. The diastereomeric mixture was separated by application of medium pressure chromatography. The reaction of the (–)-bornyl (*Rp*)-phosphinate with (*o*-bromomethoxyphenyl)magnesium bromide furnished (*R*)-(*o*-methoxyphenyl) methyl(phenyl) phosphine oxide.

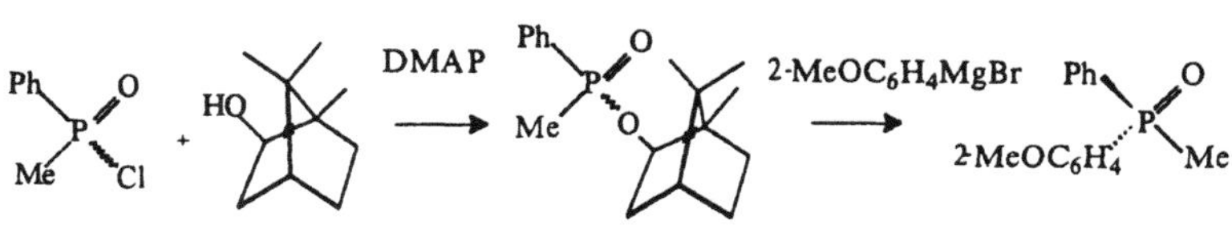

74

Diastereomeric phosphinate esters were also formed from racemic methyl(phenyl)phosphinic chloride with the terpene alcohols: (–)-menthol, (–)-isopinocampheol, and (+)-isoborneol in the ratios 1:1, 1:1, and 74:26, respectively.[259] Synthesis of menthyl amidophosphonates was described by Michalsky.[205]

The reaction of phosphorus(IV) chlorides with 2-arylethylamines proceeds analogously, but with less stereoselectivity. Thus, the diastereoselectivity of the reaction of two-fold excess of 1,1′-binaphtyl-2,2′-diylphosphonyl chloride with chiral amines depends on the reaction temperature, on the increased steric hindrance by the substituent on the carbon atom, α to the nitrogen atom, and on the

Table 2. Reaction of Racemic Chlorophosphinates with Chiral Secondary Alcohols: $RR'P(O)Cl + R^{*}OH \rightarrow RR'P(O)OR^{*}$

R	*R'*	*Nu*	*B*	*Ratio*	*Ref.*
Ph	Me	GF	Et_3N	90:10	187
Ph	Me	GF	Et_3N	97:3	22
Ph	Et	GF	Et_3N	96:4	187
Ph	*i*-Bu	GF	Et_3N	95:5	187
Ph	$PhCH_2$	GF	Et_3N	~100:0	181
Ph	Me	(1*S*)-borneol	DMAP	1:4	259
Ph	Me	(–)-menthole	DMAP	1:1	259
				2:1	285
Ph	Me	(–)-isopinocampheol	DMAP	1:1	259
Ph	Me	(+)-isoborneol	DMAP	74:26	259
4-$MeOC_6H_4$	Ph	menthol	Et_3N	4:1	193, 290

ortho-substitution of the benzene ring of the chiral amine. The rate of formation of one enantiomer of (*S*,*R*)-1,1′-binaphtyl-2,2′-diyl *N*-(*S*)-(α-methylbenzyl)phosphoramidate **76** is greater than that of the other (Table 3). The reaction is kinetically resolved, therefore half of the phosphoryl chloride remains unreacted.[291]

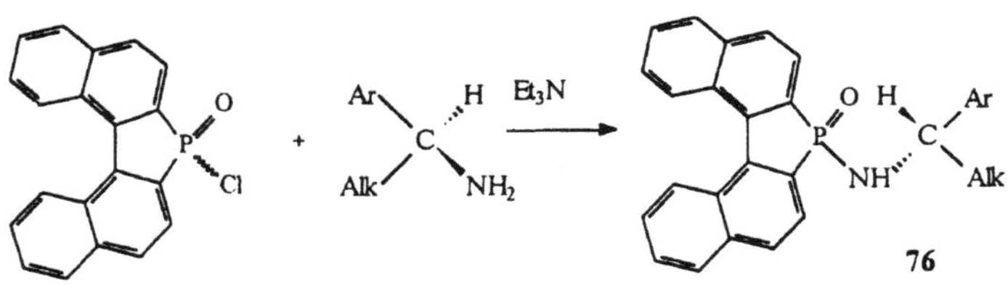

76

Table 3.

Alk.	*Ar*	*T* °C	de *of* **76** *(%)*
Me	Ph	0	27
		10	24
		20	20
i-Pr	4-ClC_6H_4	0	38
		10	31
		20	24
i-Pr	2-MeO-5-MeC_6H_3	0	82
		10	68
		20	33

Certain chlorophosphates react apparently stereospecifically with nucleophiles, without asymmetric induction at the phosphorus atom. For example, Hulst and coworkers propose a reagent **75** for the determination of enantiomeric excesses of alcohols, amines, and amino acids.[142] The reaction of **77** with nucleophiles proceeds stereospecifically.

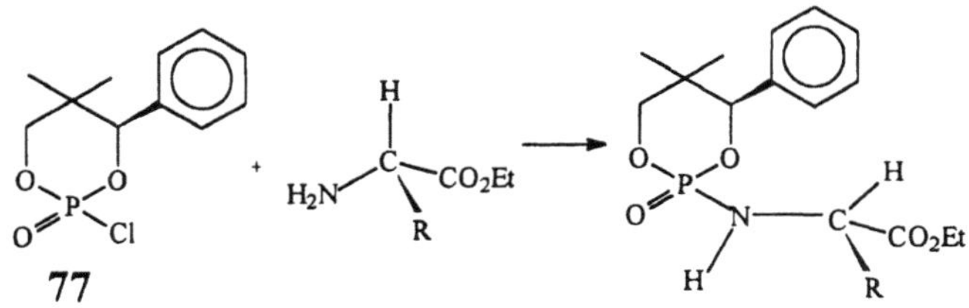

77

Reaction of phenyl(2,4,6-trimethylphenyl)phosphinic chloride with (+)-(*R*)-1-phenylethylamine in the presence of the triethylamine yields a mixture of diastereomers in the ratio 2.2:6.5. Each of the diastereomers was separated by column chromatography and was then converted by acidic methanolysis into enantiomerically pure methyl esters of phosphinic acids (Scheme 12).[114]

Reaction of the 2,2′-biphenyl diol with equimolecular quantities of thiophosphoryl chloride and *(S)*-methylbenzylamine in pyridine gives in virtually quantitative yield a enantiopure phosphorothioamidate, methanolysis, or ethanolysis of which in the presence of 4MH_2SO_4 provides alkylarylphosphorothioates **78** in quantitative yield. These were tested as agrochemicals.[72]

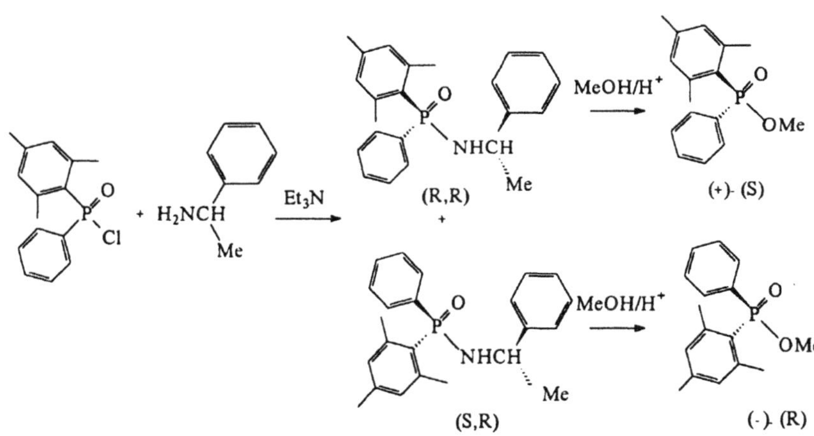

Scheme 12.

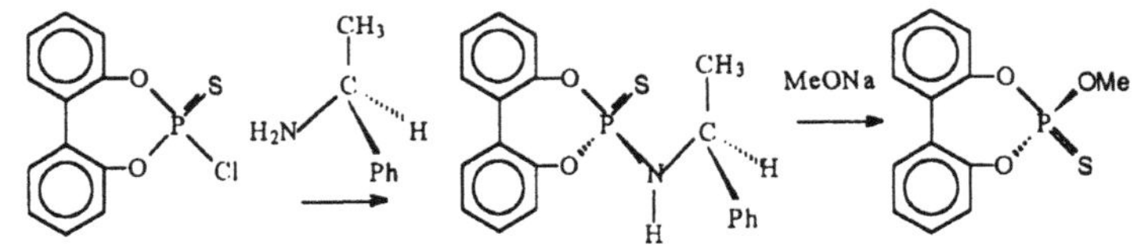

78

Reaction of the axialy substituted (±)-chlorides **79** with *O*- and *S*-nucleophiles in the presence DBN preferentially proceeds with retention of configuration at the phosphorus atom, whereas the epimer ratio is reversed with DBU as a base. *N*-Nucleophiles react exclusively with inversion. In the presence of DBU a pentacoordinate intermediate **80** was isolated as the main product (Scheme 13).[222]

Inch and coworkers showed that pyranoside phosphonates react stereospecifically in a two-step displacement sequence with different Grignard reagents giving optically pure phosphine oxides.[63,64]

Asymmetric Michaelis–Arbuzov Reaction. The Arbuzov approach has been adopted by many authors, using a range of chiral auxiliaries. Suga and coworkers have reacted the cyclic phenylphosphonite **81**, prepared from (*S*)-1,3-butanediol with phosphorus halides and demonstrated a regioselective ring opening by cleavage of the primary P–O bond.[164,263] The same group has utilized cyclic phosphonites **82** and **83** in a similar sequence, but here the enantiomeric purity of the resulting phosphine oxides was low. In this area the most successful development is due to Juge, who has developed an approach based on the previously observed

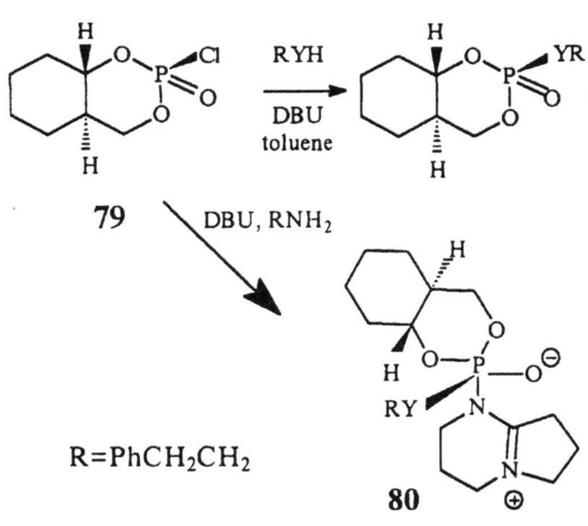

Y=	de
O	78
S	60
NH	100

Scheme 13.

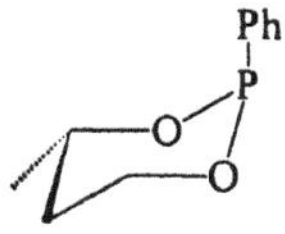

81

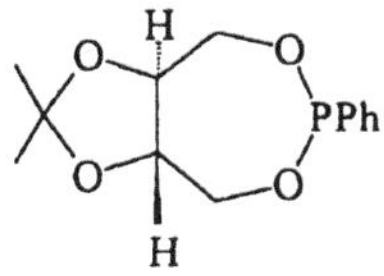

82

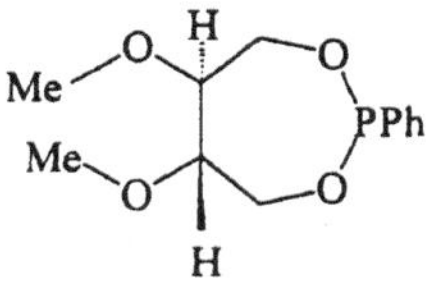

83

diastereoselectivity in formation of the oxazaphiospholidine from ephedrine.[155,164,257,263] Thus, oxazaphospholidines undergoes an efficient reaction with MeI, EtI, or PrI to give an average 9:1 ratio of phosphinic amide **84** to its diasteromer separable by fractional crystallization. The product can be converted into diarylalkylphosphine oxides by succesive acid-catalyzed methanolysis and Grignard displacement. In this way tertiary phosphine oxides were prepared in 95 and 92% e.e., respectively.

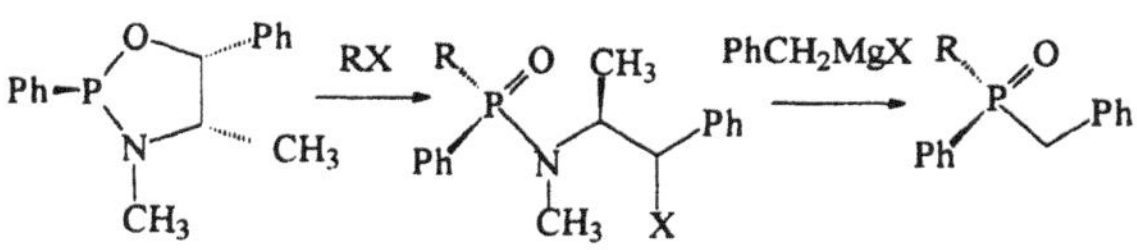

84

Methoxycarbonylalkylmethyl side chains have been utilized for homochiral phosphine oxide synthesis by two groups. Pietrusiewicz and coworkers demonstrated that a vinylphosphine reacted with (–)-menthyl bromoacetate giving a diastereomeric mixture of the Arbuzov products from which one diastereomer **85** could be crystallized.[34,243–246]

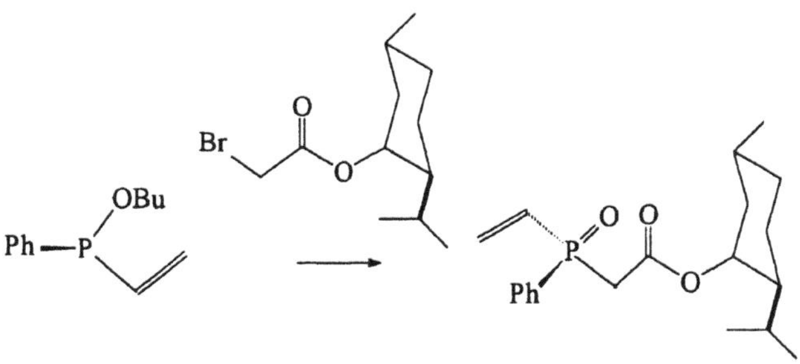

85

Johnson and Imamoto, demonstrated that menthyl ester **86** prepared in the above manner could be readily separated into its diastereomers, which were separately subjected to hydrolysis and decarboxylation to give a tertiary phosphine oxide. This approach was successfully used to prepare a variety of phosphine oxides such as **87**.[144,150]

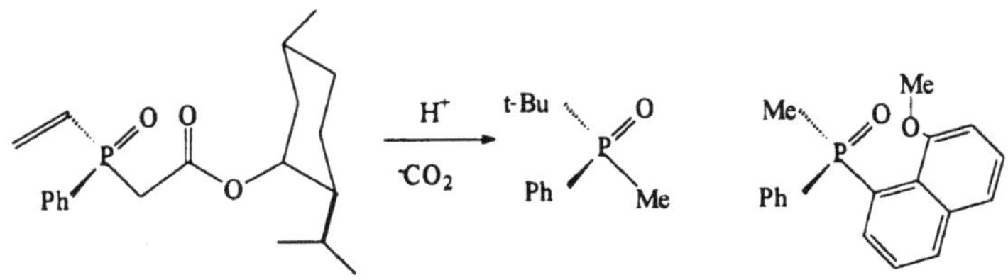

86 87

A general method for the asymmetric synthesis of phosphinates and phosphine oxides using the Michaelis–Arbuzov rearrangement of chiral diheterophophosphacycloalkanes was proposed by Juge and Genet.[156] They showed that it is possible to prepare various organophosphorus compounds of known absolute configuration from only one heterocyclic compound. The reaction of a 1,3,2-oxazaphospholidine with an alkylhalide gave the phosphinamide via the Michaelis–Arbuzov rearrangement. The reaction proceeds with retention of the chirality at the phosphorus atom; however its stereoselectivity depends on the steric hindrance of the R′ group and experimental conditions. Acid methanolysis of the phosphinamides gives R(+)-methylphenylphosphinate **88** with an ee of more than 96%.[159] The reaction was

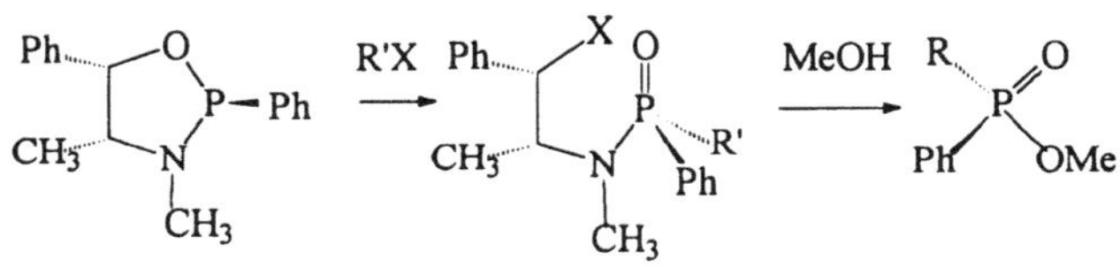

88

monitored by NMR and clearly showed the formation of two quasiphosphonium intermediates with the same diastereomeric ratio as that of the final products (Scheme 14). No NMR signals in the P(V) phosphorane region were observed.[156,159] On the basis of this methodology, enantiomerically pure derivatives of thio- and selenophosphonium acids have been obtained.[63,113]

The Michaelis–Arbuzov reaction of propionyl chloride with the diastereoisomeric mixture of a 1,3,2-oxazaphosphorinane led stereoselectively to the formation of a cyclic acylphosphonate **89** as a 92:8 ratio of diastereomers (81% yield).[107]

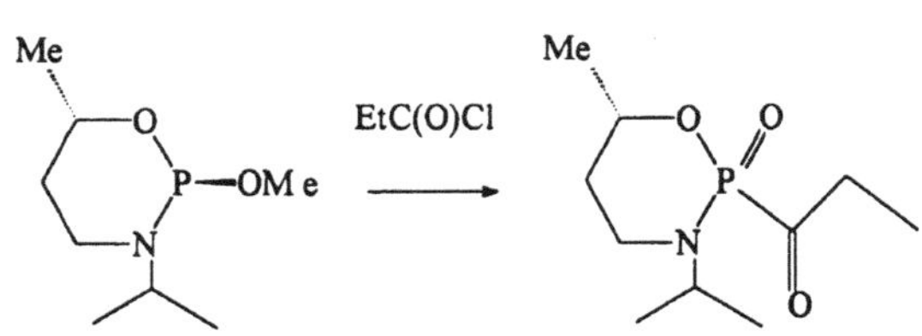

89

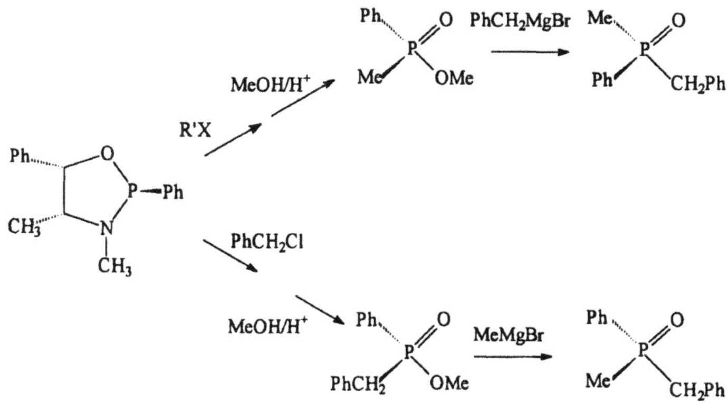

Scheme 14.

An alternative synthesis leading predominantly to the *cis*-P-alkyl oxazaphosphorinanes **90** (ratio 5:1–11:1) involved conversion of the amino alcohols to the cyclic ethyl phosphite followed by subsequent Arbuzov reaction with an appropriate alkyl tosylate.[76]

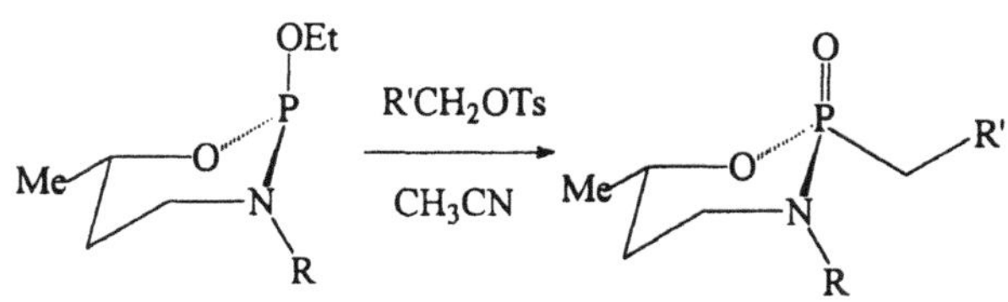

90

R = t-Bu, R'=Ph (8:1); R=t-Bu, R'=Me (5:1); R= CEt_3, R'=Ph (11:1)

Enantioselective Oxidation and Sulfurization of Phosphines. An attractive route to the «chemical» synthesis of homochiral phosphine oxides is the enantioselective oxidation of prochiral phosphines. Usually oxidation or sulfurization of tervalent phosphorus compounds proceeds stereospecifically, with a high degree of retention or inversion of configuration at the phosphorus atom.[140,242] However, in the last few years interesting methodology of enantioselective oxidation and sulfurization of tervalent phosphorus compounds have been developed. Thus, the oxidation of *N*-phosphino amino acids with the pair of tetrachloromethane–alcohol or water proceeds with high stereoselectivity and furnishes mainly one of two possible diastereomers. In some cases the stereoselectivity achieved was 100%. This reaction is especially of interest for the preparation of stereochemically pure derivatives of *N*-phosphorylated amino acids **91** having important practical significance, because existing methods for their synthesis are not stereoselective.

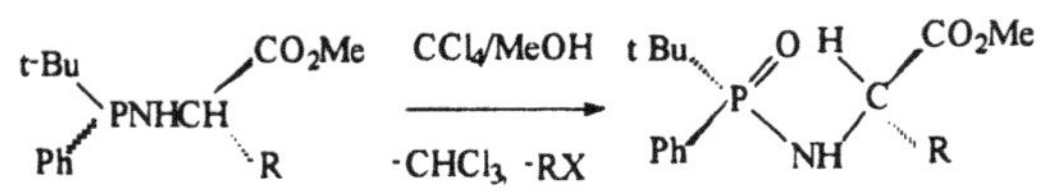

91

NMR spectroscopic studies showed that the reaction proceeds via the formation of a alkoxyhalogenophosphorane, which results in pseudorotation to give the most thermodynamically stable diastereomer. The latter converts into the amidophosphinate via the alkoxyphosphonium salt. The ratio 94:6 (P -56 and -58 ppm) shows high thermodinamical advantage of the one of diastereomers.[161,292]

Alkoxyhalogenophosphorane **92** exists in equilibrium with alkoxyphosphonium salt **93** and is converted into an amidophosphate. The position of equilibrium depends on substituents at the phosphorus atom. Alkoxyhalogenophosphorane bearing the five-membered 1,3,2-oxazophospholane cycle, appear in the more stable pentacoordinate state.[184]

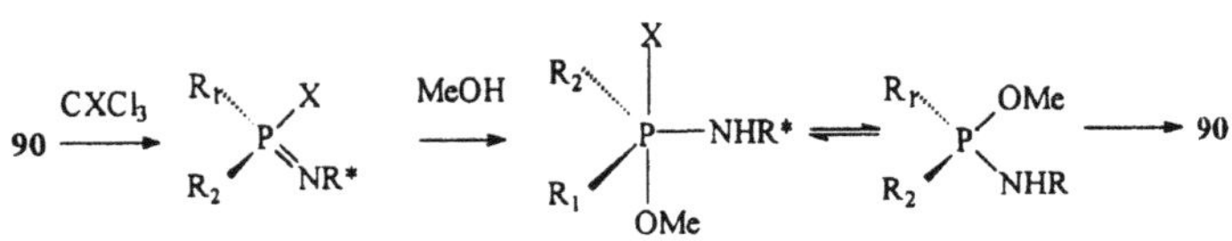

92 **93**

The reaction of tervalent phosphorus compounds with elemental sulfur is stereospecific and proceeds with inversion of configuration at the phosphorus atom and with complete retention of the enantiomeric ratio of isomers.[140] At the same time certain aminophosphines react stereoselectively with elementary sulfur. The reaction proceeds with asymmetric induction at the phosphorus atom to give predominantly one of the two possible diastereomers of compounds **94**.[188]

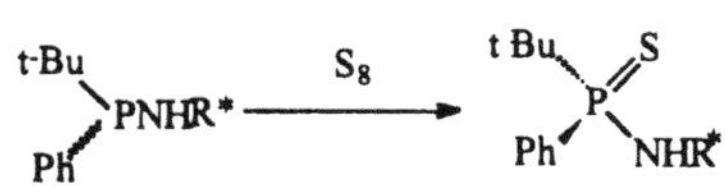

94

R*=*(S)*-CH(Me)Ph; *(S)*-CH(Bu-i)CO_2Me

Diastereochemically pure diethylamidine phosphoramidates have been site-selectively incorporated into synthetic oligonucleotides by a phosphoramidate technique. By using the terminal amino residues bound to the chiral phosphoramidates, various functional residues have been attached to the oligonucleotides in stereospecific ways. The dependence of the duplex- and triplex-forming activities of these functionalized oligonucleotides on the diastereochemistry of the phosphoramidate has been reported.[92]

C. Addition of Tervalent Phosphorus Compounds to Multiple Bonds

Reaction of addition of trivalent phosphorus compounds to C–C and C-heteroatom multiple bonds is important and versatile tool in the formation of new compounds containing P–C bonds. These reactions have been studied and generalized in several reviews.[219b]

The dichlorophenylphosphine on treatment with 2,4,4-trimethyl-2-pentane forms a cyclic phosphonium salt, which on hydrolysis gives a mixture of diastereomeric phosphetane 1-oxides. The ratio of diastereomers depends on the method used to quench the intermediate salt.[70]

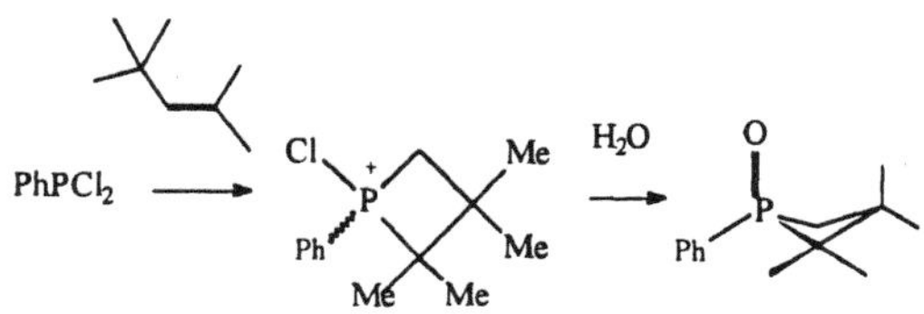

In many cases addition of chlorophosphines to dienes proceeds stereoselectively. For example, the addition of phenyldichlorophosphines to cycloheptadiene provides stereochemically pure *cis*-product **95**.[10b]

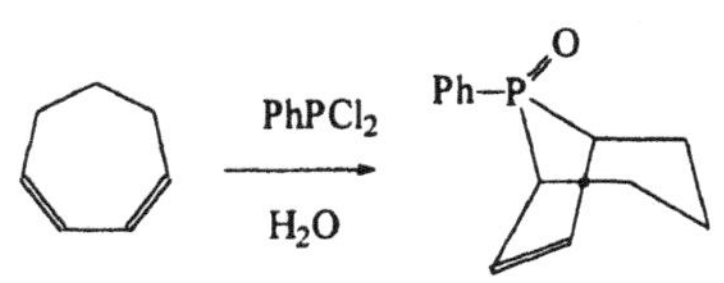

95

Substituted phosphetanes can be easily made from a branched olefin and phosphenium cation $[MePCl]^+$, $AlCl_4^-$, in a reaction which involves a 1,2-Me shift. Asymmetric reaction of a-pinene with $MePCl^+$,$AlCl_4^-$ gives a bridged bicyclic phosphetane **96** whose structure has been unequivocally assigned on the basis of chemical, spectral, and X-ray analysis.[288,289] The formation of this compound can be rationalized by the opening of the cyclobutane ring in the primary adduct allowed by a 1,2-H migration and phosphetane ring closure. An analogous reaction was realized with the camphene.[289]

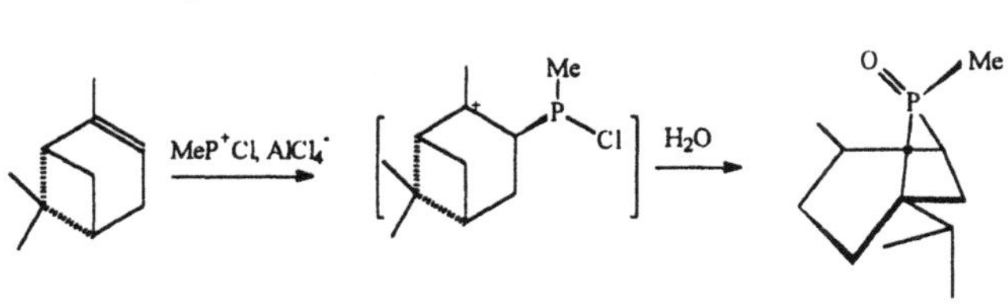

96

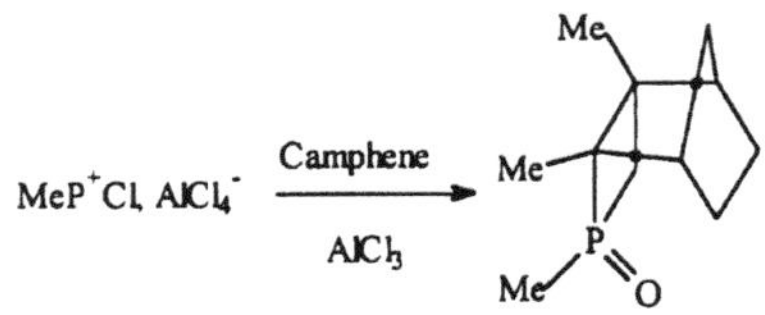

The Michael addition of secondary phosphine oxides to a,b-unsaturated esters proceeds with high stereoselectivity. Thus, in the reaction of benzylphenylphosphine oxide with methyl (*E*)-2-methyl-2- butenoate, only two of the four possible (*Rp*)- and (*Sp*)-diastereomers **97** are formed in a 1:1 ratio of distereomers.

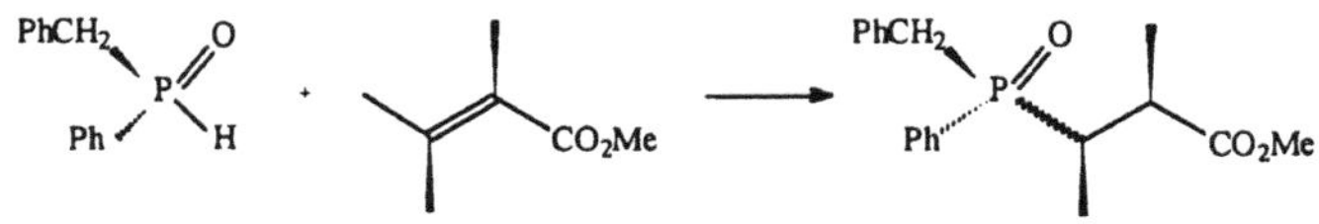

97

The Michael addition of secondary phosphine oxides to cyclohexane-1-carboxylates (both flexible and biased) gives exclusively the axialy substituted products **98**.[220b]

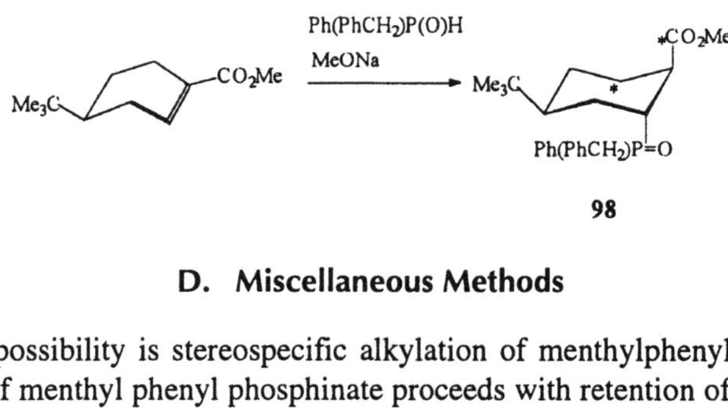

98

D. Miscellaneous Methods

Another possibility is stereospecific alkylation of menthylphenylphosphinate. Akylation of menthyl phenyl phosphinate proceeds with retention of the configuration at the phosphorus atom. Many chiral phosphinates are potentially available via this method.[43–47,96]

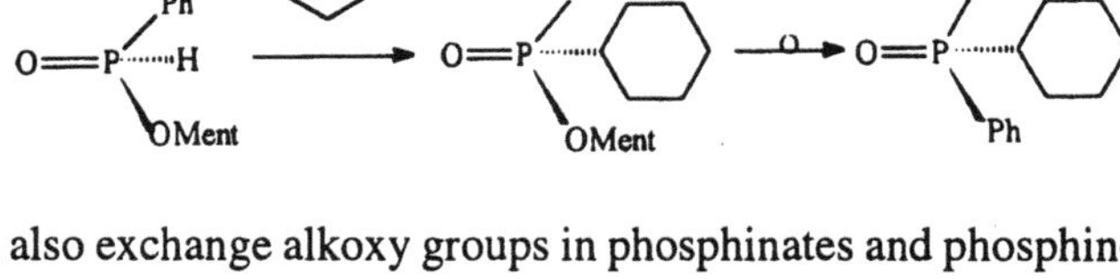

One can also exchange alkoxy groups in phosphinates and phosphinothioates **99** as shown in the equation. The preferred anion is PF_6^- and the configuration at the phosphorus is inverted.[43–47,71,96,97,132]

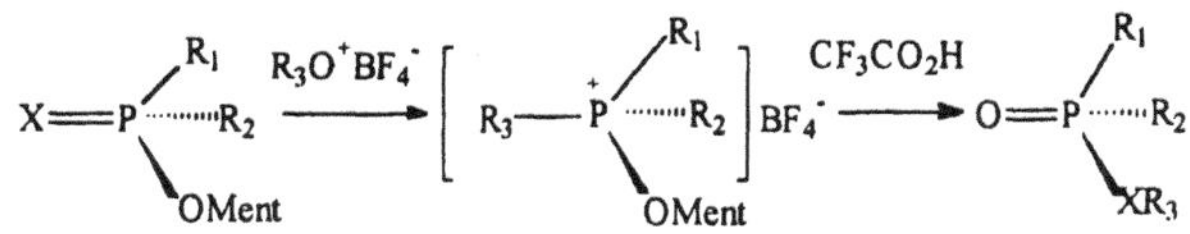

99

X=O, S; R=Me, Et

In the study involving alkaline cleavage of P-prochiral phosphonium salts, Valentine and coworkers observed that neomenthyldiphenylethylphosphonum iodide yields two diastereomeric menthylethylphenylphosphine oxides in a 7:5 ratio and that in decomposition of analogous 1,4-bis(phosphonium)butane there was ca. 2:1 preference for formation of one epimeric phosphorus configuration as compared to the other.[284]

Kabachnik and coworkers showed that *O*-(trifluoromethylalkyl)methylphosphonochloridate can be prepared from 1-trifluoromethylalkanole and methyl-phosphonicdichloride in good yield in 2:1 diastereomeric ratio:[160]

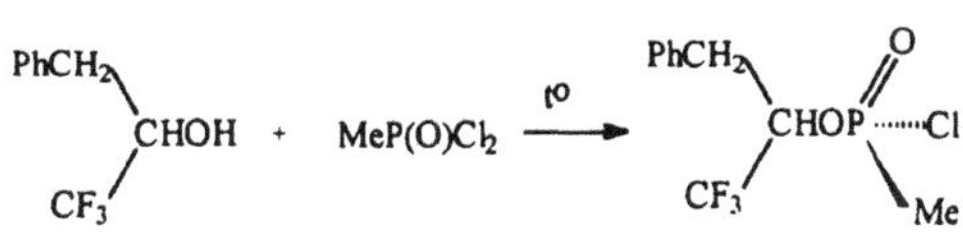

The synthesis of mixed anhydrides of chiral phosphinic and camphor sulfonic acids was elaborated by Michalski.[293]

E. Optically Active Compounds

In this section some the most interesting and perspective types of chiral tetracoordinate phosphorus compounds are discussed.

Phosphine Oxides

Phosphine oxides to be effective synthetic tools for symmetric carbon–carbon bond formation are required in optically active form. In general, phosphine oxides are configurationally stable under normal conditions. A variety of reactions of phosphine oxides will be discussed in later sections of this chapter.

Cyclic Compounds

Structures in which phosphorus is included in a ring, particularly a strained ring, show stereochemical behavior that is strongly influenced by the geometry of the substrate. A practical method for the preparation of enantiomerically pure 1,1′-binaphthalene-2,2′-diol and dithiol derivatives **99** was proposed De Lucchi.[95] These molecules may be considered prototypes of the larger class of atropoisomeric chiral molecules with C_2 symmetry.[1]

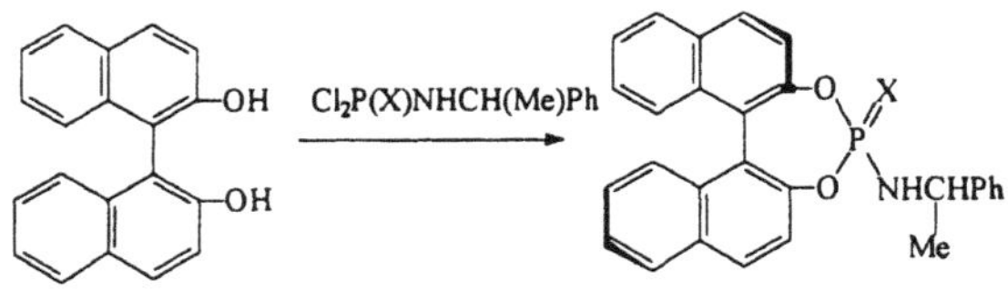

High asymmetric induction (96%) was achieved in the diastereoselective alkylation of the cyanhydrin carbanion generated from the chiral phosphate **100**. The synthesis of the optically active (*R*)-*tert*-cyanohydrins **101** demonstrates how P-chiral auxiliaries temporarily linked to the substrate through heteroatoms can be used in asymmetric synthesis.[261]

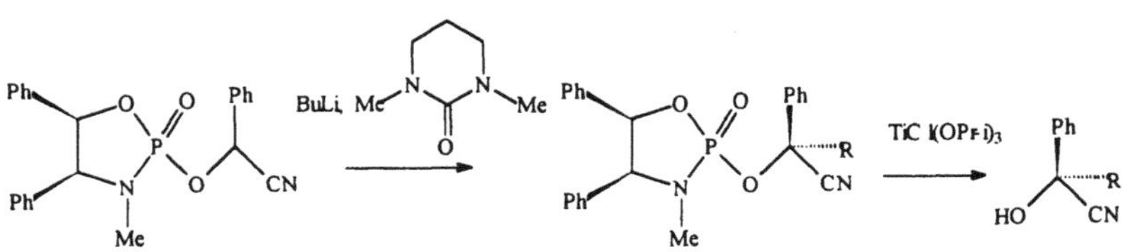

100 **101**

Lithiated optically active bis(dialkylamido)phosphites react with epoxides yielding 2-hydroxyalkylphosphonic diamides **102** with poor diastereoselectivity (de ~33%), but this opened a potential route to optically active 2-substituted alkylphosphonic acids.[149] Derivatives of ephedrine and chiral glycols have been used as

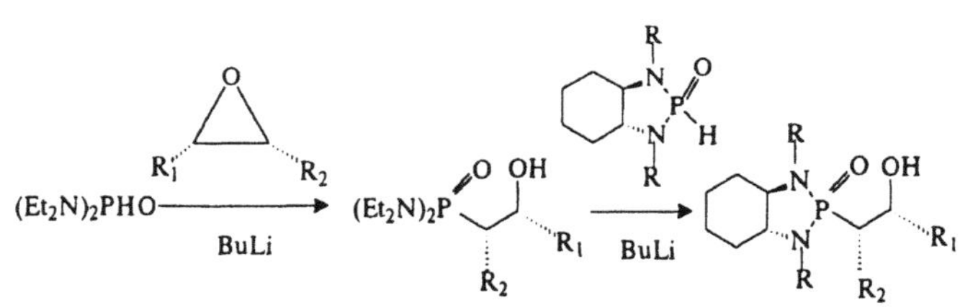

102

reagents for a very efficient determination of enantiomeric purity of alcohols by NMR analysis.[8,151]

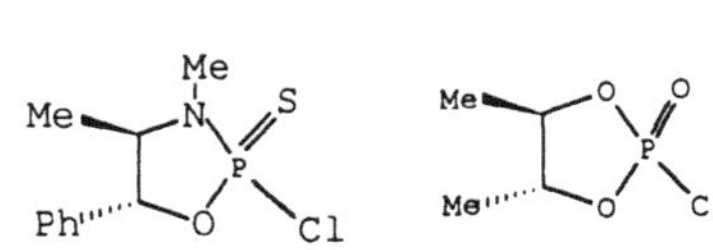

A stereoselective method to insert a heteroatom into an existing sugar framework was developed by Gallagher.[105] The sequence provides access to reasonable amounts of these phosphasugars in fair yields and excellent stereochemical purity. Alkylation of isopropyl phosphoric acid with allyl bromide in the presence of the triethylamine affords isopropyl allylphosphonite **103** (65%), which in the presence of triethylamine and *tert*-butyldimethylchlorosilane, adds smoothly to (*R*)-2,3-*O*-isopropylideneglyceraldehyde to give alternatively the dialkyl phosphinates **104,105**, as a mixture of (81%), readily separable by chromatography (de of >95%).

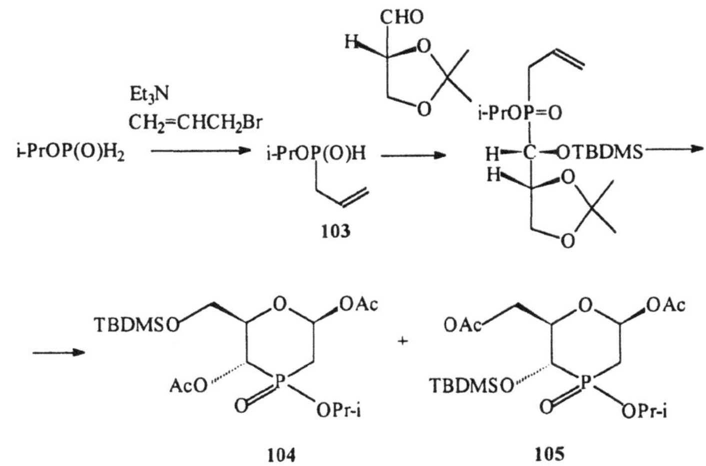

Stereospecific reactions of 3-seleno-1,3,2-dioxaphosphorinanes,[265] as well as thermally and acid-catalyzed rearrangements of sulfur and selenium compounds to isomeric selenoesters proceeding with complete retention of configuration at phosphorus have been discussed.[215,219,220]

Hydrothiolysis of the stereoisomeric mixture of amines **106** provides a single phosphorothioate, whereas the reaction of the corresponding chlorides **108** gives both isomers of **107**.[249]

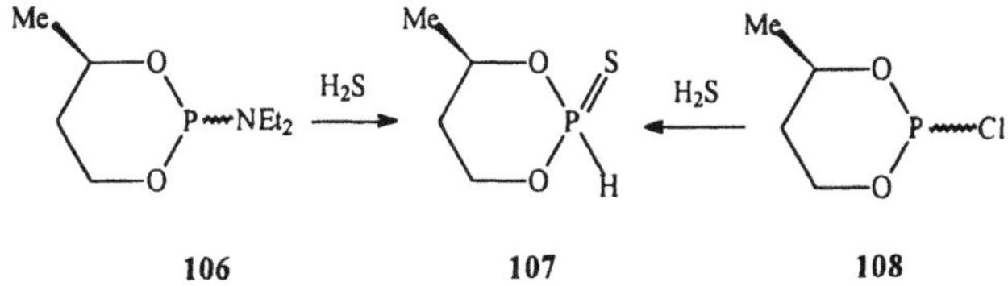

Chiral phosphoric acid diamides **110** were prepared by hydrolysis of **109** with one equivalent of water. Deprotonation of these acids with *n*-butyl lithium or LDA gives the lithium salt which reacted smoothly with alkyl halides to give substituted phosphonamides in good yield.[170]

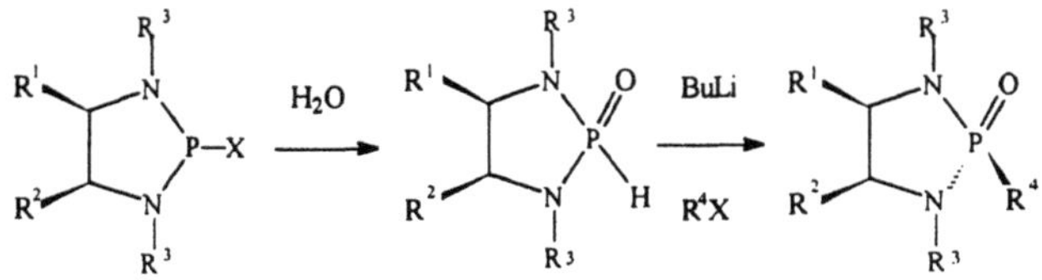

109 110

Ylides

In contrast to amines, tertiary phosphines are enantiomerically stable, and can be alkylated to chiral phosphonium salts. The chiral phosphonium salts have been converted to ylides using aqueous base or phenyllithium in ether, and the ylides were configurational stable at phosphorus. This configurational stability at phosphorus persisted through alkylation, hydrolysis, reduction and Wittig reactions as well as in reactions with epoxides.[140,216,217]

Bestmann and Tomoskozi studied the kinetic resolution in "umylidierung" reaction of racemic ylides with chiral acid chlorides; however the optical yield in this case was only 11–15.4%.[27]

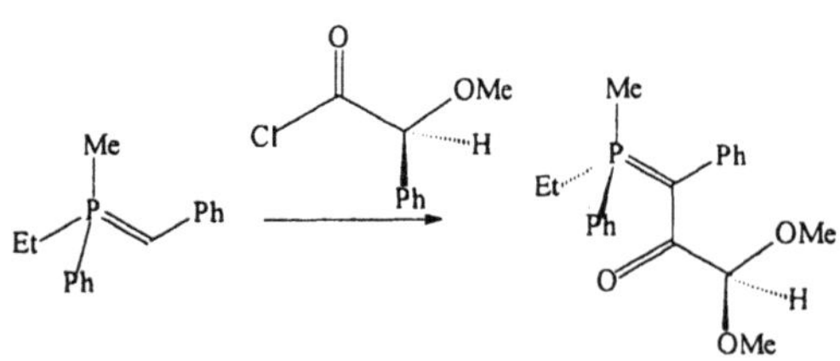

Fabbri and coworkers prepared chiral P-ylides **112** from binaphthophospholes **111**. The reaction of binaphtophospholes **111** with methyl iodide gives phosphonium salts, which react with butyllithium to produce ylides **112.**[94]

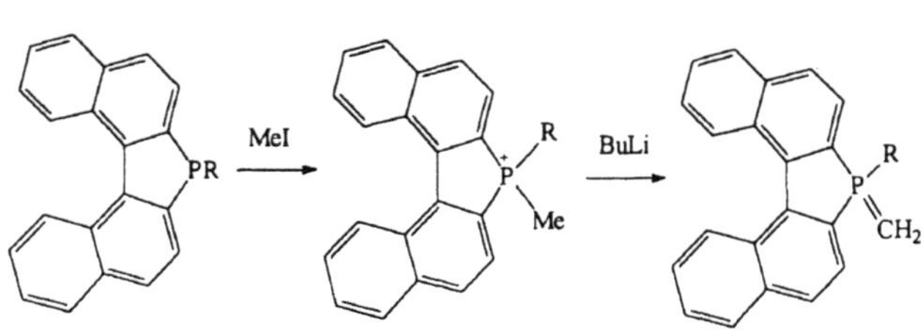

111 112

Hanessian developed a chiral phosphonamide ylide based on a C_2 cyclohexyl diamine for olefination that provides good control of stereochemistry at "remote" centers.[19]

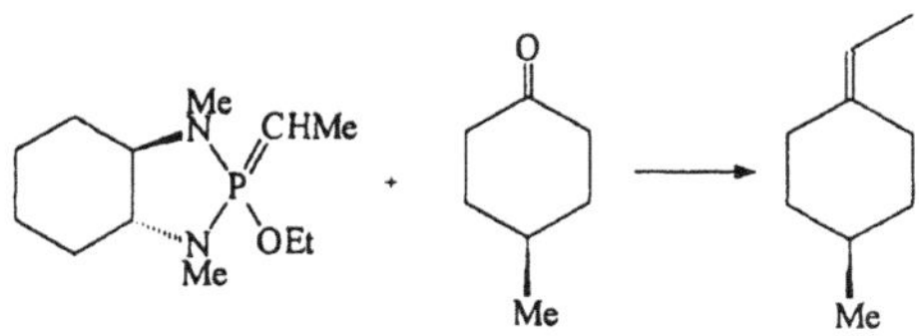

An interesting example of thermodynamically controlled asymmetric synthesis is the dehydrofluorination of alkoxyfluorophosphoranes **113** bearing chiral ligands and resulting in a mixture of diastereomers of *P*-fluoroylids **114a** and **114b** in a 1:1 ratio. However in the presence of lithium fluoride the epimerization of *P*-fluoroylids **114** takes place. As a result, the ratio of diastereomers of **114** changes strongly in favor of the thermodynamically more stable one. *P*-Fluoroylids reversibly add the lithium salt to form fluorophosphorane intermediate **115** and undergo pseudorotation to provide the most thermodynamically stable diastereomer. Determination of heats of formation for ylides bearing the *(S)*-diethylamino-2-propanol group by means of CNDO calculation revealed that the diastereomer with the configuration *S,R* is energetically more advantageous.[189]

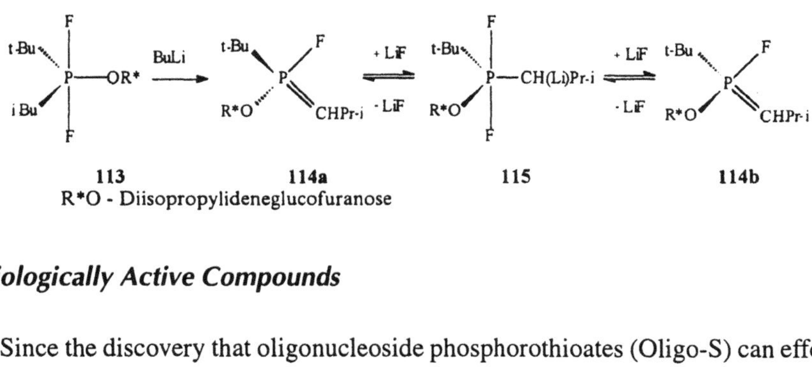

113 114a 115 114b

R*O - Diisopropylideneglucofuranose

Biologically Active Compounds

Since the discovery that oligonucleoside phosphorothioates (Oligo-S) can effectively protect cells from the lethal action of the HIV-1 virus, a number of reports have appeared concerning improvement in their synthesis.[266] It was recognized that stereodefined mixtures of Oligo-S possess interesting pharmaceutical properties. A variety of examples of stereoselective/stereospecific syntheses of Oligo-S were reviewed by Stec and Wilk.[266] Here we show some examples of asymmetric synthesis of biologically active compounds.

Cosstick and Williams demonstrated that reaction of P-prochiral substrates with 3′-*O*-acetylthymidine leads to the formation of the corresponding dinucleoside **116**, which contains the diastereomers in 79:21 ratio.[68]

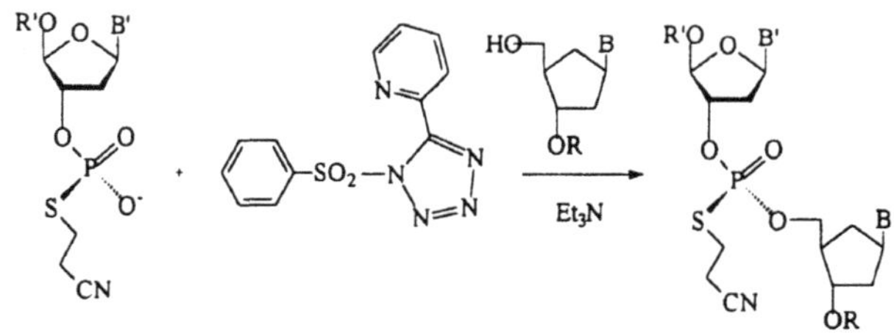

116

Diastereomerically pure dithymidine phosphoramidates have been site-selectively incorporated into synthetic oligonucleotides by a phosphoramidate technique. By using the terminal amino residue bound to the chiral phosphoramidates, various functional residues have been attached to the oligonucleotides in stereospecific ways. No racemization takes place during these procedures.[92] DNA therapeutics show great potential for gene-specific nontoxic therapy of a wide variety of diseases. It was found that the all-*R* methylphosphonate oligomers **117** show significantly greater potency as antisense or antigene therapeutics than do racemic oligomers or all-*S* oligomers.

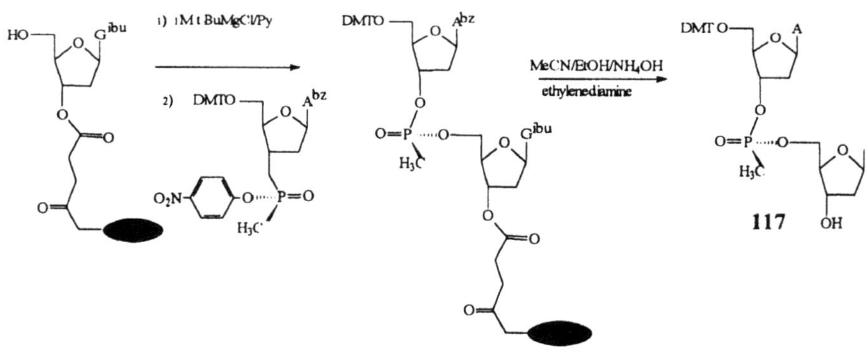

117

Le Bec and Wickstrom, using stereospecific Grignard-activated solid-phase methodology, synthesized stereoregular (*Rp*)- and (*Sp*)-DNA methylphosphonate dimers. The dimers were cleaved from the solid support, deprotected, and purified yielding methylphosphonate DNA dimers of defined stereochemistry.[21] Iyer and coworkers proposed a stereoselective synthesis of the nucleoside by reaction of the chiral oxazaphospholidine with 5′-DMT-T. The phosphoramidite synthon was stereoselectively converted by oxidative sulfurization into new (*Rp*)-nucleoside synthons for the synthesis of oligonucleotides.[147]

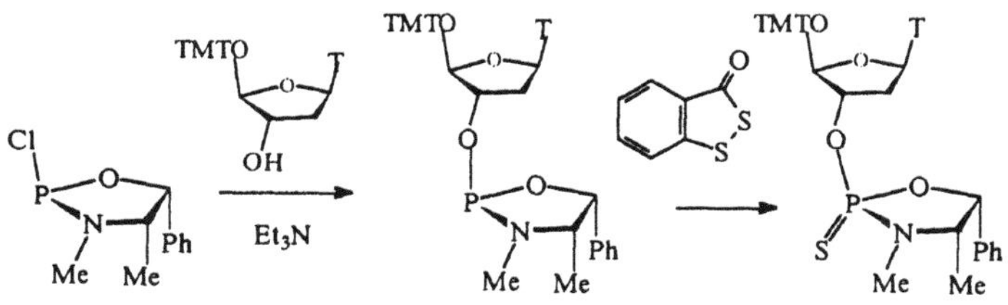

Rp Rp

Phosphoramidites of dithymidine have been synthesized in diastereomerically pure forms and have been incorporated into oligonucleotides. The duplex- and triplex-forming activities of these modified nucleotides are dependent on the stereochemistry of phosphoramidates, since chirality govern the orientation of the functional residues.

VI. ASYMMETRIC SYNTHESIS AT HYPERVALENT PHOSPHORUS ATOM

The ability of the phosphorus atom to increase its coordination number was recognized many years ago; penta- and hexacoordinate species have been isolated. However it was more or less accepted that this expansion of coordination was implicated only when the phosphorus atom is surrounded by electronegative groups, and all attempts to access hypervalent phosphorus species were designed accordingly.

In the past few years, numerous experimental results have illustrated the fundamental importance of penta- and hexacoordinate phosphorus species in reaction at the phosphorus atom. The importance of pentacoordinate intermediates in substitution reaction at phosphorus is now well accepted. In this context the existence of stable penta- and hexacoordinate phosphorus derivatives and their structure have elicited considerable interest. Isomerization processes of these compounds are also of importance.

A. Pentacoordinated Phosphorus Species

Pentacoordinated phosphorus compounds have attracted attention as models for the intermediate of the transition state in nonenzymatic and enzymatic phosphoryl transfer reactions.[135,294] Phosphorus compounds exhibit a marked tendency to expand their coordination, leading to five- or six-coordinate species. Very often the compounds of tetracoordinate phosphorus are found in equilibrium with compounds of pentacoordinate phosphorus, for which pseudorotation with fluctuations of axial and apical bonds is characteristic (Scheme 15).

Thus as shown previously, enantioselective oxidation of racemic aminophosphines **90** by the pair ROH/$CXCl_3$ procceds via the formation of a alkoxyhalogeno-

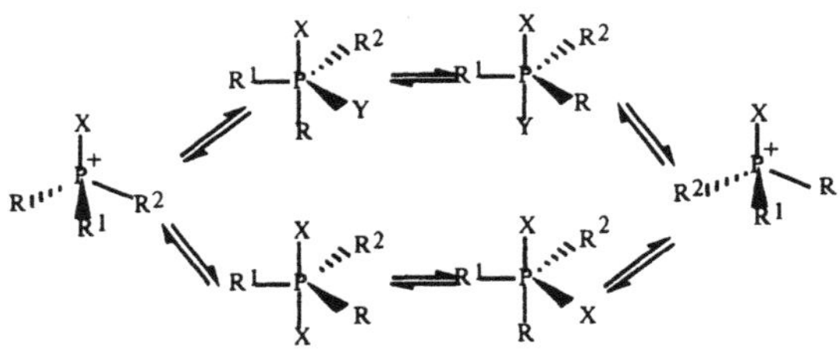

Scheme 15.

phosphorane **92**, which exists in equilibrium with alkoxyphosphonium salt **93**, ultimately converted into into the amidophosphate **91**. The position of equilibrium depends on substituents at the phosphorus atom. Alkoxyhalogenophosphoranes have been obtained for compounds **118** bearing the five-membered 1,3,2-oxazaphospholane cycle, thus stabilizing the pentacoordinate state of the diastereomers. Bicyclic alkoxyhalogenophosphoranes **118** were obtained as a 94:6 diastereomer mixture. Phosphoranes convert gradually at room temperature into the amidophosphate **120**.[184]

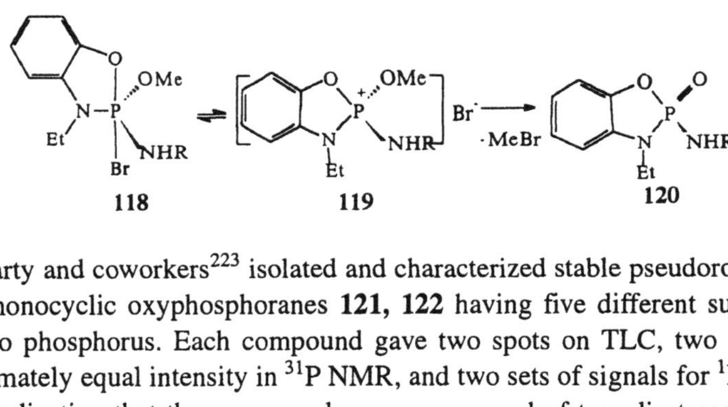

118 **119** **120**

Moriarty and coworkers[223] isolated and characterized stable pseudorotamers of chiral monocyclic oxyphosphoranes **121, 122** having five different substituents bound to phosphorus. Each compound gave two spots on TLC, two signals of approximately equal intensity in ^{31}P NMR, and two sets of signals for ^{1}H and ^{13}C NMR, indicating that the compounds were composed of two diastereomerically related isomers with different configuration around phosphorus. The diastereomers **121,122** were separated by silica gel column chromatography or fractional crystallization. Stereoisomers have been studied by NMR spectra, and X-Ray analysis.

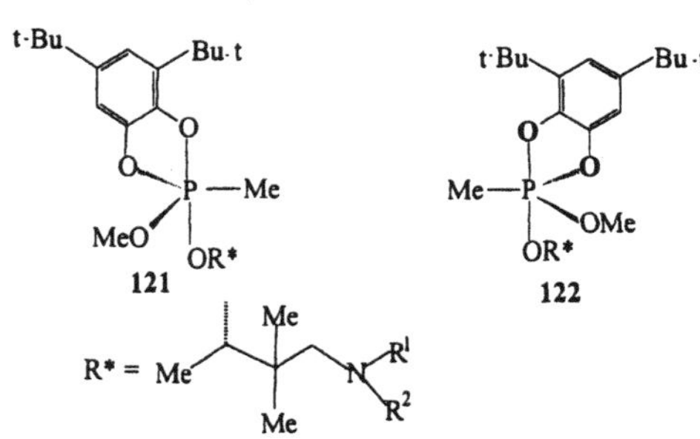

121 **122**

B. Hexacoordinated Phosphorus Species

At the present time several organic compounds of hexacoordinate phosphorus are known and in some cases such compounds act as intermediate products or transition states in nucleophilic displacement at tetrahedral phosphorus. Kinetic studies of the decomposition of 3-hydroxypropyltriphenylphosphonium chloride catalyzed by ethoxide ions showed the presence of a hexacoordinate intermediate or transition state formed in case of attack of the ethoxide ion on pentacoordinate intermediate (Scheme 16).[139]

The octahedral geometry of pentavalent hexacoordinate phosphorus allows the formation of chiral phosphate anions by complexation with three identical, symmetrical, bidentate ligands.[130,135] Enantiopure anions of D_3 symmetry can be used in several fields of chemistry that involve chiral or prostereogenic cationic species. Resolution of enantiomeric cations, determination of their enantiomeric purity, or asymmetric synthesis of cationic species have several applications. Examples of preparation of optical active pentaarylphosphorane **123–125** have been given by Hellwinkel,[131] and other authors.[56,168,171,172,204]

VII. ASYMMETRIC INDUCTION IN TRANSFER OF CHIRALITY FROM PHOSPHORUS TO OTHER CENTERS

In the last few years optically active organophosphorus compounds have found wide application in asymmetric synthesis. This is mainly because organophospho-

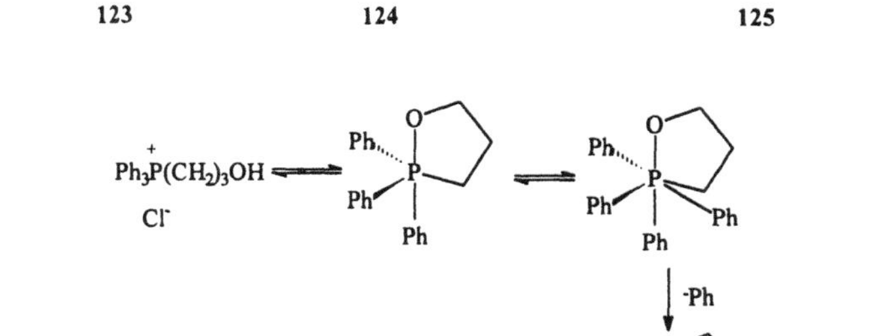

123 **124** **125**

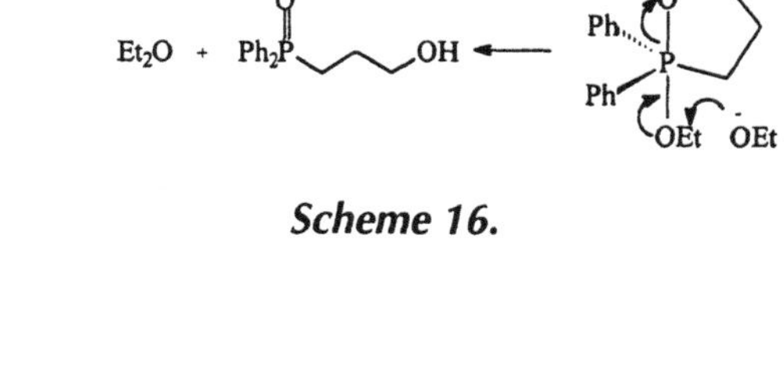

Scheme 16.

rus compounds are quite readily available in optically active form. Moreover the chiral phosphorus grouping that induces optical activity can be easily removed from the molecule, thus presenting an additional advantage in the asymmetric synthesis of chiral compounds. This section concerns reactions in which asymmetric induction in transfer of chirality from phosphorus to other center was observed. Various criteria may be used to classify such reactions. In our survey the most general criterion was used; that is, the further fate of the chiral phosphorus moiety inducing optical activity. According to this criterion all the reactions, which can be defined as the processes of transfer of chirality from the chiral phosphorus atom to the newly formed chiral carbon or heteroatomic center, can be divided into two groups:

1. Reactions resulting in the formation of diastereomeric organophosphorus compounds containing the induced new chiral center as well as the optical active phosphorus atom. Here, it is necessary to take into consideration the number of newly formed chiral centers, i.e. to consider reactions proceeding with 1,2- or with 1,4-asymmetric induction.
2. Reactions in the course of which the formation of a new chiral center is accompanied by elimination of the phosphorus atom.

A. The Abramov Reaction

Silyl phosphine esters $(RO)_2POSiR'_3$ have been shown to be remarkably versatile phosphorylating reagents. Chiral silylated organophosphorus reagents can be used in the asymmetric phosphorylation of prochiral unsaturated organic substrates such as aldehydes. One such example is the asymmetric variant of the Abramov reaction. This was applied to compounds possessing C_2 symmetry where only a single stereoisomeric form exists for the silylated organophosphorus(III) reagent and hence only two possible isomers can be produced in the reaction with benzaldehyde. The silylphosphine compounds **126** have been synthesized by the reaction of binaphtalatochlorophosphites with R_3SiOH.[110] These chiral compounds are good aymmetric phosphonylating reagents. They undergo the Abramov reaction with benzaldehyde at room temperature to afford new silyloxy esters **127** in high yield and with good stereoselectivity.

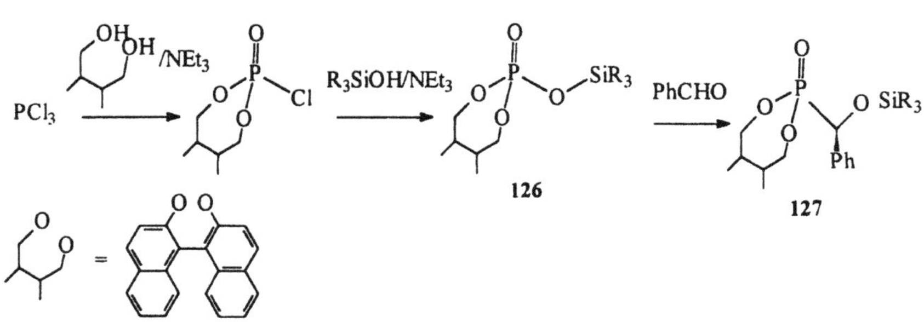

Silylphosphites derived from chiral (+)-dimethyl L-tartaric esters **128** as auxiliaries have been obtained. However, they are not sufficiently useful because they do not react very cleanly with benzaldehyde.[247]

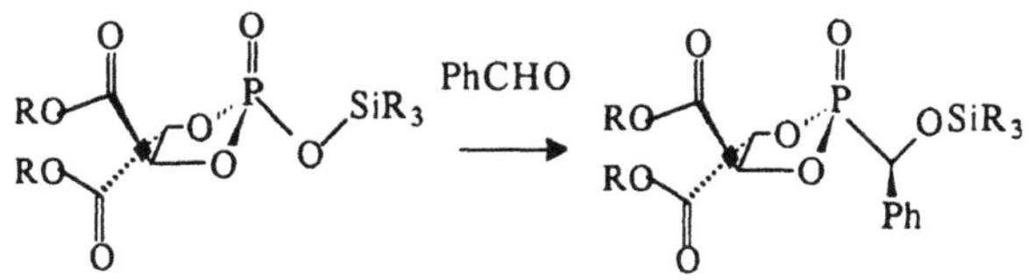

128

The 2-triorganosiloxy-1,3,2-oxazaphospholidines **129** undergo the Abramov reaction with benzaldehyde at room temperature to afford new esters in high yield and with good stereoselectivity. Recrystallization of diastereomeric mixtures from pentane affords α-siloxy phosphonate esters **130** as white crystalline solids in up to 88% isolated yield and 95% isomeric purity.[267,270] The reaction is kinetically controlled and the transfer of the silyl group to the carbonyl oxygen is intramolecular, which results in retention of relative configuration at the phosphorus atom.[108]

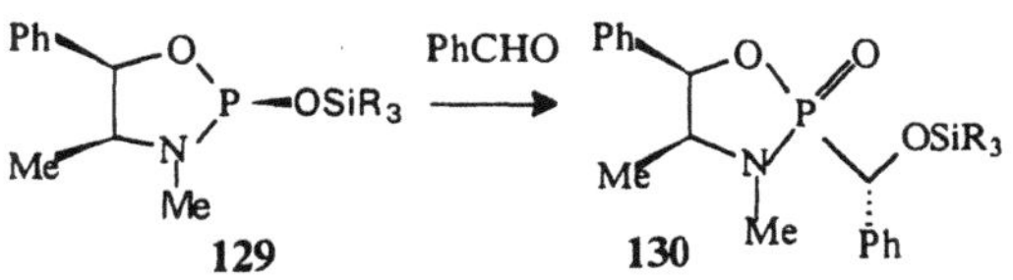

The successful development in this area is due to Juge, who has developed an approach based on the previously observed diastereoselectivity in reactions of oxazaphospholidines from ephedrine.[156]

B. The Pudovik Reaction

The reaction of dialkyl phosphite anions with aldehydes (the Pudovik reaction) is a method of synthesis of α-hydroxy phosphonamides, many of which are biologically active and have been shown to inhibit different enzymes. The absolute configuration at the α-position in substituted phosphonic acids is very important for biological activity.

Spilling and coworkers investigated the use of chiral phosphorus compounds as reagents for asymmetric synthesis, and have developed a method for the asymmetric synthesis of α-hydroxy acid derivatives formed by addition of a chiral phosphoric acid diamide to aldehydes. Addition of aldehydes to chiral phosphoric acid anions in THF solution proceeds stereoselectively to give α-hydroxy phosphonamides **131** in good yield and good stereoselectivity (54–93% de). The diastereo-selectivity was strongly dependent upon the diamide used and ranged from poor to good.[32,33]

It was found that the reaction proceeds irreversibly and under kinetic control The phosphonamides were hydrolyzed with aqueous HCl in dioxane to provide α-hydroxyphosphonic acids **132**.[31–33,169,253]

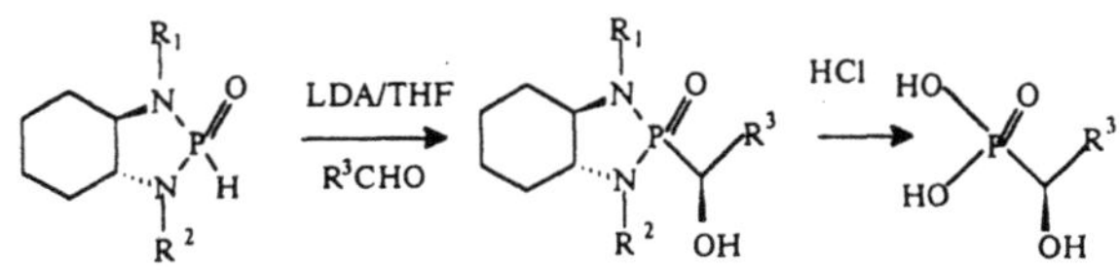

131 132

Reaction of a lithiated optically active bis-dialkylamidophosphite with 1,2-epoxybutane gave the corresponding product **133** with poor (2:1) stereoselectivity, but opened a potential route to the optically active alkylphosphonic acids.[149]

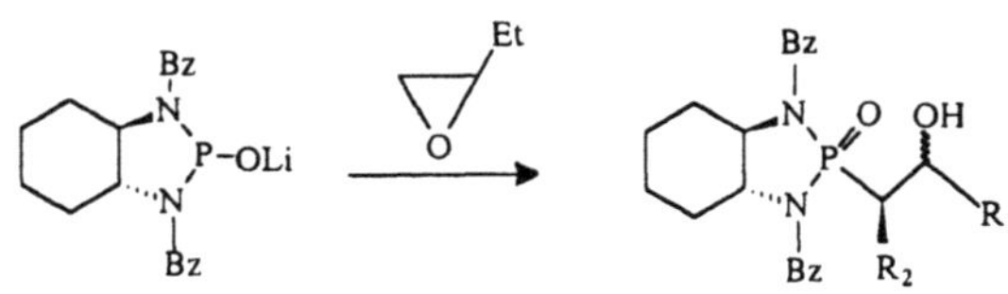

133

Gordon and Evans described the condensation of diastereoisomeric 2-methoxy-1,3,2-oxazophosphites **134** with a variety of aldehyde–boron trifluoride–etherate complexes resulting in diastereoisomerically enriched mixtures of α-hydroxy-2-oxo-1,3,2-oxazaphosphorinanes **135** (75:25 diastereomer ratio).[108]

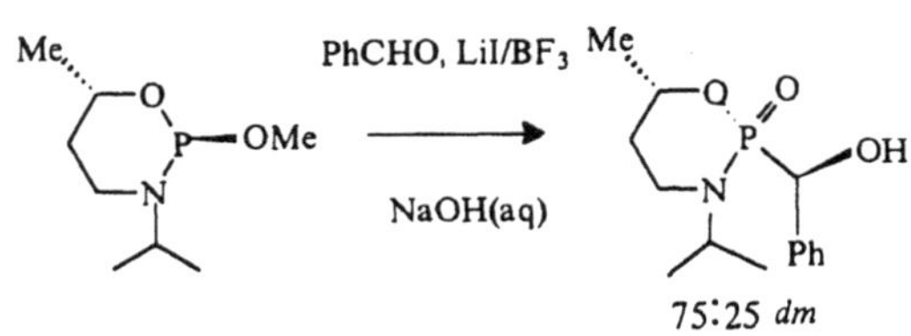

134 135

C. Enantioselective Cycloadditions

The 1,3-dipolar cycloaddition reaction, particularly the one leading to chiral isoxazolidine cycloadducts, can provide valuable building blocks for stereoselective synthesis of functionalized molecules with control of stereochemistry in acyclic systems. Brandi and Pietruciewicz were able to demonstrate that 1,3-dipolar cycloaddition of prochiral divinylphosphine **136** derivatives to a five-membered ring nitrone led to highly diastereomerically enriched (up to 92%) cycloadducts **137** of predictable stereochemistry at phosphorus.[37,38,109,241] The 1,3-dipolar cy-

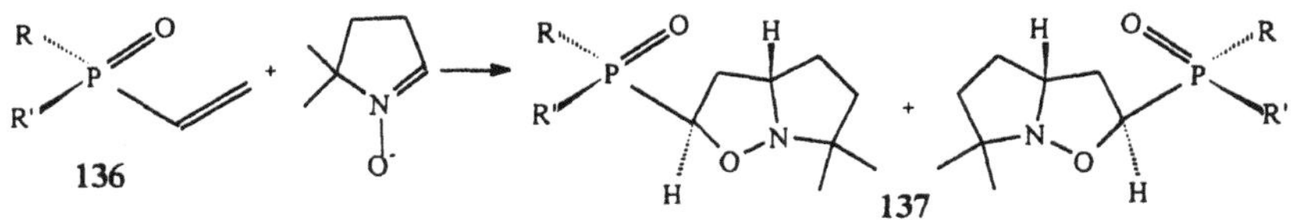

136 137

cloaddition of nitrones to diphenylvinylphosphine oxides, sulfides or selenides proceeds stereoselectively to form adducts of type A under conditions that avoid cycloreversion. The selectivity decreases with an increase in the electron-withdrawing ability of the substituent according to the sequence:

$$Ph_2P > PhMeP(O) > Ph_2P(S) \geq Ph_2P(O) > (EtO)_2P(O)$$

The stereoselectivity indicates that for both regiochemical orientations the cycloadditions prefer *exo* transition states for (*E*)-dipoles and *endo* transition states for (*Z*)-dipoles. The predominant formation of the 5-*exo* products in cycloadditions involving (*E*)-dipolarophiles suggests additionally that this preference is considerably stronger for large substituents destined for position 5 (isoxazolidine numbering) in the product than for those directed to position 4.[39]

Enantiomerically pure five-membered ring nitrones derived from L-tartaric acid via C_2-symmetric O,O′-protected 3,4-dihydroxy pyrrolidines undergo highly regio- and stereoselective cycloaddition reactions with racemic 2,3-dihydro-1-phenyl-1*H*-phosphole 1-oxide and 1-sulfide **138**. In all cases formation of only two diastereomeric cycloadducts is observed and their ratio (up to 10:1) is dependent on the size of the protecting group at the nitrone and on the extent of conversion. The tricyclic cycloadducts **139** feature a 2,2-connection of the pyrrolidine and phospholane rings and six contiguous stereogenic centers, of which three are created and the one at phosphorus is due to kinetic resolution during the cycloaddition process. It was established that during the kinetic resolutions the stereoselectivity factor $s = kS/kR$ exceeds the value of 10 (up to 14) in the most favorable cases.

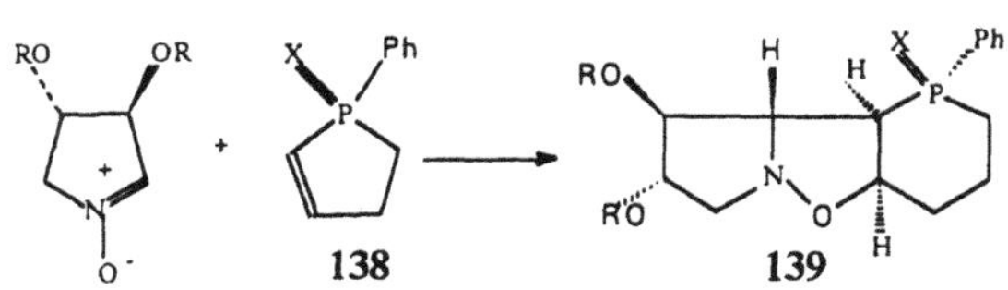

138 139

D. Chiral Phosphorus Stabilized Anions

Today an increasing number of known bond-forming processes have their asymmetric counterparts, usually with impressive levels of enantiomeric or diastereomeric purities.[3,20,23,120,260,277] The chemistry of phosphorus stabilized anions in the domain of structure or of reaction stereoselectivity has attracted the attention of many chemists.[69,73,74,78,79,82,84–86,88] One property of these chirally modified reagents is their capacity to undergo highly stereoselective asymmetric transformations of the anion (e.g. rearrangements, electrophilic substitution) and thus to afford chiral phosphorus and non-phosphorus containing compounds.

Structure

Theoretical investigations of phosphorus(V) stabilized carbanions have been performed by Cremer and coworkers.[70] The rotational coordinates about the P–C bond have been studied at the HF/3-21G* level, with stationary points characterized at levels equivalent to MP3/6-31+G*//HF/6-31+G*. The locations of local minima on the rotational coordinate were found to be dependent on opportunities for hyperconjugative stabilization. When amino substituents on phosphorus were geometrically unconstrained, two local minima were found. By contrast, when the amino groups were constrained to localized geometries similar to those found in diazaphosphorinanes, the effects on the rotation coordinate were considerable. A single minimum and a higher rotation barrier were noted.[70]

Denmark and Dorrow studied NMR spectra and performed X-ray analysis of phosphorus-stabilized carbanions **140**, **141**. X-ray crystallography of the related P-isopropyl anion showed that the Li–C distance (3.88 Å) is greater then the sum of the van der Waals radii. The carbanionic carbon is planar and at nearly 0° angle to the P=O to maximize P-type interaction and the preferred conformation of the anion is parallel. The barrier to rotation about the P(1)–C(b) bond is very low (<8 kcal/mol). The lithium anions exist in general as dimers in THF solution and in the solid state there is no metal–carbon contact.[84] There is a considerable shortening of the P–C (1′) bond (0.1 Å) and a lesser lengthening of the P–O bond (0.04 Å) and P–N bond (0.02 Å) bond compared to average literature values for neutral phosphonamides. The most remarkable, feature of the anion **140** is the pyramidality of the nitrogen, clearly disposing the methyl groups to axial and equatorial positions. The downfield shift of the ^{31}P NMR resonance is indicative of the polarization of the phosphoryl group to stabilize the anion.[79]

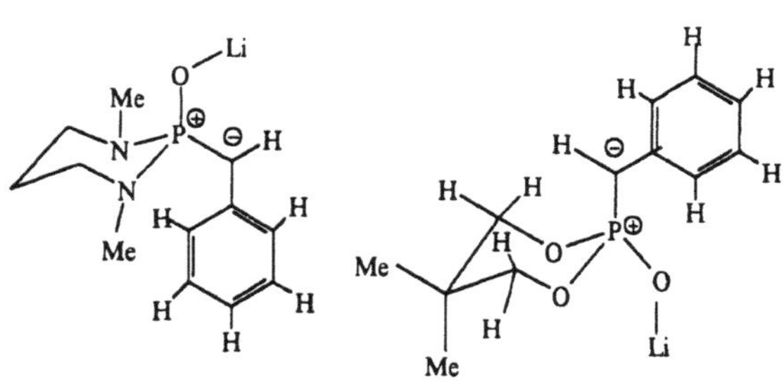

140 **141**

In summary the preferred conformation for phosphondiamide anion **140** is that which maximizes opportunities for hyperconjugative stabilization. When the amino substituents in the phosphondiamide are constrained to a diazaphosphorinane-like arrangement, the rotational barrier about the P–C bond is increased and a single local minimum structure is observed.[69]

NMR studies of oxazaphosphorinane anion **142** indicate that it exists as a rapidly interconverting mixture of conformers of the oxazaphosphorinane ring. Approach of the electrophile *syn* to the P–O bond in **142a** and away from the *t*-butyl group in **142b** would lead to the observed stereochemistry of amination. Based on the fact that anions of this type are planar, the observed trend in selectivity upon variation of the P-alkyl group is also consistent with either an orthogonal or a parallel anion in which reduced interactions between the P-alkyl group and the solvated lithium cluster lead to a loss of conformational preference in the anion.[76]

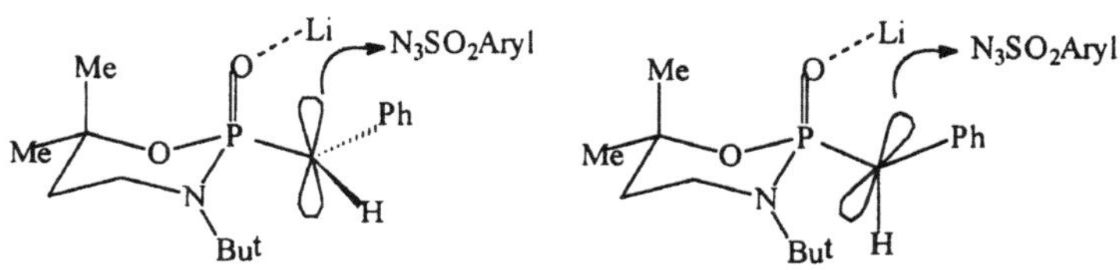

142a **142b**

The planarity of the carbanion is not primarily due to phenyl conjugation; since the barrier to rotation about the P(1)–C(6) bond is very low, the preferred conformation of the anion is parallel to the P=O bond and the anion exists as a dimer in THF solution.The intrinsic planarity and low rotational barrier for phosphonyl anions place stringent requirements on the design of chiral ligands for effective asymmetric reagents.[80,84]

Aggarval analyzed the mechanism of racemization of heteroatom substituted organolithium compounds **143** and arrived at the conclusion that rotation of the C–P bond plays a dominant role in racemization, which in turn is reduced with increasing concentration of P-stabilized carbanions.[2] The rate of racemization of

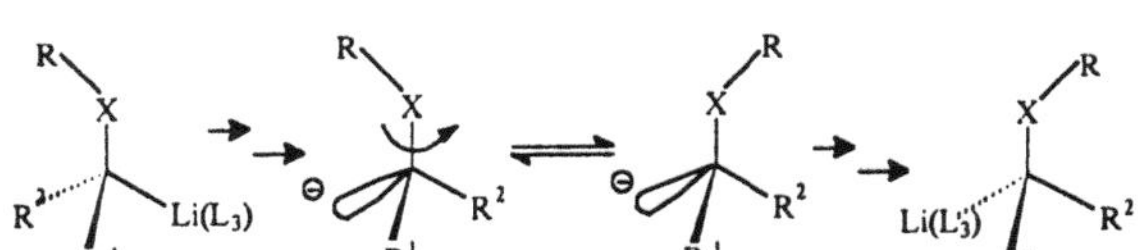

143

X= heteroatom, including P atom

phosphonate carbanions **143** is slowed down in the presence of HMPA, because HMPA increases the barrier of rotation around the P–C bond.[89]

Reactions with Electrophiles

Reaction of chiral P-stabilized carbanions with electrophiles can proceed with formation of diastereomeric organophosphorus compounds, with the newly induced chiral center formed as a consequence of 1,2 - or 1,4-asymmetric induction. Reaction of chiral P-stabilized carbanions with carbonyl compounds, possessing axial asymmetry (enantioselective Wittig or Horner–Wadsworth–Emmons reactions) result in the formation of chiral alkenes with the elimination of the phosphorus atom.

1,2-Asymmetric Induction. This section is concerned with the reaction of chiral phosphorus stabilized carbanions that involves 1,2-asymmetric induction and embraces amination, alkylation, carboxylation, and acylation reactions.

Aminoalkyl phosphonic acids have been extensively studied because of their close relationship to amino acids and their bioactive properties.[9,15–17,36,55,133] It has been shown that the biological activity of α-aminophosphonic acids is dependent upon their absolute configuration, which makes the asymmetric synthesis of this class of compounds both interesting and of practical significance.[10]

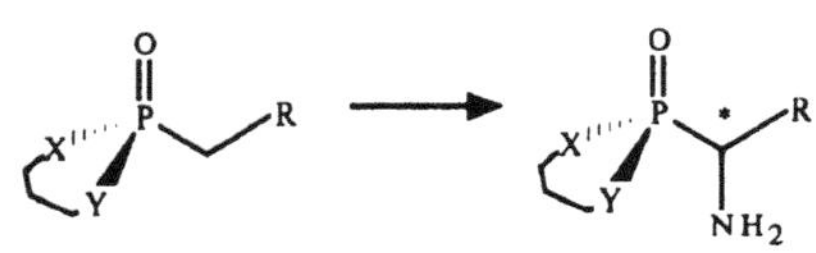

A general method for the asymmetric synthesis of enantiomerically pure or enriched α-amino-α-alkyl,[5] α,α-dialkyl,[6,7] and α-chloro-α-alkyl phosphonic acids **147**[7,85,159,179] in either enantiomeric form is based on the alkylation of chiral bicyclic phosphonamides **145** derived from (*R,R*)- and (*S,S*)-1,2-diaminocyclohexanes—a readily available C_2 symmetrical template via the chloromethyl derivative.[24,76] The enantiomeric purity of such α-amino-α-alkyl phosphonic acids **147** is 84–98%.

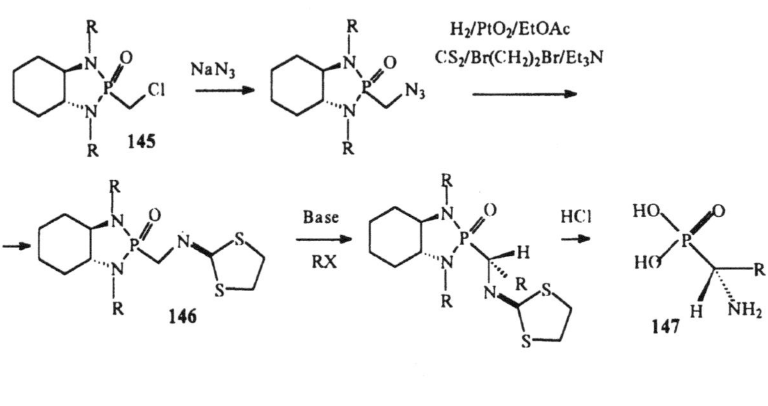

Another route to α-amino phosphonic acids **147** is possible from the α-chloro-α-alkylamides **148** which are easily obtained by the asymmetric alkylation of chloromethylphosphonamide (**145**). Nucleophilic displacement with azide ion gives the corresponding α-azido derivatives **149**. Mild acid hydrolysis followed by hydrogenation of the azido group led in almost quantitative yield to the corresponding α-amino phosphonic acids **147**. The enantiomeric purity of these α-amino-α-alkyl phosphonic acids **147** is 78-98%.[19,23,24]

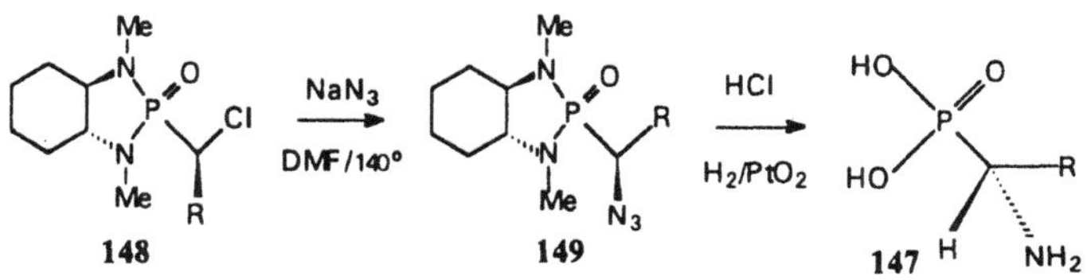

Hannesian and coworkers[116] also described the synthesis of α- and β-amino phosphonic acids based on the stereoselective addition of carbanions of chiral nonracemic α-chloromethyl and methyl alkyl phosphonamides to imines. The structure and absolute configuration were determined by single crystal X-ray analysis. Acid hydrolysis followed by esterification gave the corresponding dimethyl phosphonate derivative **150** in 80% yield and with greater than 95:5 diastereoselectivity. Stereoselective electrophilic amination of anions of chiral

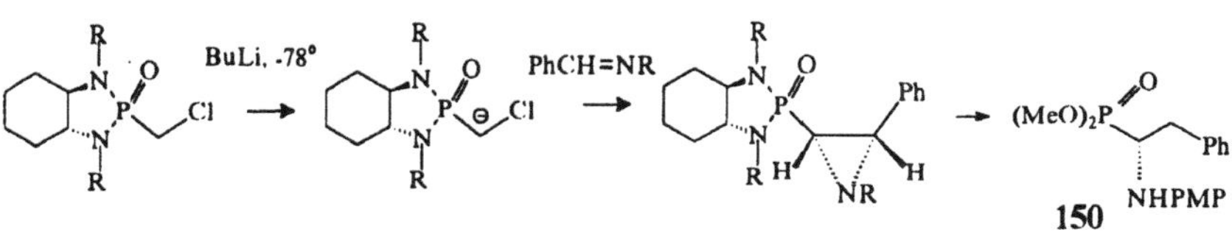

nonracemic α-alkyl phosphonamides **151**, derived from *N,N*-dimethyl (*R,R*)-1,2-diaminocyclohexane, by azocompounds proceeds with moderate to excellent enantioselectivities. The products **152** were hydrolyzed and reduced to the corresponding α-alkyl-α-aminophosphonic acids **153**, which are being extensively studied in view of their relations to amino acids and bioactive peptides.[208]

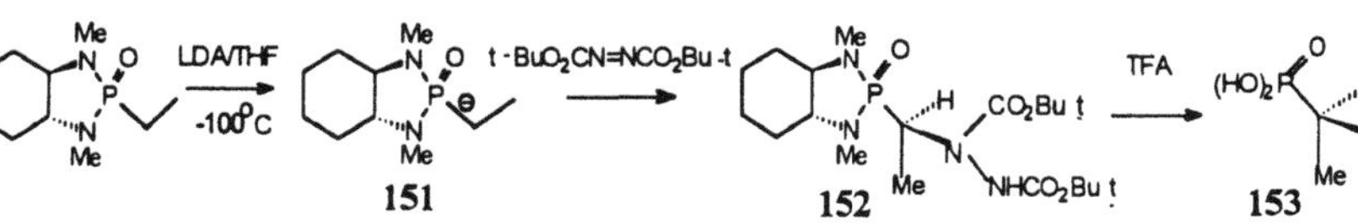

Denmark[76] proposed asymmetric electrophilic amination by trimethylbenzenesulfonyl azide (trisyl azide) of phosphorus-stabilized anions from chiral oxazaphosphorinanes and diazaphospholidines. The reaction was shown to be dependent on auxiliary structure, nature of the P-alkyl substituent and amination conditions. In the best cases, (*S*)-α-aminophosphonic acids **147** with 92% ee have been obtained. Deprotonation of racemic oxazaphosphorinanes with KHMDS was followed by the addition of trisyl azide and then by acetic acid to decompose the triazine. This protocol affords the α-azido phosphonoamidate **154** with excellent diastereoselectivity. The minor diastereomer could not be detected by NMR spectroscopy or HPLC.

In other method proposed by Denmark[76] oxazaphosphorinane was deprotonated with BuLi, treated with trisyl azide and the intermediate lithiosulfonyltriazine was captured with acetic anhydride to afford after decomposition a single isomer **154** in excellent yield. Again the minor diastereomer could not be detected by NMR or HPLC. Reduction of compound **154** was accomplished by hydrogenation (1 atm, Pd/C) to give the α-aminophosphonamidate **155**. The selectivity in azide transfer was at least 50:1. The free α-aminophosphonic acids **147** was obtained in good yield by acid hydrolysis.[1] The reaction is shown to be dependent on the structure

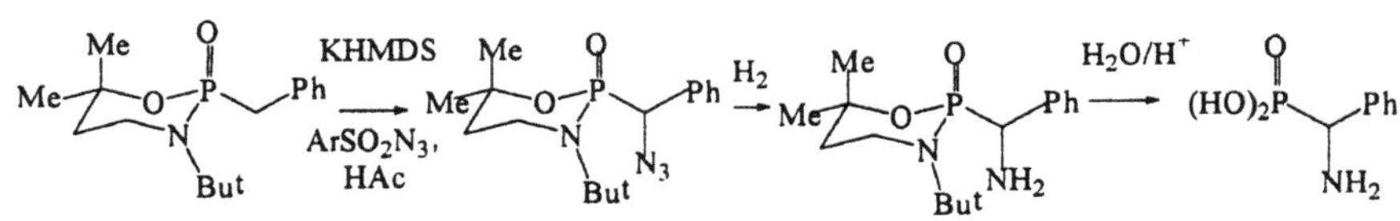

154 **155** **147**

of auxiliary, the nature of the P-alkyl substituent, and the choice of amination procedure. Using the oxaphosphorinanes, a high level of asymmetric induction can be achieved. *cis*-Oxazaphosphorinanes provide the α-aminophosphinic acids **147** with 92% ee in the (*S*)-antipode.

Hanessian[20] studied amination of carbanions derived from (*R,R*)-diaminocyclohexane also using trisyl azide for stereoselective electrophilic amination. Enantiomeric purity of **156** was 68–80%. Subsequent hydrolysis of **156** and hydrogenation afforded aminoacids **157** with 63–99 % ee.

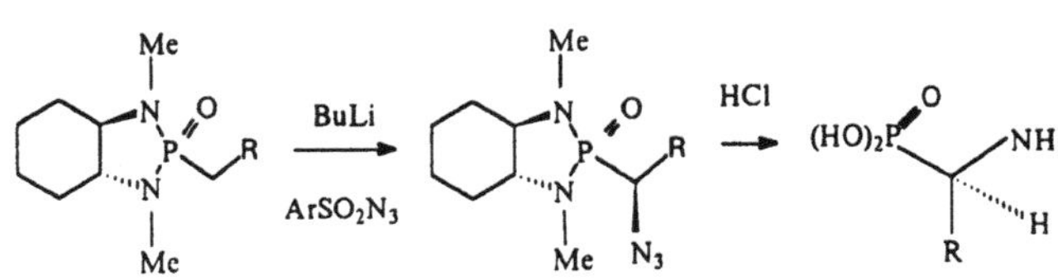

156 **157**

Alkylation of a series of enantiomerically pure *cis*- and *trans*-3-substituted 2-benzyl-6-methyl-1,3,2-oxazaphosphorinane-2-oxides was found to be sensitive to the bulk of the N-substituents. The original design of the auxiliary was made on the basis of the strongly dissymmetric environment around the anionic center due to the sterically disparate groups, i.e., N-*tert*-butyl and O-electron pair and a strong rotator bias in the anion.

Denmark and Amburgey studied enantioselective alkylation of P-carbanions. The quarternary stereocenter was created by alkylation of the various β-keto phosphonamidates as their potassium or sodium enolates. Alkylated β-keto phosphonamidates **159** could be prepared with excellent diastereoselectivities by complementary alkylation of the series **158** with EtI or MeI. Stereoselective reduction gives diastereomers of β-hydroxy phosphonamidates **160** in the ratio 1:14–1:150. Thermal cycloelimination afforded the olefins **161** in excellent yield and stereospecificity independent of configuration and substituent R_3.[73]

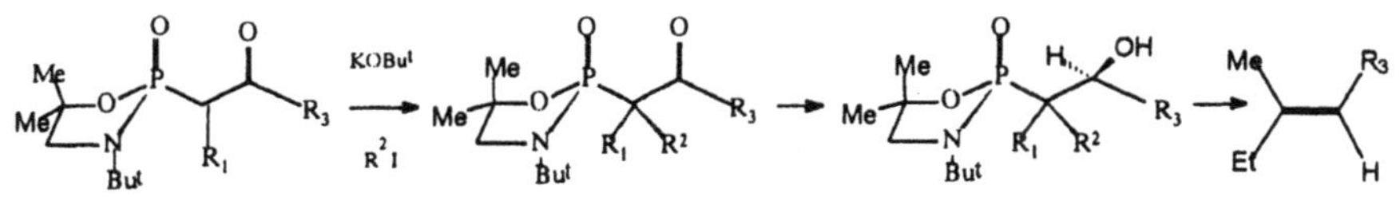

158 **159** **160** **161**

The alkylation of a series of enantiomerically pure *cis*- and *trans*-3-substituted 2-benzyl-6-methyl-1,3,2-oxazaphosphorinane 2-oxides was found to be sensitive to the bulk of the N-substituents.[75]

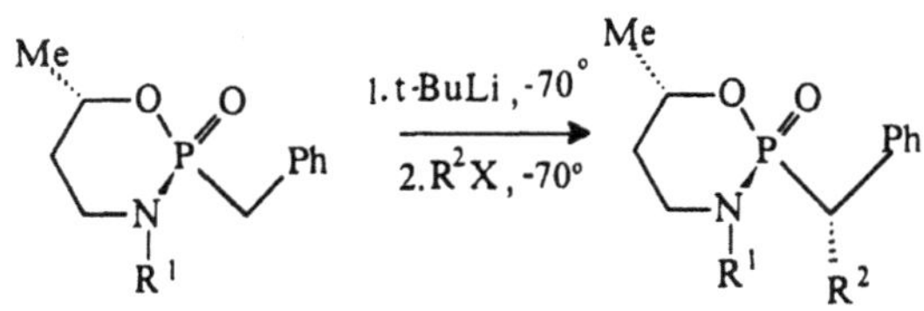

Vineyard and coworkers generated carbanions **162** by action of R_2NLi. Then this carbanion was oxidatively coupled with copper salts. The biphosphine oxide **163** was converted into biphosphine **164** with inversion at the phosphorus center using a mixture of trichlorosilane and triethylamine in acetonitrile solution. This biphosphine **164** was used as a ligand in rhodium complexes.[290]

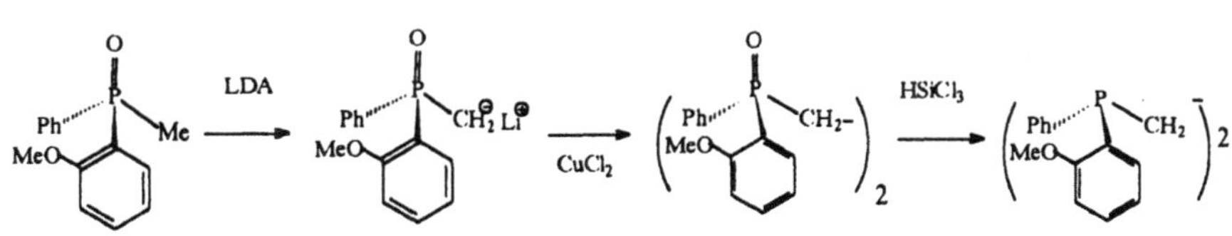

162 **163** **164**

Interesting results have been obtained in the alkylation of chiral nonracemic bicyclic phosphonamides derived from enantiomerically pure 1,2-diaminocyclohexanes.[117] The enantiomerically pure or enriched α-chloro-α-alkylphosphonic acids **166** in both enantiomeric forms have been obtained by asymmetric alkylation of bicyclic C_2 symmetric phosphondiamides. The process consists in the treatment of the chloromethyl or ethyl phosphonamides with a base such as BuLi or LDA in THF followed by addition of an appropriate alkyl halide at –100 °C. The resulting products **165** were obtained in high yield and they were of excellent optical purities. Hydrolysis of alkylated phosphonamides **165** under mild conditions gives the corresponding α-chloro-α-alkyl phosphonic acids **166** in high yield and optical purity (de 90:10–99:1). The authors assumed that attack of the initially formed

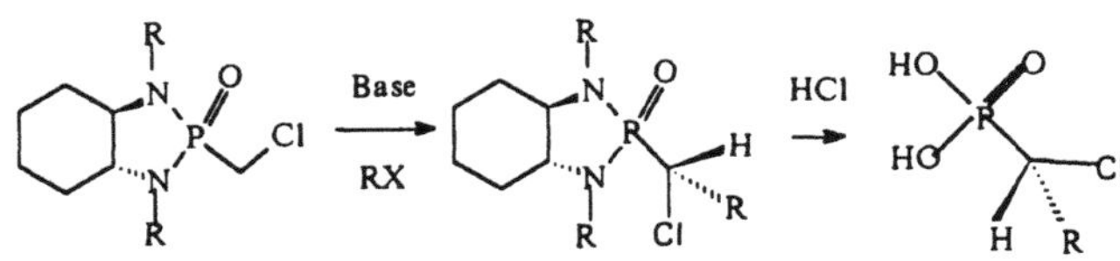

165 **166**

anion **167** on the electrophile will take place preferentially from the side facing the lone pair of one of the nitrogen atoms than the side facing the N-methyl group.[24]

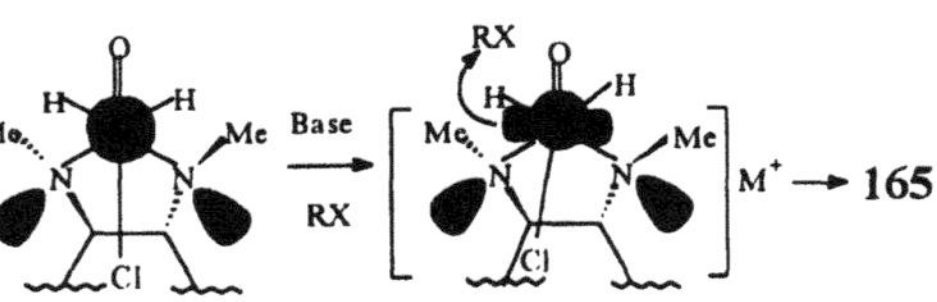

167

1,4-Asymmetric Induction. An efficient and versatile protocol for the synthesis of diastereomerically pure or highly enriched substituted cyclopropane derivatives has been described. The method involves the highly stereocontrolled conjugate 1,4-addition of the anion of a *trans*-chloroallyl phosphonamide reagent to α,β-unsaturated carbonyl compounds **168,** with the concomitant formation of the corresponding cyclopropanes **170** as a result of an intramolecular attack of the enolate upon the intermediate allylic chloride **169**. The method gave the crystalline *endo,endo*-isomer of cyclopropane **170** in 90% yield.[208–215,256b] Stereoselective reduction of the carbonyl group ($NaBH_4$/MeOH), protection, and oxidative cleavage by ozonolysis afforded the aldehyde **171,** which can be epimerized to the *exo, endo*-isomer **172**. Alternatively, utilization of the *cis*-chloroallyl phosphonamide reagent with the same enone **168** via the intermediate **173** led to the isomeric *exo, endo*-product **172** as the major isomer (>90:10)[115,123,124]

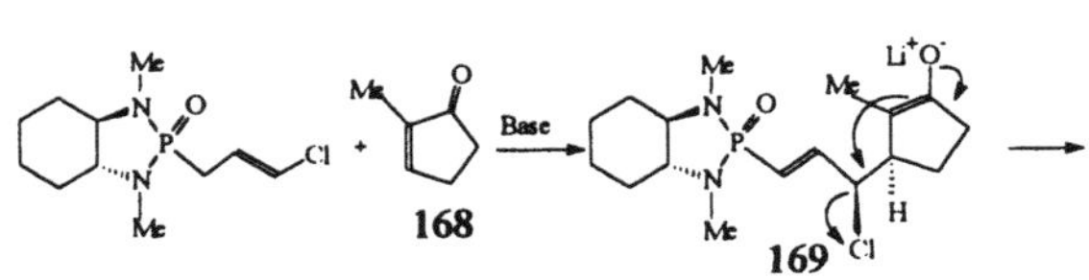

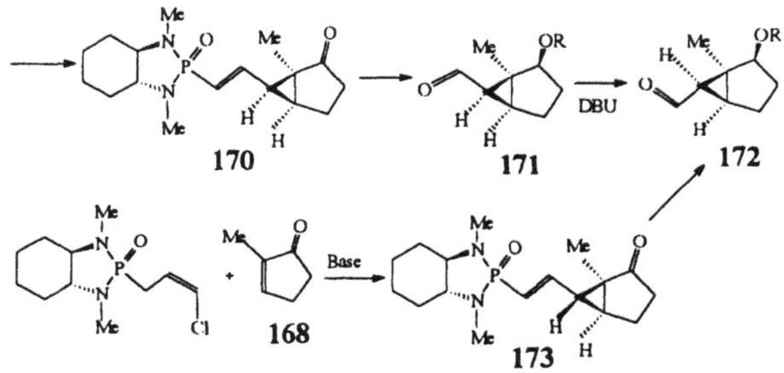

Reaction of anions derived from chiral nonracemic allyl and crotyl bicyclic phosphonamides **174** with α,β-unsaturated cyclic ketones takes place at the γ-position of the reagents and led to diastereomerically pure or highly enriched products of conjugate addition **175**. Oxidative cleavage led to products corresponding to the formal conjugative addition of an acetaldehyde or a propionaldehyde anion equivalent to α,β-unsaturated carbonyl compounds **176**. The inclusion of HMPA was found to enhance the ratio of γ-1,4-addition and to improve the stereoselectivity in the case of 3-methylcyclopentenone. Mono-, di-, and trisubstituted cyclopentanones are obtained as single diastereomers.[210]

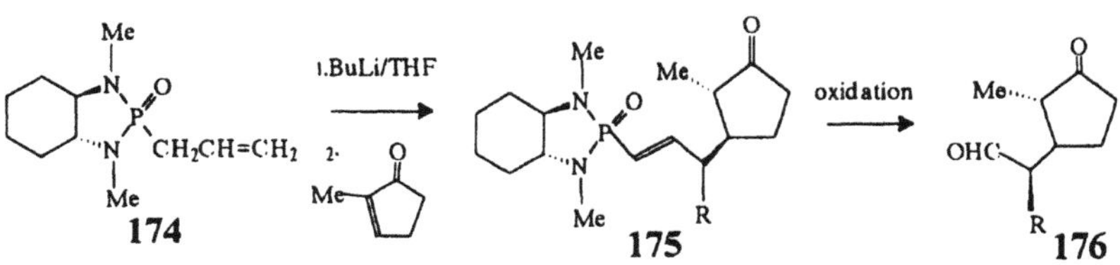

Haynes has since described the enantioselective γ-1,4-addition of individual enantiomers of (*E*)-2-butenyl-*tert*-butylphenylphosphine oxide to 2-methyl-2-cyclopentenone and formulated a model for the asymmetric induction. *t*-Butyl(methyl)phenylphosphine oxide was lithiated with butyllithium and then treated with propylene oxide and next with BF_3-ether to provide a 1:2 mixture of diastereomers of the -hydroxyphosphine oxide **177**, which was converted into (*S*)- and (*R*)-allyl-*tert*-butylphenylphosphine oxides **178**. The lithiated reagent derived from the phosphine oxide **178** was treated with 2-methylcyclopent-2-enone to give unsaturated diketone **179**, which was converted into corresponding enantiomers of hydrindenones **180**, suitable for conversion into vitamin D analogues and their enantiomers.[128,129]

The carbanion-accelerated Claisen rearrangement (CACR) of allyl vinyl ethers has proven to be a reaction of synthetic potential. The utility of various phos-

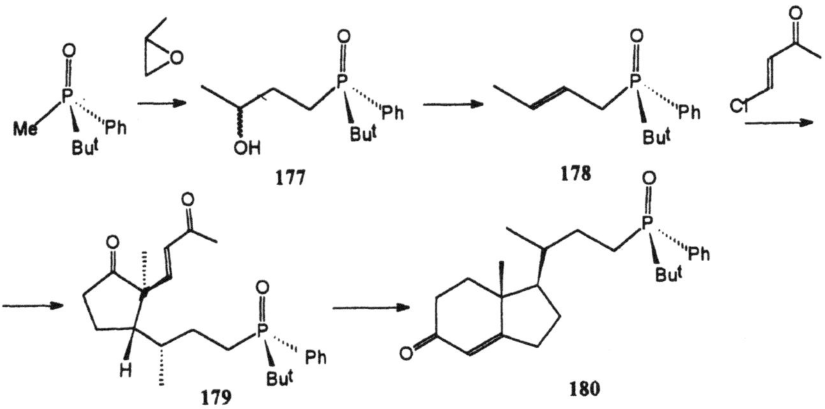

phonamide groups has been examined in the context of CACR. The *N,N'*-dibenzyl-1,3,2-diazaphospholidine group is the most optimal for the construction of the CACR precursor and the stereoselectivity of its rearrangement. Using butyllithium as the base, phosphonamides **181** rearranged readily at –20 °C into **182** with complete regioselectivity, in good yield, and, compared to carbanion accelerated at a high level of diastereoselectivity (>95% de). However in a simple Claisen rearrangement of the chiral *N,N'*-dibenzyl-1,3,2-diazaphospholidine **181** the relative asymmetric induction was poor (~20% de)[82,85] The carbanion-accelerated

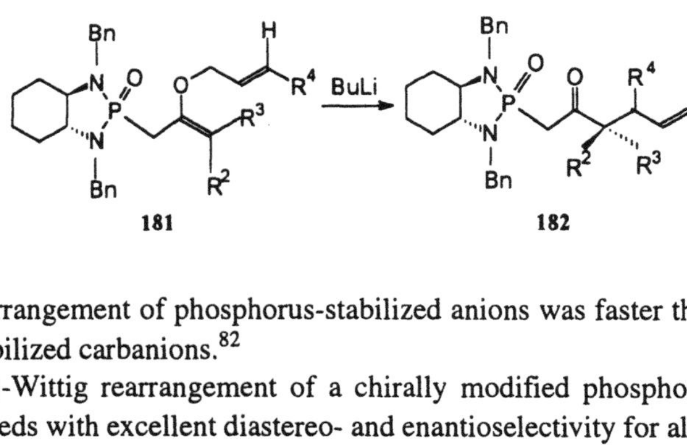

Claisen rearrangement of phosphorus-stabilized anions was faster than in case of sulfone-stabilized carbanions.[82]

The [2,3]-Wittig rearrangement of a chirally modified phosphorus-stabilized anion proceeds with excellent diastereo- and enantioselectivity for allyloxymethyl and (*Z*)-2-butenyloxymethyl derivatives.

Deprotonation of 1,3,2-oxazaphosphorinate with buthyllithium in THF at –70 °C generated the phosphorus-stabilized anion **183,** which underwent the [2,3]-Wittig rearrangement to afford a single diastereomer of hydroxy 3-butenyl-1,3,2-oxazaphosphorinanes **184** in good yield. The configuration of the hydroxy bearing stereocenters of compounds (*S*)-(+)- and (*R*)-(–)-**185** was determined by comparison of the sign of optical rotation with compounds prepared by independent synthesis.

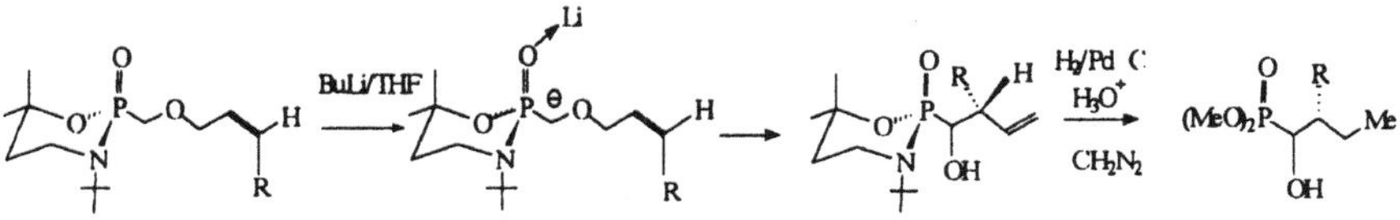

183 **184** **185**

Another interesting example of the [2,3]-Wittig rearrangement was described by Collignon. A diastereoselectivity of up to 90% was achieved in the rearrangement of lithiated allyloxymethylphosphonates **186** by using the chiral dimenthylphosphinyl ester group as stereodirecting auxiliary. After treatment with an excess of butyllithium in THF at –78 °C, dimethyl allyloxymethylphosphonate **186** underwent complete [2,3]-Wittig rearrangement giving, after low-temperature acid hydrolysis of the reaction mixture of **187** and subsequent work up (1-hydroxy-3-buten)-1-yl phosphonate (**188**), isolated in 95% as a mixture of two diastereomers in a 96:4 ratio. Esterification of **188** with diazomethane in dioxane led to formation of dimethyl 1-hydroxy butylphosphonate (**189**). Comparison of the sign of optical rotation with that a similar phosphonate of known configuration allowed the assignment of the *R* configuration to the (–)-phosphonate **189** derived from (–)-menthol and the *S* configuration to the phosphonate **189** derived from (+)-menthol.[111]

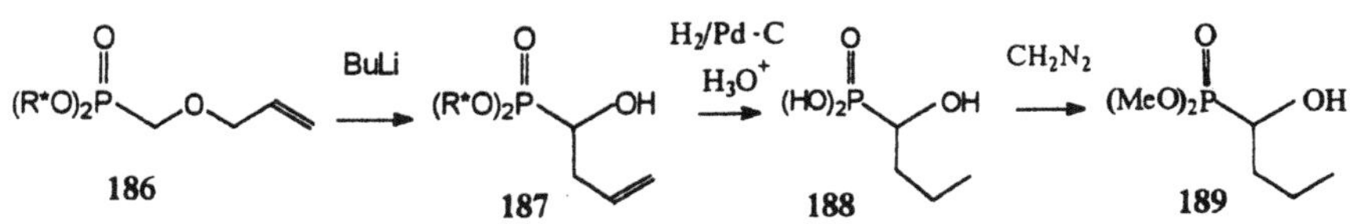

186 **187** **188** **189**

R*=(-)-Menthol

Chiral nonracemic 1-hydroxyphosphonates are interesting precursor for other α-functional phosphonates.

Enantioselective Olefination. The Horner–Wadsworth–Emmons (HWE), as well as Wittig reactions between phosphorus ylides or phosphorus-stabilized carbanions and aldehydes or ketones are an important, practical, method for the construction of the carbon–carbon double bond. Many attempts to develop an asymmetric version of the Wittig and Horner–Wittig reactions have been reported over the past three decades[104,255] since the pioneering work by Bestmann, which employed chiral P-ylides for the synthesis of optically active allenes and cycloalkylidenes.[26,27] Bestmann also described asymmetric catalysis in the Wittig reaction.[28]

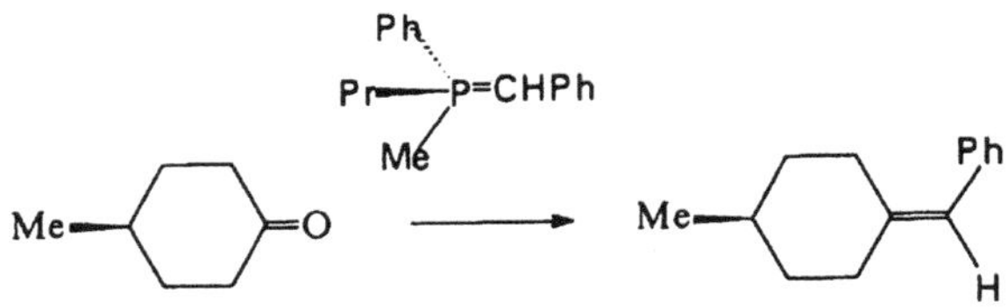

The carbonyl olefination reaction employing phosphoryl-stabilized carbanions (Horner–Wittig reaction) is now a well-established and useful alternative to the Wittig olefination. The Horner–Wittig reaction of anions derived from phosphine oxides, phosphonates, phosphonamides, and their thiono counterparts have well-documented advantages in many situations.[78] Different chirally modified phosphinates,[225,226] phosphonates,[101,104,231,254,255,276,279] phosphonamides,[17,118,121,122] phosphinothionic amides,[152] phosphine oxides,[126,248] oxazaphosphorinanes,[78] oxathiaphospho-rinanes,[174] and phosphoranes[26–28,30,278,280,281] have been employed for asymmetric olefination with variable succes. Optically active (*ortho*-methoxyphenyl) phenylphosphine oxide **190** was used to create three new chiral carbon atoms, as the carbon framework of **191** was built up by stereoselective electrophilic attack on a cyclic allylic phosphine oxide. Elimination of the chiral auxiliary in a Horner–Wittig reaction gave compound **192.** Optically active phosphine oxide was converted into the optically active alcohol (*S*)-(+)-**193** and into the optically active sulfide **194** as shown in the Scheme 16.[292]

It was shown that the reaction of chiral ylides with chiral acyl halides resulted in partial kinetic resolution and the production of optically active allenes.[154,192] The reaction of carbanion obtained from optically active *O*-methylphenylphosphinyl

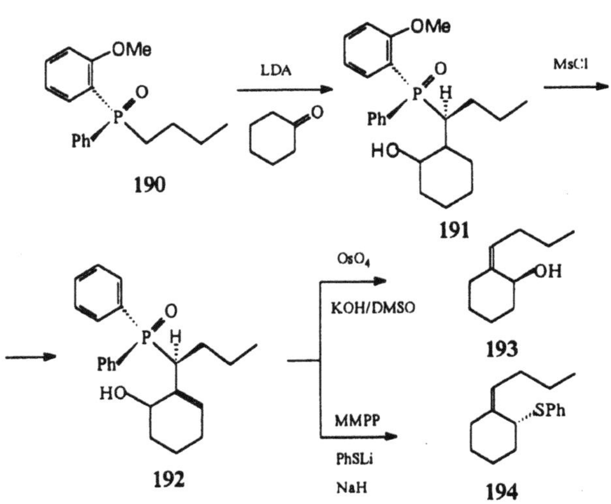

Scheme 16.

acetate with unsymmetric ketenes and with racemic 2-substituted cycloalkanones was obtained by Musierowicz and coworkers. The ketenes reacted smoothly with (*Sp*)-phosphorylated carbanion in benzene-ether solution to induce the formation of the allenes of configuration *R*. The ee was only 12–17%.[26,27,30,226,278,281]

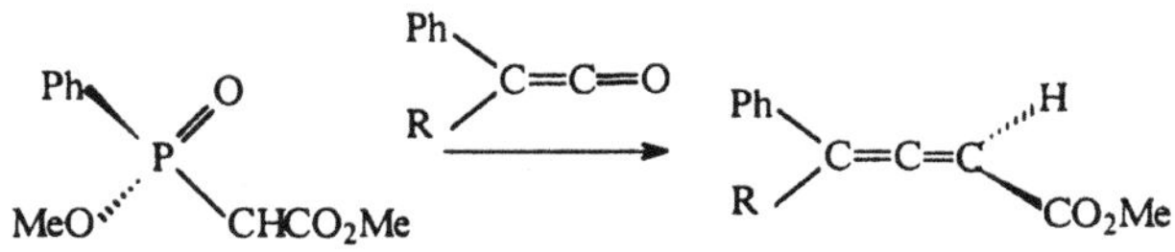

The most effective successes in the application of the Wittig–Horner reaction for the preparation of chiral alkenes were achieved with cyclic chiral phosphondiamides. Since the Wittig reaction does not create a new sp^3 carbon center, efforts to develop an asymmetric version of the Wittig reaction have been focused mainly on alkylidenecycloalkanes with axial chirality. The synthesis of topologically unique, enantiomeric chiral bicyclic phosphonamide reagents was reported by Hanessian and coworkers.[120,121] Their remarkable stereodifferentiating reactivity toward alkylcyclohexanones as well as towards alkyl halides was demonstrated. Paramount in the design of these reagents was the inherent C_2 symmetry of the parent chiral diamines and the stereoelectronic consequences resulting from the spatial disposition of the heteroatom. By far the most selective direct transformations employ the chiral phosphondiamides and menthyl phosphonoacetates with enantioselectivities of up to 90% for the elimination–coupling process with chiral sulfoximines.[102,103]

Treatment of 4-*tert*-butylcyclohexanone with the anion of **195** generated with KDA gave a 82% yield of (*R*)-4-*tert*-butylcyclohexylidene)ethane with an optical purity of 90 ± 2%.[120,121]

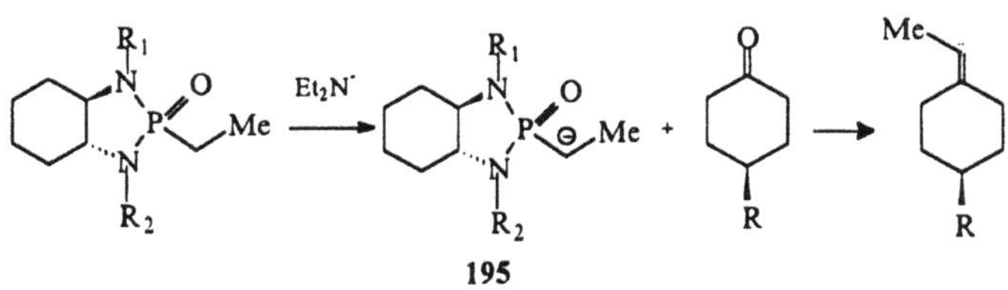

195

Hanessian et al. have described a highly selective synthesis of dissymmetric olefins (86–100% ee) that employs a novel electrophilic activation of the Horner–Wittig process. Notable results were achieved in which axially dissymmetric (alkylcyclohexylidene)ethanes were produced with up to 90% optical purity from cyclohexanones derivatives and a chiral bicyclic phosphonamide.[26,29,78,117–119,121,152,292,197,202,259] Treatment of alkyl cyclohexanones with topologically unique phosphonamides derived from the C_2 symmetrical (*R,R*)- and (*S,S*)-*N,N'*-dimethyl-

1,2-*trans*-cyclohexanediamine leads to enantiomerically pure allylidene, benzylidene and propylidene alkylcyclohexanes.[117]

A series of chiral (arylmethylene)cycloalkanes **196** was synthesized in opticaly active form for incorporation into a liquid crystal-based optical switch.[202]

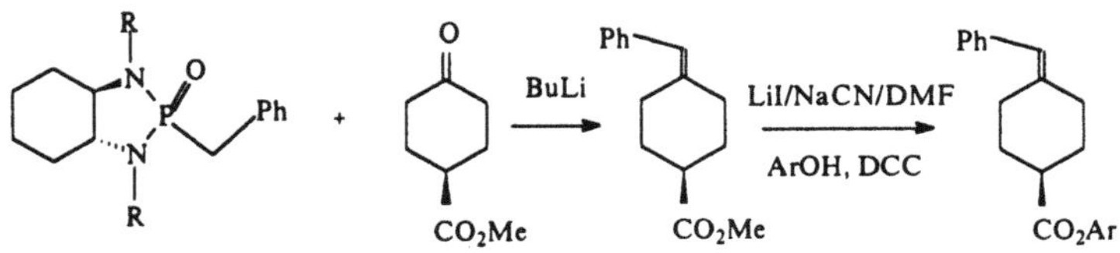

196

Wittig–Horner reactions in the solid state of the inclusion compound of 4-methyl- or 3,5-dimethylcyclohexanone and an optically active host compound with (carbethoxymethylene) triphenylphosphorane gave optically active 4-methyl- and 3,5-dimethyl-1-carbethoxymethylene-cyclohexane.[275,276] A selectivity of 82% de for 4-substituted cyclohexanones was obtained with carbanions derived from menthol phosphonates.[255]

The methodology developed by Hanessian includes isolation of β-hydroxyl intermediates, purification, and transformation into 4-substituted cyclohexanones. The treatment of alkyl cyclohexanones with anions of chiral (*R,R'*)- and (*S,R*)-bicyclic allyl or benzylphosphondiamides led to corresponding β-hydroxyphosphonamide intermediates **197** in excellent yield and enantioselectivity.

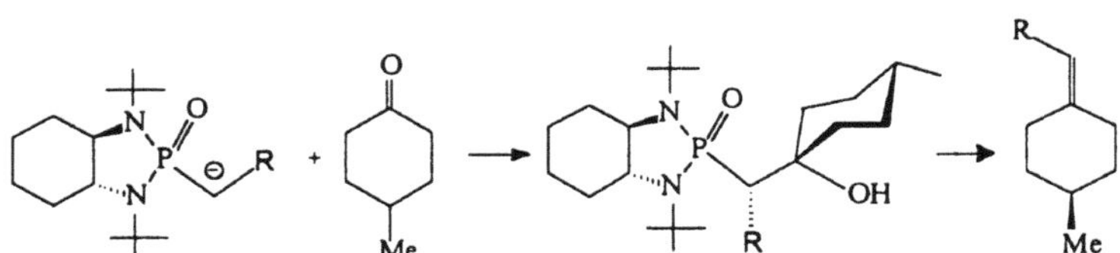

197

The isolation of an intermediate of β-hydroxyphosphonamide **197** allowed the clarification of the mechanism of reaction. The disposition of the phosphoryl appendage reflects a preferential "equatorial" attack on the carbonyl group and the orientation of the α-phenyl group corresponds to the attack on the *pro-S* face of the anion by the electrophile.

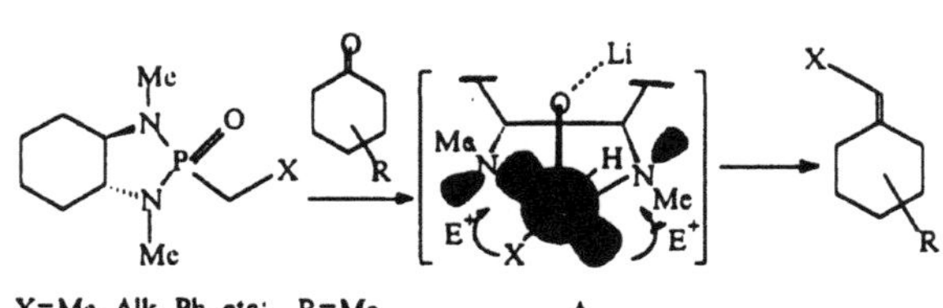

A

Hanessian and Chen applied chiral phosphonamidate anions derived from 1,3,2-oxazaphosphorinane 2-oxides to the enantioselective olefination of 4-substituted cyclohexanones. In all cases the adducts **198** were obtained in high yields and excellent diastereoselectivity (88–100%). The triflate-induced olefination also proceeded smoothly to afford dissymmetric olefins with complete stereospecificity.[18,78]

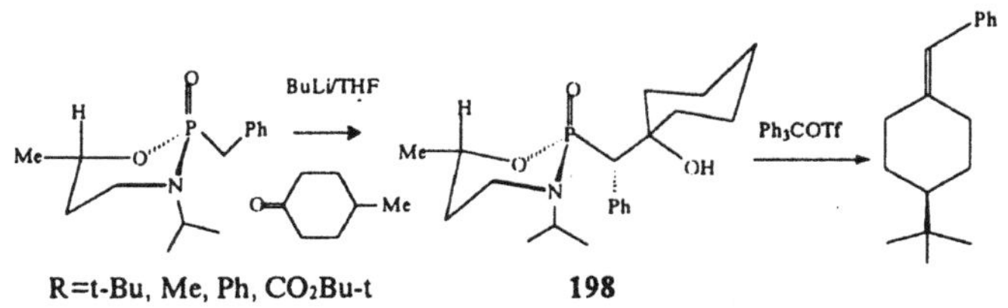

198

Optimization (brief warming at 60 °C in acetonitrile) led to the highest enantiomeric excesses. Readily available camphor-derived amino alcohol auxiliaries were used for the preparation of oxazaphospholidine 2-oxides **199**, which were converted into phosphorus stabilized carbanions, generated by action of lithium bis(trimethylsilyl)amide. Optimization of olefination conditions led to 4-*tert*-butylcarbalkoxycyclohexylidene in high yield and excellent enantioselectivity. The phosphorus isomers were formed in ratios between 1:1–1:12 (*cis/trans*) depending on the N-substituents and could be easily separated by column chromatography. The asymmetric carboalkoxyalkylidenation of 4-substituted cyclohexanones was effected by the use of chirally modified Horner–Wadsworth–Emmons reagents in good yields. The reaction was conducted in THF at –35 °C and after 48 h good yields and respectable levels of enantioselectivity were obtained (78–86%). The selectivity of the reaction depends on the anion, temperature, solvent and substituent at the nitrogen atom. Electron-withdrawing N-substituents led to a slower reaction but higher selectivities, while increasing solvent polarity (Et_2O < DME < DMF) decreases the rate and increases stereoselectivity.

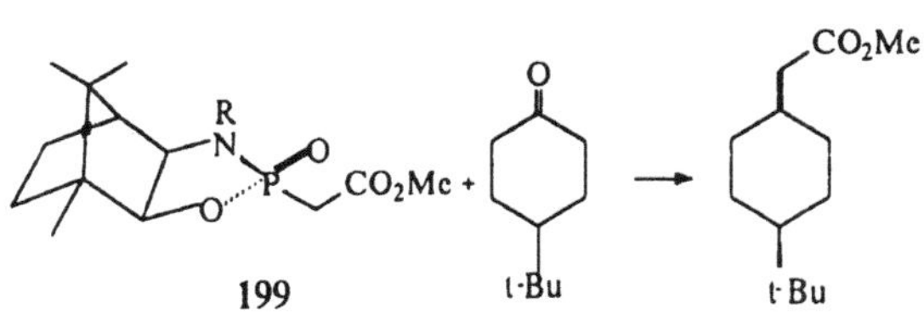

199

Asymmetric carboalkoxyalkylidenation with a chiral Horner–Wadsworth–Emmons reagents was also conducted by Japanese authors.[87,231] Natural mannitol was converted into a C_2 axis symmetrical dimethylene acetal by transacetalation with dimethoxymethane. Chiral tricyclic phosphonate **200** was prepared from the above chiral diol and a dichlorophosphonate in the presence of triethylamine. Kinetic resolution of racemic α-substituted carbonyl compounds by the Wittig reaction with the chiral tricyclic phosphonate **200** led to the formation of carboalkoxyalkylidene product **201** in high yield and good stereoselectivity.[127,235b] The absolute configuration of **201** was determined as *S*.[231]

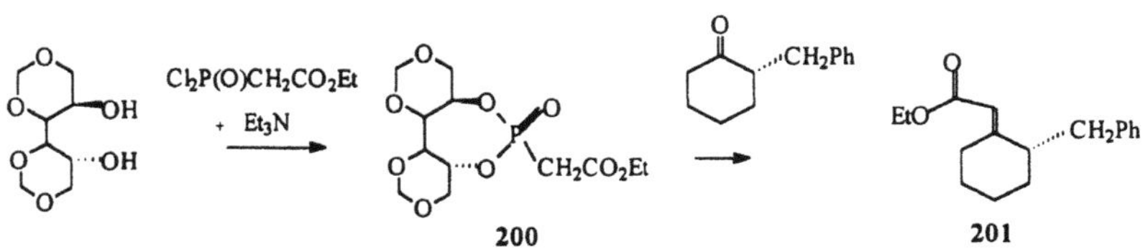

Reaction between racemic α-benzylcyclohexanone and the chiral dinaphthyl phosphonate **202** leads to chiral olefins **203** with good stereoselectivity.[231]

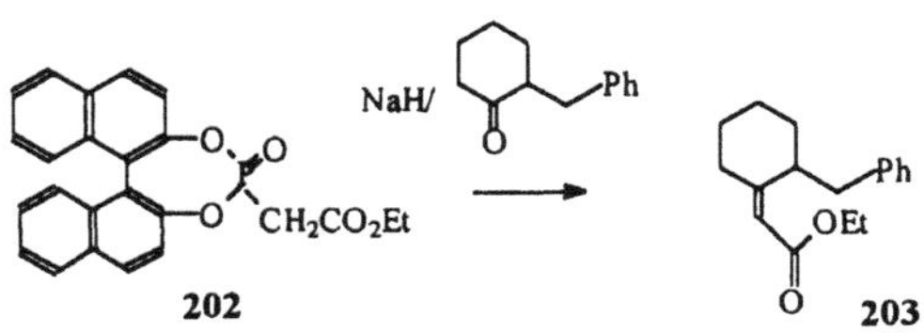

Fugi and coworkers studied asymmetrization of a *meso*-dicarbonyl compound **204** by the Horner–Wittig reaction utilizing a chiral phosphonoacetate reagent **205** bearing binaphthol as a chiral auxiliary. Treatment of *meso*-α-dicarbonyl compound with the anion **205** generated with NaH at –78 °C in THF gave β-alkoxycarbonyl-α,β-unsaturated ketone **206** with nearly a 98% enantiomeric excess. The chiral cyclic phosphonoacetate was very effective for enantiogroup differentiation in *meso*-α-diketone **204** to give the (*Z*)-olefin in which complete transfer of chirality from the binaphthol moiety to product was realized. The absolute stereochemistry was determined by X-ray analysis.[272]

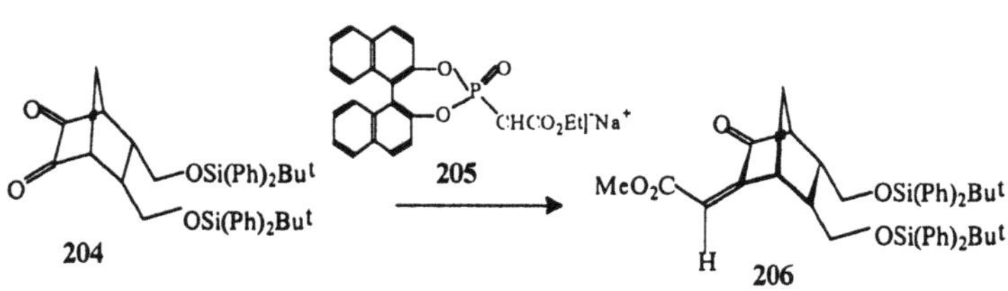

Hanessian and Beaudoin elaborated an elegant chemical method based on the construction of chiral alicyclic cyclohexenes **207** and involving the application of sequential asymmetric olefination and ene reaction. Oxidative cleavage of **207** led to their acyclic counterparts **208** containing stereochemically defined C-methyl groups (Scheme 17)[117,124]

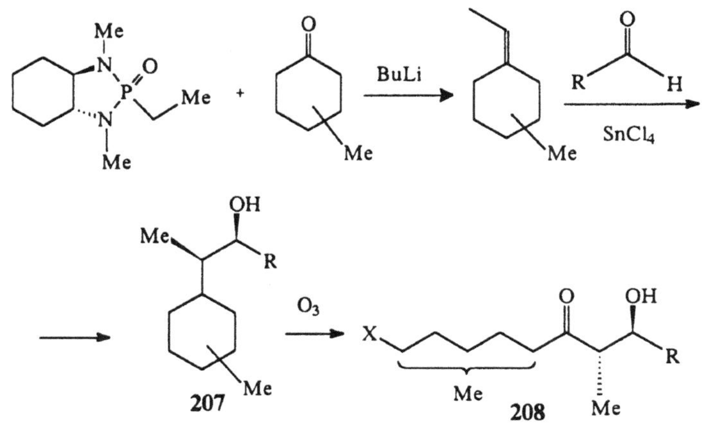

Scheme 17.

1,4-Asymmetric Addition to P-Stabilized Carbanions: Michael Reactions. One of the most fundamental questions surrounding the use of stabilized allyl anions in their reaction with electrophiles is the α/γ-regioselectivity. This type of reactivity has been found to be highly dependent on the steric and electronic nature of stabilizing groups as well as the nature of the electrophile. The synthetically most useful application of such allyl anions is the Michael addition reaction. Highly stereocontrolled sequential asymmetric Michael addition reactions with cinnamate esters which led to generation of three and four contiguous stereogenic centers was described by Hanessian. An asymmetric Michael reaction of the anion **209**, generated by BuLi from the crotyl phosphonamide with *t*-butyl cinnamate followed by addition of methyl iodide, led to formation of adducts **210** in a ratio of 92:8 and in very good yield. Ozonolysis and $NaBH_4$ reduction gave the hydroxy ester derivative **211** as a single isomer. Treatment with TFA in dichloromethane led to the lactone **212** containing three contiguous carbon substituents.[124]

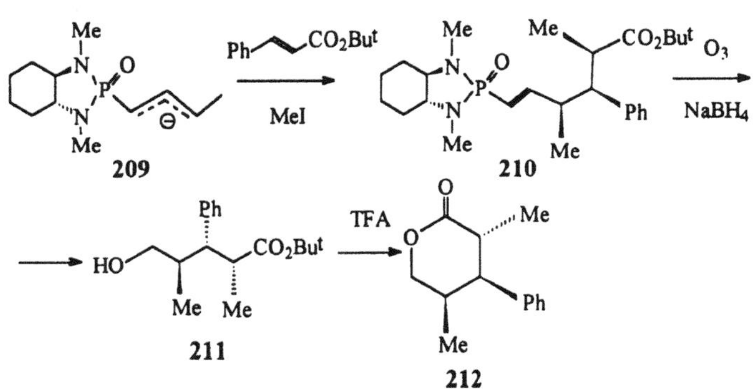

Following the same protocol as above and quenching the enolate with methyl triflate the same authors obtained a 90:10 mixture of diastereomers **213**. Ozonolysis, reduction, and chromatographic separation gave the hydroxy ester **214** as a single isomer, harboring four contiguous stereogenic centers.[123,209]

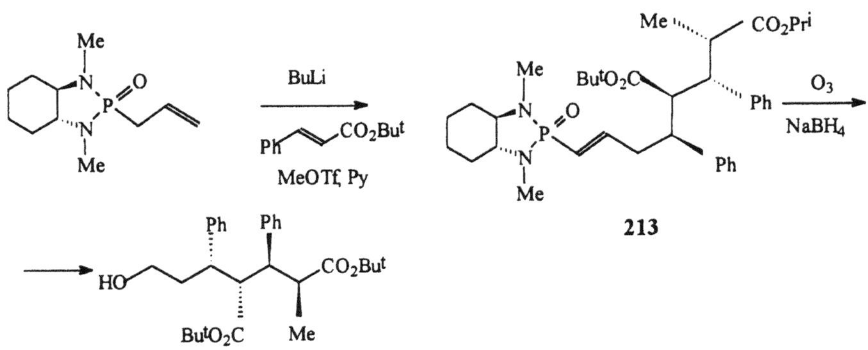

213

214

Reaction of anions derived from chiral nonracemic allyl and crotyl bicyclic phosphonoamides with α,β-unsaturated cyclic ketones, esters, lactones, and lactams takes place at the γ-position of the reagents and leads to diastereomerically pure or highly enriched products of conjugate addition. High diastereoselectivity was observed for the asymmetric Michael addition reaction of racemic 2-allyl-1,3,2-oxazaphosphorinane 2-oxide with cyclic enones (88–90% de).[81,85]

VIII. CONCLUSION AND FINAL REMARKS

Compounds with optical activity due to the presence of an asymmetric phosphorus center are accessible and can be relatively easily obtained. Certain resolved organophosphorus compounds have played a pivotal role in the preparation of other phosphorus-centered asymmetric compounds, primarily by nucleophilic substitution reactions at phosphorus. It is with this class of reactions that phosphorus chemistry clearly departs from typical carbon or nitrogen chemistry. Thus, bimolecular nucleophilic displacements at saturated carbon proceeds with inversion of configuration and does not involve a pentacoordinate intermediate, whereas S_N2 reaction at phosphorus occurs with retention or with inversion that can be explained by the intermediacy of a pentacoordinate phosphorus.

We hope that this review of chiral organophosphorus compounds will be useful to chemists interested in various aspects of chemistry and stereochemistry. The facts and problems discussed provide numerous possibilities for the study of additional stereochemical phenomena at phosphorus.

Looking in the future it may be said that asymmetric synthesis of organophosphorus compounds will be and should be the subject of further studies. It remains to await the further development of the application of prochiral low-coordinate phosphorus compounds, and the study of stereochemistry of their reactions. Further studies might be dedicated to the stereochemistry of phosphorus in living processes, and of biologically active organophosphorus compounds. It remains to be seen that biological properties of the epimers of biological active organophosphorus compounds differ in their properties from their racemates.

Further development must include studies of the stereochemistry of hypervalent organophosphorus compounds. In spite of the successes in the area of basic research of hypervalent phosphorus compounds, their synthetic application in asymmetric synthesis is still deficient. Further applications of optical active organophosphorus reagents in asymmetric organic synthesis remain to be explored.

Finally, we note that this chapter was intended to be illustrative, not exhaustive; therefore, we apologize to the authors whose important studies could not be included.

ABBREVIATIONS

BINAP - 2,2′-bis(diphenylphosphino)-1,1′-binaphthyl
CAMP - cyclohexylanisylmethylphosphine
CHIRAPHOS - 1,2-bis(diphenylphosphino)butane
DABCO - 1,4-diazabicyclo[2,2,2]-octane
DBN - 1,5-diazabicyclo[4,3,0]non-5-ene
DBU - 1,8-diazabicyclo[5,4,0]undec-7-ene
DIOP - 2,3-isopropylidene-2,3-dihydroxy-1,4-bis(diphenylphosphino)butane
DIPAMP - 1,2-bis(*o*-anisylphenylphosphino)ethane
DIPHOS - 1,2-bis(diphenylphosphino)ethane
DME - 1,2-dimethoxyethane
DMF - dimethylformamide
dr - diastereomeric ratio
de - diastereomeric excess
ee - enantiomeric excess
GF - (–)-1,2:5,6-diisopropylidene-D-glucofuranose;
HMPA - hexamethylphosphosphorous triamide
KHMDS - *N*-potassiohexamethyldisylazane
KDA - potassium diethylamide
NaHMDS - *N*-sodiohexamethyldisylazane
NEA - (–)-2-naphthylethylamine;
PAMP - phenylanisylmethylphosphine
PEA- (–)-2-phenylethylamine.
THF -tetrahydrofuran

REFERENCES

1. Abel, A. D.; Hoult, D. H.; Morris, R. M. *J. Org. Chem.* **1993**, *58*, 1531.
2. Aggarval, V. K. *Angew. Chem.* **1994**, *33*, 175.
3. Ahlbrecht, H. *Chem. Ber.* **1986**, *117*, 1239.
4. Allen, D. G.; Wild, S. B.; Wood, D. L. *Organometallics* **1986**, *5*, 1009.
5. Alexakis, A.; Mutti, S.; Normand, J. F.; Mangeney, P. *Tetrahedron: Asymmetry* **1990**, *1*, 437.
6. Alexakis, A.; Mutti, S.; Mangeney, P. *J. Org. Chem.* **1992**, *57*, 1224.
7. Alexakis, A.; Frutos, J. C.; Mutti, S.; Mangeney, P. *J. Org. Chem.* **1994**, *59*, 3326.
8. Anderson, R. C.; Shapiro, M. J. *J. Org. Chem.* **1984**, *49*, 1304.
9. Atherton, F. R.; Hassal, C. H.; Lambert, R. W. *J. Med. Chem.* **1986**, *29*, 29.
10. (a) Atherton, F. R.; Hall, M. J.; Hassal, C. H.; Lambert, R. W.; Lloid Ringrose, W. J. P. S. *Antimicrob. Agents Chemother.* **1979**, *15*, 696; (b) Awerbuch, O.; Koshman, Y. *Tetrahedron* **1975**, *31*, 33.
11. Bader, A.; Salem, G.; Willis, A. C.; Wild, S. B. *Tetrahedron: Asymmetry* **1992**, *3*, 1227.
12. Baechler, R. D.; Mislow, K. *J. Am. Chem. Soc.* **1970**, *92*, 3090.
13. Baechler, R. D.; Mislow, K. *J. Am. Chem. Soc.* **1971**, *93*, 773.
14. Baker, M. J.; Pringle, P. G. *J. Chem. Soc., Chem. Commun.* **1993**, 314.
15. Barltett, P. A.; Jacobsen, N. E. *J. Am. Chem. Soc.* **1981**, *103*, 654.
16. Barltett, P. A.; Kezer, W. B. *J. Am. Chem. Soc.* **1984**, *106*, 4282.
17. Barltett, P. A.; Giannousis, P. P. *J. Med. Chem.* **1987**, *32*, 119.
18. Bartoli, G.; Marcantoni, E.; Sambri, L.; Tamburini, M. *Angew. Chem., Int. Ed. Engl.* **1995**, *34*, 2046.
19. Bongini, A.; Camerini, R.; Pamenzio, H. *Tetrahedron: Asymmetry* **1996**, *37*, 1467.
20. le Bec, C.; Wickstrom, E. *Tetrahedron Lett.* **1994**, *35*, 9525.
21. le Bec, C.; Wickstrom, E. *J. Org. Chem.* **1996**, *61*, 510.
22. Benabra, A.; Alcudia, A., Khiar, N.; Fernandez, I.; Alcudia, F. *Tetrahedron: Asymmetry* **1996**, *7*, 3353.
23. Bennani, Y. L.; Hanessian, S. *Tetrahedron Lett.* **1990**, *31*, 6465.
24. Bennani, Y. L.; Delorme, D.; Hanessian, S. *Tetrahedron Lett.* **1990**, *31*, 6461.
25. Benson, S. C.; Snyder, J. K. *Tetrahedron Lett.* **1991**, *32*, 5885.
26. Bestmann, H. J.; Heid, E.; Ryschka, W.; Lienert, J. *Lieb. Ann. Chem.* **1974**, 1684.
27. Bestmann, H. J.; Tomoskozi, I. *Tetrahedron* **1968**, *24*, 3299.
28. Bestmann, H. J.; Lienert, J. *Chem. Ztg.* **1970**, *94*, 487.
29. Bestmann, H. J. *Pure Appl. Chem.* **1980**, *52*, 771.
30. Bestmann, H. J.; Lienert, J. *Angew. Chem., Int. Ed. Engl.* **1969**, *8*, 763.
31. Blazis, V. J.; De la Cruz, A.; Koeller, K. J.; Spilling, C. D. *Phosph. Sulf. and Silicon* **1993**, *75*, 159.
32. Blazis, V. J.; Koeller, K. J.; Spilling, S. D. *J. Org. Chem.* **1995**, *60*, 931.
33. Blazis, V. J.; Koeller, K. J.; Spilling, S. D. *Tetrahedron: Asymmetry* **1994**, *5*, 499.
34. Bodalski, R.; Rutkowska-Olm, E.; Pietrusiewicz, K. M. *Tetrahedron* **1980**, *36*, 2353.
35. Bodalsky, R.; Koszuk, J. *Phosph., Sulf., and Silicon* **1989**, *44*, 87.
36. Boutin, J. A.; Cudennec, C.A.; Hautefaye, P.; Lavielle, G.; Pierre, A.; Schaeffer, C. *J. Med. Chem.* **1991**, *34*, 1998.
37. Brandi, A.; Cichi, S.; Goti, A.; Pietrusiewicz, K. M.; Zablocka, M.; Wisniewski W. *J. Org. Chem.* **1991**, 56, 4383.
38. Brandi, A.; Cichi, S.; Goti, A.; Pietrusiewicz, K. M. *Tetrahedron: Asymmetry* **1991**, *2*, 1063.
39. Brandi, A.; Cichi, S.; Goti, A.; Pietrusiewicz, K. M.; Wisniewski, W. *Tetrahedron* **1990**, *46*, 7093.
40. Brandi, A.; Cichi, S.; Goti, A.; Koprowski, M.; Pietrusiewicz, K. M. *J. Org.* Chem. **1994**, *59*, 1315.
41. Brown, J. M.; Carey, J. V.; Russell, M. J. H. *Tetrahedron* **1990**, *46*, 4877.
42. De Bruin, K. E.; Zon, G.; Naumann, K.; Mislow, K. *J. Am. Chem. Soc.* **1969**, *91*, 7027.
43. De Bruin, K. E.; Mislow, K. *J. Am. Chem. Soc.* **1969**, *91*, 7393.

44. De Bruin, K. E.; Petersen, D. M. *J. Org. Chem.* **1972,** *37,* 2772.
45. De Bruin, K. E.; Johnson, D. M. *J. Am. Chem. Soc.* **1973,** *95,* 4675.
46. De Bruin, K. E.; Perrin, E. D. *J. Org. Chem.* **1975,** *40,* 1523.
47. De Bruin, K. E.; Padilla, A. G., Campbell, M. I. *J. Am. Chem. Soc.* **1973,** *95,* 4681.
48. Brunel, J.-M.; Pardigon, O.; Maffei, M.; Buono, G. *Tetrahedron: Asymmetry* **1992,** *3,* 1243.
49. Brunel, J.-M.; Faure, B. *Tetrahedron: Asymmetry* **1995,** *6,* 2353.
50. Brunner, H. *Topics in Stereochemistry*; Eliel E.L.; Wilen; S. H., Eds.; John Wiley: New York, 1988, Vol. 18, pp. 129–248.
51. Buisman, G. J. H.; Martin, M. E.; Vos, E. J.; Klootwijk, A.; Kamer, P. C. J.;. van Legwen P. W. N. M. *Tetrahedron: Asymmetry* **1995,** *6,* 719.
52. Burgess, K.; Ohlmeyer, M. J.; Whitmire, K. H. *Organometallics* **1992,** *11,* 3588.
53. Carey, J. V.; Barker, M. D.; Brown, J. M.; Russell, M. J. H. *J. Chem. Soc., Perkin Trans. 1* **1993,** 831.
54. Carrelhas, A. C.; Brazier, J.-F.; Houalla, D.; Wolf, R. *Phosphorus* **1977,** *8,* 61.
55. Cassaigne, A.; Neuzel, E. *Exp. Ann. Biochim. Med.* **1980,** *34,* 165.
56. Cavezzan, J.; Etemad-Moghadam; Koenig M.; Klaebe A. *Tetrahedron Lett.* **1979,** 795.
57. Cernia, E.; Giongo, G. M.; Marcati, F.; Marconi, W.; Palladino, N. *Inorg. Chim. Acta* **1974,** *11,* 195.
58. Chodkiewicz, W.; Guilerm, D. *Tetrahedron Lett.* **1979,** 3573.
59. Chodkiewicz, W.; Jore, D., Wodski, W. *Tetrahedron Lett.* **1979,** 1069.
60. Chodkiewicz, W. *J.Organometal. Chem.* **1984,** *173,* C55.
61. Crisp, G. T.; Salem, G.; Stephens, F. E.; Wild, S. B. *J. Chem. Soc., Chem. Commun.* **1987,** 600.
62. Crisp, G. T.; Salem, G.; Wild, S. B.; Stephens, F. S. *Organometallics* **1989,** *8,* 2360.
63. Cooper, D. B.; Hall, C. R.; Inch, T. D. *Tetrahedron Lett.* **1974,** 2697.
64. Cooper, D. B.; Inch, T. D.; Lewis, G. J. *J. Chem. Soc., Perkin Trans 1* **1974,** 1043.
65. Cordillo, B.; Garduno, C.; Guadarrama, G.; Hernandez, J. *J. Org. Chem.* **1995,** *60,* 5180.
66. Corey, E. J.; Chen, Z.; Tanoury, G. J. *J. Am. Chem. Soc.* **1993,** *115,* 1101.
67. Corey, E. J.; Cane D. E. *J. Org. Chem.* **1969,** *34,* 3053.
68. Costick, R.; Williams, D. M. *Nucleic Acids Res.* **1987,** *15,* 9921.
69. Cramer, C. J.; Denmark, S. E.; Miller, P. C.; Dorow, R. L.; Swiss, K. A.; Wilson, S. R. *J. Am. Chem. Soc.* **1994,** *116,* 2437.
70. (a) Cremer, S. E.; Chorvat, R. J.; Trivedi, B. C. *J. Chem. Soc., Chem. Commun.* **1969,** 769; (b) Cremer, S. E.; Chorvat, R. J. *J. Org. Chem.* **1967,** *32,* 4066.
71. De'ath, N. J.; Ellis, K.; Smith, D. J. H.; Trippett, S. *J. Chem. Soc., Chem. Commun.* **1971,** 714.
72. Delogu, G.; Fabbri, D.; Dettori, M. A. *Tetrahedron: Asymmetry* **1996,** *7,* 413.
73. Denmark, S. E.; Amburgey, J. *J. Am. Chem. Soc.* **1993,** *115,* 10386.
74. Denmark, S. E.; Marlin, J. E. *J. Org. Chem.* **1987,** *52,* 5742.
75. Denmark, S. E.; Chen, C. T. *J. Org. Chem.* **1994,** *59,* 2922.
76. Denmark, S. E.; Chatani, N.; Pansare, S. V. T. *Tetrahedron* **1992,** *48,* 2191.
77. Denmark, S. E.; Cramer, C. J. *J. Org. Chem.* **1990,** *55,* 1806.
78. Denmark, S. E.; Chen, C. T. *J. Am. Chem. Soc.,* **1992,** *114,* 10674.
79. Denmark, S. E.; Dorow, R. C. *J. Org. Chem.* **1990,** *55,* 5926.
80. Denmark, S. E.; Dorow, R. L. *J. Am. Chem. Soc.* **1990,** *112,* 864.
81. Denmark, S. E.; Kim, J. H. *J. Org. Chem.* **1995,** *60,* 7535.
82. Denmark, S. E.; Marlin, J. E. *J. Org. Chem.* **1991,** *56,* 1003.
83. Denmark, S. E.; Miller, P. R. *Tetrahedron. Lett.* **1995,** *6,* 6631.
84. Denmark, S. E.; Miller P. C.; Wilson S. R. *J. Am. Chem. Soc.* **1991,** *113,* 1468.
85. Denmark, S. E.; Stadler, H.; Dorow, R. L.; Kim, J. H. *J. Org. Chem.* **1991,** *56,* 5063.
86. Denmark, S. E.; Rajendra, G.; Marlin, J. E. *Tetrahedron Lett.* **1989,** *30,* 2469.
87. Denmark, S. E.; Rivera, J. *J. Org. Chem.* **1994,** *59,* 6887.
88. Denmark, S. E.; Dorow, L. R. *J. Org. Chem.* **1990,** *55,* 1806.

89. Denmark, E. S.; Swiss, K. A.; Wilson, S. R. *J. Am. Chem. Soc.* **1993**, *115*, 3826.
90. Dore, A.; Fabbri, S.; Gladiali, S.; De Lucchi, O. *J. Chem. Soc., Chem. Commun.* **1993**, 1124.
91. Emsley, J.; Hall, C. D. *The Chemistry of Phosphorus;* Harper and Row: New York, 1976, Chapter 8.
92. Endo, M.; Komiyama, M. *J. Org. Chem.* **1996**, *61*, 1994.
93. Evans, D. A.; Tacacs, J. M.; Hurst, K. M. *J. Am. Chem. Soc.* **1979**, *101*, 371.
94. (a) Fabbri, D.; Gladiali, S.; De Lucchi, O. *Synth. Commun.* **1994**, *24*, 1271; (b) Gladiali, S.; Fabbri, D. *Chem. Ber.* **1997**, *130*, 543.
95. Fabbri, D.; Delogu, G.; De Lucchi, O. *J. Org. Chem.* **1993**, *58* , 1748.
96. Farnham, W. B.; Murray, Jr., R. K.; Mislow, K. *J. Chem. Soc., Chem. Commun.* **1971**, 146.
97. Farnham, W. B.; Murray, Jr., R. K.; Mislow K. *J. Am. Chem. Soc.* **1970**, *92*, 5809.
98. Fisher, C.; Mosher H. *Tetrahedron. Lett.* **1977**, 2487.
99. Freeman, S.; Harger, M. J. P. *J. Chem. Soc., Chem. Commun.* **1985**, 1394.
100. Freeman, R. N.; Haynes, R. K.; Loughlin, W. A.; Mitchell, C. R.; Stokes, C. *Pure Appl. Chem.* **1993**, *65*, 647.
101. Fuji, K.; Tanaka, K.; Ohta, Y.; Taga, T. *Tetrahedron Lett.* **1993**, *34*, 4071.
102. Gais, H-J.; Bulow, G. *Tetrahedron Lett.* **1992**, *33*, 461.
103. Gais, H-J.; Bulow, G. *Tetrahedron Lett.* **1992**, *33*, 465.
104. Gais, H. E.; Schmiedl, G.; Ball, W. A.; Bund, J.; Hellmann, G.; Erdelmeier, J. *Tetrahedron Lett.* **1988**, *29*, 1773.
105. Gallagher, M. J.; Ranasinghe, M. G. *J. Org. Chem.* **1996**, *61*, 436.
106. Ganci, W.; Meier, E. J. M.; Mercling, F. A.; Przibille, G.; Ringeisen, U.; Ruedl, P. *Helv. Chim. Acta* **1997**, *80*, 421.
107. Gordon, N. J.; Evans, S. A., Jr. *J. Org. Chem.* **1993**, *58*, 5295.
108. Gordon, N. J., Evans, S. A., Jr. *J. Org. Chem.* **1993**, *58*, 5293.
109. Goti, A.; Cichi, S.; Brandi, A.; Pietruchiewicz, K. M. *Tetrahedron: Asymmetry* **1991**, *2*, 1371.
110. Greene, N.; Kee, T. P. *Synth.Commun.* **1993**, *23*, 1651.
111. Gulea-Purcarescu, M.; About-Jaudet, E.; Collignon, N. *Tetrahedron Lett.* **1995**, *36*, 6635.
112. Hall, C. R.; Inch, T. D. *J. Chem. Soc., Perkin Trans. 1* **1981**, 2368; **1979**, 1104; **1979**, 1646.
113. Hall, C. R.; Inch, T. D. *Tetrahedron Lett.* **1976**, 3645.
114. Hamor, T. A.; Jennings, W. B.; Lovely, C. J.; Reeves, K. A. *J. Chem. Soc., Perkin Trans 2* **1992**, 843.
115. Hanessian, S.; Andreotti, D.; Gomstyan, A. *J. Am. Chem. Soc.* **1995**, *117*, 10393.
116. Bennani, Y. L.; Hanessian, S.; Herve, Y. *Synlett* **1993**, *1*, 35.
117. Hanessian, S.; Beaudoin, S. *Tetrahedron Lett.* **1992**, *33*, 7655.
118. (a) Hanessian, S.; Beaudoin, S. *Tetrahedron Lett.* **1992**, *33*, 7659; (b) Hanessian, S.; Cention, L. D.; Roy, S.; Andreotti, D.; Gomstyan, A. *Tetrahedron Lett.* **1997**, *38*, 1103.
119. Hanessian, S.; Bennani, Y. L. *Tetrahedron Lett.* **1990**, *31*, 6465.
120. Hanessian, S.; Bennani, Y. L.; Delorme, D. *Tetrahedron Lett.* **1990**, *31*, 6461.
121. Hanessian, S.; Delorme, D.; Beaudoin, S.; Leblanc, Y. *J. Am. Chem. Soc.* **1984**, *106*, 5754.
122. Hanessian, S.; Delorme, D.; Beaudoin, S.; Leblanc, Y. *Chem. Scr.* **1985**, *88*, 1419.
123. Hanessian, S.; Gomstyan, A. *Tetrahedron Lett.* **1994**, *35*, 7509.
124. Hanessian, S.; Gomtsyan, A.; Payne, A.; Herve, Y.; Beaudoin, S. *J. Org. Chem.* **1993**, *58*, 5032.
125. Harger, M. J. P. *J. Chem. Soc., Chem. Commun.* **1988**, 1256.
126. Harmat, N. J. S.; Warren, S. *Tetrahedron Lett.* **1990**, *31*, 2743.
127. Hayasi, T.; Konishi, H.; Ito, H.; Kumada, M. *J. Org. Chem.* **1987**, *51*, 3772.
128. Haynes, R. K.; Stokes, J. P.; Hambley, T. W. *J. Chem. Soc., Chem. Commun.* **1991**, 58.
129. Haynes, R. K.; Freeman, R. N.; Mitchell, C. R.; Vonwiller, S. C. *J. Org. Chem.* **1994**, *59*, 2919.
130. Hellwinkel, D. In *Organic Phosphorus Compounds*; Kosolapoff, Maier L., Eds.; Wiley: New York, 1973, Vol. 3, p. 185.
131. Hellwinkel, D. *Chem. Ber.* **1966**, *99*, 3642.

132. Herriot, A. *J. Am. Chem. Soc.* **1971**, *93*, 3304.
133. Hilderbrand, R. L. *In The Role of Phosphonates in Living System*; CRC Press: Boca Raton, Fl., 1984.
134. Holmes, R. *Pentacoordinated Phosphorus ACS Monograph 176*, ACS: Washington, DC, 1980, Vol. II.
135. Holmes, R. *Chem. Rev.* **1996**, *96*, 927.
136. Horner, L. *Pure Apple Chem.* **1980**, *52*, 843.
137. Horner, L.; Jordan, M. *Phosphorus and Sulfur* **1980**, *8*, 225.
138. Horner, L. *Pure Apple Chem.* **1964**, *9*, 225.
139. Horner, L.; Simons, G. *Phosphorus and Sulfur* **1984**, *19*, 65.
140. Horner, L.; Winkler, H. *Liebigs Ann. Chem.* **1965**, *685*, 1.
141. Huong, Y.; Arif, A. M.; Bentrude, W. G. *J. Org. Chem.* **1993**, *58*, 6235.
142. Hulst, R.; Zijlstra, R. W. J.; Feringa, R. L.; Koen de Vries, N.; ten Hoeve, W.; Wynberg H. *Tetrahedron Lett.* **1993**, *34*, 1339.
143. Imamoto, T.; Kusumoto, Y.; Suzuki, N.; Sato, K. *J. Am. Chem. Soc.* **1985**, *107*, 5301.
144. Imamoto, T.; Sato, K.; Johnson, C. R. *Tetrahedron Lett.* **1985**, *26*, 783.
145. Imamoto, T.; Oshiki, T.; Onazawa, T.; Kusumoto, Y.; Sato, K. *J. Am. Chem. Soc.* **1990**, *112*, 5244.
146. Imamoto, T.; Oshiki, T. *Bull. Chem. Soc. Jpn.* **1990**, *63*, 3719.
147. Iyer, R. P.; Yu, D.; Ho, N.-H.; Tan, W.; Agraval, S. *Tetrahedron: Asymmetry* **1995**, *6*, 51; (b) Iyer, R. P.; Regan, J. B.; Egan, W.; Beaucage, S. L. *J. Am. Chem. Soc.* **1990**, *112*, 1253.
148. Jacobus, J.; Mislow, K. *J. Chem. Soc., Chem. Commun.* **1968**, 253.
149. de Jonch, C. P.; Modro, T. A. *Phosphorus and Sulfur* **1993**, *79*, 161.
150. Johnson, C. R.; Imamoto, T. *J. Org. Chem.* **1987**, *52*, 210.
151. Johnson, C. R.; Elliot, R. C.; Penning, T. D. *J. Am. Chem. Soc.* **1984**, *106*, 5019.
152. Johnson, C. R.; Elliot, R. C.; Meanwell, N. A. *Tetrahedron Lett.* **1982**, *23*, 5005.
153. Johnson, C. R.; Imamoto, T. *J. Org. Chem.* **1987**, *52*, 2170.
154. Johnson, A. W. *Ylides and imines of Phosphorus*; John Wiley, NY, 1993, p. 147.
155. Juge, S.; Stephan, M.; Achii, S.; Genet, J. P. *Phosphorus and Sulfur* **1990**, *50*, 267.
156. Juge, S.; Genet, J. P. *Tetrahedron Lett.* **1989**, *30*, 2783.
157. Juge, S.; Stephan, M.; Merdes, R.; Genet, J. P.; Halut-Desportes, S. *J. Chem. Soc., Chem. Commun.* **1993**, 531.
158. (a) Juge, S.; Stephan, M.; Laffitte, J. A.; Genet, J. P. *Tetrahedron Lett.* **1990**, *31*, 6357; (b) Juge, S.; Uziel, J.; Riegel, N.; Aka, B. *Tetrahedron Lett.* **1997**, *38*, 3405.
159. Juge, S.; Wakselman, M.; Stephan, M.; Genet, J. P. *Tetrahedron Lett.* **1990**, *31*, 4443.
160. Kabachnik, M. I.; Zakharov, L. S.; Kudryavtsev, I. Yu.; Petrovskii, P. V.; Scherbina, T. M.; Laretina, A. P. *Izv. AN* **1994**, *10*, 1840.
161. Kafarski, P.; Lejczak, B. *Phosph. Sulf. and Silicon* **1991**, *63*, 193.
162. (a) Kagan, H. B. *Asymmetric Synthesis*; Morrison, J. D., Ed.; Academic Press: Orlando, FL, 1985, Vol. 5, pp. 1–39; (b) Kagan, H. B.; Fiaud, J. C. *Top. In. Stereochem.* **1988**, *18*, 249.
163. Kagan, H. B. *Comprehensive Organometallic Chemistry*; Wilkinson, G., Ed.; Pergamon Press: Oxford, 1982, Vol. 8, p. 463.
164. (a) Kato, T.; Kobayashi, K.; Masuda, S.; Segi, M.; Nakajima, T.; Suga, S. *Chem. Lett.* **1987**, 1915; (b) Kielbasinski, P.; Zurawinski, R.; Pietrusiewich, M. K.; Zastocka, M.; Mikolajczyk, M. *Tetrahedron Lett.* **1994**, *35*, 7081.
165. King, R. B.; Bakos, J.; Hoff, C. D.; Marco, L. *J. Org. Chem.* **1979**, *44*, 3095.
166. Kinoshita, I.; Kashiwabara, K.; Fujita, J. *Chem. Lett.* **1977**, 831.
167. Klaebe, A.; Brazier, J. F.; Carrelas, A. C.; Garrigues, B.; Marre, M. R. *Tetrahedron* **1982**, *38*, 2111.
168. Klaebe, A.; Koenig, M.; Wolf, R.; Ahlberg, P. *J. Chem. Soc., Dalton Trans.* **1977**, 570.
169. Koeller, K. J.; Spilling, C. D. *Tetrahedron Lett.* **1991**, *32*, 6297.
170. Koenig, M.; Klaebe, A.; Munoz, A.; Wolf, R. *J. Chem. Soc., Dalton Trans.* **1976**, 955.
171. Koenig, M.; Klaebe, A.; Munoz, A.; Wolf, R. *J. Chem. Soc., Perkin Trans 2* **1979**, 40.

172. Kojima, S.; Nakamoto, M.; Yamazaki, K.; Akiba, K. *Tetrahedron Lett.* **1997**, *38*, 4107.
173. Koizumi, T.; Yanada, R.; Takagi, H.; Hirai, H.; Yoshii, E. *Tetrahedron Lett.* **1981**, *22*, 571; **1981**, *22*, 477.
174. Koizumi, T.; Takahashi, T.; Matsui, M.; Maeno, N. *Heterocycles* **1990**, *30*, 353.
175. Kolodiazhnyi, O. I. *Zh. Obshch. Khim.* **1995**, *65*,1926.
176. Kolodiazhnyi, O. I.; Grishkun, E. V. *Tetrahedron: Asymmetry* **1996**, *7*, 967.
177. Kolodiazhnyi, O. I.; Grishkun, E. V.; Golovatyi, O. R.; Ustenko, S. N. *Phosph., Sulf., and Silicon* **1995**, *103*, 183.
178. Kolodiazhnyi, O. I. *Rus. Chem. Rev. (Uspekhi Khimii)* **1997**, *66*, 246.
179. Kolodiazhnyi, O. I., Grishkun, E. V. *Phosph., Sulf., and Silicon* **1996**, *115*, 115.
180. Kolodiazhnyi, O. I. *Tetrahedron* **1996**, *52*, 1855.
181. Kolodiazhnyi, O. I.; Grishkun, E. V.; Golovatyi, O. R.; Ustenko, S. N. *Phosph., Sulf., and Silicon* **1996**, *109/110*, 485.
182. Kolodiazhnyi, O. I. *Theses of Symp. on Organic.Chemistry*; St-Petersbourg, 1995, Mai 16–20, p. 137.
183. Kolodiazhnyi, O. I.; Ustenko, S. N.; Grishkun, E. V.; Golovatyi, O. R. *Abstract of Lectures of XIII Int.Conf. on Phosphorus Chem.*; ICPC, Jerusalem 1995, p. 34.
184. Kolodiazhnyi, O. I. *Tetrahedron Lett.* **1995**, *36*, 3921.
185. Kolodiazhnyi, O. I. *Zh. Obshch. Khim.* **1995**, *65*, 1926.
186. Kolodiazhnyi, O. I.; Grishkun, E. V.; Golovatyi, I. R. *Zh. Obshch. Khim.* **1995**, *65*, 1214.
187. Kolodiazhnyi, O. I.; Grishkun, E. V.; Otzaluk, V. M. *Zh. Obshch. Khim.* In press.
188. (a) Kolodiazhnyi, O. I.; Sheiko, S. *Zh. Obshch. Khim.* In press; (b) Kolodiazhnyi, O. I.; Grishkun, E. V.; Sheiko, S. *Zh. Obshch. Khim.*; In press.
189. Kolodiazhnyi, O. I.; Ustenko, S. N.; Golovatyi, O. R. *Tetrahedron Lett.* **1994**, *35*, 1755.
190. Kolodiazhnyi, O. I.; Ustenko, S. N. *Zh. Obshch. Khim.* **1993**, *63*, 2149.
191. Kolodiazhnyi, O. I.; Zemlianoy, A. I.; Kukhar, V. P. *Zh. Obshch. Khim.* **1989**, *59*, 1451.
192. Kolodiazhnyi, O. I. *Phosphorus Ylides Chemistry*; Naukova Dumka: Kiev, 1994, pp. 1–560.
193. Knowles, W. S.; Sabacky, M. J.; Vineyard, B. D.; Weinkauff, D. J. *J. Am. Chem. Soc.* **1975**, *97*, 2567.
194. Korpiun, O.; Lewis, R. A.; Chickos, J.; Mislow, K. *J. Am. Chem. Soc.* **1968**, *90*, 4842.
195. Korpiun, O.; Lewis, R. A.; Chickos, J.; Mislow, K. *J. Am. Chem. Soc.* **1968**, *90*, 4847.
196. Korpiun, O.; Mislow, K. *J. Am. Chem. Soc.* **1967**, *89*, 4784.
197. Koshiwabara, K.; Hanaki, K.; Fujita, J. *Bull. Chem. Soc. Jpn.* **1980**, *53*, 2275.
198. Kumli, K. F.; McEwen, W. E.; Van der Werf, C. A. *J. Am. Chem. Soc.* **1959**, *81*, 3805.
199. Lewis, R. A.; Mislow, K. *J. Am. Chem. Soc.* **1969**, *91*, 7009.
200. Lewis, R. A.; Naumann, K.; De Bruin, K. E.; Mislow, K. *J. Chem. Soc., Chem. Commun.* **1969**, 1010.
201. Lemieux, R. P.; Schuster, G. B. *J. Org. Chem.* **1993**, *58*, 100.
202. Lambert, J. B. *Topics in Stereochemistry*; Allinger, N. L.; Eliel, E. L., Eds.; Wiley-Int: 1971, Vol. 6, p. 19.
203. (a) Li, Y.-F., Hammerschmidt F. *Tetrahedron: Asymmetry* **1993**, *4*, 109; (b) Dresher, M.; Hammerschmidt F. *Tetrahedron* **1997**, *53*, 4627.
204. Lacour, J.; Ginglinger, C.; Grivet, C.; Bernardinelli, G. *Angew. Chem., Int. Ed. Engl.* **1997**, *36*, 608.
205. Lopusinski, A.; Luczak, L.; Michalski, J. *J. Heteroat. Chem.* **1995**, *6*, 365.
206. Lu, Y.; Le Goffic, F. *Synth. Commun.* **1993**, *23*, 1943.
207. Macpherson, A. J.; Smith, D. J. H. *J. Chem. Res. (S)*, **1984**, 32.
208. Maffre, D.; Dumi, P.; Vidal, J.-P.; Escale, P.; Girard, J.-P. *J. Chem. Res. Synop.* **1994**, 30.
209. (a) Marinetti, A.; Ricard, L. *Tetrahedron* **1993**, *49*, 10291; (b) Marinetti, A.; Buzin, F.-X.; Ricard, L. *Tetrahedron* **1997**, *53*, 4627.
210. Marinetti, A.; Ricard, L. *Organometallics* **1994**, *13*, 3956.

211. Marinetti, A., Mathey, F. J.; Ricard, L. *Synthesis* **1992**, 157.
212. Marinetti, A. J.; Ricard, L.; Mathey, F. *Organometallics* **1993**, *12*, 1207.
213. Markovski, L. I.; Romanenko, V. D.; Ruban, A. V.; Drapailo, A. B.; Mikolajczyk, M.; Omelanczuk, J.; Perlikovska, V. *Zh. Obshch. Khim.* **1988**, *58*, 2627.
214. Markwald, W. *Ber.* **1904**, *37*, 1368.
215. Maryanoff, B. E.; Hutchins, R. O.; Maryanoff, C. A. *Top. Stereochem.*; John Wiley & Sons: Year, City, Vol. 11, p. 187.
216. McEwen, W. E.; Kumli, K. F.; Blade-Font, A.; Zanger, M.; Vanderwerf, C. A. *J. Am. Chem. Soc.* **1964**, *86*, 2378.
217. McEwen, W. E.; Blade-Font, A.; Vanderwerf, C. A. *J. Am. Chem. Soc.* **1962**, *84*, 677.
218. McEwen, W. E.; Berlin, K. D. *Organophosphorus Stereochemistry;* John Wiley and Sons: Strassbourg, 1975, Pt. I and II.
219. (a) Mikolajczyk, M. *J. Chem. Soc., Chem. Commun.* **1969**, 1221; (b) Mikolajczyk, M.; Omelanczuk, J.; Perlikovska, W.; Markovski, L. I.; Romanenko, V. D.; Ruban, A. V.; Drapailo, A. B. *Phosph., Sulf., and Silicon* **1988**, *36*, 267; (c) Mikolajczyk, M.; Omelanczuk, J.; Perlikovska, V. *Tetrahedron* **1979**, *35*, 1531.
220. (a) Mikolajczyk, M. *Angew.Chem., Int. Ed. Engl.* **1969**, *8*, 511; (b) Mikolajczyk, M.; Drabowicz, J.; Kielbasinski, P. *Houben-Weil;* Vol. E21, S. 5701.
221. Mikolajczyk, M.; Drabowicz, J.; Omelanczuk, J.; Fluck, E. *J. Chem. Soc., Chem. Commun.* **1975**, 382.
222. Merckling, F. A.; Ruedl, P. *Tetrahedron Lett.* **1996**, *37*, 2217.
223. Moriarty, R. M.; Hiratake, J.; Liu, K.; Wendler, A.; Awasthi, A. K.; Gilardi, R. *J. Am. Chem. Soc.* **1991**, *113*, 9374.
224. Morrison, J. D.; Mosher, H. S. *Asymmetric Organic Reaction*; Englewood Cliffs, NJ, 1971.
225. Mussierowicz, S.; Wroblewsli, A. E. *Tetrahedron* **1980**, *96*, 1375.
226. Mussierowicz, S.; Wroblewsli, A. E.; Krawczyk, K. K. *Tetrahedron Lett.* **1975**, 437.
227. (a) Nagel, U.; Krink, T. *Chem. Ber.* **1993**, *126*, 1091; (b) Nakai, T.; Tomooka, K. *Pure and Appl.Chem.* **1997**, *68*, 595.
228. Nagel, U.; Rieger, B. *Chem. Ber.* **1988**, *121*, 1123.
229. Nagel, U.; Bublewitz, A. *Chem. Ber.* **1992**, *126*, 1061.
230. Nagel, U.; Rieger, B., Bublewitz, A. *J. Organomet. Chem.* **1989**, *370*, 223.
231. Narasaka, K.; Hidai, E.; Hayashi, Y.; Gras, J. L. *J. Chem. Soc., Chem. Commun.* **1993**, 102.
232. Naylor, R. A.; Walker, B. *J. Chem. Soc., Chem. Commun.* **1975**, 45.
233. Neuffer, J.; Richter, W. J. *J. Organomet. Chem.* **1986**, *301*, 289.
234. Nifant'ev, E.; Gratshev, M. K.; Vasianin, L. K. *Zh. Obshch. Khim.* **1995**, *65*, 1926.
235. (a) Noyori, R. *Asymmetric Catalysis in Organic Synthesis*; John Wiley and Sons: New York, 1994; (b) Noyori, R.; Ikeda, T.; Ohkuma, T. *J. Am. Chem. Soc.* **1989**, *111*, 9134.
236. (a) Nudelmann, A.; Cram, D. J. *J. Am. Chem. Soc.* **1968**, *90*, 3869; (b) Nudelmann, A.; Cram, D. J. *J. Org. Chem.* **1971**, *36*, 335.
237. Omelanchuk, J. *J. Chem. Soc., Chem. Commun.* **1992**, 1718.
238. Ourupin, A. B.; Kadyko, M. I.; Petrovskii, P. V.; Fedin, E. I. *Tetrahedron: Asymmetry* **1995**, *6*, 1813.
239. Pabel, M.; Willis, A. C.; Wild, S. B. *Tetrahedron: Asymmetry* **1995**, *6*, 2369.
240. Pastor, S. D.; Shum, S. P.; Rodebaugh, R. K.; Debellis, A. D.; Clarke, F. H. *Helv. Chim. Acta* **1993**, *76*, 900.
241. Pietrushiewicz, K. M.; Wieczorek, W.; Cichi, S.; Brandi, A. *Heteroat. Chem.* **1991**, *2*, 661.
242. Pietrusiewicz, K. M.; Zablocka, M. *Chem. Rev.* **1994**, *94*, 1375.
243. Pietrusiewicz, K. M.; Zablocka, M.; Monkiewicz, J. *J. Org. Chem.* **1984**, *49*, 1522.
244. Pietrusiewicz, K. M.; Zablocka, M. *Tetrahedron Lett.* **1989**, *30*, 477.
245. Pietrusiewicz, K. M.; Zablocka, M. *Tetrahedron Lett.* **1988**, *29*, 937.
246. Pietrusiewicz, K. M.; Zablocka, M. *Tetrahedron Lett.* **1988**, *29*, 1987.

247. Pickersgill, F.; Devitt, P. G.; Kee, T. P. *Synth.Commun.* **1993**, *23*, 1643.
248. Polniazek, R. P.; Foster, A. L. *J. Org. Chem.* **1991**, *56*, 3137.
249. Predvotitelev, D. A.; Afanas'eva, D. N.; Fillipovich, Y. B.; Nifant'ev E. E. *Zh. Obshch. Khim.* **1974**, *44*, 1697.
250. Quin, L. D.; Sadani, N. D.; Wu, Xiao-Ping. *J. Am. Chem. Soc.* **1989**, *111*, 6853.
251. Reissig, H.-U. *Angew. Chem., Int. Ed. Engl.* **1996**, *35*, 971.
252. Rauk, A.; Allen, L. C.; Mislow, K. *Angew. Chem. Int. Ed. Engl.* **1970**, *9*, 400.
253. Rath, P.; Spilling, C. D. *Tetrahedron Lett.* **1994**, *35*, 227.
254. Rein, T.; Kann, N. *J. Org. Chem.* **1993**, *58*, 3802.
255. Rewinkel, H.; Skupsch, J.; Vorbruggen, H. *Tetrahedron Lett.* **1988**, *29*, 1775.
256. Richter, W. J. *Angew. Chem., Int. Ed. Engl.* **1982**, *21*, 292.
257. (a) Richter, W. J. *Chem. Ber.* **1984**, *117*, 2328; (b) Romanenko, V. D.; Reitel, G. V.; Kachkovskaaya, L. S.; Mikolajczyk, M.; Omelanczuk, J.; Perlikovska, V. *Zh. Obshch. Khim.* **1993**, *63*, 1182.
258. Savelova, V. A.; Solomoychenko, T. N.; Ved, T. V.; Sadovski, Yu. S.; Simonenko, Yu. S. *Zh. Org. Chem.* **1993**, *29*, 666.
259. Schmidt, U.; Riedl, B.; Griesser, H.; Fitz, C. *Synthesis* **1991**, 655.
260. Schollkopf, U. *Lieb. Ann. Chem.* **1987**, 45.
261. Schrader, T. *Angew. Chem.* **1995**, *107*, 1001.
262. Seebach, D.; Hayakawa, M.; Sakaki, J.; Schweizer, W. B. *Tetrahedron* **1993**, *49*, 1711.
263. Segi, M.; Nakamura, Y.; Nakajima, T.; Suga, S. *Chem. Lett.* **1983**, 913.
264. Seiferth, D.; Welch, D. E.; Heeren, J. K. *J. Am. Chem. Soc.* **1964**, *86*, 1100.
265. Stec, W. J.; Uznanski, B.; Bruzik, K.; Michalski, J. *J. Org. Chem.* **1976**, *41*, 1291.
266. Stec, W. J.; Wilk, A. *Angew. Chem., Int. Ed. Engl.* **1994**, *33*, 709.
267. Sum, V.; Kee, T. P. *J. Chem. Soc., Perkin Trans. 1* **1993**, 1778.
268. Sum, V.; Kee, T. P. *J. Chem. Soc., Perkin Trans. 1* **1993**, 2701.
269. (a) Sum, V.; Patel, M. T.; Kee, T. P.; Thornton-Pett, M. *Polyhedron* **1992**, *11*, 1743; (b) Sum, V.; Kee, P.; Thornton-Pett, M. *J. Chem. Soc., Chem. Commun.* **1994**, 743.
270. Sum, V.; Davies, A. J.; Kee, T. P. *J. Chem. Soc., Chem. Commun.* **1992**, 1771.
271. Takahaschi, T.; Matsui, M.; Maeno, N.; Koizumi, T. *Heterocycles* **1990**, *30*, 353.
272. Tanaka, K.; Ohta, Y.; Fuji, K.; Taga, T. *Tetrahedron Lett.* **1993**, *34*, 4071.
273. Tasz, M. K.; Gamliel, A.; Rodriguez, O. P.; Lane, T. M.; Cremer, S. E.; Bennett, D. W. *J. Org. Chem.* **1995**, *60*, 6281.
274. Thatcher, C. R. J.; Kluger, R. *Adv. Phys. Org. Chem.* **1989**, *25*, 99.
275. Toda, F. *Top. Cur. Chem.* **1988**, *149*, 211.
276. Toda, F.; Akai, H. *J. Org. Chem.* **1990**, *55*, 3446.
277. Togni, A.; Pastor, A. *Tetrahedron Lett.* **1989**, *30*, 1071.
278. Tomoskozi, I.; Bestmann, H. J. *Tetrahedron Lett.* **1964**, 1293.
279. Tomoskozi, I.; Janzo, G. *Chem. Ind. (London)* **1962**, 2085.
280. Trost, B. M.; Curran, D. P. *J. Am. Chem. Soc.* **1980**, *102*, 5699.
281. Trost, B. M.; Curran, D. P. *Tetrahedron Lett.* **1981**, 22, 4929.
282. Tripett S. *Pure Appl. Chem.* **1974**, *40*, 595.
283. Uznanski, B.; Grazkowski, A.; Krzyzanovska, B. *J. Am. Chem. Soc.* **1992**, *114*, 10197.
284. Valentine, D. Jr.; Blount, J. F.; Toth, K. *J. Org. Chem.* **1980**, *45*, 3691.
285. Valentine, D. Jr.; Scott, J. W. *Synthesis* **1978**, 329.
286. Valentine, D. *Asymmetric Synthesis*; Morrison, J. D.; Scott, J. W., Eds.; Academic Press: Orlando, FL, 1984, Vol. 3, Chap. 3.
287. de Vaumas, R.; Marinetti, A.; Ricard, L.; Mathey, F. *J. Am. Chem. Soc.* **1992**, *114*, 261.
288. Vilkas, E.; Vilcas, M.; Joniaux, D.; Pascard-Billy, C. *J. Chem. Soc., Chem. Commun.* **1978**, 125.
289. Vilkas, E., Vilcas, M., Sainton, J., Meunier, B., Pascard, C. *J. Chem. Soc. Perkin Trans.* **1979**, 2136.
290. Vineyard, B. D.; Knowles, W. S.; Bachman, G. L.; Weinkauff, D. J. *J. Am. Chem. Soc.* **1977**, *99*, 5946.

291. Wangi, M.; Liu, S. Z.; Hu, B. F. *J. Org. Chem.* **1995**, *60*, 7364.
292. Warren, S.; Harmat, N. J. S. *Tetrahedron Lett.* **1990**, *31*, 2743.
293. Wasiak, J.; Michalski, J. *Tetrahedron Lett.* **1994**, *35*, 9473.
294. Westheimer, F. H. *Acc. Chem. Res.* **1968**, *1*, 70.
295. Wittig, G.; Cristau, H. J.; Braun, H. *Angew. Chem., Int. Ed. Engl.* **1967**, *6*, 700.
296. (a) Whitsell, J. K. *Chem. Rev.* **1989**, *89*, 1581; (b) Wozniak, L.; Choinowski, J. *Tetrahedron* **1989**, *45*, 2465.
297. Yamamoto, K.; Tomita, A.; Tsui, J. *Chem. Lett.* **1978**, 3.
298. Zon, G.; Mislow, K. *Fortschr. Chem. Forsch.* **1971**, *19*, 61.
299. Zon, G.; De Bruin, K. E.; Naumann, K.; Mislow, K. *J. Am. Chem. Soc.* **1969**, *91*, 7023.

INDEX

www.ingramcontent.com/pod-product-compliance
Ingram Content Group UK Ltd.
Pitfield, Milton Keynes, MK11 3LW, UK
UKHW020233290926
15470UKWH00002B/515

9 780762 301065